Banach Spaces and their Applications in Analysis

Nigel J. Kalton

Banach Spaces and their Applications in Analysis

Proceedings of the International Conference
at Miami University
May 22–27, 2006

In Honor of Nigel Kalton's 60th Birthday

Editors

Beata Randrianantoanina
Narcisse Randrianantoanina

Walter de Gruyter · Berlin · New York

Editors

Beata Randrianantoanina
Department of Mathematics and Statistics
Miami University
Oxford, OH 45056, USA
E-mail: randrib@muohio.edu

Narcisse Randrianantoanina
Department of Mathematics and Statistics
Miami University
Oxford, OH 45056, USA
E-mail: randrin@muohio.edu

Keywords: applications of Banach space theory, approximation theory, functional calculus, algebraic methods in Banach spaces, homological methods in Banach spaces, isomorphic theory, nonlinear theory

Mathematics Subject Classification 2000: primary 46-06, 46N10, 46N20, 46N30, 46N40; secondary 46A22, 46B10, 46B20, 46E39, 47H09

∞ Printed on acid-free paper which falls within the guidelines of the ANSI to ensure permanence and durability.

Library of Congress Cataloging-in-Publication Data

A CIP catalogue record for this book is available from the Library of Congress.

ISBN 978-3-11-019449-4

Bibliographic information published by the Deutsche Nationalbibliothek

The Deutsche Nationalbibliothek lists this publication in the Deutsche Nationalbibliografie; detailed bibliographic data are available in the Internet at http://dnb.d-nb.de.

Printed in Germany.
Cover design: Thomas Bonnie, Hamburg.
Printing and binding: Hubert & Co. GmbH & Co. KG, Göttingen.

Preface

Stefan Banach once said:

> "A mathematician is a person who can find analogies between theorems; a better mathematician is one who can see analogies between proofs; and the best mathematician can notice analogies between theories. One can imagine that the ultimate mathematician is one who can see analogies between analogies."

According to this definition, Nigel Kalton is one of the ultimate mathematicians. In his work, Kalton finds underlying connections between seemingly unrelated areas of mathematics. He has been immensely successful in applying Banach space methods to numerous problems in analysis. Thus, we honor him on the occasion of his 60th birthday in 2007.

As evidenced by the participation of over 160 mathematicians from around the world, it is clear that our community sees the power and potential of Banach space methods in solving a broad array of analysis problems. Indeed, in recent years there has been a surge of profound new developments in analysis – developments whose connecting thread is the use of Banach space methods. Many problems seemingly far from classical geometry of Banach spaces have been solved using Banach space techniques.

In this conference, specialists who have been instrumental in these new developments were brought together. The emphasis of the conference was on applications of Banach space methods in the following areas:

1. Nonlinear theory (Lipschitz classifications of Banach/metric spaces, linear programming methods and related topics);
2. Isomorphic theory of Banach spaces including connections with combinatorics and set theory;
3. Algebraic and homological methods in Banach spaces;
4. Approximation theory and algorithms in Banach spaces (greedy algorithms, interpolation etc.);
5. Functional calculus and applications to partial differential equations.

At the conference there were 15 plenary talks giving a broad overview of various areas where Banach space methods found applications. In addition, 105 talks were delivered in specialized sessions. These *Proceedings* reflect the conference. They include 11 papers by plenary speakers and 16 specialized papers by participants of the conference. We especially thank Gilles Godefroy for writing an excellent article surveying the vast work of Nigel Kalton. Godefroy describes many of the important breakthroughs in different areas of analysis and presents open problems for further research.

We thank Miami University for hosting the conference and providing substantial support for a successful meeting. We especially thank Mark A. Smith, chair of the Department of Mathematics and Statistics at Miami University for both financial and logistical support. We thank the following units of Miami University for financial grants in support of the conference: the Office of the Dean of Arts and Science, the Office of the Dean of Engineering and Applied Science, the Office of the Provost and the International Visiting Scholar Exchange Fund.

We thank Mark Ashbaugh, chair of the Department of Mathematics at the University of Missouri-Columbia, and Curator's Professor Fund from University of Missouri-Columbia for their financial support.

We thank the American Mathematical Society and a private donor for their financial contributions.

We thank the following publishers who generously provided books for display at the conference: Cambridge University Press, Princeton University Press and Springer-Verlag. We also thank Brill Science Library of Miami University for lending additional books for display.

We thank the National Science Foundation, whose grant provided travel support for many conference participants.

We thank the support staff, especially Linda Ferriell, for helping to run the conference smoothly.

Last, but most important of all, we thank all the speakers and participants of the conference who made it a success.

Oxford, Ohio, June 2007

Beata Randrianantoanina
Narcisse Randrianantoanina

Table of Contents

Banach Spaces and their Applications in Analysis, 1–35

A glimpse at Nigel Kalton's work

Gilles Godefroy

Abstract. This article is a non exhaustive survey of Nigel Kalton's contribution to functional analysis. We focus on geometry and structure of Banach spaces and quasi-Banach spaces, non linear isomorphisms, isometric theory, interpolation theory, differentiation of interpolation lines and twisted sums, basis theory and applications.

Key words. Banach spaces, quasi-linear maps, interpolation, entropy functions.

AMS classification. 46A16, 46B20, 46B70.

1 Introduction

Nigel Kalton's work concerns every domain of functional analysis, and goes in fact much further than that and even beyond the limits of analysis. His research spans over more than forty years, and the published part of his work represents thousands of dense, original and deep pages of concisely written mathematics.

Pretending to outline such an amount of mathematics in a few pages is simply pointless. The present survey therefore focuses on Nigel Kalton's work on the geometry and structure of Banach spaces and quasi-Banach spaces. But I should make it clear that even with this restriction, the present sketch is very far from comprehensive, and that many important articles (or Memoirs) will not even be mentioned.

Indeed a choice has been made, in order to provide the reader with some pieces of information that would not be too easily found elsewhere. This choice consisted into picking some works with which the author is relatively familiar, and on which he could provide, if not an original point of view, at least some facts which may not be commonly known, or recent references on related topics. I should make it clear that there is no claim on my side that this part of Nigel Kalton's contribution is the most important one. It is simply a part I understand, and which is clearly exciting enough to deserve a presentation, and hard enough to deserve an explanation.

This outline contains some sketches of proof and a few technical arguments, but it mainly consists into a glance at Nigel Kalton's work from a distance, in order to provide the reader with some intuition on what goes on. Obviously, this outline cannot be used as a substitute for an actual reading of the original articles. On the contrary, it is an invitation to dwell into Nigel Kalton's work, and to enjoy every line of it, and even all what can be found between the lines.

Let us now describe the contents of this note. Section 2 deals with non locally convex analysis, and the important concepts which Nigel Kalton discovered and used in this particular theory: extensions, quasi-linear maps, $\mathcal{K}$-spaces. Various objects from classical functional analysis, such as Maharam submeasures, entropy functions or distorted norms, turn out to be related with what happens when p is less than 1 and this shows in our presentation. Section 3 is devoted to some recent progress in non-linear geometry. This topic is relatively new and many basic questions are still unanswered. However asymptotic structures happen to provide usable invariants for Lipschitz and even uniform isomorphisms and this is displayed here. Section 4 is devoted to isometric theory, and to the wealth of information that can be deduced from the existence of special norms on a given space. Complex and real spaces behave quite differently there, and Hermitian operators, which are so useful but only in the complex case, partly explain why. We will also see that isometric theory sometimes has a quite algebraic flavor. Interpolation theory, differentiation of interpolation lines and the corresponding calculus on the "manifold" of Banach spaces are displayed in section 5, where unexpected connections with commutators and the trace class are shown. Finally, section 6 contains some applications of basis theory to the solution of important problems of classical analysis, showing that some of the work which has been done for its own sake some thirty years ago provides powerful tools and examples when geometry of Banach spaces joins forces with harmonic analysis or semigroup theory.

These sections have been created for the reader's convenience, but tight relations exist between them and the versatility of some of the concepts which are displayed below underlines the profound unity of mathematical analysis, when seen from Nigel Kalton's point of view. Quite exotic tools, such as non-linear liftings, discontinuous linear functionals, conditional bases of the Hilbert space, non locally convex twisted sums, and so on, are used below in such a way that they provide useful information on classical and main-stream analysis. And why dispensing with such tools when they can be so powerful?

2 The Kalton zone: $0 < p < 1$.

Hahn–Banach theorems are cornerstones of functional analysis, to such a point that analysts may get nervous when they have to dispense with them. But it turns out that non-locally convex spaces show up very naturally in many cases when there is no reason to "stop at $p = 1$": for instance, if $f : \Omega \to \mathbb{C}$ is an holomorphic function on some open subset Ω of $\mathbb{C}$ then $|f|^p$ is subharmonic for all $p > 0$. This very elementary but important example is an invitation to visit what I suggest to call the Kalton zone: $0 < p < 1$.

We recall that metrizable complete topological vector spaces (on $\mathbb{K} = \mathbb{R}$ or $\mathbb{C}$) are called F-spaces. Their topology is induced by an *F-norm*, that is, a map Λ from the space X to $\mathbb{R}^+$ such that:

(i) $\Lambda(x) > 0$ if $x \neq 0$.
(ii) $\Lambda(\alpha x) \leq \Lambda(x)$ if $|\alpha| \leq 1$.
(iii) $\lim_{\alpha \to 0} \Lambda(\alpha x) = \Lambda(0) = 0$.
(iv) $\Lambda(x + y) \leq \Lambda(x) + \Lambda(y)$ for all $(x, y) \in X^2$.

The space X is locally bounded if and only if its topology can be generated by a *quasi-norm* $\|\cdot\|$, namely a map $\|\cdot\| : X \to \mathbb{R}^+$ such that:

(i) $\|x\| > 0$ if $x \neq 0$
(ii) $\|\alpha x\| = |\alpha|\|x\|$ for all $x \in X$ and $\alpha \in \mathbb{K}$
(iii) $\|x + y\| \leq C(\|x\| + \|y\|)$ for all $(x, y) \in X^2$

where $C \geq 1$ is the "modulus of concavity" of the quasi-norm. An F-space is called a *quasi-Banach space* when its topology is generated by a quasi-norm, or equivalently by the Aoki–Rolewicz theorem, by a p-subadditive quasi-norm $|||\cdot|||$, which satisfies the condition

(iv) $|||x + y|||^p \leq |||x|||^p + |||y|||^p$

for all $(x, y) \in X^2$ and $p > 0$ given by $p = (1 + \log_2(C))^{-1}$.

We refer to [83] for an authoritative book on F-spaces. It is clear that the classical applications of Baire's lemma are not sensitive to local convexity assumptions, while Hahn–Banach theorem is, and in fact to the point where it leads to a characterization:

Theorem 2.1 ([56]). *A quasi-Banach space X is locally convex (i.e. is a Banach space) if and only if every continuous linear functional defined on a closed subspace E of X has an extension to a continuous linear functional on X.*

In other words, a quasi-Banach space is a Banach space if and only if the weak and quasi-norm topologies have the same closed subspaces.

The proof of Theorem 2.1 relies on the construction of Markushevich basic sequences (see [71, Proposition 3.4]), obtained by refining Mazur's classical argument. For being able to do so, one needs however a weaker topology, even if it is not "weak" in the classical sense. A quasi-Banach space is *minimal* if it does not have any weaker Hausdorff vector topology. A separable quasi-Banach space is minimal exactly when it contains no basic sequence. It turns out that quite general assumptions force the existence of basic sequences. Let us say that a p-subadditive quasi-norm $\|\cdot\|$ is *plurisubharmonic* (in short, $p.s.$) if

$$\|x\| \leq \frac{1}{2\pi} \int_0^{2\pi} \|x + e^{i\theta} y\| d\theta$$

for all $(x, y) \in X^2$. It is shown in [131] that a complex quasi-Banach space X which has an equivalent $p.s.$ norm is *not* minimal, and this applies to subspaces of $L^p(0 < p < 1)$ and more generally to all *natural* quasi Banach spaces, where "natural" means "subspace of a lattice-convex quasi-Banach lattice".

However, minimal quasi-Banach spaces do exist and this is shown in [69]:

Theorem 2.2. *There is a quasi-Banach space X which contains a one-dimensional subspace E such that every infinite-dimensional closed subspace Y of X contains E. In particular, X contains no basic sequence and it is minimal.*

The reader may find it amusing to think of that space as a book: it may have many pages but they all meet on the one-dimensional binding.

The proof of Theorem 2.2 is the culmination of several works, which we now outline.

If X and Y are quasi-Banach spaces, Y is a *minimal extension* of X if there is a one-dimensional subspace E of Y such that $Y/E \simeq X$. The reader should be warned that the word "minimal" is used here with two different meanings. A minimal extension is usually *not* a minimal quasi-Banach space, but the above theorem asserts that this may happen. A minimal extension is said to be *trivial* if it splits, i.e. if $Y \simeq X \oplus E$. The case when X is actually a Banach space is important, and indeed Kalton [59], Ribe [119], Roberts [123] independently constructed non-trivial minimal extensions of $X = \ell_1$, thus solving negatively the three-space problem for local convexity.

All minimal extensions can be obtained in the following way [58]: let X be a quasi-Banach space over the field $\mathbb{K}$ and X_0 a dense linear subspace of X. A map $F : X_0 \to \mathbb{K}$ is *quasilinear* if:

(i) $F(\alpha x) = \alpha F(x)$ for $x \in X_0$ and $\alpha \in \mathbb{K}$.
(ii) There is $C \in \mathbb{R}$ such that $|F(x+y) - F(x) - F(y)| \leq C(\|x\| + \|y\|)$ for all $(x, y) \in X^2$.

Then one can define a quasi-norm on $\mathbb{K} \oplus X_0$ by

$$\|(\alpha, x)\|_F = |\alpha - F(x)| + \|x\|$$

and the completion of $\mathbb{K} \oplus X_0$ for this quasi-norm if a minimal extension of X denoted $\mathbb{K} \oplus_F X$. Conversely any minimal extension of X is obtained in this way, and $\mathbb{K} \oplus_F X$ splits if and only if there is a linear map $G = X_0 \to \mathbb{K}$ such that

$$|F(x) - G(x)| \leq C'\|x\| \tag{2.1}$$

for all $x \in X_0$. This approximation is related with Hyers–Ulam functional stability, for which we refer to [6, Chapter 15]. The Ribe space is then obtained by considering the quasilinear functional

$$F(x) = \sum_{k=1}^{\infty} x_k \log |x_k| - (\sum_{k=1}^{\infty} x_k) \log |\sum_{k=1}^{\infty} x_k|$$

on the dense subspace c_{00} of finitely supported sequences in ℓ_1. For showing that (2.1) fails for this F, it suffices to note that when X is a Banach space, (2.1) is equivalent to an estimate

$$|F(\sum_{i=1}^{n} x_i) - \sum_{i=1}^{n} F(x_i)| \leq C'' \sum_{i=1}^{n} \|x_i\|$$

for all $x_1, \ldots, x_n \in X_0$, and to compute

$$F(\sum_{i=1}^{n} e_i) - \sum_{i=1}^{n} F(e_i) = -n \log(n).$$

Ribe's function F is closely related to Shannon's entropy function from information theory. This motivates the following terminology. Let X be a Banach sequence space.

The quasilinear map Φ_X defined on c_{00}^{+} by

$$\Phi_X(x) = \sup_{\|t\|_X \leq 1} \sum_{k=1}^{\infty} x_k \log |t_k| \tag{2.2}$$

and extended to c_{00} by

$$\Phi_X(x) = \Phi_X(x^+) - \Phi_X(x^-)$$

is called the *entropy function* of X [107]. For instance

$$\Phi_{\ell_1}(x) = \sum_{k=1}^{\infty} x_k \log \frac{|x_k|}{\|x\|_1},$$

while $\Phi_{\ell_p} = \frac{1}{p}\Phi_{\ell_1}$.

In order to construct a minimal extension Y of ℓ_1 with no basic sequence and such that $Y/E \simeq \ell_1$, we need that every infinite-dimensional closed subspace of Y contains E. This is reminiscent of Gowers–Maurey's construction of a Banach space X_{GM} without unconditional basic sequence [45], which is such that for any infinite-dimensional subspaces U and V of X_{GM}

$$\inf\{\|u - v\|; u \in U, v \in V, \|u\| = \|v\| = 1\} = 0.$$

And indeed, Gowers' modification [43] of the original construction, used in his solution of the hyperplane problem, provides a space X whose entropy function Φ_X yields to a minimal extension $\mathbb{K}\oplus_{\Phi_X} \ell_1$ with no basic sequence [69], and which is therefore a minimal quasi-Banach space. Note that this function $\Phi_X = F$ fails to satisfy (2.1) when restricted to any infinite dimensional subspace J of c_{00} or equivalently

$$\sup\{|F(x)| : x \in J, \|x\| \leq 1\} = \infty,$$

hence $\Phi_X = F$ is distorted in the sense of [108].

Minimal quasi-Banach spaces M are pretty strange objects: every one-to-one continuous linear map from M into a Hausdorff topological vector space is actually an isomorphism on its range! However existing examples are "non-isotropic" in the sense where they contain a distinguished line, namely the orthogonal of the dual space. It in not known whether an even stranger example exists which would exhibit this behaviour everywhere.

Open problem. *Does there exist an "atomic" quasi-Banach space, that is, a quasi-Banach space which contains no infinite-dimensional proper closed subspace?*

The Ribe space is a non-trivial minimal extension of ℓ_1. However there exist infinite-dimensional quasi-Banach spaces X which are such that every minimal extension of X is trivial. Such spaces are called $\mathcal{K}$- *spaces* in [81] (but in my opinion Kalton spaces would sound perfect) and it is shown in [59] that if $0 < p < 1$, the spaces ℓ_p and L_p are $\mathcal{K}$-spaces, from which it follows in particular that if E is a one-dimensional subspace of $L_p(0 < p < 1)$ then L_p/E is not isomorphic to L_p [81].

Some Banach spaces are $\mathcal{K}$-spaces: it is shown in [86] that every quotient space of a $\mathcal{L}_\infty$-space is a $\mathcal{K}$-space, and in [59] that a Banach space with non-trivial type is a $\mathcal{K}$-space. In fact, Kalton conjectures that a Banach space is a $\mathcal{K}$-space exactly when its dual space has non-trivial cotype.

Minimal extensions of $\mathcal{L}_\infty$-spaces are trivial, in other words it is so that quasi-linear maps on "cubes" are close to linear ones. This creates a link between this field and the "Maharam problem", which we now outline.

The Maharam problem originates in von Neumann's question (Problem 163 in the Scottish book) of characterizing measure algebras. D. Maharam's conditional negative answer [104] to von Neumann's question (provided the Souslin hypothesis fails) lead her to consider what is now called a *Maharam submeasure*: if Σ is a σ-algebra of subsets of some set Ω, a Maharam submeasure is a map $\phi = \Sigma \to \mathbb{R}^+$ such that:

(M1) $\phi(\emptyset) = 0$.
(M2) $\phi(A) \leq \phi(A \cup B) \leq \phi(A) + \phi(B)$ for $A, B \in \Sigma$.
(M3) If $(A_n) \in \Sigma$ is a disjoint sequence then $\lim \phi(A_n) = 0$.

The *Maharam problem* asks if for every Maharam submeasure ϕ, there is a (countably additive, positive, finite) measure λ *equivalent* to ϕ, in the sense where $\phi(A) = 0$ if and only if $\lambda(A) = 0$.

This problem has an equivalent form which turns it into a functional analysis problem: let X be an F-space and $G : \Sigma \to X$ be a countably additive vector measure. Does there exist a measure λ on Σ which *controls* G, that is $\lambda(A) = 0$ implies $G(A) = 0$? The answer to this "control measure problem" is positive when X is a Banach space by a classical result of Bartle–Dunford–Schwartz (see [65]), when $X = L_p$ $(0 < p < 1)$ through a change of density argument due to Maurey [105], and when $X = L_0$ as a consequence of the fact that L_0-valued vector measures are bounded [129, 82].

The equivalence between the Maharam problem and the control measure problem relies in part on the simple fact that is $G : \Sigma \to X$ is a vector measure, then its semi-variation $\|G\|$ defined by

$$\|G\|(A) = \sup\left\{\|G(B)\| : B \subseteq A,\ B \in \Sigma\right\}$$

is a Maharam submeasure.

The Maharam problem was investigated in [49] where a negative answer was shown to be equivalent to the existence of a *pathological* nonzero Maharam submeasure, where ϕ "pathological" means: for all $\varepsilon > 0$, there exists $B_1, B_2, \ldots, B_m$ in Σ such that

(P1) $\displaystyle \frac{1}{m}\sum_{i=1}^{m} \mathbf{1}_{B_i} > (1-\varepsilon)\mathbf{1}_\Omega$
(P2) $\phi(B_i) < \varepsilon$ for $i = 1, 2, \ldots, m$.

At this point, we should relate this line of thought with Roberts' construction of a compact convex set without extreme points [122]. This construction relies on the notion of a *needle point*: a point x in a quasi-Banach space X is a needle point if for all $\varepsilon > 0$, there exist $x_1, x_2, \ldots, x_n \in X$ such that $\|x_i\| < \varepsilon$ for all $i = 1, 2, \ldots, n$, and if $y \in \operatorname{conv}\{x_1, x_2, \ldots, x_n\}$ then $\|y - tx\| < \varepsilon$ for some $t \in [0, 1]$.

We already met an example of needlepoint: if $(\mathbb{K} \oplus_F Y) = X$ is a non-trivial minimal extension of a Banach space Y, then $(1,0)$ is a needle point in X. It is also relevant to compare the definition of a needle point with the conditions (P1) and (P2) which characterize pathological submeasures.

A *needle space* is a quasi-Banach space such that every point is a needle point, and Roberts showed in [122] that every needle space contains a compact convex set without extreme points, and in [121] that L_p is a needle space if $0 < p < 1$.

Let us observe in passing a feature of quasi-Banach spaces which is quite unusual for those who usually work in Banach spaces where "all non-zero vectors are equivalent": given $x \neq 0$ and y in a quasi-Banach space X, there is in general *no* continuous linear operator T such that $Tx = y$. It may even be so that the *only* continuous linear map are scalar multiplications [85], and such "rigid spaces" are highly non-isotropic.

We now translate the Maharam problem into geometric terms. When a Maharam submeasure ϕ is equivalent to some measure λ, this submeasure satisfies a stronger uniform version of (M3), namely:

(M4) for all $\varepsilon > 0$, there is $N = N(\varepsilon)$ so that if $A_1, A_2, \ldots, A_N \in \Sigma$ are N disjoint sets then

$$\min_{1 \leq i \leq N} \phi(A_i) < \varepsilon.$$

Such submeasures are called *uniformly exhaustive*. When $\phi = \|G\|$ for a vecor measure $G : \Sigma \to X$, the relative compactness of $G(\Sigma)$ (or equivalently of $\mathrm{conv}(G(\Sigma)) \subseteq \overline{G(\Sigma)}$ by Liapunoff's theorem) provides, when available, the requested uniformity and ϕ is uniformly exhaustive.

The link between Roberts' construction and the Maharam problem is provided by [57] where it is shown that if $\overline{\mathrm{conv}}(G(\Sigma))$ is compact, then $\|G\|$ is equivalent to a measure (alternatively, G has a control measure) if and only if $\overline{\mathrm{conv}}(G(\Sigma))$ is locally convex.

This rises hope to obtain a negative solution to Maharam's problem (even with a uniformly exhaustive submeasure) by modifying Robert's approach to construct a non-locally convex compact *zonoid* $\overline{\mathrm{conv}}(G(\Sigma))$.

This however, cannot be done, and this is the content of the following Kalton–Roberts' theorem:

Theorem 2.3 ([86]). *A Maharam submeasure is equivalent to a measure if and only if it is uniformly exhaustive.*

The proof relies on the existence of remarkable bipartite graphs called concentrators (see [33]).

Therefore every compact zonoid is locally convex, every vector measure valued in an F-space with relatively compact range has a control measure, and every minimal extension of $Y = c_0$ is trivial (in other words, c_0 is a $\mathcal{K}$-space). Back to the set function side, Theorem 2.3 provides:

Theorem 2.4 ([86]). *There is a universal constant K such that if Σ is an algebra of subsets of some set Ω and $\phi : \Sigma \to \mathbb{R}$ is a set function such that*

$$|\phi(A \cup B) - \phi(A) - \phi(B)| \leq 1$$

if $A \cap B = \emptyset$, then there is an additive set function λ such that

$$|\phi(A) - \lambda(A)| \leq K$$

for all $A \in \Sigma$.

The best value of K belongs to the interval $[3/2, 45]$ but its precise value seems to be unknown.

The Maharam problem has recently been solved by Talagrand [130] who constructed a pathological Maharam submeasure. Talagrand's theorem provides a final negative answer to von Neumann's original question, and answers also negatively the control measure problem. On the other hand it shows that Kalton–Roberts' Theorem 2.3 is essentially optimal.

Let us note that if $G : \Sigma \to X$ is a vector measure with no control measure (which exists by Talagrand's theorem), then there is no one-to-one continuous linear operator $T : X \to L_0(\mu)$ with μ a measure, since $(T \circ G)$ would then be an L_0-valued vector measure for which the existence of a control measure is known. It is however simpler to show the existence of such "pathological" spaces, and in fact Kalton shows in [60] that if $0 < p < 1$ and

$$T = L_p(\mathbb{T})/H^p(\mathbb{T}) \to L_0$$

is a continuous linear operator then $T = 0$. The above space $L_p(\mathbb{T})/H^p(\mathbb{T})$ is a simple example of a quasi-Banach space which is not "natural" as defined above.

To conclude on the Maharam problem, let us mention that it has been shown by Louveau [103] that if ϕ is a pathological submeasure of an algebra Σ of subsets of Ω and if we denote for $B \in \Sigma$

$$t^{\omega}_{\phi}(B) = \sup \{t \geq 0 : \text{for all } n, \text{ there is a partition } \mathcal{P}_n \text{ of } B \text{ such that } \mathcal{P}_n \subseteq \Sigma, \ \operatorname{card} \mathcal{P}_n \geq n, \text{ and } \phi(P) \geq t \text{ for all } P \in \mathcal{P}_n\}$$

then $t^{\omega}_{\phi} \geq \phi/3$. In particular $t^{\omega}_{\phi}(\Omega) \geq \phi(\Omega)/3$ and Kalton–Roberts' theorem follows since it means that $\phi(\Omega) = 0$ when $t^{\omega}_{\phi}(\Omega) = 0$. Louveau's argument is an adaptation of Kalton–Roberts' proof.

The "Ribe–Roberts" circle of ideas lead to a number of results at the right of the Kalton zone, in the locally convex case $p = 1$. Indeed, the Ribe space is isomorphic [61] to the subspace of L^p $(0 < p < 1)$ generated by the constant function $\mathbf{1}$ and the sequence $(|\xi_n|)$, where (ξ_n) is a sequence of independent random variables with the Cauchy distribution (in other words, 1-stable variables), such that

$$\int e^{is\xi_n(t)}\,dt = e^{-|s|}.$$

It can be shown [61] that

$$\|\alpha_0 + \sum_k \alpha_k |\xi_k| \|_p \sim |\alpha_0 - \phi_{\ell_1}((\alpha_t))| + \sum_k |\alpha_k|$$

where ϕ_{ℓ_1} is the entropy function of the space ℓ_1 defined by (2.2). Through this construction, the constant function in the Ribe space appears as the most natural example of needle point.

Of course $\xi_n \notin L_1$, but a similar construction yields an interesting subspace of L_1 [39, Example 4.1.1]. If $p \in (1, 2]$, a random variable Y is called *p-stable* if for some constant C

$$\int e^{isY(t)} dt = e^{-C|s|^p}.$$

Let (Z_j) be a sequence of independent symmetric p_j-stable random variables such that $\|Z_j\|_1 = 1$, let $U_j = |Z_j| - 1$ and let

$$X = \overline{\text{span}}\{\mathbf{1}, (U_j)_{j\geq 1}\}$$

be the closed subspace of L_1 generated by $\mathbf{1}$ and the sequence (U_j). Then, for a proper choice of the sequence (p_j) with $\lim_{j\to\infty} p_j = 1$, the unit ball B_X of the space X is compact for the topology τ_m of convergence in measure, although $x^*(\mathbf{1}) = 0$ for every $x^* \in X^*$ whose restriction to B_X is τ_m-continuous; in particular (B_X, τ_m) is not locally convex.

The space X is the first example of a separable Banach space whose unit ball is compact but not locally convex for a Hausdorff t.v.s. topology. However it falls short from being the satisfactory example one could hope for, and indeed the following question, which could be related with the atomic space problem, remains open.

Open problem. *Does there exist a separable Banach space X such that B_X is compact for some Hausdorff topological vector space topology, but* $\text{Ext}(B_X) = \emptyset$*?*

It is not even known whether there is a subspace Y of L_1 such that B_Y is τ_m-compact but such that $\text{Ext}(B_Y) = \emptyset$, or merely such that Y fails the Radon–Nikodym property. Along these lines, let us mention the important Bourgain–Rosenthal example [9] of $Z \subseteq L_1$ failing RNP but such that every *dyadic* bounded tree converges in Z. The closed unit ball B_Z is τ_m*-relatively* compact and the proof relies in part on Robert's argument.

If instead of taking absolute values of p_j-stable random variables one takes the variables themselves and the space $Z = \overline{\text{span}}^{\|\cdot\|}(\mathbf{1}, (Z_j)_{j\geq 1})$, one obtains another interesting subspace of L_1 [40, Section V], which shows for instance that Dor's Theorem B [22] does not extend from bases to FDD's: more precisely, for every $\varepsilon > 0$ the space Z is $(1+\varepsilon)$-isomorphic to an ℓ_1-sum of finite dimensional spaces; however, if $J : Z \to L_1$ denotes the canonical injection, then

$$\|\mathbb{E}^S - J\|_{L(Z, L_1)} \geq 1$$

where $\mathbb{E}^S$ denotes the conditional expectation with respect to the σ-algebra generated by an arbitrary measurable partition. A similar construction provides in [74] an example of a Banach space which isometrically embeds into L^p but not into L^r with $p < r < 1$, namely the span in L^p of a sequence of 1-stable Cauchy independent random variables and of the constant function. Note that the isomorphic embedding into L^r follows from the Maurey–Nikishin factorization theorem.

It is fairly natural that what happens in the Kalton zone $0 \leq p \leq 1$ overspills to $p = 1$. But what could there be at the *left* of the Kalton zone, where $p < 0$? For an answer to this question, we refer to [96, Section 4], and we state that an n-dimensional space X "embeds into L_p" $(-n < p < 0)$ if $\|\cdot\|_X^{-p}$ is a positive definite distribution. With this terminology, every n-dimensional space imbeds into L_{-n+3}. In particular 4-dimensional spaces embed into L_{-1} and this is important since embedding into L_{-1} characterizes intersection bodies [95, 94]. It follows, with more work, that the Buseman–Petty problem: "if K and L are symmetric convex bodies in $\mathbb{R}^n$ such that

$$\mathrm{vol}_{n-1}(K \cap \xi^{\perp}) \leq \mathrm{vol}_{n-1}(L \cap \xi^{\perp})$$

for all $\xi \in S^{n-1}$, does it follow that $\mathrm{vol}_n(K) \leq \mathrm{vol}_n(L)$?", has a positive answer if $n \leq 4$ (and a negative answer if $n \geq 5$) [34], [135].

Of course, going left of the Kalton zone has a price. When $p \in [0, 1)$, we loose the dual space. But when $p < 0$, the space itself disappears.

3 Non-linear geometry

Banach spaces are in particular metric spaces, and it is natural to wonder if two spaces are linearly isomorphic when they are Lipschitz isomorphic. We refer to [6, Chapter 7] for an account of this part of non-linear functional analysis.

It turns out that local theory provides non-trivial Lipschitz invariants [118, 27, 48] which permit to show that the moduli of uniform convexity or uniform smoothness are invariant under Lipschitz (and even uniform) homeomorphisms. Of course, alternative Lipschitz invariants are needed for dealing with non super-reflexive spaces, and it is shown in [38] and [52] that asymptotic structures can help.

Up to the notation, the following *modulus of asymptotic smoothness* is defined in [106]: if $x \in S_X$ and $Y \subseteq X$ is a closed linear subspace of X, let

$$\overline{\rho}(x, Y, \tau) = \sup\left\{\|x + y\| - 1; y \in Y, \|y\| = \tau\right\}$$

and let

$$\overline{\rho}(\tau) = \sup_{x \in S_X} [\inf_{\substack{Y \subseteq X \\ \dim(X/Y) < \infty}} (\overline{\rho}(x, Y, \tau))].$$

The space X is *asymptotically uniformly smooth* (in short, a.u.s.) if

$$\lim_{\tau \to 0} \frac{\overline{\rho}(\tau)}{\tau} = 0.$$

It is clear that uniformly smooth spaces are a.u.s. but the converse completely fails: for instance, the natural norm on $c_0(\mathbb{N})$ is such that $\overline{\rho}(\tau) = 0$ for all $\tau \in [0,1]$. This example points towards an interesting feature of the asymptotic modulus: if ℓ_p $(1 \leq p < \infty)$ is equipped with its natural norm then

$$\overline{\rho}_{\ell_p}(\tau) \sim \tau^p$$

and thus it distinguishes between the ℓ_p's when $2 \leq p$, while the modulus of uniform smoothness fails to do that. What happens is that well-chosen finite-codimensional subspaces Y permit to avoid the spherical subspaces provided by Dvoretzky's theorem, which show that the optimally smooth Banach space is ℓ_2, while the optimally a.u.s. Banach space is c_0.

The dual notion reads as follows: the norm of X is a.u.s. if and only if for every $\tau > 0$, there exists $\theta_X(\tau) > 0$ such that

$$\liminf_{n\to\infty} \|x^* + x_n^*\| \geq 1 + \theta_X(\tau)$$

whenever $\|x^*\| = 1$, $x_n^* \xrightarrow{w^*} 0$ and $\|x_n^*\| \geq \tau$.

Using this duality, the following result follows from [38, Theorem 5.4]:

Theorem 3.1. *Let X be an asymptotically uniformly smooth separable Banach space. If a Banach space Y is Lipschitz-isomorphic to X and $M > d_L(X,Y)$, then there is an equivalent norm on Y such that for all $\tau \in (0,1]$*

$$\overline{\rho}_Y(\tau/4M) \leq 2\overline{\rho}_X(\tau).$$

Here $d_L(X,Y)$ denotes the Lipschitz distance between X and Y.

If $U : X \longrightarrow Y$ is a Lipschitz isomorphism, the norm we need on Y is predual to the equivalent norm defined on Y^* by

$$\|y^*\| = \sup\left\{\frac{y^*(U(x_1) - U(x_2))}{\|x_1 - x_2\|} \;:\; x_1 \neq x_2\right\}$$

The rate of change which appears in this definition somehow provides a substitute to the lack of points of differentiability for Lipschitz map between a.u.s. spaces. We note that Theorem 3.1 also has a quite similar uniform version [38, Theorem 5.3] involving a sequence of norms.

The difficult part of the proof of Theorem 3.1 consists into controlling $\theta_Y(\tau)$ from below, since the natural "adjoint" of the Lipschitz map $U : X \longrightarrow Y$ maps the space $\mathrm{Lip}(Y)$ of real-valued Lipschitz functions on Y to $\mathrm{Lip}(X)$, but does not map Y^* to X^*. A proper use of the Gorelik principle ([42]; see [38, Prop. 5.1]) overcomes this obstruction by showing that if X_0 is a finite-codimensional subspace of X, then $U(B_{X_0})$ asymptotically norms the weak*-null sequences in Y^*. This principle therefore says that Lipschitz isomorphisms retain some restricted form of weak continuity, although they are of course not weakly continuous in general.

The quantitative Lipschitz invariance of a.u.s. norms leads (through [93] and [38]) to the Lipschitz invariance of Szlenk indices – at least when they are minimal, see [23] for more – and it also applies to the "optimal" case: indeed it is shown in [37] that a separable Banach space X is isomorphic to a subspace of $c_0(\mathbb{N})$ if and only if it has an equivalent norm such that $\rho_X(\tau_0) = 0$ for some $\tau_0 > 0$. Then Theorem 3.1 implies that the class of subspaces of $c_0(\mathbb{N})$ is stable under Lipschitz isomorphisms, from which it follows that if X is Lipschitz-isomorphic to $c_0(\mathbb{N})$ then it is actually linearly isomorphic to $c_0(\mathbb{N})$. [37, Theorem 2.2].

This looks like the right place to state an important question.

Open problem. *Let X and Y be two Lipschitz-isomorphic separable Banach spaces. Are X and Y linearly isomorphic?*

This problem is open even if $X = \ell_1(\mathbb{N})$, or if X and Y are assumed to be super-reflexive. Counterexamples are available in the non-separable case [1], and the relevance of separability is displayed in [36]. Let $\mathrm{Lip}_0(X)$ be the space of real-valued Lipschitz functions on a Banach space X which vanish at 0, and let $\mathfrak{F}(X)$ be the natural predual of $\mathrm{Lip}_0(X)$, whose weak*-topology coincide on the unit ball of $\mathrm{Lip}_0(X)$ with the pointwise convergence on X. The Dirac map $\delta : X \longrightarrow \mathfrak{F}(X)$ defined by $\langle g, \delta(x)\rangle = g(x)$ is an isometric (non-linear!) embedding from X to a subset of $\mathfrak{F}(X)$ which generates a dense linear subspace. This Dirac map has a linear left inverse $\beta : \mathfrak{F}(X) \longrightarrow X$ which is the quotient map such that $x^*(\beta(\mu)) = \langle x^*, \mu\rangle$ for all $x^* \in X^*$; in other words, β is the extension to $\mathfrak{F}(X)$ of the barycenter map. This setting provides canonical examples of Lipschitz-isomorphic spaces. Indeed, if we let $Z_X = Ker(\beta)$, it follows easily from $\beta\delta = Id_X$ that $Z_X \oplus X = \mathfrak{G}(X)$ is Lipschitz-isomorphic to $\mathfrak{F}(X)$.

Following [36], let us say that a Banach space X has the *lifting property* if when Y and Z are Banach spaces and $S : Z \to Y$ and $T : X \to Y$ are continuous linear maps, the existence of a Lipschitz map $\mathcal{L}$ such that $T = S\mathcal{L}$ implies the existence of a continuous linear operator L such that $T = SL$.

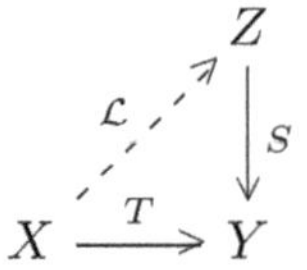

A diagram-chasing argument shows that $\mathfrak{G}(X)$ is linearly isomorphic to $\mathfrak{F}(X)$ if and only if X has lifting property.

It turns out that non-separable reflexive spaces, and also the spaces $\ell_\infty(\mathbb{N})$ and $c_0(\Gamma)$ when Γ is uncountable, fail the lifting property and this provides canonical examples of pairs of Lipschitz but not linearly isomorphic spaces. On the other hand, every *separable* space X has the lifting property: to prove it, one can pick a Gaussian measure γ whose support is dense in X and use the fact that $(\mathcal{L} * \gamma)$ is weakly Gâteaux-differentiable. Then $L = (\mathcal{L} * \gamma)'(0)$ satisfies $T = SL$.

The lifting property for separable spaces forbids the existence of canonical pairs of non-isomorphic separable spaces, but on the other hand it leads to

Theorem 3.2. *Let X be a separable Banach space. If there exists an isometric embedding from X into a Banach space Y, then Y contains a linear subspace which is isometric to X.*

Indeed a theorem due to Figiel [32] states that if $J : X \longrightarrow Y$ is an isometric embedding such that $J(0) = 0$ and $\overline{\text{span}}[J(X)] = Y$ then there is a linear quotient map with $\|Q\| = 1$ and $QJ = Id_X$, and then one applies the lifting property with $J = \mathcal{L}, S = Q$ and $T = Id_X$.

Theorem 3.2 does not extend to the non-separable case: if X is reflexive and non-separable then X embeds isometrically but not linearly into $\mathfrak{F}(X)$ [36, Proposition 4.1].

Non-linear theory is also an invitation to enter the Kalton zone $0 < p < 1$, following [73]. If $(X, \|\cdot\|)$ is a metric space and $\omega : [0, \infty) \longrightarrow [0, \infty)$ is a subadditive function such that $\lim_{t \to 0} \omega(t) = \omega(0) = 0$ and $\omega(t) = t$ if $t \geq 1$, then the space $\text{Lip}_\omega(X)$ of $(\omega \circ d)$-Lipschitz functions on X which vanish at 0 has a natural predual denoted $\mathfrak{F}_\omega(X)$, and the barycentric map $\beta_\omega : \mathfrak{F}_\omega(X) \longrightarrow X$ (whose adjoint is the canonical embedding from X^* to $\mathfrak{F}_\omega(X)$) is still a linear quotient map such that $\beta_\omega \delta = Id_X$. However, the Dirac map $\delta : X \longrightarrow \mathfrak{F}_\omega(X)$ is now uniformly continuous with modulus ω – e.g. α-Hölder when $\omega(t) = \max(t^\alpha, t)$ with $0 < \alpha < 1$.

Uniformly continuous functions usually fail the differentiability properties that Lipschitz functions may enjoy, and thus one can expect that this part of the theory is more "distant" from the linear theory than the Lipschitz one. It is indeed so, and [73, Theorem 4.6], reads as follows.

Theorem 3.3. *If ω satisfies $\lim_{t \to 0} \omega(t)/t = \infty$, then $\mathfrak{F}_\omega(X)$ is a Schur space – that is, weakly convergent sequences in X are norm convergent.*

It follows from Theorem 3.3 that the uniform analogue of the lifting property fails unless X has the (quite restrictive) Schur property. Moreover, $\mathfrak{F}_\omega(X)$ is (3ω)-uniformly homeomorphic to $[X \oplus \text{Ker}(\beta_\omega)]$ and as soon as $\lim_{t \to 0} \omega(t)/t = 0$ and X fails the Schur property we obtain canonical pairs of uniformly homeomorphic separable Banach spaces which are not linearly isomorphic. We refer to [120, 53] for other examples of such pairs.

Along with Hölder maps between Banach spaces, one may as well consider Lipschitz map between quasi-Banach spaces, and this is done in [2] where similar methods provide examples of separable quasi-Banach spaces which are Lipschitz but not linearly isomorphic.

It is also shown in [2, Theorem 3.2] that Theorem 3.2 drastically fails for quasi-Banach spaces: if $0 < p < 1$, there exists separable p-normed quasi-Banach spaces X and Y such that X embeds isometrically in Y, but if $T : X \longrightarrow Y$ is a continuous linear operator then $T = 0$. The proof relies in particular on the previously seen fact that if $0 < p < 1$, the space $X = L_p(\mathbb{T})/H^p(\mathbb{T})$ is not "natural" in Kalton's sense [62].

We conclude this section by mentioning a very recent embedding result which provides the optimal quantitative version of Aharoni's embedding theorem: if M is a separable metric space, there exists a 2-Lipschitz embedding from M into $c_0(\mathbb{N})$ and the constant 2 is best possible [75]. It is not known if a separable Banach space X which contain a Lipschitz copy of every separable metric space (or equivalently, a Lipschitz copy of $c_0(\mathbb{N})$) contains a closed subspace isomorphic to $c_0(\mathbb{N})$. However, differentiation shows that such an X fails the Radon–Nikodym property, and it is also known that $c_0(\mathbb{N})$ does not embed uniformly or coarsely into a reflexive space.

Embedding spaces, trees or groups in the Hilbert space, in a Lipschitz, uniform or coarse way, has recently become a major field of research, which happens to be related with deep geometric questions investigated in particular by M. Gromov and A. Naor. In order to remain within our present topic, let us simply mention the article [55] where it is shown that if $p > 2$, the space l_p does not coarsely embed into a Hilbert space.

4 Isometric theory

It is sometimes useful to work with special norms on Banach spaces: they might be canonical or easy to compute, or they can be tightly related with the structure of operators of the space, or they can provide isomorphic information on the space.

All this motivates the investigation of isometric theory, that is, the study of Banach spaces equipped with a given norm.

It should be pointed out that the real and complex isometric theory are quite different. On a complex Banach space X, one can define the notion of a *Hermitian* linear operator T by: $\|e^{ixT}\| = 1$ for all $x \in \mathbb{R}$. Let us say that $x \in X$ is Hermitian if there exists $x^* \in X^*$ such that $(x^* \otimes x)$ is a Hermitian operator. It is not difficult to check that a projection P is Hermitian if and only if $\|P + \lambda(I - P)\| = 1$ for all $\lambda \in \mathbb{C}$ with $|\lambda| = 1$. In other words, Hermitian projections are "orthogonal". It follows that a complex Banach space with a 1-unconditional basis is the closed linear span of its Hermitian elements. A remarkable result of Kalton and Wood ([92]; see [126]) states the converse:

Theorem 4.1. *A complex Banach space X which is the closed linear span of its Hermitian elements has a* 1*-unconditional basis.*

An important corollary of Theorem 4.1 is that if X is a 1-complemented subspace of a complex Banach space Y with a 1-unconditional basis, then X has a 1-unconditional basis: indeed let us write

$$I_Y = \sum_{j=1}^{\infty} y_j^* \otimes y_j$$

where (y_j) is a 1-unconditional basis of Y, and let $P : Y \longrightarrow X$ be such that $P^2 = P$ and $\|P\| = 1$. If $T_j = y_j^* \otimes P(y_j)$ then $x = \sum_{j=1}^{\infty} T_j(x)$ for all $x \in X$ and

$$\Big\|\sum_{j=1}^{\infty} \alpha_j T_j(x)\Big\| \leq \|x\|$$

if $|\alpha_j| = 1$ for all j. But if $\|x^*\| = \|x\| = x^*(x) = 1$, we have

$$1 = \sum_{j=1}^{\infty} \langle x^*, T_j(x)\rangle$$

and

$$|\sum_{j=1}^{\infty} \alpha_j \langle x^*, T_j(x) \rangle| \leq 1$$

if $|\alpha_j| = 1$, and this implies that $\langle x^*, T_j(x) \rangle$ is real (and ≥ 0) for all j. It follows that $P(y_j) \in X$ is Hermitian for all j, and Theorem 4.1 concludes the proof.

It should be noted that in the *real* case, the existence of 1-unconditional bases does not pass to 1-complemented subspaces [101]. We refer to [117] for important results on 1-complemented subspaces of spaces with 1-unconditional bases. The Kalton–Wood theorem is one of the few available positive results, while the following question is still open:

Open problem. *Let X be a complemented subspace of a space Y with an unconditional basis. Does X have an unconditional basis?*

A negative answer looks plausible, since for instance it is observed in [14, §4] that a space with BAP but no FDD constructed by C. Read (unpublished) is complemented in a space with an unconditional FDD. On the other hand [14] contains a major positive result.

Theorem 4.2. *Let X be a separable Banach space. If X has the metric approximation property (MAP) then X has the commuting metric approximation property (CMAP).*

The proof goes as follows: if X has MAP then there exists a sequence of finite rank operators (T_n) such that $T_k T_n = T_n$ if $k > n$ and $\|T_n\| \leq 1 + \varepsilon_n$ with $\sum \varepsilon_n = \beta < \infty$. For $t > 0$ and $n \geq 1$, let

$$V_n(t) = e^{-nt} \exp(t \sum_{k=1}^{n} T_k).$$

Then $\|V_n(t)\| \leq e^{\beta t}$ and

$$\lim_{n \to \infty} V_n(t)x = S(t)x$$

exists for all $x \in X$. The operator $S(t)$ is compact for all $t > 0$, $\lim_{t \to 0} \|S(t)\| = 1$ and $\lim_{t \to 0} S(t)x = x$ for all $x \in X$. Now $S(t+s) = S(t)S(s)$ and thus of course $S(t)S(s) = S(s)S(t)$. Now the compact operators $S(1/n)$ can be approximated by finite rank operators which "nearly commute" and the conclusion follows through a perturbation argument. □

An alternative proof of Theorem 4.2 (with a slightly better control on the commuting sequence) is given in [35] for separable spaces which do not contain $\ell_1(\mathbb{N})$, and it is also shown there that the unconditional MAP is equivalent to its commutative version for all separable spaces (with a simpler proof in the complex spaces, using the Hermitian trick which is displayed above).

The following important problem is, however, still open.

Open problem. *Let X be a separable Banach space with the bounded approximation property. Does X have the commuting bounded approximation property?*

By Theorem 4.2 an equivalent formulation of the problem is to know if a Banach space with BAP has an equivalent norm with MAP. It follows from results of Casazza [11] and Johnson [51] that still another formulation of this problem is to know if every "π-space" (that is, every space which has BAP with a sequence of finite rank *projections*) has a FDD. We refer to [12] for an updated survey on approximation properties.

An isometric concept which turned out to be very useful was defined by Alfsen and Effros in 1972 [3]: a closed subspace X of a Banach space Y is called an *M-ideal* in Y if there is a subspace V of Y^* such that

$$Y^* = V \oplus_1 X^\perp$$

where $\oplus_1$ means that $\|y^* + z^*\| = \|y^*\| + \|z^*\|$ for all $y^* \in V$ and all $z^* \in X^\perp$. We refer to [46] for the theory as it was in 1993, immediately after Kalton's breakthrough [68].

Although the notion of M-ideal is independent of any algebraic structure, it turns out to be tightly related with the notion of "ideal" from operator theory, and in fact ideals $K(X)$ of compact operators in spaces $L(X)$ of bounded operators provide a wealth of examples of M-ideals (see [46, Chapters V and VI]). It is shown in [68] that if X is a separable Banach space then $K(X)$ is an M-ideal in $L(X)$ (in short, $K(X) \subset_M L(X)$) if and only if there is a sequence (K_n) of compact operators such that

$$\lim_{n\to\infty} \|x - K_n(x)\| = 0$$

$$\lim_{n\to\infty} \|x^* - K_n^*(x^*)\| = 0$$

for all $x \in X$ and $x^* \in X^*$, such that

$$\lim_{n\to\infty} \|I - 2K_n\| = 1$$

and moreover X has *Propery (M)*, which means that for any sequence $(x_n) \subseteq X$ with $w - \lim_{n\to\infty}(x_n) = 0$, one has

$$\overline{\lim}\|x + x_n\| = \overline{\lim}\|y + x_n\|$$

for all $(x, y) \in X^2$ with $\|x\| = \|y\| = 1$. In other words, the norm of X is "asymptotically isotropic" and all vectors of the sphere "look the same" when seen from infinity. It is shown in this same paper that property (M) forbids distortion: if X has (M), there is some $p \in [1, +\infty]$ such that for every $\varepsilon > 0$, X contains a subspace $(1+\varepsilon)$-isomorphic to ℓ_p if $p < \infty$ and to c_0 if $p = \infty$.

Note that it is obvious that ℓ_p spaces equipped with their natural norms have (M), while L_p spaces fail to have it: this is related to the fact that they contain L_2 – and incidentally to the fact that the "asymptotic modulus of smoothness" (see Section 3 above) is better for ℓ_p than for L_p $(p > 2)$. In fact, if X is a separable order-continuous nonatomic Banach lattice and X has an equivalent norm with (M) then X is lattice-isomorphic to L_2 [68], and this is in particular the case when X is isomorphic to a subspace of an Orlicz sequence space h_F. In the article [4], twisted sums meet property

(M) and it is shown in particular that the Kalton–Peck space Z_2 has an equivalent norm with (M). On the other hand, it is shown in the recent work [24] that norms with (M) are optimally asymptotically uniformly smooth among all equivalent norms on a given Banach space. Since Theorem 3.1 above asserts that the corresponding modulus is a Lipschitz invariant, it follows for instance that Lipschitz-isomorphic Orlicz spaces contain the same l_p spaces.

The main result of [68] was soon improved in [91] where it was shown that the unconditionality condition (3) actually follows from the isotropy condition (M). Thus [91, Theorem 2.13] reads:

Theorem 4.3. *Let X be a separable Banach space. Then $K(X)$ is an M-ideal in $L(X)$ if and only if X does not contain a copy of ℓ_1, X has (M) and X has the metric compact approximation property.*

In this same paper, property (M) which, as seen before, yields to "discrete" spaces when applied to subspaces of L_p, is used to prove [91, Theorem 4.4].

Theorem 4.4. *Let $1 < p < \infty, p \neq 2$, and X be a closed infinite-dimensional subspace of L_p. Then B_X is $\|\cdot\|_1$-compact if and only if for any $\varepsilon > 0$, there is a subspace X_ε of ℓ_p such that $d(X, X_\varepsilon) < 1 + \varepsilon$.*

Note that if $p > 2$ the assumption amounts to say that X contains no subspace isomorphic to ℓ_2, and Theorem 4.4 improves a result of [54]. We mention at this point that it has been recently shown in [47] that if $p > 2$, $X \subseteq L_p$ and $\ell_p(\ell_2)$ does not embed in X, then X embeds into $\ell_p \oplus \ell_2$.

Theorem 4.4 has been pushed to the case $p = 1$ in [39], to characterize subspaces of L_1 which ε-embed into ℓ_1, but the non-reflexivity of L_1 (and the non-convexity of L_p if $p < 1$) somehow complicates the matter, and [39, Theorem 3.3] states that if X is a closed subspace of L^1 with the approximation property, then B_X is L_p-compact locally convex $(0 \leq p < 1)$ if and only if for every $\varepsilon > 0$, there is a quotient space E_ε of c_0 such that $d(X, E_\varepsilon^*) < 1 + \varepsilon$. We already saw in Section 2 above that the "Ribe–Roberts" approach provides examples where B_X is L_p-compact with $p < 1$ (or equivalently, τ_m-compact) but not locally convex.

Following [14], we say that $X \subseteq Y$ is a *u-ideal* in Y if

$$Y^* = V \oplus X^\perp$$

in such a way that $\|y^* + z^*\| = \|y^* + \lambda z^*\|$ for all $y^* \in V, z^* \in X^\perp$ and $|\lambda| = 1$.

The article [41] is devoted to this notion which, thanks again to Hermitian operators, is quite nicer in the complex case. Let us denote $Ba(X)$ the subspace of X^{**} consisting of weak*-limits of weak*-convergent sequences of elements of X. With this notation, it follows from [41, Theorem 6.5] that we have:

Theorem 4.5. *Let X be a separable complex Banach space which is a u-ideal in its bidual. Then there exists an Hermitian projection from X^{**} onto $Ba(X)$. Moreover for every $x^{**} \in Ba(X)$ and every $\varepsilon > 0$, there is a sequence (x_n) in X such that*

$$x^{**} = w^* - \lim_{n\to\infty} \sum_{j=1}^{n} x_j \quad \textit{and} \quad \|\sum_{j=1}^{n} \theta_j x_j\| \leq (1+\varepsilon)\|x^{**}\|$$

for all $n \geq 1$ and all $|\theta_j| = 1$.

If one thinks of $Ba(X)$ as the "band" generated by X in X^{**}, one can express Theorem 4.5 by saying that the embedding of a u-ideal in its bidual looks very much like the embedding of an order-continuous Banach lattice. However, u-ideals (such as $K(X)$ spaces, with X reflexive with the unconditional compact approximation property) bear in general no usable order structure. We refer to [87] for (order) ideal properties of (algebraic) ideals of operators between Banach lattices.

5 Interpolation, twisted sums, and the Kalton calculus

The Banach–Mazur functional d_{BM} is a classical tool for estimating the "distance" between two isomorphic Banach spaces – or equivalently, the distance between two equivalent norms and similar functionals such as the Lipschitz distance d_L can be defined when more general notions of isomorphisms are taken into consideration. In [80], "distances" are defined between spaces which are not isomorphic but are somewhat similar, such as ℓ_p and ℓ_q when p and q are close to each other. If X and Y are two subspaces of a Banach space Z, let $\Lambda(X,Y)$ be the Hausdorff distance between B_X and B_Y, that is

$$\Lambda(X,Y) = \max\{\sup_{x\in B_X} \inf_{y\in B_Y} \|x-y\|, \sup_{y\in B_Y} \inf_{x\in B_X} \|y-x\|\}.$$

The Kadets distance $d_K(X,Y)$ is the infimum of $\Lambda(\tilde{X},\tilde{Y})$ over all Banach spaces Z containing isometric copies $\tilde{X}$ and $\tilde{Y}$ of X and Y respectively. The Kadets distance is a pseudo-metric which is controlled from above by d_{BM}, but there are non-isomorphic Banach spaces X and Y such that $d_K(X,Y) = 0$.

The Gromov–Hausdorff distance d_{GH} is the non-linear analogue of the Kadets distance, defined along the same lines, except that the infimum is taken over all metric spaces containing isometric copies of X and Y. Of course, $d_{GH} \leq d_K$ and for instance $d_{GH}(\ell_p,\ell_1) \to 0$ as $p \to 1$ while $d_K(\ell_p,\ell_1) = 1$ for all $p > 1$. However, if X is a $\mathcal{K}$-space then $d_{GH}(X_n,X) \to 0$ implies that $d_K(X_n,X) \to 0$. This can again be understood as an "approximation by linear maps" on $\mathcal{K}$-spaces.

Interpolation theory provides families of Banach spaces which are not isomorphic but tightly related, and Kadets distance will make this remark precise – and usable. Moreover, interpolation leads to a "differential calculus" on the "manifold" of Banach spaces. We will outline how N. Kalton's vision created links between this calculus, twisted sums, and quasi-linear maps.

Complex interpolation studies analytic families of Banach spaces. Let us restrict our discussion to a very important special case: let W be some complex Banach space and let X_0 and X_1 be two closed subspaces of W. We denote

$$S = \{z \in \mathbb{C}; 0 < \mathrm{Re}(z) < 1\}$$

and $\mathfrak{F}$ is the space of analytic functions $F : S \longrightarrow W$ which extend continuously to $\bar{S}$ and such that $\{F(it); t \in \mathbb{R}\}$ is a bounded subset of X_0 and $\{F(1+it); t \in \mathbb{R}\}$ is a bounded subset of X_1. The space $\mathfrak{F}$ is normed by

$$\|F\|_{\mathfrak{F}} = \max_{j=0,1} \sup\{\|F(j+it)\|_{X_j}; t \in \mathbb{R}\}.$$

For $\theta \in (0,1)$ and $x \in W$, we define

$$\|x\|_\theta = \inf\{\|F\|_{\mathfrak{F}}; F(\theta) = x\}$$

and

$$X_\theta = \{x \in W; \|x\|_\theta < \infty\}$$

If $W_0 = \mathrm{span}\{X_\theta; \theta \in (0,1)\}$, a linear map $T : W_0 \longrightarrow W_0$ is called *interpolating* if $F \longmapsto T \circ F$ is defined and bounded on $\mathfrak{F}$. If T is interpolating, then $T(X_\theta) \subseteq X_\theta$ for all $\theta \in (0,1)$.

The above space $X_\theta = [X_0, X_1]_\theta$ is said to be obtained from X_0 and X_1 by the *complex interpolation method* and one has [80] for $0 < \theta < \phi < 1$

$$d_K(X_\theta, X_\phi) \leq 2 \frac{\sin[\pi \frac{(\phi-\theta)}{2}]}{\sin[\pi \frac{(\phi+\theta)}{2}]}.$$

This continuity of the interpolation method with respect to the Kadets distance permits to apply connectedness arguments. Indeed, let us call a property (P) *stable* if there exists $\alpha > 0$ so that if X has (P), and $d_K(X,Y) < \alpha$, then Y has (P). For instance, each of the following properties (P) is stable: separability, reflexivity, $X \supseteq \ell_1$, super-reflexivity, type$(X) > 1$. Connectedness thus shows that if $0 < \theta < 1$ and $X_\theta = [X_0, X_1]_\theta$ has (P), then X_φ has (P) for every $\varphi \in (0,1)$.

This line of thought opens an exciting field of research. It can be shown that the connected component of any separable Banach space X contains all isomorphic copies of X. It follows from [111] that the connected component of ℓ_2 contains all super-reflexive Banach lattices, and it is not known whether it contains all super-reflexive spaces. It is conjectured that the component of c_0 consists of all spaces isomorphic to a subspace of c_0.

These concepts are also relevant to non-linear isomorphisms: it follows from instance from Sobczyk's theorem that if $d_{GH}(X_n, c_0) \to 0$ then we have not only that $d_K(X_n, c_0) \to 0$ (since c_0 is a $\mathcal{K}$-space), but actually $d_{BM}(X_n, c_0) \to 0$ [80]. This implies for instance that if the uniform distance between X and c_0 is small then X is linearly isomorphic to c_0 [38, Theorem 5.7]. It is not known whether a space which is uniformly homeomorphic to c_0 is linearly isomorphic to c_0.

The Kadets and Gromov–Hausdorff distances clearly are topological notions, but interpolation points to some kind of differential structure, which we now outline.

Minimal extensions have been displayed in Section 2. Let us say more generally that if X, Y and Z are quasi-Banach spaces, Z is an extension of X by Y if

$$Z/Y \simeq X.$$

An extension Z is also called a *twisted sum* of X and Y (a non-trivial twisted sum if Y is not complemented in Z), and we refer to [17] for a comprehensive survey of this matter.

A map $\Omega : X \longrightarrow Y$ is called *quasi-linear* (see Section 2) if $\Omega(\lambda x) = \lambda\Omega(x)$ for all $x \in X$ and $\lambda \in \mathbb{K}$, and if there is $C > 0$ such that

$$\|\Omega(x_1 + x_2) - \Omega(x_1) - \Omega(x_2)\| \leq C(\|x_1\| + \|x_2\|)$$

for all $x_1, x_2 \in X$. The extension $X \oplus_\Omega Y$ of X by Y is the space $X \oplus Y$ equipped with the quasi-norm

$$\|(x, y)\| = \|x\| + \|y - \Omega(x)\|$$

Even when X and Y are Banach spaces $X \oplus_\Omega Y$ is not, unless Ω actually satisfies for all $n \geq 1$ and all (x_k)

$$\|\sum_{k=1}^{n} \Omega(x_k) - \Omega(\sum_{k=1}^{n} x_k)\| \leq C \sum_{k=1}^{n} \|x_k\|.$$

It will always be so when X is a $\mathcal{K}$-space [59]. Every extension can actually be obtained with such an Ω: if $q : Z \longrightarrow X$ is the quotient map. Take $\Omega = S - R$ where $qS = qR = Id_X$, S is homogenous (but not necessarily linear) such that $\|S(x)\| \leq 2\|x\|$, and R is linear (but not necessarily continuous). As in the case of minimal extensions, the existence of a bounded linear projection from $X \oplus_\Omega Y$ onto Y is equivalent to the existence of a linear map $L : X \longrightarrow Y$ such that

$$\|\Omega(x) - L(x)\| \leq C\|x\|$$

for all $x \in X$. When $X = Y$, the space $X \oplus_\Omega X$ is called a *self-extension* of X and it is denoted by

$$X \oplus_\Omega X = d_\Omega X.$$

When $X = \ell_2$, a non-trivial self-extension of ℓ_2 is called a twisted Hilbert space, and it was shown in [28] that such spaces exist. An alternative example, the Kalton–Peck space Z_2, is constructed in [81] with the help of the Ribe functional (see Section 2): let $\Omega = \ell_2 \longrightarrow \mathbb{R}^{\mathbb{N}}$ be defined by

$$\Omega((\xi_n)) = \Big(\xi_n \log(\frac{|\xi_n|}{\|\xi\|_2})\Big)_{n \geq 1}$$

(and $\Omega(0) = 0$). The space $Z_2 = d_\Omega \ell_2$ is therefore the space of pairs of sequences $((\xi_n), (\eta_n))$ such that

$$\|(\xi, \eta)\| = \Big(\sum_{n=1}^{\infty} |\xi_n|^2\Big)^{\frac{1}{2}} + \Big(\sum_{n=1}^{\infty} \Big|\eta_n - \xi_n \log \frac{|\xi_n|}{\|\xi\|_2}\Big|^2\Big)^{\frac{1}{2}} < \infty.$$

Of course Z_2 is a Banach space since ℓ_2 is a $\mathcal{K}$-space. The Kalton–Peck space Z_2 exhibits remarkable features, which are not yet fully understood although that space was constructed 30 years ago. It is plain that Z_2 has an unconditional FDD consisting of 2-dimensional spaces; however it has no unconditional basis and no local unconditional structure [53]. Actually, an unconditional FDD with spaces of bounded dimension provides an unconditional basis which can be chosen from the subspaces *if* the space has local unconditional structure [15]. It is unknown, however, if a twisted Hilbert space can have local unconditional structure; the best result so far is that it has no unconditional basis in full generality [72]. The space Z_2 is also an example of a symplectic Banach space which is not the direct sum of two isotropic subspaces [88]. In fact, intuition suggests that the space Z_2 is "even-dimensional" and thus that it should not be isomorphic to its hyperplanes: this 30-years old conjecture is still open, although examples of infinite-dimensional Banach spaces which are not isomorphic to their hyperplanes are now known [43]. In fact, there are even (necessarily non-separable) $C(K)$ spaces which are indecomposable and not isomorphic to their hyperplanes [98, 114]. These spaces are obtained by techniques which are completely independent from the Gowers–Maurey approach. The complexity of their construction relies on the underlying compact space K, which is obtained through an elaborate transfinite induction. Coming back to Z_2, we note that spaces with 2-dimensional unconditional FDD but no unconditional basis show up in the classification results shown in [97], which play a crucial role in Gowers' homogeneous space theorem [44].

As so often in N. Kalton's work, the conceptual frame in which the construction is completed provides flexibility and leads to more results. If $F : \mathbb{R} \longrightarrow \mathbb{C}$ is any Lipschitz map and E is a Banach sequence space, let

$$\Omega_F(\xi) = \big(\xi_n F(\log \frac{|\xi_n|}{\|\xi\|_E})\big)_{n\geq 1}$$

and let

$$d_{\Omega_F} E = E \oplus_{\Omega_F} E.$$

Taking $E = \ell_2$ and $F(t) = t^{1+i\alpha}$ ($\alpha \neq 0$) provides a complex Banach space $Z(\alpha)$ which is not complex-isomorphic to its conjugate space $\overline{Z(\alpha)} = Z(-\alpha)$. The existence of such spaces had been shown in [8] and [128] by probabilistic methods. We refer to [30] for recent work on this topic, and the existence of Banach spaces with exactly n complex structures, for any given integer n, and also to [31].

The notation $d_\Omega X$ is reminiscent of differential calculus, and this is not a chance. With the above notation of the complex interpolation method, and following [125], we define a derived space $dX_\theta \subseteq W \times W$ by $dX_\theta = \{(x_1, x_2) : \|(x_1, x_2)\|_{dX\theta} < \infty\}$ where

$$\|(x_1, x_2)\|_{dX_\theta} = \inf\big\{\|F\|_{\mathfrak{F}} : F(z) = x_1, F'(z) = x_2\big\}.$$

The space $Y = \{(x_1, x_2) \in dX_\theta : x_1 = 0\}$ is isometric to X_θ and so is dX_θ/Y. Hence dX_θ is a self-extension of X_θ. By the above, one has

$$dX_\theta = d_\Omega X_\theta$$

for some quasi-linear map $\Omega : X_\theta \longrightarrow W$ and $\Omega(x) = F'(z)$, where $F \in \mathfrak{F}$ is such that $\|F\|_{\mathfrak{F}} \leq C\|x\|_\theta$ and $F(z) = x$, does the work. Now, if T is an interpolating operator then $(x_1, x_2) \longrightarrow (Tx_1, Tx_2)$ is bounded on dX_θ and this translates into "commutator estimates":

$$\|T(\Omega(x)) - \Omega(T(x))\|_\theta \leq C\|x\|_\theta$$

for all $x \in X_\theta$.

The sequence spaces ℓ_p $(1 \leq p \leq \infty)$ provide a first example of interpolation line and the above calculations applied to $X_0 = \ell_1$ and $X_1 = \ell_\infty$ provide the Kalton–Peck space

$$Z_2 = dX_{1/2}.$$

Similar calculations are possible for the function spaces $L_p(\mathbb{T})$. For this interpolation scale, the Hilbert transform H is a very important example of interpolating operator and the Rochberg–Weiss commutator estimate becomes in this case

$$\|H(f \log|f|) - H(f)\log|H(f)|\|_p \leq C_p\|f\|$$

for $1 < p < \infty$ and some $C_p < \infty$.

Following [67] and [64], we now relate this differential calculus with entropy functions (see Section 2). If X is a sequence space, its entropy function Φ_X is defined for positive sequences u by

$$\Phi_X(u) = \sup_{\|x\|_X \leq 1} \sum_{k=1}^{\infty} u_k \log|x_k|.$$

If X_0 and X_1 are separable sequence spaces, the interpolation spaces X_θ are given by the Calderon formula $X_\theta = X_0^{1-\theta} X_1^\theta$, that is:

$$\|x\|_\theta = \inf\{\|x_0\|_0^{1-\theta}\|x_2\|_1^\theta \ ; \ |x| = |x_0|^{1-\theta}|x_1|^\theta\}$$

and the entropy function Φ_X can conveniently be described as the *logarithm* of the sequence space X. Indeed one has

$$\Phi_{X_\theta} = (1-\theta)\Phi_{X_0} + \theta\Phi_{X_1}$$

and by the Lozanovsky factorization theorem,

$$\Phi_X + \Phi_{X^*} = \Phi_{\ell_1}$$

where Φ_{ℓ_1} is the Ribe functional while (see Section 2)

$$\Phi_{\ell_p} = \frac{1}{p}\Phi_{\ell_1}$$

and $\Phi_{\ell_\infty} = 0$. It now becomes natural to see the Hilbert space as the geometric mean between any sequence space X and its dual X^*. The map "$X \mapsto \Phi_X$" is logarithmic-like, but in order to complete the picture we conversely need an *exponential* function which maps a quasi-linear map Φ to a Banach space. If $\Phi : c_{00}^{+} \longrightarrow \mathbb{R}$ is any functional, there exists a Banach sequence space X such that $\Phi_X = \Phi$ if and only if Φ and $(\Phi_{\ell_1} - \Phi)$ are convex functions and Φ is positively homogeneous [67], and the space X has unit ball

$$B_X = \big\{(x_k);\ \sum_{k=1}^{\infty} u_k \log|x_k| \le \Phi(u) \text{ for all } u \ge 0\big\}.$$

This exponential map leads to what I suggest to call the *Kalton calculus*, which bears an uncanny resemblance with the exponentation from a Lie algebra to its Lie group, and creates "lines" from infinitesimals; in other words, yields to *extrapolation*. For instance, if X is p-convex $(1 < p < 2)$ and p^*-concave, then $X = Y^{1/p}$ for some sequence space Y and so $(\frac{1}{p}\Phi_{\ell_1} - \Phi_X)$ is convex. Similarly p^*-concavity means that $(\frac{1}{p}\Phi_{\ell_1} - \Phi_{X^*})$ is convex. Now the equation

$$\begin{aligned}\Phi_X &= (1-\theta)\Phi + \theta\Phi_{\ell_2}\\ &= (1-\theta)\Phi + \frac{\theta}{2}\Phi_{\ell_1}\end{aligned}$$

provides a convex function Φ such that $(\Phi_{\ell_1} - \Phi)$ is also convex and thus $\Phi = \Phi_Z$ for some Z. Exponentiating, we find $X = Z^{1-\theta}\ell_2^{\theta}$ (a result from [111]).

To close the circle of ideas relating the entropy functions with derived spaces, we note that if $X_\theta = X_0^{1-\theta}X_1^{\theta}$ then $dX_\theta = d_\Omega X_\theta = X_\theta \oplus_\Omega X_\theta$ where the quasi-linear map Ω satisfies

$$|\langle x^*, \Omega(x)\rangle - \Phi(xx^*)| \le C\|x\|_{X_\theta}\|x^*\|_{X_\theta^*}$$

where $\Phi = \Phi_{X_1} - \Phi_{X_0}$ and (xx^*) denotes the pointwise product of the sequences x and x^*.

Special properties of the derived space $d_\Omega X_\theta$ can "spread out" by exponentation to a segment $\{X_\varphi;\ |\varphi - \theta| < \varepsilon\}$. Indeed, if X_0 and X_1 are acceptable function spaces on $\mathbb{T}$ and R is the vector-valued Riesz transform, then R is bounded on X_θ for $|\theta - \theta_0| < \delta$ if and only if $\|R\Omega - \Omega R\|_{X_{\theta_0}} < \infty$. It follows that there exist twisted Hilbert spaces which are not UMD [67] although Z_2 is UMD [63]. We note at this point that higher order derivatives can be considered, and this has been done e.g. in [10, 106, 124].

As seen before, differentiating interpolation lines yield quasi-linear maps Ω such that $dX_\theta = d_\Omega X_\theta$. If for instance X_1 is obtained from X_0 through a change of weight, the map Ω enjoys a commutation property, namely

$$\|\Omega(ax) - a\Omega(x)\|_{X_\theta} \le C\|a\|_\infty \|x\|_{X_\theta}.$$

These special maps are called *centralizers* in [64] and the corresponding space $d_\Omega X_\theta$ is a *lattice twisted sum*.

Centralizers yield to extrapolation results: if for instance X is a super-reflexive sequence space and Ω is a real centralizer on X, then there exist super-reflexive Banach sequence spaces X_0 and X_1 such that $X = X_0^{1/2}X_1^{1/2} = X_{1/2}$ and moreover $dX_{1/2} \simeq d_\Omega X$. Note that the Kalton calculus, which we displayed here (following [79]) for sequences, is designed for function spaces and this is what is done in [67] and [64].

We now recall the Rochberg–Weiss commutator estimates: if $X_\theta = X_0^{1-\theta}X_1^\theta$ and $dX_\theta = d_\Omega X_\theta$ then

$$\|T(\Omega(x)) - \Omega(T(x))\|_{X_\theta} \le C\|x\|_{X_\theta}$$

for interpolating operators T. When, for instance, Ω is a centralizer, this estimate says that Ω nearly commutes not only with multiplication operators, but with all interpolating operators.

Now the extrapolation technique allows a change of perspective: starting from an operator T on X, we may consider all pairs (X_0, X_1) such that $X = X_0^{1-\theta}X_1^\theta$ and T is interpolating between X_0 and X_1 and get a whole family of estimates on T. Doing this for $X = \ell_p$ yields to the family of quasi-linear maps

$$\Phi_G(u) = \sum_{n=1}^{\infty} u_n G(\log|u_n|)$$

where G runs through the family $\mathcal{G}$ of 1-Lipschitz maps from $\mathbb{R}$ to $\mathbb{R}$ whose derivative is compactly supported. This leads to considering the quasi-Banach space h_1^{sym} defined by

$$\|\xi\|_{h_1^{\text{sym}}} = \sum_{k=1}^{\infty} |\xi_k| + \sup_{G\in\mathcal{G}} \Phi_G(\xi) < \infty.$$

This "tangent space" h_1^{sym} is conveniently described as the space of sequence (ξ_k) in ℓ_1 such that

$$\sum_{n=1}^{\infty} \frac{1}{n}|\xi_1 + \xi_2 + \cdots + \xi_n| < \infty.$$

The same steps applied to function spaces lead to the symmetric Hardy function space $H^1_{\text{sym}}(\mu)$ of all functions $f \in L^1(\mu)$ such that

$$\|f\|_{H^1_{\text{sym}}} = \int |f|d\mu + \sup_{G\in\mathcal{G}} \int |f|G(\log|f|)d\mu < \infty.$$

Commutator estimates on interpolating operators then show the following theorem [64]:

Theorem 5.1. *Suppose* $1 < p_0 < p < p_1 < \infty$ *and* $p^{-1} + q^{-1} = 1$. *Suppose that* $T : L_{p_j} \longrightarrow L_{p_j}$ *is linear bounded for* $j = 0, 1$. *Then the bilinear form*

$$B_T(f, g) = f.T^*g - g.Tf$$

is bounded from $L_p \times L_q$ *to* H^1_{sym}.

The above dot denotes of course the pointwise product of functions. This theorem can be applied to a variety of interpolating operators. When applied to the Riesz projection on $L_2(\mathbb{T})$, it gives when f and g belongs to $H^1(\mathbb{T})$ and $g(0) = 0$ the inequality

$$\|fg\|_{H^1_{\text{sym}}} \le C\|f\|_2\|g\|_2$$

and thus since $H_0^1 = H^2.H_0^2$, one gets for every function $h \in H^1$ with $h(0) = 0$

$$\|h\|_{H^1_{\text{sym}}} \le C\|h\|_1.$$

This result, first shown in [18] and [19], somehow means that functions in $H^1(\mathbb{T})$ have a quite symmetric behavior around their singularities. Conversely, real-valued functions in $H^1_{\text{sym}}(\mathbb{T})$ are real parts of functions in $H^1(\mathbb{T})$.

The ideas developed above have non-commutative analogues, and the bridge which brings to the non-commutative world is the concept of trace. If X is a symmetric Banach sequence space, we denote $\mathcal{C}_X$ the space of all operators T on ℓ_2 whose sequence $(s_n(T))_{n\ge 1}$ of singular numbers belongs to X. When X_0 and X_1 are reflexive then

$$[\mathcal{C}_{X_0}, \mathcal{C}_{X_1}]_\theta = \mathcal{C}_{X_0^{1-\theta} X_1^\theta} = \mathcal{C}_{X_\theta}$$

and interpolation tools apply to the spaces $\mathcal{C}_X$.

Let $\mathcal{C}_{\ell_1} = \mathcal{C}_1$ be the ideal of trace-class (or nuclear) operators on ℓ_2. A *trace* on $\mathcal{C}_1$ is a linear map τ such that $\tau(AB) = \tau(BA)$ for all $A \in \mathcal{C}_1$ and all bounded operators B. We denote $\text{Comm}(\mathcal{C}_1)$ the linear span of all commutators

$$[A, B] = AB - BA$$

with $A \in \mathcal{C}_1$ and B bounded. Clearly, if $S \in \mathcal{C}_1$ then $S \in \text{Comm}(\mathcal{C}_1)$ if and only if $\tau(S) = 0$ for every trace τ. It was shown in [133] that $\text{Comm}(\mathcal{C}_1)$ is strictly contained in $\{T \in \mathcal{C}_1; tr(T) = 0\}$, or equivalently that there exist *discontinuous* traces on $\mathcal{C}_1$. The precise description of $\text{Comm}(\mathcal{C}_1)$ was obtained in [66] by interpolation arguments and it reads as follows:

Theorem 5.2. *Let $T \in \mathcal{C}_1$ be a trace-class operator. Then $T \in \text{Comm}(\mathcal{C}_1)$ if and only if its eigenvalue sequence $\lambda_n(T))_{n\ge 1}$ belongs to h^1_{sym}.*

It was shown in [66] that every $T \in \text{Comm}(\mathcal{C}_1)$ is the sum of 6 commutators, but this number has now been put down to 3 and the case of general ideas of operators is also treated in [25, 26, 70]. We refer to [29] for characteristic determinants of trace-class operators and their use in this context.

6 Multipliers and some of their uses

Bases and their various substitutes provide coordinate systems in which some computations can take place, and this is important for understanding the geometry of the Banach spaces these bases generate, and for applied mathematics as well: let us refer for instance to frame theory (see [16, 13] and Casazza's subsequent work), and to wavelets and greedy bases (see [134, 21]).

Bases allow in particular to define "diagonal" operators, shift operators and more generally Toeplitz-like $Q(S)$ operators, where

$$S(\sum_j x_j e_j) = \sum_j x_j e_{j+1}$$

is the shift and $Q(\xi) = \sum_{n\in\mathbb{Z}} a_n \xi^n$ is a formal bilateral series. We stress that such operators can be bounded or not, and their norm can be controlled or not by some norm of Q, and this is where the fun begins. Of course, more general "matrices" can be considered which would define operators, but it seems to be a fact of life that if something significant happens far away from the diagonal, then it is very hard to control.

Let us now see how diagonal operators, when properly used, actually suffice for obtaining important examples. We begin with Hilbertian theory. A linear operator T on a complex Hilbert space H is called *power bounded* if $\sup_{n\geq 0} \|T^n\| < \infty$ and *polynomially bounded* if there is $C < \infty$ such that

$$\|P(T)\| \leq C\|P\|_\infty = C \sup\{\|P(z)\|; |z| \leq 1\}$$

for every $P \in \mathbb{C}[\xi]$. J. von Neumann's classical inequality shows that every operator T which is similar to a contraction (that is, there exists A invertible such that $\|A^{-1}TA\| \leq 1$) is polynomially bounded, and an example due to Pisier (see [112]) shows that the converse is false. A weaker requirement would be to show that T is similar to an operator such that $\sup_n \|T^n\|$ is close to 1 if it is power-bounded, and this question is asked in [110]. Note that a classical theorem of Rota ([127]) shows that power-bounded operators are similar to operators of norm close to 1.

Basis theory joins forces with harmonic analysis to provide a negative answer to Peller's question in [78]. A *weight* w on $\mathbb{T}$ is a non-zero function from $L_1(\mathbb{T})$ with $w \geq 0$. We denote by $L_2(w)$ the corresponding weighted L_2-space, and

$$H^2(w) = \overline{\text{span}}\{e^{in\theta}; n \geq 0\} \subseteq L_2(w).$$

The Riesz projection is formally defined on $L_2(w)$ by

$$R(\sum_{n\in\mathbb{Z}} \widehat{f}(n)e^{in\theta}) = \sum_{n\geq 0} \widehat{f}(n)e^{in\theta}$$

and w is called an A_2-*weight* if R is a bounded projection from $L_2(w)$ onto $H^2(w)$. If $\|R\|_w$ denotes the norm of the Riesz projection and $0 \leq \varphi < \pi/2$, then [78, Proposition 2.2] proves the following Helson–Szegö estimate: $\|R\|_w \leq (\cos(\varphi))^{-1}$ if and only if there exists $h \in H^1(\mathbb{T})$ such that $|w - h| \leq w \sin(\varphi)$.

If $\alpha \in (0, 1)$ and $w(e^{i\theta}) = |\theta|^\alpha$ for $\theta \in (-\pi, \pi]$, this estimate leads to

$$\inf\{\|R\|_v; v \sim w\} = (\cos(\frac{\pi\alpha}{2}))^{-1}$$

where $v \sim w$ means that (w/v) and (v/w) both belong to $L_\infty(\mathbb{T})$.

Following [78], let us consider a basis $(e_n)_{n\geq 0}$ of H and call $T : H \longrightarrow H$ a *fast monotone multiplier* (with respect to (e_n)) if

$$T(\sum_{k=0}^{\infty} a_k e_k) = \sum_{k=0}^{\infty} \lambda_k a_k e_k$$

where $(\lambda_k) \in (0,1)$ is an increasing sequence such that

$$\lim_{k\to\infty} \frac{1-\lambda_k}{1-\lambda_{k-1}} = 0.$$

Easy computations then show that $\sup_{n\geq 0} \|T^n\|$ is at most equal to the basis constant b of (e_n).

If we now consider the (usually conditional!) basis $(e_k)_{k\geq 0}$ of $H^2(w)$, where $e_k(\theta) = e^{ik\theta}$, then its basis constant is $b = \|R\|_w$. The main result of [78] implies in particular the following:

Theorem 6.1. *Let $\alpha \in (0,1)$ and $w_\alpha(e^{i\theta}) = |\theta|^\alpha$ for $\theta \in (-\pi,\pi]$. Let T be a fast monotone multiplier with respect to the basis of exponential functions $(e_n)_{n\geq 0}$ of $H^2(w_\alpha)$. Then T is power bounded, and*

$$\inf_A \{ \sup_n \|(A^{-1}TA)^n\| \} = (\cos(\frac{\pi\alpha}{2}))^{-1}$$

where the infimum is taken over all invertible operators A.

Showing Theorem 6.1 amounts to proving

$$\inf_A\{\sup_n\|(A^{-1}TA)^n\|\} = \inf\{\|R\|_v; v \sim w\}$$

and "$\geq$" is the hard part of the above equation. If $S(f) = e_1 f$ is the shift operator on $H^2(w)$ and if $\sup_n\|(A^{-1}TA)^n\| = \sigma$ then consider a Banach limit

$$\langle f,g\rangle = \lim_{n\to\mathcal{U}}(AS^n f, AS^n g)$$

where (.,.) is the inner product of $H^2(w)$ and $\mathcal{U}$ is a free ultrafilter on $\mathbb{N}$. This defines an equivalent inner-product norm

$$|||f||| = \langle f,f\rangle^{\frac{1}{2}}$$

on $H^2(w)$. Since T is a fast monotone multiplier, its powers provide good approximations of the "tails" $Q_n = I - P_n$ of the basis (e_n), and it follows that the basis constant of (e_n) in the new norm $|||\cdot|||$ is at most σ.

Now the sequence $C_k = \langle e_0, e_k\rangle$ if $k \geq 0$ and $c_k = \overline{c_{-k}}$ if $k < 0$ is positive definite. Then Bochner's theorem gives a weight $v \sim w$ which defines $\|\cdot\|_v = |||\cdot|||$, and $\|R\|_v \leq \sigma$ since $\|R\|_v$ is equal to the basis constant of (e_n) in the norm $\|.\|_v$.

Hence a negative answer to Peller's question is obtained with fast monotone multipliers with respect to Babenko's conditional basis of the Hilbert space [5]. More general weights are also considered in [78], which show that the infimum is usually *not* attained in Theorem 6.1. On the other hand, polynomially bounded operators which are obtained by this method are actually similar to contractions. It is conjectured that there exist polynomially bounded operators which are not similar to operators T such that $\sup_n \|T^n\|$ is close to 1. Along these lines, we mention the existence of uniformly bounded representations of the free group with 2 generators which are not similar to representations with small norm ([113, Corollary 5.10]).

Multipliers can as well be unbounded, and such objects provide a negative answer to another important open question. Let us consider the following Cauchy problem:

$$u'(t) + B(u(t)) = f(t)$$

with the initial condition $u(0) = 0$, where $t \in [0, T)$, $-B$ is the closed, densely defined infinitesimal generator of a bounded analytic semi-group on a complex Banach space X, and u and f are X-valued functions on [0,T). One says that B satisfies *maximal regularity* if $u' \in L_2(X)$ as soon as $f \in L_2(X)$. The importance of this notion lies in the fact that it permits to solve quasi-linear parabolic P.D.E. through fixed point arguments.

When X is the Hilbert space, every such B has maximal regularity [20], and the question of the converse occurs, in particular when $X = L_p$ with $1 < p < \infty$. A quite general answer is obtained in [76]:

Theorem 6.2. *If a Banach space X has an unconditional basis, then every closed densely defined operator B such that $(-B)$ generates a bounded analytic semi-group on X has maximal regularity if and only if X is isomorphic to the Hilbert space.*

The proof goes as follows: if B satisfies maximal regularity, then solving the Cauchy problem for well-chosen functions $f \in L_2(X)$ amounts to showing that for every X-valued trigonometric polynomial $g(t) = \sum_{n=-N}^{N} \widehat{g}(n)e^{int}$ one has

$$\|\sum_{n\in\mathbb{Z}} in(in + B)^{-1}\widehat{g}(n)e^{int}\|_{L^2(X)} \leq C\|g\|_{L^2(X)}.$$

This is applied to closed density defined operators B of the form

$$B(\sum_{n=1}^{\infty} a_n e_n) = \sum_{n\geq 1} a_n b_n e_n$$

where (e_n) is an unconditional basis of X and (b_n) is an increasing sequence of real numbers such that $b_n > 0$. Any such multiplier $B = M((b_n))$ is *sectorial* of type ω for any $\omega \in (0, \pi)$ and in particular $(-B)$ generates an analytic bounded semigroup.

Note that when the sequence $(\widehat{g}(n))$ is unconditional, then

$$\|g\|_{L_2(X)} \simeq \|\sum_{-N}^{N} \widehat{g}(n)\|_X.$$

Now the maximal regularity inequality applied with proper choices of scalars (b_n) shows after some work that for any block basis (u_j) of any permutation $(e_{\pi(n)})$ of (e_n), the space $\overline{\text{span}}((u_j))$ is complemented in X. It then follows from [102] that (e_j) is equivalent to the canonical basis of c_0 or ℓ_p for some $p \in [1, \infty)$, and since $\ell_p \simeq (\sum_{n\geq 1} \oplus \ell_2^n)_p$ has non-equivalent unconditional bases if $1 < p < \infty, p \neq 2$ [109], it follows that X is c_0, ℓ_1 or ℓ_2. Finally, c_0 and ℓ_1 can be discarded with the help of a multiplier relative to the (conditional) summing basis of c_0.

Since the Haar system is an unconditional basis of L_p, Theorem 6.1 answers negatively the maximal regularity problem for L_p $(1 < p < \infty, p \neq 2)$. The case of L_1 had been done in [99]. The unconditionality assumption in Theorem 6.2 can be released, but it is not known if the conclusion holds for every space X with a basis (see [77]). We refer to [100] for examples of quasi Banach spaces which have unique unconditional bases up to permutation.

Theorem 6.2 started, together with [132], an important and very active field of research. The above discussion shows how basis theory, developed for its own sake in the 60's and 70's, turns out to be very useful in abstract P.D.E. problems, since (via multipliers), bases – or merely Schauder decompositions – provide sectorial operators. More elaborate concepts, such as R-bounded families of operators, allow an intensive application of geometric tools to maximal regularity and Kato's square root estimates, for which I refer to the joint works of N. Kalton and L. Weis [90, 89] and to [50]. Let us mention that the discrete time analogue of the maximal regularity problem, which presents some specific difficulties since there is no differentiation with respect to time in this case, is investigated in [7, 115, 116] and [84].

Acknowledgments. I am very grateful to Beata and Narcisse Randrianantoanina, for the splendid and perfectly organized Oxford meeting, and for the help they provided in the preparation and typing of this article. My deepest gratitude goes to Nigel Kalton, to whom I owe more than I can express.

References

[1] I. Aharoni and J. Lindenstrauss, *Uniform equivalence between Banach spaces*, Bull. Amer. Math. Soc. 84 (1978), pp. 281–283.

[2] F. Albiac and N. J. Kalton, *Lipschitz strucure of quasi-Banach spaces*, to appear.

[3] E. M. Alfsen and E. G. Effros, *Structure in real Banach spaces. I, II*, Ann. of Math. (2) 96 (1972), pp. 98–128; ibid. (2) 96 (1972), pp. 129–173.

[4] G. Androulakis, C. D. Cazacu, and N. J. Kalton, *Twisted sums, Fenchel-Orlicz spaces and property (M)*, Houston J. Math. 24 (1998), pp. 105–126.

[5] K. I. Babenko, *On conjugate functions*, Doklady Akad. Nauk SSSR (N. S.) 62 (1948), pp. 157–160.

[6] Y. Benyamini and J. Lindenstrauss, *Geometric nonlinear functional analysis. Vol. 1*, American Mathematical Society Colloquium Publications 48. American Mathematical Society, Providence, RI, 2000.

[7] S. Blunck, *Maximal regularity of discrete and continuous time evolution equations*, Studia Math. 146 (2001), pp. 157–176.

[8] J. Bourgain, *Real isomorphic complex Banach spaces need not be complex isomorphic*, Proc. Amer. Math. Soc. 96 (1986), pp. 221–226.

[9] J. Bourgain and H. P. Rosenthal, *Martingales valued in certain subspaces of L^1*, Israel J. Math. 37 (1980), pp. 54–75.

[10] M. J. Carro, J. Cerdà, and J. Soria, *Higher order commutators in interpolation theory*, Math. Scand. 77 (1995), pp. 301–319.

[11] P. G. Casazza, *The commuting B.A.P. for Banach spaces*, Analysis at Urbana, Vol. II (Urbana, IL, 1986–1987), London Math. Soc. Lecture Note Ser., 138, Cambridge Univ. Press, Cambridge, 1989, pp. 108–127.

[12] ______, *Approximation properties*, Handbook of the geometry of Banach spaces, Vol. I, North-Holland, Amsterdam, 2001, pp. 271–316.

[13] ______, *An introduction to irregular Weyl-Heisenberg frames*, Sampling, wavelets, and tomography, Appl. Numer. Harmon. Anal., Birkhäuser, Boston, MA, 2004, pp. 61–81.

[14] P. G. Casazza and N. J. Kalton, *Notes on approximation properties in separable Banach spaces*, Geometry of Banach spaces (Strobl, 1989), London Math. Soc. Lecture Note Ser. 158, Cambridge Univ. Press, Cambridge, 1990, pp. 49–63.

[15] ______, *Unconditional bases and unconditional finite-dimensional decompositions in Banach spaces*, Israel J. Math. 95 (1996), pp. 349–373.

[16] ______, *Roots of complex polynomials and Weyl-Heisenberg frame sets*, Proc. Amer. Math. Soc. 130 (2002), pp. 2313–2318.

[17] J. M. F. Castillo and M. González, *Three-space problems in Banach space theory*, Lecture Notes in Mathematics 1667. Springer-Verlag, Berlin, 1997.

[18] O. D. Cereteli, *A metric characterization of the set of functions whose conjugate functions are integrable*, Sakharth. SSR Mecn. Akad. Moambe 81 (1976), pp. 281–283.

[19] B. Davis, *Hardy spaces and rearrangements*, Trans. Amer. Math. Soc. 261 (1980), pp. 211–233.

[20] L. de Simon, *Un'applicazione della teoria degli integrali singolari allo studio delle equazioni differenziali lineari astratte del primo ordine*, Rend. Sem. Mat. Univ. Padova 34 (1964), pp. 205–223.

[21] S. J. Dilworth, D. Kutzarova, and P. Wojtaszczyk, *On approximate l_1 systems in Banach spaces*, J. Approx. Theory 114 (2002), pp. 214–241.

[22] L. E. Dor, *On projections in L_1*, Ann. of Math. (2) 102 (1975), pp. 463–474.

[23] Y. Dutrieux, *Lipschitz-quotients and the Kunen-Martin theorem*, Comment. Math. Univ. Carolin. 42 (2001), pp. 641–648.

[24] S. Dutta and A. Godard, *Property (M) and asymptotic uniform smoothness*, to appear.

[25] K. J. Dykema, T. Figiel, G. Weiss, and M. Wodzicki, *Commutator structure of operator ideals*, Adv. Math. 185 (2004), pp. 1–79.

[26] K. J. Dykema and N. J. Kalton, *Spectral characterization of sums of commutators. II*, J. Reine Angew. Math. 504 (1998), pp. 127–137.

[27] P. Enflo, *On the nonexistence of uniform homeomorphisms between L_p-spaces*, Ark. Mat. 8 (1969), pp. 103–105 (1969).

[28] P. Enflo, J. Lindenstrauss, and G. Pisier, *On the "three space problem"*, Math. Scand. 36 (1975), pp. 199–210.

[29] C. M. Eoff and N. J. Kalton, *Zeros of entire functions and characteristic determinants*, Complex Variables Theory Appl. 14 (1990), pp. 53–63.

[30] V. Ferenczi, *Uniqueness of complex structures and real heredetarily indecomposable Banach spaces*, to appear.

[31] V. Ferenczi and E. M. Galego, *Even infinite-dimensional Banach spaces*, to appear.

[32] T. Figiel, *On nonlinear isometric embeddings of normed linear spaces*, Bull. Acad. Polon. Sci. Sér. Sci. Math. Astronom. Phys. 16 (1968), pp. 185–188.

[33] O. Gabber and Z. Galil, *Explicit constructions of linear-sized superconcentrators*, J. Comput. System Sci. 22 (1981), pp. 407–420.

[34] R. J. Gardner, A. Koldobsky, and T. Schlumprecht, *An analytic solution to the Busemann-Petty problem on sections of convex bodies*, Ann. of Math. (2) 149 (1999), pp. 691–703.

[35] G. Godefroy and N. J. Kalton, *Approximating sequences and bidual projections*, Quart. J. Math. Oxford Ser. (2) 48 (1997), pp. 179–202.

[36] ______, *Lipschitz-free Banach spaces*, Studia Math. 159 (2003), pp. 121–141.

[37] G. Godefroy, N. J. Kalton, and G. Lancien, *Subspaces of $c_0(\mathbf{N})$ and Lipschitz isomorphisms*, Geom. Funct. Anal. 10 (2000), pp. 798–820.

[38] ______, *Szlenk indices and uniform homeomorphisms*, Trans. Amer. Math. Soc. 353 (2001), pp. 3895–3918.

[39] G. Godefroy, N. J. Kalton, and D. Li, *On subspaces of L^1 which embed into l_1*, J. Reine Angew. Math. 471 (1996), pp. 43–75.

[40] ______, *Operators between subspaces and quotients of L^1*, Indiana Univ. Math. J. 49 (2000), pp. 245–286.

[41] G. Godefroy, N. J. Kalton, and P. D. Saphar, *Unconditional ideals in Banach spaces*, Studia Math. 104 (1993), pp. 13–59.

[42] E. Gorelik, *The uniform nonequivalence of L_p and l_p*, Israel J. Math. 87 (1994), pp. 1–8.

[43] W. T. Gowers, *A solution to Banach's hyperplane problem*, Bull. London Math. Soc. 26 (1994), pp. 523–530.

[44] ______, *A new dichotomy for Banach spaces*, Geom. Funct. Anal. 6 (1996), pp. 1083–1093.

[45] W. T. Gowers and B. Maurey, *The unconditional basic sequence problem*, J. Amer. Math. Soc. 6 (1993), pp. 851–874.

[46] P. Harmand, D. Werner, and W. Werner, *M-ideals in Banach spaces and Banach algebras*, Lecture Notes in Mathematics 1547. Springer-Verlag, Berlin, 1993.

[47] R Haydon, E Odell, and T Schlumprecht, *On small subspaces of L_p with $p > 2$*, to appear.

[48] S. Heinrich and P. Mankiewicz, *Applications of ultrapowers to the uniform and Lipschitz classification of Banach spaces*, Studia Math. 73 (1982), pp. 225–251.

[49] W. Herer and J. P. R. Christensen, *On the existence of pathological submeasures and the construction of exotic topological groups*, Math. Ann. 213 (1975), pp. 203–210.

[50] T Hytönen, A McIntosh, and P. Portal, *Kato's square root problem in Banach spaces*, to appear.

[51] W. B. Johnson, *Factoring compact operators*, Israel J. Math. 9 (1971), pp. 337–345.

[52] W. B. Johnson, J. Lindenstrauss, D. Preiss, and G. Schechtman, *Almost Fréchet differentiability of Lipschitz mappings between infinite-dimensional Banach spaces*, Proc. London Math. Soc. (3) 84 (2002), pp. 711–746.

[53] W. B. Johnson, J. Lindenstrauss, and G. Schechtman, *Banach spaces determined by their uniform structures*, Geom. Funct. Anal. 6 (1996), pp. 430–470.

[54] W. B. Johnson and E. Odell, *Subspaces of L_p which embed into l_p*, Compositio Math. 28 (1974), pp. 37–49.

[55] W. B. Johnson and N. L. Randrianarivony, *l_p $(p > 2)$ does not coarsely embed into a Hilbert space*, Proc. Amer. Math. Soc. 134 (2006), pp. 1045–1050.

[56] N. J. Kalton, *Basic sequences in F-spaces and their applications*, Proc. Edinburgh Math. Soc. 19 (1974/75), pp. 151–167.

[57] ———, *Linear operators whose domain is locally convex*, Proc. Edinburgh Math. Soc. 20 (1976/77), pp. 293–299.

[58] ———, *Quotients of F-spaces*, Glasgow Math. J. 19 (1978), pp. 103–108.

[59] ———, *The three space problem for locally bounded F-spaces*, Compositio Math. 37 (1978), pp. 243–276.

[60] ———, *Linear operators on L_p for $0 < p < 1$*, Trans. Amer. Math. Soc. 259 (1980), pp. 319–355.

[61] ———, *Sequences of random variables in L_p for $p < 1$*, J. Reine Angew. Math. 329 (1981), pp. 204–214.

[62] ———, *Convexity conditions for nonlocally convex lattices*, Glasgow Math. J. 25 (1984), pp. 141–152.

[63] ———, *Applications of an inequality for H_1-functions*, Texas Functional Analysis Seminar 1985–1986 (Austin, TX, 1985–1986), Longhorn Notes, Univ. Texas, Austin, TX, 1986, pp. 41–51.

[64] ———, *Nonlinear commutators in interpolation theory*, Mem. Amer. Math. Soc. 73 (1988), pp. iv+85.

[65] ———, *The Maharam problem*, Séminaire d'Initiation à l'Analyse, Exp. No. 18, 13 94 (1989).

[66] ———, *Trace-class operators and commutators*, J. Funct. Anal. 86 (1989), pp. 41–74.

[67] ———, *Differentials of complex interpolation processes for Köthe function spaces*, Trans. Amer. Math. Soc. 333 (1992), pp. 479–529.

[68] ———, *M-ideals of compact operators*, Illinois J. Math. 37 (1993), pp. 147–169.

[69] ———, *The basic sequence problem*, Studia Math. 116 (1995), pp. 167–187.

[70] ———, *Spectral characterization of sums of commutators. I*, J. Reine Angew. Math. 504 (1998), pp. 115–125.

[71] ———, *Quasi-Banach spaces*, Handbook of the geometry of Banach spaces, Vol. 2, North-Holland, Amsterdam, 2003, pp. 1099–1130.

[72] ———, *Twisted Hilbert spaces and unconditional structure*, J. Inst. Math. Jussieu 2 (2003), pp. 401–408.

[73] ———, *Spaces of Lipschitz and Hölder functions and their applications*, Collect. Math. 55 (2004), pp. 171–217.

[74] N. J. Kalton and A. Koldobsky, *Banach spaces embedding isometrically into L_p when $0 < p < 1$*, Proc. Amer. Math. Soc. 132 (2004), pp. 67–76 (electronic).

[75] N. J. Kalton and G. Lancien, *Lipschitz embeddings into $c_0(\mathbf{N})$*, to appear.

[76] ———, *A solution to the problem of L^p-maximal regularity*, Math. Z. 235 (2000), pp. 559–568.

[77] ———, *L^p-maximal regularity on Banach spaces with a Schauder basis*, Arch. Math. (Basel) 78 (2002), pp. 397–408.

[78] N. J. Kalton and C. Le Merdy, *Solution of a problem of Peller concerning similarity*, J. Operator Theory 47 (2002), pp. 379–387.

[79] N. J. Kalton and S. Montgomery-Smith, *Interpolation of Banach spaces*, Handbook of the geometry of Banach spaces, Vol. 2, North-Holland, Amsterdam, 2003, pp. 1131–1175.

[80] N. J. Kalton and M. I. Ostrovskii, *Distances between Banach spaces*, Forum Math. 11 (1999), pp. 17–48.

[81] N. J. Kalton and N. T. Peck, *Quotients of* $L_p(0,1)$ *for* $0 \le p < 1$, Studia Math. 64 (1979), pp. 65–75.

[82] N. J. Kalton, N. T. Peck, and J. W. Roberts, L_0*-valued vector measures are bounded*, Proc. Amer. Math. Soc. 85 (1982), pp. 575–582.

[83] ———, *An F-space sampler*, London Mathematical Society Lecture Note Series 89. Cambridge University Press, Cambridge, 1984.

[84] N. J. Kalton and P. Portal, *Some remarks on* ℓ_1 *and* ℓ_∞ *maximal regularity for power bounded operators*, to appear.

[85] N. J. Kalton and J. W. Roberts, *A rigid subspace of* L_0, Trans. Amer. Math. Soc. 266 (1981), pp. 645–654.

[86] ———, *Uniformly exhaustive submeasures and nearly additive set functions*, Trans. Amer. Math. Soc. 278 (1983), pp. 803–816.

[87] N. J. Kalton and P. Saab, *Ideal properties of regular operators between Banach lattices*, Illinois J. Math. 29 (1985), pp. 382–400.

[88] N. J. Kalton and R. C. Swanson, *A symplectic Banach space with no Lagrangian subspaces*, Trans. Amer. Math. Soc. 273 (1982), pp. 385–392.

[89] N. J. Kalton and L. Weis, *Hilbertian structures*. To appear.

[90] ———, *The* H^∞*-calculus and sums of closed operators*, Math. Ann. 321 (2001), pp. 319–345.

[91] N. J. Kalton and D. Werner, *Property* (M)*,* M*-ideals, and almost isometric structure of Banach spaces*, J. Reine Angew. Math. 461 (1995), pp. 137–178.

[92] N. J. Kalton and G. V. Wood, *Orthonormal systems in Banach spaces and their applications*, Math. Proc. Cambridge Philos. Soc. 79 (1976), pp. 493–510.

[93] H. Knaust, E. Odell, and T. Schlumprecht, *On asymptotic structure, the Szlenk index and UKK properties in Banach spaces*, Positivity 3 (1999), pp. 173–199.

[94] A. Koldobsky, *Intersection bodies in* $\mathbf{R}^4$, Adv. Math. 136 (1998), pp. 1–14.

[95] ———, *Intersection bodies, positive definite distributions, and the Busemann-Petty problem*, Amer. J. Math. 120 (1998), pp. 827–840.

[96] A. Koldobsky and H. König, *Aspects of the isometric theory of Banach spaces*, Handbook of the geometry of Banach spaces, Vol. I, North Holland, Amsterdam, 2001, pp. 899–939.

[97] R. A. Komorowski and N. Tomczak-Jaegermann, *Banach spaces without local unconditional structure*, Israel J. Math. 89 (1995), pp. 205–226.

[98] P. Koszmider, *Banach spaces of continuous functions with few operators*, Math. Ann. 330 (2004), pp. 151–183.

[99] C. Le Merdy, *Counterexamples on* L_p*-maximal regularity*, Math. Z. 230 (1999), pp. 47–62.

[100] C. Leránoz, *Uniqueness of unconditional bases of* $c_0(l_p)$, $0 < p < 1$, Studia Math. 102 (1992), pp. 193–207.

[101] D. R. Lewis, *Ellipsoids defined by Banach ideal norms*, Mathematika 26 (1979), pp. 18–29.

[102] J. Lindenstrauss and L. Tzafriri, *On the complemented subspaces problem*, Israel J. Math. 9 (1971), pp. 263–269.

[103] A. Louveau, *Progrès récents sur le problème de Maharam d'après N. J. Kalton et J. W. Roberts*, Séminaire d'Initiation à l'Analyse 20 (1983/84), pp. 1–8.

[104] D. Maharam, *An algebraic characterization of measure algebras*, Ann. of Math. (2) 48 (1947), pp. 154–167.

[105] B. Maurey, *Théorèmes de factorisation pour les opérateurs linéaires à valeurs dans les espaces L^p*. Astérisque 11, Société Mathématique de France, Paris, 1974.

[106] V. D. Milman, *Geometric theory of Banach spaces. II. Geometry of the unit ball*, Uspehi Mat. Nauk 26 (1971), pp. 73–149. English translation: Russian Math. Surveys 26 (1971), pp. 79-163.

[107] E. Odell and T. Schlumprecht, *The distortion of Hilbert space*, Geom. Funct. Anal. 3 (1993), pp. 201–207.

[108] ———, *The distortion problem*, Acta Math. 173 (1994), pp. 259–281.

[109] A. Pełczyński, *Projections in certain Banach spaces*, Studia Math. 19 (1960), pp. 209–228.

[110] V. V. Peller, *Estimates of functions of power bounded operators on Hilbert spaces*, J. Operator Theory 7 (1982), pp. 341–372.

[111] G. Pisier, *Some applications of the complex interpolation method to Banach lattices*, J. Analyse Math. 35 (1979), pp. 264–281.

[112] ———, *Similarity problems and completely bounded maps*, expanded ed, Lecture Notes in Mathematics 1618. Springer-Verlag, Berlin, 2001.

[113] ———, *Are unitarizable groups amenable?*, Infinite groups: geometric, combinatorial and dynamical aspects, Progr. Math. 248, Birkhäuser, Basel, 2005, pp. 323–362.

[114] G. Plebanek, *A construction of a Banach space $C(K)$ with few operators*, Topology Appl. 143 (2004), pp. 217–239.

[115] P. Portal, *Discrete time analytic semigroups and the geometry of Banach spaces*, Semigroup Forum 67 (2003), pp. 125–144.

[116] ———, *Maximal regularity of evolution equations on discrete time scales*, J. Math. Anal. Appl. 304 (2005), pp. 1–12.

[117] B. Randrianantoanina, *1-complemented subspaces of spaces with 1-unconditional bases*, Canad. J. Math. 49 (1997), pp. 1242–1264.

[118] M. Ribe, *On uniformly homeomorphic normed spaces*, Ark. Mat. 14 (1976), pp. 237–244.

[119] ———, *Examples for the nonlocally convex three space problem*, Proc. Amer. Math. Soc. 73 (1979), pp. 351–355.

[120] ———, *Existence of separable uniformly homeomorphic nonisomorphic Banach spaces*, Israel J. Math. 48 (1984), pp. 139–147.

[121] J. W. Roberts, *Pathological compact convex sets in $L_p(0,1)$, $0 < p < 1$*, The Altgeld book, University of Illinois Functional Analysis Seminar, 1975-76.

[122] ———, *A compact convex set with no extreme points*, Studia Math. 60 (1977), pp. 255–266.

[123] ———, *A nonlocally convex F-space with the Hahn-Banach approximation property*, Banach spaces of analytic functions (Proc. Pelczynski Conf., Kent State Univ., Kent, Ohio, 1976). Lecture Notes in Math. 604, Springer, Berlin, 1977, pp. 76–81.

[124] R. Rochberg, *Higher order estimates in complex interpolation theory*, Pacific J. Math. 174 (1996), pp. 247–267.

[125] R. Rochberg and G. Weiss, *Derivatives of analytic families of Banach spaces*, Ann. of Math. (2) 118 (1983), pp. 315–347.

[126] H. P. Rosenthal, *On one-complemented subspaces of complex Banach spaces with a one-unconditional basis, according to Kalton and Wood*, Israel seminar on geometrical aspects of functional analysis (1983/84), Tel Aviv Univ., Tel Aviv, 1984, pp. IX, 16.

[127] G.-C. Rota, *On models for linear operators*, Comm. Pure Appl. Math. 13 (1960), pp. 469–472.

[128] S. J. Szarek, *On the existence and uniqueness of complex structure and spaces with "few" operators*, Trans. Amer. Math. Soc. 293 (1986), pp. 339–353.

[129] M. Talagrand, *Les mesures vectorielles à valeurs dans L^0 sont bornées*, Ann. Sci. École Norm. Sup. (4) 14 (1981), pp. 445–452 (1982).

[130] ———, *Maharam's problem*, C. R. Acad. Sci. Paris 342 (2006), pp. 501–503.

[131] S.-C. Tam, *The basic sequence problem for quasi-normed spaces*, Arch. Math. (Basel) 62 (1994), pp. 69–72.

[132] L. Weis, *Operator-valued Fourier multiplier theorems and maximal L_p-regularity*, Math. Ann. 319 (2001), pp. 735–758.

[133] G. Weiss, *Commutators of Hilbert-Schmidt operators. II*, Integral Equations Operator Theory 3 (1980), pp. 574–600.

[134] P. Wojtaszczyk, *A mathematical introduction to wavelets*, London Mathematical Society Student Texts 37. Cambridge University Press, Cambridge, 1997.

[135] G. Zhang, *A positive solution to the Busemann-Petty problem in $\mathbf{R}^4$*, Ann. of Math. (2) 149 (1999), pp. 535–543.

Author information

Gilles Godefroy, Institut de Mathématiques de Jussieu, 175, rue du Chevaleret, 75013 Paris, France.
Email: godefroy@math.jussieu.fr

Banach Spaces and their Applications in Analysis, 37–57

Multivariate functions of bounded (k, p)-variation

Yuri Brudnyi

Abstract. The paper is a survey of the basic facts of the theory of functions of bounded (k, p)-variation considered from the point of Local Approximation Theory. An overview of the later theory, now separated among quite a few papers of the author and his students is also presented.

Key words. Multivariate function, (k, p)-variation, local approximation spaces, Besov spaces, Sobolev spaces, embedding, duality, N-approximation.

AMS classification. 26B30, 41A46, 46E35, 46B70.

Dedicated to Nigel Kalton on the occasion of his 60-th birthday.

1 Introduction

Functions of bounded variation play a considerable role in some central problems of analysis (convergence of Fourier series, differential properties of univariate functions, parameterization of rectifiable curves, F. Riesz duality theorem etc). Similar problems for functions of several variables have been stimulated attempts of finding an appropriate concept for multivariate functions. Since the situation in this case is much more complicated, the modern definition of the space $BV(\mathbb{R}^n)$ was gradually introduced within about fifty years since the Vitali and B. Levi papers at the very beginning of the 20th century (see [1] for the history and references). This was firstly due to Tonelli (1926) for continuous functions and extended by Cesari (1936) to integrable functions. The modern definition appears only in 1954 (Di-Giorgi, Fichera). However, none of these is based on the behaviour of an appropriately defined oscillation which is the main point of Jordan's definition. Such a definition is contained in the author paper [3] and looks as follows.

Let $f \in L_p^{\mathrm{loc}}(\mathbb{R}^n)$. Its *p-oscillation* is a set function given for a measurable set $S \subset \mathbb{R}^n$ by

$$\operatorname{osc}_p(f; S) := \sup_h \Big\{ \int_{S \cap (S+h)} |f(x+h) - f(x)|^p \, dx \Big\}^{\frac{1}{p}}. \tag{1.1}$$

Definition 1.1. A function $f \in L_p^{\mathrm{loc}}(\mathbb{R}^n)$ is said to be of bounded (λ, p)-variation ($\lambda \in \mathbb{R}$) if

$$V_p^\lambda(f) := \sup_\pi \Big\{ \sum_{Q \in \pi} \frac{\operatorname{osc}_p(f; Q)}{|Q|^\lambda} \Big\} \tag{1.2}$$

is finite. Here π runs over all disjoint families of cubes $Q \subset \mathbb{R}^n$.

The following special case of Theorem 12 from [3, §4] motivates this definition.

Theorem 1.2. *Let* $1 \le p \le n/(n-1)$ *and* $\lambda := 1/p - n/(n-1)$. *Then uniformly in* $f \in L_p^{\text{loc}}(\mathbb{R}^n)$

$$|f|_{BV(\mathbb{R}^n)} \approx V_p^\lambda(f). \tag{1.3}$$

Hereafter we conventionally write $F \approx G$, if for some constants $C_1, C_2 > 0$ and all common arguments of the functions F, G it is true that

$$C_1 F \le G \le C_2 F.$$

The theorem shows that there is, in fact, even a family of Jordan's type variations for the definition of BV-functions. As we will see below definitions of this kind may be as useful for the study of properties of BV-functions as the classical Tonelli–Cesari definition. As the only illustration we for now present a precise form of the Gagliardo embedding theorem [20] stating, in particular, that

$$BV([0,1]^n) \subset L_{\frac{n}{n-1}}([0,1]^n).$$

Using (1.3) for $p := n/(n-1)$ (and $\lambda = 0$) one can prove, see [17] for $n = 2$, that for every $N \ge 1$ there is a subdivision π_N of $[0,1]^n$ into at most N dyadic cubes Q such that

$$\inf_{\{a_Q\} \subset \mathbb{R}} \Big\| f - \sum_{Q \in \pi_N} a_Q \, 1\!\!1_Q \Big\|_{\frac{n}{n-1}} \le O(1) N^{-\frac{1}{n}} |f|_{BV([0,1]^n)}. \tag{1.4}$$

Here $1\!\!1_S$ is the indicator of a set $S \subset \mathbb{R}^n$.

The similar result can be also proved for $1 \le p < n/(n-1)$. In this case we have an additional information on the subdivision π_N; e.g., π_N is uniform, if $p = 1$.

Finally, we modify the above definition replacing osc_p by a more flexible object, *local polynomial approximation*. In this special case, one uses the equivalence

$$\text{osc}_p(f; Q) \approx \inf_{a \in \mathbb{R}} \Big\{ \int_Q |f - a|^p \, dx \Big\}^{\frac{1}{p}} \tag{1.5}$$

with the constants independent of f and Q. In other words, instead of (1.2) we may define the variation by

$$\sup_\pi \Big\{ \sum_{Q \in \pi} \frac{E_1(f; Q; L_p)}{|Q|^\lambda} \Big\} \tag{1.6}$$

where $E_1(f; Q; L_p)$ stands for the right-hand side in (1.5).

Now the way to further generalization is clear; we simply replace constants in (1.5) by polynomials and the weighted ℓ_1-norm in (1.6) by a suitable ℓ_p-norm.

The spaces of (generalized) BV-functions introduced in this way are specialization of more general objects, *local polynomial approximation spaces* (briefly, l.a.- spaces), It is natural to overview some basic points of the theory of l.a.- spaces.

2 Topics in local approximation theory

First we describe the basic ingredients used for the definition of l.a.- spaces. They consist of

(1) *family of carriers*,

(2) *space of polynomials*,

(3) *underlying space of integrable functions*,

(4) *local polynomial approximation*,

(5) *functional parameter*.

We restrict our account to a special case concerning the main theme of the paper and then briefly outline the general situation.

A. As a family of carriers we use the set $K(\mathbb{R}^n)$ of open n-cubes homothetic to the unit cube $Q_0 := [0,1]^n$. In other words, $Q \in K(\mathbb{R}^n)$ is an open ball of the space ℓ_∞^n. We write $Q = Q_r(x)$ for the cube centered at $x \in \mathbb{R}^n$ and of radius $r > 0$. We also denote by c_Q and r_Q the center and radius of Q; then λQ stands for the cube centered at c_Q and of radius λr_Q, $\lambda > 0$.

The next two facts together with the classical Bescovich and Vitali covering lemmas are the main geometric tools of the theory.

To formulate the first one we recall that *order of the family of subsets* F is defined by

$$\operatorname{ord} F := \sup_x \Big(\sum_{S \in F} 1\!\!1_S \Big).$$

In other words, every point is contained in at most ord F elements of F.

Theorem 2.1 ([11]). *(a) Let F be a subfamily of $K(\mathbb{R}^n)$ of order at most N. Then F can be presented as union of at most $2^n(N-1)+1$ disjoint subfamilies.*

(b) If F consists of congruent cubes, then the upper estimate is $2^{n-1}(N-1)+1$.

Remark 2.2. The second estimate is sharp for $n = 1, 2$. Is it sharp for $n \geq 3$?

The next result was, in essence, due to Krugljak (its proof can extracted from the argument of [23, Theorem 1.9]). For its formulation we need

Definition 2.3. Let $F \subset K(\mathbb{R}^n)$ be a cover of a set $S \subset \mathbb{R}^n$ by cubes. A family $F_0 \subset K(\mathbb{R}^n)$ is said to be an *extended (contracted) subcover* of F if

(a) F_0 covers S;

(b) for every cube $Q \in F_0$ there is a constant $\lambda = \lambda(Q) \geq 1$ (respectively, $0 < \lambda(Q) \leq 1$) such that

$$\lambda Q \in F.$$

Hereafter we write $c = c(\alpha, \beta, \ldots)$ if the constant depends *only* on the arguments in the brackets.

Definition 2.4. Let $S_1, S_2 \subset \mathbb{R}^n$ be measurable sets and a number $\varepsilon \in (0,1)$. The sets are *ε-linked* if the following is true:

$$|S_1 \cap S_2| \ge \varepsilon \min(|S_1|, |S_2|).$$

S_1, S_2 are *strongly ε-linked* if they satisfies the inequality

$$|S_1 \cap S_2| \ge \varepsilon \max(|S_1|, |S_2|).$$

Theorem 2.5. *Let $F \subset K(\mathbb{R}^n)$ cover a bounded set $S \subset \mathbb{R}^n$ by the centers of its cubes (i.e., $S = \{c_Q : Q \in F\}$). Then there exists an extended (contracted) subcover F_0 of F such that:*

(a) F_0 is at most countable;

(b) for some $c = c(n)$

$$\operatorname{ord} F_0 \le c(n);$$

(c) for every $Q \in F$ there is a cube $K \in F_0$ such that $Q \subset K$ (respectively, $K \subset Q$);

(d) for some $\varepsilon = \varepsilon(n) \in (0,1)$ every pair of intersecting cubes from F_0 is strongly ε-linked.

B. As a space of polynomials we use the space P_k of polynomials in $x_1, \dots, x_n$ of degree $k-1$. Conventionally, $P_0 := \{0\}$.

The next result is of importance for many applications.

Theorem 2.6 ([9]). *Let S be a measurable subset of a convex bounded set $V \subset \mathbb{R}^n$ such that for some $\ell \in (0,1)$*

$$|S| \big/ |V| \ge \ell.$$

Then for every polynomial $p \in P_k$ it is true that

$$\max_V |p| \le t_{k-1}\left(\frac{1 + \sqrt[n]{1-\ell}}{1 - \sqrt[n]{1-\ell}}\right) \max_S |p|.$$

Here $t_m(x) := \cos(m \arccos x)$ is the m-th Chebyshev polynomial.

C. We most frequently use the space $L_p^{\mathrm{loc}}(\mathbb{R}^n)$, $0 < p \le \infty$, as an underlying space for the construction of l.a.- spaces. However, sometimes we need for this goal the space of locally bounded continuous functions, Lorentz spaces $L_{pq}^{\mathrm{loc}}(\mathbb{R}^n)$ and $BMO(\mathbb{R}^n)$.

D. Using the concepts introduced we now present the basic

Definition 2.7. *Local polynomial (best) approximation E_k* is a function

$$E_k : \left(L_p^{\mathrm{loc}} \times K\right)(\mathbb{R}^n) \to \mathbb{R}_+$$

given for $f \in L_p^{\mathrm{loc}}(\mathbb{R}^n)$ and $Q \in K(\mathbb{R}^n)$ by

$$E_k(f; Q; L_p) := \inf_{m \in P_k} \|f - m\|_{L_p(Q)}. \tag{2.1}$$

We also define *normalized local polynomial approximation* by

$$\mathcal{E}_k\left(f;Q;L_p\right) := |Q|^{-\frac{1}{p}} E_k\left(f;Q;L_p\right). \tag{2.2}$$

Note that for $k=0$ the right-hand side equals $\left\{\frac{1}{|Q|}\int\limits_Q |f|^p\,dx\right\}^{1/p}$.

Replacing in (2.1) cube Q by an arbitrary bounded measurable subset $S \subset \mathbb{R}^n$ we extended E_k to this class of subsets. The following result describes basic characteristics of the E_k regarded as a set function.

Theorem 2.8. *(a) Assume that a sequence $\{S_j\}_{j\ge 1}$ of bounded measurable sets of $\mathbb{R}^n$ converges in measure to a bounded measurable set V. Then for $p<\infty$*

$$\lim_{j\to\infty} E_k\left(f;S_n;L_p\right) = E_k\left(f;S;L_p\right). \tag{2.3}$$

For $p=\infty$ relation (2.3) *holds if $\{S_j\}_{j\ge1}$ converges to S in the Hausdorff metric.*

(b) Assume that for some $\varepsilon \in (0,1)$ measurable bounded subsets $S_1, S_2 \subset \mathbb{R}^n$ are ε-linked. Then for some $c = c(n,k) > 0$

$$E_k\left(f;S_1\cup S_2;L_p\right) \le c\cdot\varepsilon^{-k+1}\left\{E_k\left(f;S_1;L_p\right)+E_k\left(f;S_2;L_p\right)\right\}.$$

Now we present the basic fact of the theory, the relation of E_k to difference characteristics of functions. It order to present the result let us recall that *k-modulus of continuity* of $f\in L_p^{\mathrm{loc}}\left(\mathbb{R}^n\right)$ is a function

$$\omega_k\colon L_p^{\mathrm{loc}}\left(\mathbb{R}^n\right)\times(0,+\infty)\longmapsto(0,+\infty]$$

given by

$$\omega_k\left(f;t;L_p\right) : = \sup_{|h|\le t}\left\|\Delta_h^k f\right\|_{L_p(\mathbb{R}^n)}$$

where $|h|$ is the Euclidian norm of $h\in\mathbb{R}^n$ and Δ_h^k is the k-th difference, i.e.,

$$\Delta_h^k f(x) := \sum_{j=0}^{k}(-1)^{k-j}\binom{k}{j} f(x+jh).$$

Theorem 2.9. *For every $f\in L_p^{\mathrm{loc}}\left(\mathbb{R}^n\right)$, $0<p\le\infty$, and $t>0$ it is true that*

$$\omega_k\left(f;t;L_p\right)\approx\sup_{\pi}\left\{\sum_{Q\in\pi}E_k\left(f;Q;L_p\right)^p\right\}^{\frac{1}{p}} \tag{2.4}$$

where π runs over all disjoint families of congruent cubes Q of volume t^n. The constants of equivalence depend only on k and n.

The result for $p\ge1$ was proved in [3]; its extension to $p<1$ is presented in [6].

Remark 2.10. Using monotonicity of E_k in Q and Besicovich's covering theorem one can rewrite the right-hand side of (2.4) in the integral form to get

$$\omega_k(f;t;L_p) \approx \Big\{ \int\limits_{\mathbb{R}^n} \mathcal{E}_k(f;Q_t(x);L_p)^p \, dx \Big\}^{\frac{1}{p}}.$$

As a consequence of Theorem 2.9 we obtain an estimate of $\mathcal{E}_k$ for functions from Sobolev space W_p^s with $s \le k$.

Corollary 2.11. *If* $1 \le p \le q \le \infty$ *and* $f \in W_p^s(Q)$ *then*

$$\mathcal{E}_k(f;Q;L_p) \le c(k,n)|Q|^{\frac{s}{n}} \sup_{|\alpha|=s} \mathcal{E}_{k-s}(D^\alpha f;Q;L_q). \tag{2.5}$$

The final result is a direct consequence of the Hahn–Banach theorem. To its formulation we define for a fixed $Q \in K(\mathbb{R}^n)$ a linear subspace $P_k^\perp(Q)$ of the space $L_1^{\mathrm{loc}}(\mathbb{R}^n)$ given by the conditions:

(a) $\operatorname{supp} f \subset Q$;

(b) $\int\limits_Q x^\alpha f(x)\, dx = 0$ for all $|\alpha| \le k-1$.

Theorem 2.12. *Let* $1 \le p < \infty$ *and* p' *stand for the conjugate exponent* $(1/p + 1/p' = 1)$. *Then for every* $f \in L_p(Q)$

$$E_k(f;Q;L_p) = \sup \Big\{ \int\limits_Q fg\, dx : \ g \in P_k^\perp(Q) \text{ and } \|g\|_{p'} = 1 \Big\}. \tag{2.6}$$

E. Let μ be a Borel measure on $K(\mathbb{R}^n)$.

Definition 2.13. A *functional parameter* Φ is a quasi-Banach lattice of μ-measurable (classes of) functions on $K(\mathbb{R}^n)$.

That is to say, the Φ meets the following axioms:

(P$_1$) If $F \in \Phi$ and $|G| \le |F|$, then $G \in \Phi$ and

$$\|G\|_\Phi \le \|F\|_\Phi.$$

(P$_2$) For some $c = c(\Phi) \ge 1$ and all $F, G \in \Phi$

$$\|F+G\|_\Phi \le c\{\|F\|_\Phi + \|G\|_\Phi\}.$$

(P$_3$) Φ is complete.

Example 2.14. (a) Identifying $K(\mathbb{R}^n)$ with the open half-space

$$\mathbb{R}^{n+1}_+ := \{(x,r) : x \in \mathbb{R}^n, r > 0\}$$

by the bijection $Q \longmapsto (c_Q, r_Q)$ we define a Borel measure dQ on $K(\mathbb{R}^n)$ by

$$dQ := dx \otimes r^{-1}\, dr.$$

Using this we define a (functional) parameter $L_p(L_q^\lambda)$ where $0 < p, q \le \infty$ and $\lambda \ge 0$ by the quasinorm

$$\|F\|_{L_p(L_q^\lambda)} := \Big\{ \int_0^\infty \Big(\int_{\mathbb{R}^n} \Big| \frac{F(x,r)}{r^\lambda} \Big|^p dx \Big)^{\frac{q}{p}} \frac{dr}{r} \Big\}^{\frac{1}{q}}. \tag{2.7}$$

Changing the order of integration we then define the parameter $L_p(L_q^\lambda)$.
(b) Let x_0 be a fixed point of $\mathbb{R}^n$. Define a Borel measure μ on $K(\mathbb{R}^n)$ as $\delta_{x_0} \otimes r^{-1}\, dr$. Then a parameter $L_p^\lambda(x_0)$ is defined by the quasinorm

$$\|F\|_{L_p^\lambda(x_0)} := \Big\{ \int_0^\infty \Big| \frac{F(x_0,r)}{r^\lambda} \Big|^p \frac{dr}{r} \Big\}^{\frac{1}{p}}. \tag{2.8}$$

For $p = \infty$ we will also use a separable subspace of $L_\infty^\lambda(x_0)$ denoted by $\ell_\infty^\lambda(x_0)$ which is given by the condition

$$\lim_{r \to 0} \frac{F(x_0, r)}{r^\lambda} = 0. \tag{2.8$'$}$$

(c) The next example is basic for the main part of the paper.

Let μ be now the counting measure on $K(\mathbb{R}^n)$. Then we define a parameter V_p^λ, $0 < p \le \infty$ and $\lambda \in \mathbb{R}$, by the quasinorm

$$\|F\|_{V_p^\lambda} := \sup_\pi \Big\{ \sum_{Q \in \pi} \Big| \frac{F(Q)}{|Q|^\lambda} \Big|^p \Big\}^{\frac{1}{p}} \tag{2.9}$$

where π runs over all disjoint families of cubes $Q \in K(\mathbb{R}^n)$.

In some applications ℓ_p-norm in (2.9) should be replaced the ℓ_{pq}-norm, $0 < p, q \le \infty$. Let us recall that ℓ_{pq} is the Lorentz space whose elements are sequences $s = (s_1, s_2, \dots)$ satisfying

$$\|s\|_{\ell_{pq}} := \Big\{ \sum_{n=1}^\infty \big(n^{\frac{1}{q}} s_n^*\big)^p n^{-1} \Big\}^{\frac{1}{p}} < \infty$$

where $(s_n^*)_{n \ge 1}$ is the nonincreasing rearrangement of $(s_n)_{n \ge 1}$.

The parameter V_{pq}^λ introduced in this way is therefore defined by the quasinorm

$$\|F\|_{V_{pq}^\lambda} := \sup_\pi \Big\| \big(F(Q)\big)_{Q \in \pi} \Big\|_{\ell_{pq}}. \tag{2.10}$$

It may be easily seen that all of the examples introduced satisfy axioms (P$_1$) – (P$_3$). Moreover, all of them but $\ell_\infty^\lambda(x_0)$ satisfy the *Fatou property*. Let us recall that Φ satisfies this property if every bounded in Φ sequence $\{f_j\}$ converging in measure to a function f satisfies the Fatou inequality

$$\|f\|_\Phi \le \varliminf_{j\to\infty} \|f_j\|_\Phi.$$

Finally, let us consider a couple of $\overline{\Phi} := (\Phi_0, \Phi_1)$ of parameters over measure space $(K(\mathbb{R}^n), \mu)$. It is easily proved that the spaces $\overline{\Phi}_{\theta p}$, $\Phi_0^{1-\theta}\Phi_1^\theta$ and $[\overline{\Phi}]_\theta$ where $0 < \theta < 1$ and $0 < p \le \infty$ satisfy conditions (P$_1$) – (P$_3$) and hence are functional parameters, as well. Here the first is an interpolation space defined by the real method, the second is defined by the Calderón construction and the last one is a complex interpolation space (in this case both the Φ_i are Banach); see e.g., [13] for the corresponding definitions. In general, the space $\Phi_0^{1-\theta}\Phi_1^\theta$ may be not an interpolation space of $\overline{\Phi}$, but it is the case for Φ_i possessing the Fatou property, (Ovchinnikov's theorem, see e.g., [13, Theorem 4.3.10].)

F. Using the above introduced ingredients we now define the main object of this section, a *local approximation space.*

Let Φ be a parameter on $(K(\mathbb{R}^n), \mu)$ and $0 < p \le \infty$, $k \in \mathbb{Z}_+$ be fixed.

Definition 2.15. A *homogeneous* l.a.-space $\dot{A}_\Phi^k(L_p)$ consists of $f \in L_p^{\mathrm{loc}}(\mathbb{R}^n)$ for which the seminorm

$$|f|_{A_\Phi^k(L_p)} := \|E_k(f; \cdot\,; L_p)\|_\Phi \tag{2.11}$$

is finite.

We also define a (non-homogeneous) l.a.- space $A_\Phi^k(L_p)$ by the quasinorm

$$\|f\|_{A_\Phi^k(L_p)} := \|f\|_{L_p(Q_0)} + |f|_{A_\Phi^k(L_p)} \tag{2.12}$$

where Q_0 is a fixed cube, say, $Q_0 := [0,1]^n$.

In the sequel we encounter also (for special cases) the spaces $\dot{A}_\Phi^k(L_{pq})$, $\dot{A}_\Phi^k(C)$ and $\dot{A}_\Phi^k(BMO)$ obtained by replacing L_p^{loc} by the Lorentz space $L_{pq}(\mathbb{R}^n)$, the space of continuous functions on $\mathbb{R}^n$ and by $BMO(\mathbb{R}^n)$, respectively.

The next proposition whose simple proof may be found in [3] describes some basic facts following directly from definitions.

Proposition 2.16. *(a) If Φ possesses the Fatou property, then $A_\Phi^k(L_p)$ is quasi-Banach space (Banach, if L_p and Φ are). The same is true for $\dot{A}_\Phi^k(L_p)$ after factorization by P_k.*

(b) If $\Phi \hookrightarrow \Phi'$ and $k \le \ell$, $p \ge q$, then

$$\dot{A}_\Phi^k(L_p) \hookrightarrow \dot{A}_{\Phi'}^\ell(L_q)P_k. \tag{2.13}$$

Here $X \hookrightarrow Y$ denotes linear and continuous embedding of quasi-Banach space X into quasi-Banach space Y.

(c) Let now $\overline{\Phi} = (\Phi_0, \Phi_1)$ *be a pair of parameters over* $(K(\mathbb{R}^n), \mu)$. *Then*

$$\left(\dot{A}^k_{\Phi_0}(L_{p_0}),\ \dot{A}^k_{\Phi_1}(L_{p_1})\right)_{\theta q} \subset \dot{A}^k_{\bar{\Phi}_{\theta q}}(L_{pq}) \tag{2.14}$$

where $1/p_0 := (1-\theta)/p_0 + \theta/(p-1)$.

If Φ_i *and* L_{p_i} *are Banach, then*

$$\left[\dot{A}^k_{\Phi_0}(L_{p_0}),\ \dot{A}^k_{\Phi_1}(L_{p_1})\right]_\theta \subset \dot{A}^k_{\Phi_0^{1-\theta}\Phi_1^\theta}(L_{p\theta}). \tag{2.15}$$

The basic problems of the theory concern with the inversion of embeddings (2.13)–(2.15). We will see below that in many important special cases such inversions hold under certain mild restriction on parameters.

We also briefly describe one more problem, *the trace extension problem* for l.a.-spaces. In this case we are interesting in characterization of the trace space

$$\dot{A}^k_\Phi(L_p)\Big|_S := \Big\{ f \in L_p^{\mathrm{loc}}(S) : \exists\, g \in \dot{A}^k_\Phi(L_p) \ \text{ such that } \ f = g\Big|_S \Big\}$$

where $S \subset \mathbb{R}^n$ is a subset of positive measure. To formulate the problem we define the trace $K(S) = K(\mathbb{R}^n)\Big|_S$ by

$$K(S) := \Big\{ Q \in K(\mathbb{R}^n) : \ c_Q \in S \ \text{ and } \ 0 < |S \cap Q| \le |S| \Big\}. \tag{2.16}$$

Further we set

$$E_k(f; Q; L_p(S)) := E_k(f; Q \cap S; L_p) \tag{2.17}$$

and define the space $\dot{A}^k_\Phi(L_p(S))$ using in Definition 2.15 set function (2.17) instead of $E_k(f; Q; L_p)$. It is readily seen that

$$\dot{A}^k_\Phi(L_p)\Big|_S \hookrightarrow \dot{A}^k_\Phi(L_p(S))$$

under some restrictions on Φ and on S (Ahlfors regularity) the converse is also true.

The above mentioned results have a wide range of applications, since almost all classical spaces of differentiable functions (Sobolev, Besov, Lizorkin–Triebel, BMO, etc.) can be presented as l.a.- spaces. We formulate only some of these facts.

One begins with the family of Besov spaces of positive smoothness. Among equivalent definitions of these spaces the simplest one is based on the seminorm

$$|f|_{B_p^{\sigma q}(\mathbb{R}^n)} := \Big\{ \int_{\mathbb{R}_+} \Big(\frac{\omega_k(f; t; L_p)}{t^\sigma}\Big)^q \frac{dt}{t} \Big\}^{\frac{1}{q}};$$

here $\sigma > 0$, $k = k(\sigma)$ is the smallest integer bigger than σ and $0 < p, q \le \infty$. In this way one defines the *homogeneous Besov space* denoted by $\dot{B}_p^{\sigma q}(\mathbb{R}^n)$. The corresponding Banach (quasi-Banach, if $\min(p,q) < 1$) space is then defined by (quasi-norm)

$$\|f\|_{B_p^{\sigma q}(\mathbb{R}^n)} := \|f\|_{L_p(\mathbb{R}^n)} + |f|_{B_p^{\sigma q}(\mathbb{R}^n)}.$$

For $p < 1$ there is another definition proposed by Peetre [26]. Denoting Peetre's space by $\widehat{B}_p^{\sigma q}(\mathbb{R}^n)$ we have the following embedding

$$\widehat{B}_p^{\sigma q}(\mathbb{R}^n) \subset B_p^{\sigma q}(\mathbb{R}^n)$$

which is proper only if $\sigma < n(p^{-1} - 1)$.

Theorem 2.17. *Assume that $p < r \leq \infty$ satisfy the condition*

$$\frac{\sigma}{n} > \frac{1}{p} - \frac{1}{r}.$$

Then the following equality

$$\dot{B}_p^{\sigma q}(\mathbb{R}^n) = \dot{A}_\Phi^k(L_p)$$

holds up to the equivalence of seminorms. Here the parameter $\Phi = L_r\left(L_q^{\sigma^*}\right)$, see (2.7), where $\sigma^* := \sigma - n/p$ and $k = k(\sigma)$.

For proof see [3].

Now the present an analogous result for homogeneous Lizorkin–Triebel spaces $\dot{F}_p^{\sigma q}(\mathbb{R}^n)$, see, e.g., [28] for definition. In general, these are spaces of tempered distributions and are function spaces under some restrictions, e.g., under the assumption of the result presented below.

Theorem 2.18. *Let $0 < \min(p, q) \leq \infty$ and $1 \leq r \leq \infty$. Assume that*

$$\frac{\sigma}{n} > \left(\frac{1}{\min(p,q)} - \frac{1}{r}\right)_+ .$$

Then the following equality

$$\dot{F}_p^{\sigma q}(\mathbb{R}^n) = \dot{A}_\Phi^k(L_r)$$

holds up to the equivalence of seminorms. Here $\Phi = L_q^{\sigma^*}(L_r)$, where $\sigma^* := \sigma - n/p$ and $k = k(\sigma)$.

The result for $0 < q \leq \infty$ and $1 < p < \infty$ was due to Dorronsoro [18]. The general result was due to Shvarsman and the author [14] by local approximation methods.

Finally we present local approximation presentation of Taylor spaces $T_p^\sigma(x_0)$ and $t_p^\sigma(x_0)$ introduced by Calderón and Zygmund [16]. A function $f \in L_p^{\mathrm{loc}}(\mathbb{R}^n)$ belongs to $T_p^\sigma(x_0)$ where $\sigma > 0$ and $x_0 \in \mathbb{R}^n$, if there exists a polynomial t of degree *strictly less* than σ such that

$$\left\{\frac{1}{|Q|}\int_Q |f(x) - t(x - x_0)|^p \, dx\right\}^{\frac{1}{p}} \leq c|Q|^{\frac{\sigma}{n}} \tag{2.18}$$

for all cubes $Q = Q_r(x_0)$ with $r \leq 1$ and a suitable constant $c > 0$.

If is clear that t, the *Taylor polynomial of f at x_0*, is unique. Under the (quasi-) norm

$$\|f\|_{T_p^\sigma(x_0)} := \|f\|_{L_p(Q_0)} + \inf c$$

this becomes a (quasi-) Banach space.

One then defines a subspaces $t_p^\sigma(x_0)$ of $T_p^\sigma(x_0)$ by replacing the right-hand side in (2.18) by $o\left(|Q|^{\sigma/n}\right)$ as $Q \to x_0$, and choosing the Taylor polynomial t of degree *less than or equals* to σ.

Theorem 2.19. *Assume that $0 < \sigma < k$ and σ is noninteger. Then the equalities*

$$T_p^\sigma(x_0) = A_\Phi^k(L_p) \text{ and } t_p^\sigma(x_0) = A_{\Phi'}^k(L_p)$$

hold up to the equivalence of (quasi-) norms.

Here $\Phi = L_p^\sigma(x_0)$ and $\Phi' = \ell_p^\sigma(x_0)$, see (2.8) *and* (2.8′).

G. In order to cover the bigger range of applications one should vary to large extent the basic ingredients for the construction of l.a.- spaces. The next simple example illustrates the situation.

Let $\mathrm{Var}(\mathbb{R}^2)$ be the space of bounded variations in Vitali's sense [29]. Its definition follows that of Jordan's but as the deviation of $f : \mathbb{R}^2 \to \mathbb{R}$ one uses a function of rectangle I determined by the mixed difference of f over the vertices of I, see, e.g., [1] for details. Presentation of $\mathrm{Var}(\mathbb{R}^2)$ as an l.a.- space uses the set of rectangles $\mathcal{L}(\mathbb{R}^2)$ as a family of carriers, and the linear space $\{\varphi(x) + \psi(y) : \varphi, \psi \in L_\infty^{\mathrm{loc}}(\mathbb{R})\}$ as a space of "polynomials". Let us note that $\mathcal{L}(\mathbb{R}^2)$ is a much more complicated geometric object than the set of cubes (e.g., the Vitali and Besicovich covering lemmas are untrue). Moreover, the space of approximating "polynomials" is infinite dimensional and local estimates analogous those in Theorem 2.6 do not hold. However, some basic facts of the theory remains to be true, e.g., an analog of Theorem 2.9, and the theory can be applied for the study of $\mathrm{Var}(\mathbb{R}^2)$.

To cover all kind of applications the general theory uses the class of translation and dilation invariant closed subspaces of $L_\infty^{\mathrm{loc}}(\mathbb{R})$ as spaces of "polynomials". The space P_k is clearly of this kind but there are also infinite-dimensional subspaces in this class. Nevertheless, the set of such spaces can be parameterized by finite families of points in $\mathbb{Z}_+^n$ (e.g., the set $\{\alpha \in \mathbb{Z}_+^n : |\alpha| = k\}$ is assigned to P_k).

The definition of families of carriers is more restrictive; e.g., it excludes $\mathcal{L}(\mathbb{R}^2)$ but contains the family of rectangles $I = I_1 \times I_2 \subset \mathbb{R}^2$ satisfying for fixed numbers $0 < \lambda < \mu$ the inequality $\lambda < |I_1|/|I_2| < \mu$. Another example is the family of open balls in a quasi-metric space $(\mathbb{R}^n, d)$ where a quasi-metric d is given for $x, y \in \mathbb{R}^n$ by

$$d(x, y) := \max_{1 \le i \le n} |x_i - y_i|^{\mu_i};$$

here $\mu_i > 0$ are fixed.

The latter family is of importance for the study of anisotropic spaces of smooth functions.

Finally, the general theory uses a wide class of underlying spaces including quasi-Banach lattices (e.g., Lorentz and Orlicz spaces), BMO spaces, Morrey spaces, Sobolev spaces, and Besov spaces.

3 Functions of bounded (k, p)-variation

A. Let a function $f \in L_q^{\text{loc}}(\mathbb{R}^n)$ where $1 \le q \le \infty$, and a triple $\eta := (k, \lambda, p)$ where $k \in \mathbb{N}$, $\lambda \in \mathbb{R}^n$ and $0 < p \le \infty$ be fixed.

Definition 3.1. *η-variation* of f in $L_q(\mathbb{R}^n)$ is given by

$$\operatorname{Var}_\eta(f; L_q) := \sup_\pi \Big\{ \sum_{Q \in \pi} \Big(\frac{E_k(f; Q; L_q)}{|Q|^\lambda} \Big)^p \Big\} \tag{3.1}$$

where π runs over all disjoint families of cubes $Q \in K(\mathbb{R}^n)$.

By $V_\eta(L_q)$ we denote the space of functions defined by finiteness of (3.1). This space is, clearly, equal to $\mathring{A}_\Phi^k(L_q)$ with the parameter Φ defined by (2.9). In particular, the factor-space $V_\eta(L_q)/P_k$ is quasi-Banach (Banach, if $p \ge 1$) and

$$V_\eta(L_q) \hookrightarrow V_{\widetilde{\eta}}(L_{\widetilde{q}}) \tag{3.2}$$

provided that $\widetilde{\eta} = \big(\widetilde{k}, \widetilde{\lambda}, \widetilde{p}\big)$ satisfies

$$p \le \widetilde{p}, \ \ k \le \widetilde{k}, \ \ \widetilde{q} \le q, \ \ \widetilde{\lambda} \le \lambda,$$

see Proposition 2.16.

More important characteristics of the space $V_\eta(L_q)$ relate to numerical invariants, *smoothness* σ and *limiting exponent* p^*, given by

$$\sigma := n\Big(\lambda + \frac{1}{p} - \frac{1}{q}\Big), \quad \frac{1}{p^*} := \frac{1}{q} - \lambda. \tag{3.3}$$

Note that $-\infty < p^* \le \infty$ and

$$\frac{1}{p^*} = \frac{1}{p} - \frac{\sigma}{n}. \tag{3.4}$$

In the sequel, we consider only spaces $V_\eta(L_q)$ whose smoothness σ satisfies the condition

$$0 < \sigma \le k. \tag{3.5}$$

This choice is motivated by

Theorem 3.2 ([3]). *(a) If smoothness of the space $V_\eta(L_q)$ is zero and $\lambda \ge 0$, $1 \le p < \infty$, then*

$$L_p(\mathbb{R}^n) + P_k \hookrightarrow V_\eta(L_q) \hookrightarrow L_{p\infty}(\mathbb{R}^n) + P_k \tag{3.6}$$

provided that $q < p < \infty$. *In the remaining case*[1] $q = p$ *and the left-hand side equals* $V_\eta(L_q)$.

(b) If smoothness of the space $V_\eta(L_q)$ *is strictly greater than* k, *then for* $q \le p$

$$V_\eta(L_q) = P_k \ (\textit{modulo measure zero}), \tag{3.7}$$

and for $p < q$ *this space is infinite dimensional but*

$$V_\eta(L_q) \cap C^k(\mathbb{R}^n) = P_k. \tag{3.8}$$

Remark 3.3. Apparently, $C^k(\mathbb{R}^n)$ in (3.8) may be replaced by $C^{k-1}(\mathbb{R}^n)$. The following example might justify this claim. Let $V_p(\mathbb{R})$, $0 < p < \infty$ be the space of functions of bounded p-variation in the Jordan–Wiener–L. Young sense. Since

$$E_1(f; [a,b]; L_\infty) = \frac{1}{2}\operatorname{ess\,sup}\Big\{f(x) - f(y);\ \ x, y \in [a,b]\Big\},$$

p-variation of a locally bounded function $f : \mathbb{R} \to \mathbb{R}$ can be defined as

$$\operatorname{Var}_p f = \sup_\pi \Big\{\sum_{I\in\pi} E_1(f; I; L_\infty)^p\Big\}^{\frac{1}{p}}$$

where π runs over all disjoint families of intervals.

Hence, modulo measure zero $V_p(\mathbb{R}) = V_\eta(L_\infty)$ where $\eta = (1, 0, p)$. For the latter space $\sigma = 1/p$ and for $p < 1$, $k = 1 > \sigma$; hence, (3.8) implies that

$$\dim V_p(\mathbb{R}) = \infty \quad \text{and} \quad V_p(\mathbb{R}) \cap C^1(\mathbb{R}) = \{\text{const}\}.$$

But it may be easily verified that $V_p(\mathbb{R})$ with $p < 1$ consists only of steps functions whose jumps $j_1, j_2, \ldots$ satisfies $\sum\limits_s |j_s|^p < \infty$. In particular, $V_p(\mathbb{R}) \cap C(\mathbb{R}) = \{\text{const}\}$.

The next result demonstrates the role of limiting exponent, see (3.3). To its formulation we set

$$X_{p^*} := \begin{cases} L_{p^*\infty}(\mathbb{R}^n), & \text{if } p^* > 0; \\ BMO(\mathbb{R}^n), & \text{if } p^* = 0; \\ C(\mathbb{R}^n), & \text{if } p^* < 0. \end{cases}$$

Theorem 3.4 ([3]). *Assume that limiting exponent* p^* *of the space* $V_\eta(L_q)$ *satisfies the inequality*[2] $1/p^* < 1/q$. *Then up to the equivalence of norms*

$$V_\eta(L_q) = V_{\eta^*}(X_{p^*})$$

where $\eta^* = (k, \lambda^*, p)$ *and* $\lambda^* := \min(0, 1/p^*)$.

[1] Conditions $\lambda \ge 0$ and $\sigma = 0$ imply that $q \le p$, see (3.3).

[2] Equivalently: $\lambda > 0$.

B. Now we present results concerning relations of BV-spaces to Sobolev and Besov spaces.

Theorem 3.5 ([3]). *Assume that smoothness of $V_\eta(L_q)$ satisfies: $0 < \sigma < k$. Then up to factorization by P_k*

$$\dot{B}_p^{\sigma}(\mathbb{R}^n) \hookrightarrow V_\eta(L_q) \hookrightarrow \dot{B}_p^{\sigma\infty}(\mathbb{R}^n) \tag{3.9}$$

provided $q < \infty$ or $q = \infty$ and $p < 1$.

Remark 3.6. (a) In the remaining case of $q = \infty$ and $p \geq 1$, the left-side of (3.9) should be replaced by $\dot{B}_p^{\sigma 1}(\mathbb{R}^n)$. This result for $n = 1$ was due to Bergh and Peetre [2] who also proved that this embedding is sharp.

(b) The partial case $V_p(\mathbb{R}) \subset \dot{B}_p^{1/p}(\mathbb{R}^n)$, $1 < p < \infty$, was, in essence, due to Marcinkiewicz [25].

In order to formulate the next result, let us recall that the homogeneous Sobolev space $\dot{W}_p^k(\mathbb{R}^n)$, $1 \leq p \leq \infty$, is defined by finiteness of the seminorm

$$|f|_{W_p^k(\mathbb{R}^n)} := \sup_{|\alpha|=k} \|D^\alpha f\|_{L_p(\mathbb{R}^n)} .$$

We also define the space $BV^k(\mathbb{R}^n)$ consisting of locally integrable functions on $\mathbb{R}^n$ whose k-th distributional derivatives are bounded Borel measures. We equip this space by the seminorm

$$|f|_{BV^k(\mathbb{R}^n)} := \sup_{|\alpha|=k} \operatorname{Var}_{\mathbb{R}^n}(D^\alpha f) .$$

Theorem 3.7 ([3]). *Assume that smoothness of the space $V_\eta(L_q)$ is maximal, i.e. $\sigma = k$, and the remaining parameters satisfy the conditions*

$$1 \leq p < q \leq \infty \quad \text{and} \quad \min\left(\frac{1}{p^*}, \lambda\right) > 0.$$

Then for $p > 1$

$$V_\eta(L_q) = \dot{W}_p^k(\mathbb{R}^n) \tag{3.10}$$

and for $p = 1$

$$V_\eta(L_q) = BV^k(\mathbb{R}^n) . \tag{3.11}$$

Remark 3.8. For $k = n = 1$ equality (3.10) is equivalent to the classical F. Riesz lemma, see, e.g., [27], and equality (3.11) is the Hardy–Littlewood characteristic of $V_1(\mathbb{R})$. For $k = 1$ and $n > 1$ equality (3.11) is mentioned in the introduction, see Theorem 1.2; note that if $\sigma = k = 1$ and $p = 1$, then $1/p^* = (n-1)/n$ and $0 \leq \lambda \leq (n-1)/n$.

C. Now we present several results on differentiability of BV-functions. They are taken from the author dissertation [5] still waiting its publication. However, the facts and methods required for proofs can be found in [3] and [6].

To formulate the results we recall the concepts of the *Hausdorff* (outer) *measure* $\mathcal{H}_s$ and *Hausdorff content* $\mathcal{C}_s$. Actually, set for $\Omega \subset \mathbb{R}^n$ and $\varepsilon > 0$

$$\mathcal{H}_s^\varepsilon(\Omega) := \inf\Big\{\sum_i |Q_i|^{\frac{s}{n}} : \bigcup_i Q_i \supset S \text{ and } |Q_i| \le \varepsilon\Big\};$$

here $\{Q_i\}$ are families of cubes.

Then the required notions are given by

$$\mathcal{H}_s := \lim_{\varepsilon \to 0} \mathcal{H}_s^\varepsilon \text{ and } \mathcal{C}_s = \mathcal{H}_s^\infty.$$

Further, given a function $f \in L_1^{\text{loc}}(\mathbb{R}^n)$ we define its *regularization* $\bar{f}$ by setting for a fixed x_0

$$\bar{f}(x_0) := \lim_{Q \to x_0} \int_Q f\, dx.$$

By the Lebesgue differentiation theorem the domain of $\bar{f}$ (the *Lebesgue set* of f) differs from $\mathbb{R}^n$ by a set of measure zero.

Theorem 3.9. *Let a function f belongs to the space $V_\eta(L_q)$ whose parameters satisfy for some $s \in (0, n]$:*

$$\mu := \frac{s}{p} - \frac{n}{p^*} > 0.$$

Then the following is true.

(a) If $s = n$ and $\mu = k$, then f belongs to the Taylor space $t_q^k(x)$ for almost all x.

Moreover, for every $\varepsilon > 0$ there is a C^k-function $f_\varepsilon : \mathbb{R}^n \to \mathbb{R}^n$ such that

$$\big|\{x \in \mathbb{R}^n : f_\varepsilon(x) \ne f(x)\}\big| < \varepsilon.$$

(b) If $0 < \mu < k$ and $\mu \notin \mathbb{N}$, then f belongs to the Taylor space $T_q^\mu(x)$ for all $x \in \mathbb{R}^n$ excepting a set of Hausdorff $\mathcal{H}_s$-measure zero.

Moreover, for every $\varepsilon > 0$ there is a function $f_\varepsilon : \mathbb{R}^n \to \mathbb{R}$ from the Lipschitz–Besov space $\dot{B}_\infty^\mu(\mathbb{R}^n)$ such that

$$\mathcal{C}_s\big(\{x \in \mathbb{R}^n : f_\varepsilon(x) \ne f(x)\}\big) < \varepsilon.$$

Remark 3.10. The assertions of Theorem 3.9(b) hold also for integer μ as well. For integer μ the Taylor space $T_q^\mu(x_0)$ is defined the upper complex method letting

$$T_q^\mu(x_0) := \big[T_q^{\mu-\varepsilon}(x_0), T_q^{\mu+\varepsilon}(x_0)\big]^{\frac{1}{2}}$$

where $\varepsilon > 0$ is sufficiently small. It can be shown that this space is independent of ε.

In combination with the results of subsection B, Theorem 3.9 gives a unified approach to a wide range of results on differential properties of functions from the classical function spaces (V_p, BV, W_p^k, $B_p^{\sigma q}$, etc). Here we only mention the Calderón–Zygmund generalization of the classical Lebesgue theorem to functions from $BV(\mathbb{R}^n)$ ([15]), and the Federer–Ziemer Luzin's type theorem for functions from $W_p^k(\mathbb{R}^n)$, $1 < p < \infty$ ([19]).

D. Now we present results on nonlinear N-term approximation in $L_q(\mathbb{R}^n)$ of functions from the space $V_\eta(L_q)$. It can be shown that the order of approximation for some of these functions does not become zero as $N \to \infty$. A necessary condition on a function to be *well approximating* is presented by the limit relation below.

Let $v : V_\eta(L_q) \times K(\mathbb{R}^n)$ is a function given by

$$v(f;Q) := \sup_\pi \Big\{ \sum_{k\in\pi} \Big(\frac{E_k(f;K;L_q)}{|K|^\lambda} \Big)^p \Big\}^{\frac{1}{p}}$$

where π runs over all disjoint families of cubes $K \subset Q$.

Then the subspaces of well approximating functions denoted by $\overset{\circ}{V}_\eta(L_q)$ is defined by the condition:

Uniformly in $Q \in K(\mathbb{R}^n)$

$$\lim_{|Q|\to 0} v(f;Q) = 0.$$

It can be shown that under the assumptions of Theorem 3.7 ($V_\eta(L_q)$ equals $BV^k(\mathbb{R}^n)$ or $W_p^k(\mathbb{R}^n)$ with $p > 1$ for this case) the subspace $\overset{\circ}{V}_\eta(L_q) = V_\eta(L_q)$.

Theorem 3.11. *If $f \in \overset{\circ}{V}_\eta(L_q)$, then for every integer $N \geq 1$ there is an N-term linear combination*

$$T_N(f) = \sum_Q m_Q \cdot \mathrm{1\!I}_Q \tag{3.12}$$

where Q are dyadic cubes and m_Q are polynomials of degree $k-1$ such that

$$\|f - T_N(f)\|_{L_q(\mathbb{R}^n)} \leq cN^{-\frac{\sigma}{n}} |F|_{V_\eta(L_q)}$$

where the constant $c > 0$ depends only on η and n; here $\sigma \in (0,k]$ is smoothness of the space $V_\eta(L_q)$.

An approximation result of this kind was firstly proved in [17] for $BV(\mathbb{R}^2)$: it is cited in Introduction, see (1.4). In the general case, we use the approximation algorithm described in detail in [8] for functions from Besov spaces. The proof of the result requires an additional argument, the covering theorem due to Dol'nikov, Irodova and the author, see its weaker version in [10]. To its formulation one considers a *dyadic sieve*, a subset of $\mathbb{R}^n$ of the form

$$S_N := Q_0 \setminus \bigcup_{Q\in\pi_N} Q$$

where π_N is a disjoint family of N open diadic subcubes of $Q_0 = [0,1]^n$.

Theorem 3.12. *There exists a cover of S_N by cN closed cubes such that intersecting cubes of the cover are ε-linked.*

Here the constants $\varepsilon \in (0,1)$ and $c > 1$ depend only on dimension n.

Note that if cubes of π_N are not dyadic, the cardinality of cover may be arbitrarily large.

Remark 3.13. (a) The algorithm of the paper [8] can be applied to replace in (3.12) the approximating piecewise polynomials by a linear combination of splines or compactly supported wavelets; in the latter case $q < \infty$.
(b) Replacing in Theorem 3.11 L_q-norm by an $L_{q'}$-norm with $1 \leq q' < q$ one may choose the collection of dyadic cubes in (3.12) to have the pairwise disjoint interiors. This improvement allows to estimate ε-entropy of the unit ball of the trace space $V_\eta(L_q)\Big|_{Q_0}$ in the space $L_{q'}(Q_0)$, $q' < q$. Note that this ball is not precompact in $L_q(Q_0)$ but inherits some properties of precompact sets.

E. We present now a duality result for the space $V_\eta(L_q)$.

Theorem 3.14. *Assume that the parameters of the space $V_\eta(L_q)$ satisfy*[3]

$$1 < p, q \leq \infty \text{ and } \lambda \geq 0. \tag{3.13}$$

Then this space is dual.

We outline the proof. To construct the required predual Banach space we need several notions.

Let us recall that a measurable function $a : \mathbb{R}^n \to \mathbb{R}$ is said to be a (k,λ,q)-*atom* if for some cube $Q \in K(\mathbb{R}^n)$ we have:

(a) $\operatorname{supp} a \subset Q$;
(b) $\|a\|_{L_q(Q)} \leq |Q|^{-\lambda/n}$;
(c) $\int\limits_Q a(x)x^\alpha \, dx = 0$ for all $|\alpha| < k$.

We say that a function $m : \mathbb{R}^n \to \mathbb{R}^n$ is a (k,λ,q)-*molecula* if m is a linear combination of (k,λ,q)-atoms whose supports are *pairwise disjoint*.

Let $m = \sum\limits_i c_i a_i$ be such presentation of the molecula; then for $1 \leq p \leq \infty$ we set

$$[m]_p := \left\{ \sum_i |c_i|^p \right\}^{\frac{1}{p}}.$$

Now we define a space $\mathcal{U}_p^{k,\lambda}(L_q)$ to be the linear space of functions f presented as infinite sums of (k,λ,q)-moleculas convergent in $L_q(\mathbb{R}^n)$; it is equipped by the norm

$$\|f\|_{\mathcal{U}_p^{k,\lambda}(L_q)} := \inf \Big\{ \sum_i [m_i]_p \Big\}$$

[3] Recall that $0 < \sigma \leq k$.

where infimum is take over all decomposition

$$f = \sum_i m_i \text{ (convergence in } L_q)$$

into sums of (k, λ, q)-moleculas m_i.

This construction and Theorem 2.12 allow to check easily that

$$\mathcal{U}_{p'}^{k,\lambda}(L_{q'})^* = V_\eta(L_q)\big/P_{k-1}.$$

Here p', q' are conjugate to p, q.

To formulate the next claim we define another separable subspace of $V_\eta(L_q)$ denoted by $v_\eta(L_q)$.

Let for a fixed $\varepsilon > 0$ a function $v_\varepsilon : V_\eta(L_q) \times K(\mathbb{R}^n) \to \mathbb{R}$ be given by

$$v_\varepsilon(f; K) := \sup_\pi \Big\{ \sum_{Q\in\pi} \Big(\frac{E_k(f; Q; L_q)}{|Q|^\lambda} \Big)^p \Big\}^{\frac{1}{p}}$$

where π runs over all disjoint families of cubes $Q \in K(\mathbb{R}^n)$ of *volume at most* ε.

The subspace $v_\eta(L_q)$ is defined by the condition:

$$\lim_{\varepsilon\to 0} v_\varepsilon(f; Q) = 0 \text{ for every } Q.$$

Conjecture 3.15. *If parameters of $V_\eta(L_q)$ satisfy the conditions*

$$1 < p < \infty, \quad 1 < q \le \infty \text{ and } \lambda \ge 0,$$

then it is true that

$$v_\eta(L_q)^{**} = V_\eta(L_q).$$

The result has been proved in several cases, see, in particular, [24] and [22].

F. Finally, we present several results on interpolation of Banach couples of BV-space by the real method. We begin with the result concerning K-functional of the quasi-Banach couple

$$\bar{V} := \big(V_{\eta_0}(L_q), V_{\eta_1}(L_q)\big) \tag{3.14}$$

where $\eta_i = (k, \lambda_i, p_i)$, $i = 0, 1$. For its formulation we define invariants of $\bar{V}$, setting for $p_0 \ne p_1$

$$\alpha(\bar{V}) := \frac{\lambda_0 - \lambda_1}{\frac{1}{p_0} - \frac{1}{p_1}}$$

and for $p_0 = p_1$

$$\alpha(\bar{V}) = \pm\infty$$

where the sign equals $\operatorname{sgn}(\lambda_0 - \lambda_1)$.

We also set

$$\lambda(\bar{V}) := \lambda_i + \frac{\alpha(\bar{V})}{p_i}$$

with $i = 0$ (or, what is the same, $i = 1$) for the first case, and

$$\lambda(\bar{V}) := \alpha(\bar{V})$$

for the second case.

We, further, define, for a disjoint family $\pi \subset K(\mathbb{R}^n)$, the related to $\bar{V}$ Banach couple

$$\bar{\ell}(\pi) := \left(\ell_{p_0}^{\lambda_0}(\pi), \ell_{p_1}^{\lambda_1}(\pi)\right)$$

where the space $\ell_p^\lambda(\pi)$ is given by the quasi norm

$$\|F\|_{\ell_p(\pi)}^\lambda := \Big\{ \sum_{Q \in \pi} \Big| \frac{F(Q)}{|Q|^\lambda} \Big|^p \Big\}^{\frac{1}{p}}.$$

Theorem 3.16 ([12]). *Assume that*

$$\alpha(\bar{V}) \notin \left[0, 1 - \frac{1}{n}\right] \quad \textit{and} \quad \operatorname{sgn}\lambda(\bar{V}) = \operatorname{sgn}\alpha(\bar{V}). \tag{3.15}$$

Then for every $f \in V_{\eta_0}(L_q) + V_{\eta_1}(L_q)$

$$K(t; f; \bar{V}) \approx \sup_\pi K\big(t; E_k(f; \cdot\,; L_q); \bar{\ell}(\pi)\big) \tag{3.16}$$

where π runs over all disjoint families of cubes and the constants of equivalence are independent of f and t.

The main tool of the proof is an amplified version of the Krugjak covering lemma ([23]), see Theorem 2.5 above, which imposes restrictions (3.15) on $\alpha(\bar{V})$. Apparently, Theorem 3.16 is true also for $\alpha(\bar{V}) = 0$ while the Krugjak covering lemma is untrue for the parameter corresponding to this case. To justify this claim let us consider the couple $\big(L_q, W_p^k\big)(\mathbb{R}^n)$. Presenting this in form (3.14), see Theorem 3.2 and Theorem 3.7, and using his covering lemma, Krugjak proves that for $k/n > 1/p - 1/q$

$$\big(L_q, W_p^k\big)_{\theta p_\theta} = B_{p_\theta}^{k_\theta} \tag{3.17}$$

where $1/p_\theta := (1-\theta)/q + \theta/p$ and $0 < \theta < 1$.

For this case $\alpha(\bar{V}) > 0$; however, (3.17) is also true for $k/n = 1/p - 1/q$ and $q < \infty$ when $\alpha(\bar{V}) = 0$. This result follows from the variant of Theorem 3.14 using splines or wavelets.

Presence of supremum in (3.16) makes computation of the real interpolation space $\bar{V}_{\theta p}$ for $p < \infty$ to be a very difficult problem. For some special cases ($\sigma_0 = k$ and/or $\sigma_1 = 0$ and $p = p_\theta$) the problem is solved in [23] by using a maximal function of a new kind.

However, for $p = \infty$ equality (3.16) easily implies that

$$\bar{V}_{\theta\infty} = \dot{A}_{\Phi_\theta}(L_q)$$

where the parameter Φ_θ is given by (2.10) with the corresponding parameters. More precisely, Φ_θ is defined by seminorm

$$\sup_\pi \left\| \left\{ \frac{E_k(f;Q;L_k)}{|Q|^\lambda} \right\}_{Q\in\pi} \right\|_{\ell_{p\infty}}$$

where π runs over all disjoint families of cubes and $\lambda_\theta := (1-\theta)\lambda_0 + \theta\lambda_1$.

The special case $n = 1$ and $\lambda_0 = \lambda_1 = 0$ was given in [4] by using nonlinear spline approximation.

References

[1] L. Ambrosio, N. Fusco, and D. Pallara, *Functions of bounded variation and free dicontinuity problem*, Oxford Science Publ. Clarendon Press, 2000.

[2] J. Bergh and J. Peetre, *On the spaces V_p $(0 < p < \infty)$* Boll. Un. Mat. Ital. 10 (1974), pp. 632–648.

[3] Y. Brudnyi, *Spaces defined by local approximation*(in Russian), Trudy Moskov. Mat. Obshch. 24 (1971), pp. 69–132. English translation: Transl. Moscow Math. Soc. 24 (1971), pp. 73–139.

[4] ———, *Spline approximation and functions of bounded variation* (in Russian), Dokl. Akad. Nauk SSSR 215 (1974), pp. 511–513.

[5] ———, *Investigation in the Theory of Local Approximations* (in Russian), Yaroslavl, 1977.

[6] ———, *Adaptive approximation of functions with singularities* (in Russian), Trudy Moskov. Mat. Obshch. 55 (1994), pp. 149–242. English translation: Transl. Moscow Math. Soc. (1995), pp. 123–186.

[7] ———, *Nonlinear approximation by refinable functions and piecewise polynomials*, Approximation Theory X: Splines, Wavelets and Applications, Vanderbilt University Press, Nashville, 2002.

[8] ———, *Nonlinear N-term approximation by refinable functions*, St. Petersburg Math. J. 16 (2005), pp. 143–179.

[9] Y. Brudnyi and M. Ganzburg, *On an extremal problem for polynomials in n variables* (in Russian), Izv. Akad. Nauk SSSR, Ser. Matem. 37 (1973), pp. 344–355.

[10] Y. Brudnyi and I. Irodova, *Nonlinear multivariate spline and B-spaces* (in Russian), Algebra i Analiz 4 (1992), pp. 45–79. English translation: St. Petersburg Math. J. 4 (1993), pp. 667–694.

[11] Y. Brudnyi and B. Kotliar, *A problem of combinatorial geometry* (in Russian), Sibir. Math. J. 11 (1970), pp. 1171–1173.

[12] Y. Brudnyi and N. Kruglyak, *The real interpolation for couples of spaces of smooth functions* (in Russian), Doklady RAN 349 (1996), pp. 721–723. English translation: Doklady Math. 54 (1996), pp. 601–603.

[13] ———, *Interpolation functors and interpolation spaces*, vol. 1, North-Holland Math. Library, vol. 47, North-Holland, Amsterdam, 1991.

[14] Y. Brudnyi and P. Shvartsman, *On traces of Lizorkin-Triebel function spaces to Ahlfors regular subsets of $\mathbb{R}^n$*, to appear.

[15] A. Calderón and A. Zygmund, *On the differentiability of functions which are of bounded variation in Tonelli's sense*, Rev. Un. Mat. Argentina 20 (1962), pp. 102–121.

[16] ——, *Local properties of solutions of elliptic partial differential equations*, Studia Math. 20 (1961), pp. 171–225.

[17] A. Cohen, R. DeVore, P. Petrushev, and H. Xu, *Non linear approximation and the space* $BV(R^2)$, Amer. J. Math. 121 (1999), pp. 587–628.

[18] J. R. Dorronsoro, *Poisson integrals of regular functions*, Trans. Amer. Math. Soc. 297 (1986), pp. 669–685.

[19] H. Federer and W. Ziemer, *The Lebesgue set of a function whose distribution derivatives are p-th power summable*, Indiana Univ. Math. J. 22 (1972), pp. 139–158.

[20] E. Gagliardo, *Caratterizzazioni delle tracce sulla frontiera relative ad alcune classi di funzioni in piû variabili*, Rend. Sem. Mat. Univ. Padova 27 (1957), pp. 284–305.

[21] F. John and L. Nirenberg, *On functions of bounded mean oscillation*, Comm. Pure Appl. Math. 14 (1961), pp. 415–426.

[22] S. Kislyakov, *A remark on the space of functions of bounded p-variation*, Math. Nachr. 119 (1984), pp. 137–140.

[23] N. Krugljak, *Smooth analogs of the Calderón-Zygmund decomposition, quantative covering theorems and the K-functional of the couple* (L_q, W_p^k) (in Rusian), Algebra i Analiz 8 (1996), pp. 110–160. English translation: St.Petersburg Math. J. 8 (1997), pp. 617–649.

[24] J. Lindenstrauss and C. Stegall, *Examples of separable Banach spaces which do not contain* ℓ_1 *and whose duals are non-separable*, Studia Math. 54 (1975), pp. 81–105.

[25] J. Marcinkiewicz, On a class of functions and their Fourier series, Comptes Rendus Soc. Sci. Varsovie 26 (1934), pp. 71–77.

[26] J. Peetre, *Remarques sur les espaces de Besov. Le cas* $0 < p < 1$. C. R. Acad. Sci. Paris 277 (1973), pp. 947–949.

[27] F. Riesz and B. Sz-Nagy, *Lecons d'Analyse Fonctionelle*, Akademiai Kiadó, Budapest, 1952.

[28] H. Triebel, *Theory of Function Spaces II*, Monographs in Mathematics, 84, Birkhäuser-Verlag, Basel, 1992.

[29] G. Vitali, *Sulle funzioni integrali*, Atti della Reale Accademia delle scienze di Torino 40 (1904/5), pp. 1021–1034.

Author information

Yuri Brudnyi, Department of Mathematics, Technion-Israel Institute of Technology, 32000 Haifa, Israel.
Email: ybrudnyi@math.technion.ac.il

Banach Spaces and their Applications in Analysis, 59–76

Twisted dualities in Banach space theory

Jesús M. F. Castillo and Yolanda Moreno

Abstract. We study duality theory from the point of view of exact sequences of Banach spaces. This includes: duality for z-linear maps, classical Banach duality for twisted sums, Kalton and True dualities (introduced here) for exact sequences, and Fuks duality for the associated Banach functors.

Key words. Duality, Banach spaces, twisted sums, Banach functor.

AMS classification. 46M18, 46B10, 46A20, 46M15.

Things need not have happened to be true
Tales and dreams are the shadow truth
Neil Gaiman

1 Introduction

This research is an attempt to understand a few phenomena connected with the behavior of exact sequences of Banach and quasi-Banach spaces under duality. Our departure point has been Theorem 5.1. in Kalton–Peck "groundtwisting" paper [16] in which the authors develop a method to construct a nontrivial exact sequences

$$0 \longrightarrow l_p \longrightarrow l_p(\phi) \longrightarrow l_p \longrightarrow 0$$

starting with an unbounded lipschitz map $\phi : \mathbb{R} \to \mathbb{R}$.

[Kalton–Peck duality Theorem] *The space $l_p(\phi)^*$ is equivalent to $l_{p^*}(\psi)$, where*

$$\psi(t) = -\phi\left((p-1)^{-1}t\right), \quad t \geq 0.$$

The formulation suggests that what the authors intend to say is not only that the spaces $l_p(\phi)^*$ and $l_{p^*}(\psi)$ are isomorphic, but that the exact sequences

$$\begin{array}{ccccccccc} 0 & \longrightarrow & l_{p^*} & \longrightarrow & l_p(\phi)^* & \longrightarrow & l_{p^*} & \longrightarrow & 0 \\ & & \| & & & & \| & & \\ 0 & \longrightarrow & l_{p^*} & \longrightarrow & l_{p^*}(\psi) & \longrightarrow & l_{p^*} & \longrightarrow & 0 \end{array}$$

are "the same" (in some sense, not necessarily classical equivalence of sequences)); or else, that the sequences $0 \to l_p \to l_p(\phi) \to l_p \to 0$ and $0 \to l_q \to l_q(\psi) \to l_q \to 0$ are "dual" (again, in some sense to be specified) one of each other. So, the problem we address can be formulated as:

This research has been supported in part by project BMF2004-02635.

Problem. *Construct a duality theory for exact sequences of Banach/quasi Banach spaces.*

The duality one is looking for can be treated from different levels. At the basic Banach space level one has classical Banach space duality. From this point of view the problem can be understood as how to obtain the dual $(Y \oplus_F Z)^*$ of a twisted sum space $Y \oplus_F Z$, and uncover the duality between both. This we do in Section 4 polishing the theory displayed in [3].

Since exact sequences of Banach spaces can be identified with (equivalence classes of) z-linear maps, another form of understanding the duality problem is to ask how, given a z-linear map F, to construct its dual z-linear map F^d. As one can easily guess, if F corresponds to the exact sequence $0 \to Y \to X \to Z \to 0$ the map F^d should be the map associated with the dual sequence $0 \to Z^* \to X^* \to Y^* \to 0$. This we do in Section 3 developing he ideas from [3] in combination with the methods of [11].

A duality for z-linear maps is not, however, the same as a duality theory for exact sequences: the rather involved equivalence relation that governs exact sequences must be taken into account. To do that is the purpose of Section 5 in which we introduce the so-called Kalton duality between z-linear maps. Kalton duality is put in connection with the duality for z-linear maps and the Banach duality for the twisted sum spaces. This clarifies rather convincingly the meaning of the Kalton–Peck duality theorem.

In Section 6 we focus on the idea that the heart of a duality is a bilinear map. Having this point of view in mind we introduce the so-called true duality for z-linear maps and show that it coincides with Kalton duality in a fairly general context. Besides its conceptual simplicity, true duality provides ideas that work even for exact sequences in the quasi-Banach setting.

The categorical point of view about the duality is adopted in Section 7 where we follow Fuks' definition of dual functor (see [13]) to completely identify the dual functor of "the z-linear map" functor as an appropriate tensor product.

Some open lines for research together with hints of where will they lead us are posed in the last section.

2 Preliminaries

Recall from [3, 9] that two exact sequences $0 \to Y \to X \to Z \to 0$ and $0 \to Y_1 \to X_1 \to Z_1 \to 0$ are said to be isomorphically equivalent if there exist isomorphisms $\alpha : Y \to Y_1$, $\beta : X \to X_1$ and $\gamma : Z \to Z_1$ making commutative the diagram

$$\begin{array}{ccccccccc} 0 & \longrightarrow & Y & \longrightarrow & X & \longrightarrow & Z & \longrightarrow & 0 \\ & & \downarrow{\scriptstyle \alpha} & & \downarrow{\scriptstyle \beta} & & \downarrow{\scriptstyle \gamma} & & \\ 0 & \longrightarrow & Y_1 & \longrightarrow & X_1 & \longrightarrow & Z_1 & \longrightarrow & 0. \end{array}$$

When α, γ are identities we recover the classical notion of equivalent sequences; and when α, γ are multiples of the identity we recover the notion of projectively equivalent sequences from [15, 16]. In [3] we introduced the definition of dual sequence:

we will say that $0 \to A \to B \to C \to 0$ is a dual sequence if there exists some exact sequence $0 \to Y \to X \to Z \to 0$ such that $0 \to A \to B \to C \to 0$ and $0 \to Z^* \to X^* \to Y^* \to 0$ are isomorphically equivalent. We will see that it is in this sense that the sequence $0 \to l_{p^*} \to l_{p^*}(\psi) \to l_{p^*} \to 0$ is dual of $0 \to l_p \to l_p(\phi) \to l_p \to 0$.

The quasi-linear technique introduced by Kalton in [15] and then by Kalton and Peck in [16] to study exact sequences consists of representing exact sequences of quasi-Banach spaces $0 \to Y \to X \to Z \to 0$ by a certain type of non-linear maps, called quasi-linear maps, $F : Z \curvearrowright Y$ (we use this notation to stress the fact they are not linear maps). The quasi-linear technique was further developed in [16] and adapted in [6, 3, 7] to cover the Banach case by identifying exact sequences of Banach spaces with a special type of quasi-linear maps $F : Z \curvearrowright Y$ called z-linear.

Indeed, there is a correspondence (see [15, 16, 7]) between exact sequences $0 \to Y \to X \to Z \to 0$ of Banach spaces and z-linear maps, which are homogeneous maps $F : Z \curvearrowright Y$ with the property that there exists some constant $C > 0$ such that for all finite sets $x_1, \ldots, x_n \in Z$ one has $\|F(\sum_{n=1}^N x_n) - \sum_{n=1}^N F(x_n)\| \leq C \sum_{n=1}^N \|x_n\|$. The infimum of those constants C is called the z-linearity constant of F and denoted $Z(F)$. The process to obtain a z-linear map out from an exact sequence $0 \to Y \to X \xrightarrow{q} Z \to 0$ is the following: get an homogeneous bounded selection $b : Z \to Y$ for the quotient map q, and then a linear $\ell : Z \to Y$ selection for the quotient map. Then $F = b - \ell$ is a z-linear map. A z-linear map $F : Z \curvearrowright Y$ induces the exact sequence of quasi-Banach spaces $0 \to Y \to Y \oplus_F Z \to Z \to 0$ in which $Y \oplus_F Z$ means the vector space $Y \times Z$ endowed with the quasi-norm $\|(y, z)\|_F = \|y - Fz\| + \|z\|$. By $0 \to Y \to X \to Z \to 0 \equiv F$ we mean that the exact sequences $0 \to Y \to X \to Z \to 0$ and $0 \to Y \to Y \oplus_F Z \to Z \to 0$ are equivalent. Two z-linear maps $F, G : Z \curvearrowright Y$ are said to be equivalent, and we write $F \equiv G$, if the induced exact sequences are equivalent. Two maps $F, G : Z \curvearrowright Y$ are equivalent if and only if the difference $F - G$ can be written as $B + L$, where $B : Z \to Y$ is a homogeneous bounded map and $L : Z \to Y$ a linear map. The space of all exact sequences $0 \to Y \to \diamondsuit \to Z \to 0$ modulo the equivalence relation is denoted $\operatorname{Ext}(Z, Y)$. The space of all z-linear maps $Z \curvearrowright Y$ modulo the equivalence relation is denoted $\mathcal{Z}(Z, Y)$. The vector spaces $\operatorname{Ext}(Z, Y)$ and $\mathcal{Z}(Z, Y)$ are obviously isomorphic. In [4, 11] we introduced natural locally bounded and complete vector topologies – not necessarily Hausdorff – in $\operatorname{Ext}(A, B)$ using the seminorm

$$\mathfrak{z}([F]) = \inf\{Z(G) : G \in [F]\}.$$

The naturality of these topologies stems from the fact that they are fully compatible with the homology sequence, see [4]. It is this a good moment to remark that although the problem of obtaining $[\operatorname{Ext}(A, B), \mathfrak{z}(\cdot)]^*$ seems to be a hopeless task, our approach gives us some insight about its nature (see Section 6).

All our work considers only real Banach/quasi-Banach spaces.

3 Basic duality theory for z-linear maps

From the point of view of exact sequences the existence of a duality theory means that if one has an exact sequence then the dual sequence is well defined and exact. Thus, if $F : Z \curvearrowright Y$ is a z-linear map defining the starting sequence then there must be a z-linear map $Y^* \curvearrowright Z^*$ inducing the dual sequence $0 \to Z^* \to (Y \oplus_F Z)^* \to Y^* \to 0$. This map could properly be called the transpose or dual map F^d of F. Let us quote from [3] the construction of F^d.

Let $F : Z \curvearrowright Y$ be a z-linear map. Let $y^* \in Y^*$. The composition $y^* \circ F : Z \to \mathbb{R}$ is z-linear with $Z(y^* \circ F) \le \|y^*\| Z(F)$. By [3] there must be some linear map $m(y^*F) : Z \to \mathbb{R}$ at distance at most $\|y^*\|Z(F)$ from $y^* \circ F$. The map $H : Y^* \to Z'$ defined by $H(y^*) = m(y^*F)$ is not linear, although there is no loss of generality assuming that it is homogeneous. Take now a Hamel basis (f_α) for Y^* and define a linear map $L_H : Y^* \to Z'$ by $L_H(f_\alpha) = H(f_\alpha)$ (and linearly on the rest; this is what we will call the linearization process of a map H). The map $F^d = L_H - H$ is a z-linear map $Y^* \curvearrowright Z^*$ with $Z(F^d) \le Z(F)$ and the sequences $0 \to Z^* \to Z^* \oplus_{F^d} Y^* \to Y^* \to 0$ and $0 \to Z^* \to (Y \oplus_F Z)^* \to Y^* \to 0$ are equivalent. Up to here the theory as it is shown in [3].

Now we borrow some elements from [11]. Given a space X and a Hamel basis (e_γ) for it, we will say that a z-linear map $F : X \curvearrowright \mathbb{R}$ is in its canonical form with respect to that basis if $F(e_\gamma) = 0$ (or, what is the same, it has the form $G - L_G$). Let X be a Banach space. If $Z(X, \mathbb{R})$ denotes the vector space of all z-linear maps $X \curvearrowright \mathbb{R}$ in canonical form. We define *the* z-dual of X as the Banach space

$$X^z = [Z(X, \mathbb{R}), Z(\cdot)].$$

The space X^z is actually a dual space (see [11]) and its natural predual can be realized as the space $co_z(X)$ spanned in $(X^z)^*$ by the evaluation functionals $\delta_x : X^z \to \mathbb{R}$ given by $\delta_x(F) = F(x)$. One has:

Proposition 3.1 ([11]). *There is a z-linear map $\Omega_X : X \curvearrowright co_z(X)$ with the property that given a z-linear map $F : X \curvearrowright Y$ there exists an operator $\phi_F : co_z(X) \to Y$ such that $\phi_F \Omega_X \equiv F$. Moreover, if F is in its canonical form $\phi_F \Omega_X = F$.*

Consider now the following point where the linear and nonlinear theories differ. Given an operator $T \in \mathfrak{L}(X, Y)$ and a space W, the "right composition with T" operator $T^* : \mathfrak{L}(Y, W) \longrightarrow \mathfrak{L}(X, W)$ is defined by $T^*(f) = fT$. Analogously, given $F : X \curvearrowright Y$ a quasi-linear map and a target space W we can construct the *linear* map $F^* : \mathfrak{L}(Y, W) \longrightarrow \mathcal{Z}(X, W)$ defined by right composition with F; namely, $F^*(f) = fF$. If one has a z-linear map $F : X \curvearrowright Y$ in canonical form, choosing as target space $W = \mathbb{R}$, it induces a linear map $F^* : Y^* \longrightarrow X^z$. In the linear world this is the adjoint, but not in the z-linear world!

A *metric projection* on X will be a (possibly nonlinear) map $m : X^z \to X'$ such that for some $C > 0$ one has $\|m(F) - F\| = \sup\{\|m(F)(x) - F(x)\| : \|x\| \le 1\} \le CZ(F)$. We call $\|m\|$ the infimum of the constants C above. The "nonlinear Hahn–Banach theorem" of [3] we used above yields the existence of metric projections $m : X^z \to X'$ with $\|m\| = 1$. The papers [5, 10] contain studies of Banach space properties of a space X in terms of properties of the metric projections.

To obtain the true adjoint, take a metric projection $m : X^z \curvearrowright X'$ and make the composition $mF^* : Y^* \curvearrowright X'$. This is a much better candidate for F^d, and the true z-dual map, as we will show later. For the time being, since it is required that F^d take values in X^* (instead of X') then it is necessary to push it down: if $L_{mF^*} : Y^* \to X'$ is a linearization of mF^*, then

$$F^d = L_{mF^*} - mF^*$$

is a z-linear map $Y^* \curvearrowright X^*$. This correspondence $F \to F^d$ establishes a duality between the twisted sum spaces.

4 Banach duality for twisted sum spaces

This is the classical duality between Banach/topological vector spaces.

Proposition 4.1. *There is a duality isomorphism* $\mathfrak{d} : Z^* \oplus_{F^d} Y^* \longrightarrow (Y \oplus_F Z)^*$ *induced by the bilinear form*

$$\mathfrak{d}(z^*, y^*)(y, z) = z^*(z) + y^*(y) - L_{m(y^*F)}(z).$$

This is essentially a rewording of [3, Thm. 3]. For the sake of clarity we sketch the proof.

Proof. Let us show that $\mathfrak{d}$ is continuous:

$$\begin{aligned}
|\mathfrak{d}(z^*, y^*)(y, z)| &= |z^*(z) + y^*(y) - L_{m(y^*F)}(z)| \\
&= |z^*(z) + y^*(y) - F^d(y^*)(z) + F^d(y^*)(z) - L_{m(y^*F)}(z)| \\
&= |z^*(z) + y^*(y) - F^d(y^*)(z) - m(y^*F)(z)| \\
&\le |z^*(z) - F^d(y^*)(z)| + |y^*(y) - y^*(Fz) + y^*(Fz) - m(y^*F)(z)| \\
&\le \|z^* - F^d y^*\|\|z\| + \|y^*\|\|y - Fz\| + (1+\varepsilon)Z(F)\|y^*\|\|z\| \\
&\le \max\{1, (1+\varepsilon)Z(F)\}\|(z^*, y^*)\|_{F^d}\|(y, z)\|_F.
\end{aligned}$$

That $\mathfrak{d}$ is an isomorphism follows from the commutativity of the diagram (plus the classical 3-lemma; see [7]):

$$\begin{array}{ccccccccc}
0 & \longrightarrow & Z^* & \longrightarrow & Z^* \oplus_{F^d} Y^* & \longrightarrow & Y^* & \longrightarrow & 0 \\
 & & \| & & \downarrow{\scriptstyle \mathfrak{d}} & & \| & & \\
0 & \longrightarrow & Z^* & \longrightarrow & (Y \oplus_F Z)^* & \longrightarrow & Y^* & \longrightarrow & 0
\end{array}$$

□

It is a pity, nonetheless, to have such an ugly duality formula in which the extra linear map is not natural. If one could replace F^d by $m(F^*)$ the linear map would disappear. The problem to do this is that given $y^* \in Y^*$, the approximation $m(y^*F)$ is linear, but not necessarily continuous. A way to surround this difficulty is to represent the dual $(Y \oplus_F Z)^*$ as a "derived" space instead of as a twisted sum. Recall that given Banach spaces X and Y which we assume to be vector subspaces of a vector space W, and given a map $F : X \to W$ such that for each finite family of points $x_1, \dots x_n \in X$ the Cauchy differences $F(\sum_n x_x) - \sum F(x_n)$ fall in Y and for some constant $Z(F) > 0$ verify $\|F(\sum_n x_x) - \sum F(x_n)\| \leq Z(F) \sum \|x_n\|$, the derived space Yd_FX is defined as

$$Yd_FX = \{(y,z) \in W \times W : z \in X \quad \text{and} \quad y - Fz \in Y\},$$

endowed with the quasi-norm (equivalent to a norm) $\|(y,x)\|_F = \|y - Fx\| + \|x\|$. There is a natural exact sequence $0 \to Y \to Yd_FX \to X \to 0$ with inclusion $y \to (y,0)$ and quotient map $(y,x) \to x$. That exact sequence can be transformed into a standard twisted sum sequence considering the z-linear map $F - L_F : X \curvearrowright Y$. The two sequences are *isometrically equivalent* in the language of [8]; namely, there is a commutative diagram

$$\begin{array}{ccccccccc}
0 & \longrightarrow & Y & \longrightarrow & Yd_FX & \longrightarrow & X & \longrightarrow & 0 \\
 & & \| & & \downarrow{\scriptstyle T} & & \| & & \\
0 & \longrightarrow & Y & \longrightarrow & Y \oplus_{F-L_F} X & \longrightarrow & X & \longrightarrow & 0
\end{array}$$

in which $T(y,z) = (y - L_Fz, z)$ is an isometry. In this terms, if one has an exact sequence $0 \to Y \to Y \oplus_F X \to X \to 0$ and considers the derived dual exact sequence $0 \to X^* \to X^*d_{-mF^*}Y^* \to Y^* \to 0$ then the duality between $X^*d_{-mF^*}Y^*$ and $Y \oplus_F X$ is clean:

$$\mathfrak{d}(x', y^*)(y,x) = x'(x) + y^*(y).$$

In comparison with the linear theory, recall that F^d is not uniquely defined as a map: it depends upon the metric projection m previously chosen; which is natural since the equality for z-linear maps is $\equiv$, not $=$. Of course, if F^d and F^D are two dual maps of F then $F^d \equiv F^D$ (see Proposition 5.2).

5 Kalton duality

We pass to establish the meaning of F^d by operative conditions; precisely, when are F and G dual one of each other?

Let us consider the Kalton–Peck map $F_2 : l_2 \curvearrowright l_2$ defined (usual simplifications of the problem allow one to work with finitely supported sequences only) as $F_2x = x \log \frac{|x|}{\|x\|}$. We want to obtain F_2^d. Observe that $F_2(e_j) = 0$ which means that $m(e_jF_2) = 0$. It is not easy to determine $m(yF_2)$ for $y \in l_2$ since it means to determine the values $\alpha_n = m(yF_2)(e_n)$ for all n that verify $|yF_2x - m(yF_2)x| \leq C\|x\|\|y\|$ for a predetermined $C \geq Z(F_2)$, namely

$$\left|\sum \left(\alpha_n x_n \log \frac{|x_n|}{\|x\|} - \alpha_n x_n\right)\right| \leq C\|x\|\|y\|.$$

Setting $\alpha_n = y_n \log \frac{|y_n|}{\|y\|}$, the required calculus was performed by Kalton–Peck in Lemma 5.2 in [16] through the estimate

$$\left|\sum x_n y_n \log \frac{|x_n|}{|y_n|}\right| \leq e^{-1}(\|x\|^2 + \|y\|^2),$$

that immediately gives for $\|x\| = \|y\| = 1$

$$\left|\sum y_n x_n \log \frac{|x_n|}{\|x\|} - x_n y_n \log \frac{\|y\|}{|y_n|}\right| \leq \frac{2}{e}.$$

So, $m(yF_2)(e_n) = y_n \log \frac{|y_n|}{\|y\|} = F_2 y(e_n)$. Moreover, on finitely supported sequences, $m(yF)$ is linear and continuous. Therefore, on finitely supported sequences, one gets

$$F_2^d = -F_2.$$

See Kalton and Peck [16] or Benyamini and Lindenstrauss [1]. Observe that F_2 and $-F_2$ are not equivalent maps. Thus the minus sign in the definition of F^d turns out to be essential. The maps F_2 and F_2^d are however isometrically equivalent (with the isometry given by $T(x, y) = (-x, y)$). Such is therefore the meaning of the starting Kalton–Peck duality Theorem.

For general Kalton–Peck maps $F_\phi : l_p \curvearrowright l_p$ induced by a Lipschitz function $\phi : \mathbb{R} \to \mathbb{R}$ one gets (see [16, Thm 5.1]): $F_\phi^d = -F_{\phi^d}$ where, for $t \geq 0$, one has $\phi^d(t) = \phi(\frac{1}{p-1}t)$ with the estimate

$$|\langle F_\phi z, y^*\rangle + \langle z, F_\phi^d y^*\rangle| \leq \frac{Lip(\phi)}{(p-1)e}\|y^*\|\|z\|.$$

To deduce from here the general duality relationship between a map and its dual one has to take in consideration the linear maps involved (which, for several reasons, are not present when working with Kalton–Peck maps). Precisely, recall that by the definition, F and F^d are connected by the relation

$$| < Fx, y^* > + < x, F^d y^* > - < x, L_{mF*}(y^*) > | \leq \|m\|Z(F)\|x\|\|y^*\|$$

and induce the duality $\eth$ between the corresponding twisted sum spaces. So, we will say that $F : X \curvearrowright Y$ and $G : Y^* \curvearrowright X^*$ *are in Kalton duality* when there is a bi-bounded map $\mathcal{B} : X \times Y^* \to \mathbb{R}$ (a map $b : A \times B \to C$ will be called bi-bounded if $\|b(x, y)\| \leq C\|x\|\|y\|$ for some constant $C > 0$) and a bilinear form $\Lambda : X \times Y^* \to \mathbb{R}$ such that

$$\langle Fx, y^*\rangle + \langle x, Gy^*\rangle = \mathcal{B}(x, y^*) + \Lambda(x, y^*).$$

To say that F and G are in Kalton duality we will write $\kappa(F, G)$; and to make explicit mention of the bi-bounded and bi-linear maps we will write $\kappa(F, G) = \mathcal{B} + \Lambda$. The connection with classical Banach duality is clear:

Proposition 5.1. *Two z-linear maps $F : X \curvearrowright Y$ and $G : Y^* \curvearrowright X^*$ are in Kalton duality $\kappa(F,G) = \mathcal{B} + \Lambda$ if and only if the Banach spaces $Y \oplus_F X$ and $X^* \oplus_G Y^*$ are in duality induced by the bilinear form $\langle (y,x),(x^*,y^*)\rangle = x^*x + y^*y - \Lambda(x,y^*)$.*

We now show that the dual map is as unique as it can be.

Proposition 5.2. *If F and G are in Kalton duality then $G \equiv F^d$*

Proof. What we will actually prove is that $\kappa(F,G)$ and $\kappa(F,H)$ imply $G \equiv H$. The assertion above and the uniqueness of the dual F^d in terms of $\equiv$ follow immediately. As for the proof, the combination of $\langle Fx, y^*\rangle + \langle x, Gy^*\rangle = \mathcal{B}_1(x,y^*) + \Lambda_1(x,y^*)$ with $\langle Fx, y^*\rangle + \langle x, Hy^*\rangle = \mathcal{B}_2(x,y^*) + \Lambda_2(x,y^*)$ yields

$$\langle x, Gy^* - Hy^*\rangle = \mathcal{B}_1(x,y^*) + \Lambda_1(x,y^*) - \mathcal{B}_2(x,y^*) + \Lambda_2(x,y^*)$$

from which it follows

$$Gy^* - Hy^* = \mathcal{B}(y^*) + \Lambda(y^*)$$

where $\mathcal{B}(y^*) = \mathcal{B}_1(y^*) - \mathcal{B}_2(y^*)$ is bounded and $\Lambda(y^*) = \Lambda_1(y^*) - \Lambda_2(y^*)$ is linear. □

Observe that the role of F and G is not symmetric: indeed, if $F \equiv 0$ then $F^d \equiv 0$, while $F^d \equiv 0$ does not imply $F \equiv 0$, as the many existing nontrivial exact sequences having trivial dual sequence (such as $0 \to c_0 \to l_\infty \to C(\mathbb{N}^*) \to 0$) show. Thus, if $\kappa(F,G)$ and G is trivial it does not follows that F is trivial. The reason is that one has to take into account the spaces between which the maps are defined. A particularly interesting "symmetric" situation is considered in the next section. We establish now a useful dualization formula.

Lemma 5.3. *Let $F : X \curvearrowright Y$ be a z-linear form, and let $\phi : Y \to M$ and $\psi : W \to X$ be linear continuous operators. Then*

$$(F\psi)^d \equiv \psi^* F^d, \qquad (\phi F)^d \equiv F^d \phi^*.$$

Proof. To prove $(F\psi)^d \equiv \psi^* F^d$, set $m : X^z \to X'$ to be a metric projection and define the bilinear map

$$\Lambda(w,y^*) = \langle L_{mF^*}(y^*), \psi w\rangle.$$

After Proposition 5.2, the following estimate yields the desired result:

$$|\langle \psi^* F^d y^*, w\rangle + \langle y^*, F\psi w\rangle - \Lambda(w,y^*)| \le C\|w\|\|y^*\|.$$

The other identity is analogous. □

The universal property of the z-linear map Ω_X displayed in Proposition 3.1 admits now a dual counterpart.

Proposition 5.4. *Let $G : Y \curvearrowright X^*$ be a z-linear map. There exists a linear continuous operator $\psi_G : Y \to X^z$ such that $G \equiv \Omega^d \psi_G$.*

Proof. Let $\delta_A : A \longrightarrow A^{**}$ denote the canonical into isometry. Since $G^d\delta_X = \phi_{G^d\delta_X}\Omega_X$ then $G \equiv (G^d\delta_X)^d \equiv \Omega_X^d\phi^*_{G^d\delta_X}$. □

It is on the other hand obvious that every operator $\psi : Y \to X^z$ induces a z-linear map $\Omega_X^d\psi : Y \curvearrowright X^*$.

5.1 A special case: the equality $\mathrm{Ext}(X, Y^*) = \mathrm{Ext}(Y, X^*)$

In an entire analogy with the equality $\mathfrak{L}(X, Y^*) = \mathfrak{L}(Y, X^*)$ for operators between Banach spaces, the equality of vector spaces $\mathrm{Ext}(X, Y^*) = \mathrm{Ext}(Y, X^*)$ holds. Among the several possible proofs for this identity, probably it should be enough an appeal to the homology sequence associated with the functor $\mathfrak{L}$ plus the corresponding equality for operators. This proof nevertheless requires the use of both projective and injective presentations. A diagram-based pull-back proof is in this case simpler and more effective. It does however not provide the "duality map". We need the nonlinear construction of F^d to get the isometric version:

Proposition 5.5. *The semi-Banach spaces* $[\mathrm{Ext}(X, Y^*), \mathfrak{z}(\cdot)]$ *and* $[\mathrm{Ext}(Y, X^*), \mathfrak{z}(\cdot)]$ *are isometric.*

Proof. Let $F : X \curvearrowright Y^*$ be a z-linear map and $m : X^z \to X'$ some metric projection. We set the corresponding dual map $F^d : Y^{**} \to X^*$

$$F^d(y^*) = L_{mF^*} - m(y^*F).$$

Finally, we define the map $D : \mathcal{Z}(X, Y^*) \longrightarrow \mathcal{Z}(Y, X^*)$ as $D([F]) = [F^d\delta_Y]$.

1. *The map D is well-defined.* This reduces to show that if $F - G = B + L$ then $mF^* \equiv mG^*$. Calling b, b' to some bounded maps, one has:

$$\begin{aligned} m(y^*F) - m(y^*G) &= m(y^*F) - y^*F + y^*F - y^*G + y^*G - m(y^*G) \\ &= b(y^*) + y^*B + y^*L + b'(y^*). \end{aligned}$$

2. *D is linear.* It is enough to realize that $m\,(F^* + G^*) \equiv m(F^*) + m(G^*)$.
3. *D is bijective.* The key is to realize that $\delta_X^* F^{dd}\delta_Y \equiv F$. After that, the inversion formula is clear: $D^{-1}([G]) = [G^d\delta_X]$.
4. *D is an isometry.* Since $Z(F^d) \leq Z(F)$, choosing a bounded B such that $Z(F + B) \leq \mathfrak{z}([F]) + \varepsilon$ one gets $\mathfrak{z}([F^d] \leq Z((F + B)^d) \leq Z(F + B) \leq \mathfrak{z}([F]) + \varepsilon$, which suffices to derive the result.

□

In general, the map $D([F]) = [F^d]$ establishes only an into isometry

$$D(\cdot) : [\mathcal{Z}(X, Y), \mathfrak{z}(\cdot)] \longrightarrow [\mathcal{Z}(Y^*, X^*), \mathfrak{z}(\cdot)]$$

which is not onto because (see [3]) not every z-linear map $Y^* \curvearrowright X^*$ is the dual of a z-linear map $X \curvearrowright Y$.

6 True duality

Let us revisit the Kalton duality in the symmetric situation; i.e., between two z-linear maps $F : X \curvearrowright Y^*$ and $G : Y \curvearrowright X^*$. Recall that in this case we understand that F^d is actually ${F^d}_{|Y}$; and the same about G. Proposition 5.1 yields that if $\kappa(F, G) = \mathcal{B} + \Lambda$ then $Y^* \oplus_F X$ and $X^* \oplus_G Y$ act in duality through the bilinear form

$$\mathfrak{d}\left((x^*, y), (y^*, x)\right) = x^*x + y^*y - \Lambda(x, y).$$

However, if one is thinking of duality between F and G, the idea that the heart of duality should be a bilinear form is difficult to keep apart. So, we should have a bilinear form. But, defined where?

Given a z-linear map $F : X \curvearrowright Y^*$, the analogy with the linear situation leads one quite quickly to consider the map $\mathcal{B}_F : X \times Y \curvearrowright \mathbb{R}$ defined by $\mathcal{B}_F(x, y) =< Fx, y >$. The problem with it is that $\mathcal{B}_F$ is not a bilinear map. Nevertheless, $\mathcal{B}_F$ can be made a bilinear map when defined on the appropriate spaces, as we will do next. If $F : X \curvearrowright Y^*$ is a z-linear map, then it induces a bilinear form $\mathcal{B}_F : co_z(X) \times Y \longrightarrow \mathbb{R}$ defined by

$$\mathcal{B}_F(w, y) =< \phi_F w, y > .$$

The classical theory of bilinear forms allows us to interpret a bilinear form $\mathcal{B} : A \times B \to \mathbb{R}$ in two ways: As an operator $\mathcal{B}_1 : A \to B^*$; or as an operator $\mathcal{B}_2 : B \to A^*$. In the situation we are considering, we have $(\mathfrak{B}_F)_1 = \phi_F$ and thus $F \equiv (\mathfrak{B}_F)_1\Omega_X$; therefore $F^d \equiv \Omega^d_X(\mathfrak{B}_F)^*_1 \equiv \Omega^d_X(\mathfrak{B}_F)_2$.

Let us say that two z-linear maps $F : X \curvearrowright Y^*$ and $G : Y \curvearrowright X^*$ are in *true duality* if there exists a linear continuous operator ϕ such that $F \equiv \phi\Omega_X$ and $G \equiv \Omega^d_X\phi^*$. The reason for the name and its relationship with the previous dualities is displayed in the following proposition.

Proposition 6.1. *Let $F : X \curvearrowright Y^*$ and $G : Y \curvearrowright X^*$ be z-linear maps. The following are equivalent:*

1. The z-linear maps (F, G) are in true duality.

2. The z-linear maps (F, G) are in Kalton duality.

3. There is a bilinear form $\mathfrak{B} : co_z(X) \times Y \to \mathbb{R}$ such that $F \equiv \mathfrak{B}_1\Omega_X$ and $G \equiv \Omega^d_X\mathfrak{B}_2$.

Proof. It is clear that if F, G are in true duality, $G \equiv F^d$ and therefore F, G are in Kalton duality; and conversely. It is also obvious that (3) implies that $G \equiv F^d$, and since $(\mathfrak{B}_1)^* = \mathfrak{B}_2$ it follows that (3) implies (1). □

Given a bilinear map $\mathfrak{B} : co_z(X) \times Y \to \mathbb{R}$ to recover the z-linear map $F : X \curvearrowright Y^*$ is simple:

$$F(x)(y) = \mathfrak{B}(\Omega x, y).$$

Since true duality is a symmetric relation, 3. is equivalent to: there is a bilinear form $\mathfrak{C} : X \times co_z(Y) \to \mathbb{R}$ such that $F \equiv \Omega^d_Y\mathfrak{C}_1$ and $G \equiv \mathfrak{C}_2\Omega_Y$. In these variables, however, things cannot be simply "reversed'" (indeed, $F(x)(y)$ is not a bilinear form on $X \times co_z(Y)$ since, for example, it is not linear on x). Actually, the right definition must be

$$\mathfrak{A}(x, \Omega y) = F^d(y)(x).$$

The good news about having bilinear forms at hand is their connection with tensors. Namely, if $\mathrm{Bil}(A \times B)$ denotes the space of bilinear forms on $A \times B$ then we have shown:

$$\left(co_z(X)\widehat{\otimes}_\pi Y\right)^* = \mathrm{Bil}(co_z(X) \times Y) = Z(X, Y^*) = Z(Y, X^*) = \left(X\widehat{\otimes}_\pi co_z(Y)\right)^*.$$

The duality between $Z(X, Y^*)$ and $co_z(X)\widehat{\otimes}_\pi Y$ comes thus induced by the bilinear form

$$\mathfrak{d}_{co}\left(F, \sum_i \Omega x_i \otimes y_i\right) \longrightarrow \sum_i F x_i(y_i). \tag{6.1}$$

6.1 Duality with Ext

The transport of true duality from $Z(X, Y^*) = \mathrm{Bil}(co_z(X) \times Y)$ to $\mathrm{Ext}(X, Y^*)$ requires to translate the equivalence relation between z-linear maps to an equivalence relation between bilinear forms in $\mathrm{Bil}(co_z(X) \times Y)$. Two bilinear maps $B, B' \in \mathrm{Bil}(co_z(X) \times Y)$ are said to be equivalent, written $B \equiv B'$, if there exists a bi-bounded map $b : X \times Y \to \mathbb{R}$ and a bilinear form $\Lambda : X \times Y \to \mathbb{R}$ such that $B - B' = b + \Lambda$. One has (the proof is straightforward):

Lemma 6.2. *Let $B, B' \in \mathrm{Bil}(co_z(X) \times Y)$ be two bilinear forms; let $F, F' : X \curvearrowright Y^*$ be the z-linear maps they induce; namely $Fx = B(\Omega_X x, \cdot)$, and $F'x = B'(\Omega_X x, \cdot)$. Then $B \equiv B'$ if and only if $F \equiv F'$.*

But to handle the equivalence relation is not easy. Observe that if $B(X, Y^*)$ denotes the space of homogeneous bounded maps, then $\mathrm{Ext}(X, Y^*) = Z(X, Y^*)/B(X, Y^*)$. But $B(X, Y^*)$ is usually not $Z(\cdot)$-closed in $Z(X, Y^*)$. So, it is hard to think any use for the following, otherwise obvious, assertion:

If $B(X, Y^*)$ is weak-closed in $Z(X, Y^*)$ – with respect to the duality $\mathfrak{d}_{co}$ – then*

$$\mathrm{Ext}(X, Y^*) = \left(B(X, Y^*)_\perp\right)^*.$$

Nevertheless, it has. But one has to move to the non locally convex domain since it is easier for bounded maps to be closed when linear continuous maps are scarce. Considering X a quasi-Banach space for which $X^* = 0 = X^z$ (such as the L_p spaces for $0 < p < 1$). Let Y be a closed subspace of X. It was shown in [4, Thm. 6] that $\mathrm{Ext}(X/Y, \mathbb{R})$ is actually Hausdorff; which shows us that sometimes an Ext space can be a Banach space. A natural generalization for this result is (see [4, Thm. 8]) that when $\mathfrak{L}(X, E) = 0 = \mathrm{Ext}(X, E)$ then $\mathrm{Ext}(X/Y, E) = \mathfrak{L}(Y, E)$. Actually, under the appropriate hypothesis one can show that the two functors $\mathrm{Ext}(X/Y, \cdot)$ and $\mathfrak{L}(Y, \cdot)$

are naturally equivalent). Moreover, the condition $\mathfrak{L}(X,E) = 0$ can be relaxed to: whenever $F : X \curvearrowright E$ and $G : X \curvearrowright E$ are equivalent, $F \equiv G$ then $m(Fq)_{|Y} - m(Gq)_{|Y} \in \mathfrak{L}(Y,E)$. One has:

Proposition 6.3. *Let Y be a Banach subspace of a quasi-Banach space X and let E^* a dual Banach space. Assume that $\mathfrak{L}(X,E^*) = 0 = \mathrm{Ext}(X,E^*)$ then $\mathrm{Ext}(X/Y,E^*) = \left(Y\widehat{\otimes}_\pi E\right)^*$ with duality induced by the bilinear form*

$$\mathfrak{d}_\otimes\left([F], \sum_i y_i \otimes v_i\right) \longrightarrow \sum_i m(Fq)(y_i)(v_i). \tag{6.2}$$

The duality $\mathfrak{d}_\otimes$ plays in the non-locally convex setting the same role of the duality $\mathfrak{d}_{co}$: observe that under the hypotheses $\mathrm{Ext}(X,E^*) = 0 = \mathfrak{L}(X,E^*)$ the space Y behaves "as" the space $co_z(X/Y)$ (if that would make sense).

7 Fuks duality

The last approach to duality we want to present is the application of Fuks theory of dual functors, see [13], to z-linear maps. This theory was developed in the Banach space ambient by Mitjagin and Schvarts [21], which we present now in its polished form.

Recall that a functor $\mathcal{F} : \mathbf{A} \to \mathbf{C}$ between two categories is a correspondence assigning objects to objects and arrows to arrows which respects composition and identities. The functor is called covariant if whenever $f : A \to C$ then $\mathcal{F}(f) : \mathcal{F}(A) \to \mathcal{F}(C)$. It is said contravariant if, however, $\mathcal{F}(f) : \mathcal{F}(C) \to \mathcal{F}(A)$. From now on a *Banach functor* is a functor $\mathcal{F} : \mathbf{B} \to \mathbf{B}$ acting in a certain subcategory of Banach spaces. It is often required that $\mathcal{F}$ be linear, with the meaning that $\mathcal{F}(\lambda T + S) = \lambda\mathcal{F}(T) + \mathcal{F}(S)$; also, it is also sometimes required that it is bounded, with the meaning $\|\mathcal{F}(T)\| \leq \|T\|$. For our purposes the two most important examples of linear covariant Banach functors are:

- Given a Banach space X the covariant $\mathfrak{L}_X$ functor defined by $\mathfrak{L}_X(Y) = \mathfrak{L}(X,Y)$ and $\mathfrak{L}_X(T) = T_*$ with the meaning $T_*(S) = TS$. The choice $X = \mathbb{R}$ gives the identity.
- Given a Banach space X the covariant $\otimes_X$ functor defined by $\otimes_X(Y) = X\widehat{\otimes}_\pi Y$ and $\otimes_X(T) = 1_X \otimes T$.

While the simplest example of contravariant functor is

- Given a Banach space X the contravariant $\mathfrak{L}^X$ functor defined by $\mathfrak{L}^X(Y) = \mathfrak{L}(Y,X)$ and $\mathfrak{L}^X(T) = T^*$ with the meaning $T^*(S) = ST$. The choice $X = \mathbb{R}$ gives the adjoint.

Given two Banach functors $\mathcal{F}, \mathcal{G}$ acting between the same categories $\mathbf{B} \to \mathbf{B}$, a natural transformation $\tau : \mathcal{F} \to \mathcal{G}$ is a correspondence that assigns to each object A in $\mathbf{B}$ an operator $\tau_A : \mathcal{F}(A) \to \mathcal{G}(A)$ in such a way that given an operator $f : A \to C$

there is a commutative diagram:

$$\begin{array}{ccc} \mathcal{F}(A) & \xrightarrow{\tau_A} & \mathcal{G}(A) \\ {\scriptstyle \mathcal{F}(f)}\downarrow & & \downarrow{\scriptstyle \mathcal{G}(f)} \\ \mathcal{F}(C) & \xrightarrow[\tau_C]{} & \mathcal{G}(C). \end{array}$$

A Banach natural transformation is a natural transformation $\tau : \mathcal{F} \to \mathcal{G}$ such that the quantity

$$\|\tau\| = \sup_X \|\tau_X\|$$

is finite; here the supremum is taken over all the Banach spaces of the category **B**. Addition and multiplication by scalars can be defined in an obvious way for Banach natural transformations, and it is clear that the quantity above is a complete norm. So the space $[\mathcal{F}, \mathcal{G}]$ of all Banach natural transformations between the Banach functors $\mathcal{F}$ and $\mathcal{G}$ is a Banach space. Given a Banach functor $\mathcal{F}$ consider the Banach functor $[\mathfrak{L}, \mathcal{F}]$. It assigns to a Banach space A the Banach space $[\mathfrak{L}_A, \mathcal{F}]$ of natural transformations $\mathfrak{L}_A \to \mathcal{F}$, and to an operator $T : A \to B$ the operator $T_* : [\mathfrak{L}_A, \mathcal{F}] \longrightarrow [\mathfrak{L}_B, \mathcal{F}]$ given by $T_*(\nu)_X(\phi) = \nu_X(\phi T)$. Dually, one can consider the functor $[\mathcal{F}, \otimes]$ which assigns to a Banach space A the Banach space $[\mathcal{F}A, \otimes_A]$ and to an operator $T : A \to B$ the operator $T^* : [\mathcal{F}, \otimes_A] \longrightarrow [\mathcal{F}, \otimes_B]$ given by $T^*(\nu)_X(\phi) = \nu_X(\phi T)$.

We identify now a few important spaces of natural transformations.

Lemma 7.1. *Let $\mathcal{F}$ be a covariant Banach functor, and let A be a Banach space; one has*

1. $[\mathfrak{L}_A, \mathcal{F}] = \mathcal{F}(A)$
2. $[\otimes_A, \mathcal{F}] = \mathfrak{L}(A, \mathcal{F}(\mathbb{R}))$.

Proof. To prove (1) we set a natural equivalence $\eta : \mathcal{F} \longrightarrow [\mathfrak{L}, \mathcal{F}]$ defined as follows: for each A the operator $\eta_A : \mathcal{F}A \to [\mathfrak{L}_A, \mathcal{F}]$ takes points $p \in \mathcal{F}A$ and send them into a natural transformation $\eta_A(p) : \mathfrak{L}_A \to \mathcal{F}$: for each X one must have an operator $\eta_A(p)_X : \mathfrak{L}(A, X) \to \mathcal{F}X$, which is

$$\eta_A(p)_X(T) = \mathcal{F}T(p).$$

This is well defined because if $T : A \to X$ then $\mathcal{F}T : \mathcal{F}A \to \mathcal{F}X$.

To obtain (2), recall that every element $x \in X$ can be understood as an operator $x : \mathbb{R} \to X$; therefore $\mathcal{F}x$ is an operator $\mathbb{R} \to \mathcal{F}X$. Given an operator $T : X \to \mathcal{F}\mathbb{R}$ the associated natural transformation $\nu(T) : \otimes_X \to \mathcal{F}$ comes defined by

$$\nu(T)_X(a \otimes x) = \mathcal{F}x(Ta).$$

Conversely, given a natural transformation $\eta \in [\otimes_A, \mathcal{F}]$ the associated operator is $T = \eta_{\mathbb{R}}$. □

Part (1) is due to Mitjagin and Schvarts [21, Lemma 2]; in particular, it yields $[\mathfrak{L}_X, \mathfrak{L}_Y] = \mathfrak{L}(Y, X)$. Part (2) is due to Pothoven [23], although it had been formulated by Mitjagin and Schvarts [21] as $[\otimes_X, \otimes_Y] = \mathfrak{L}(X, Y)$. Both results can be considered special instances of Yoneda's lemma (see [19]).

Definition. *The dual functor of $\mathcal{F}$ is defined to be $D\mathcal{F} = [\mathcal{F}, \otimes]$.*

Proposition 7.2. *$D\otimes_A = \mathfrak{L}_A$ and $D\mathfrak{L}_A = \otimes_A$.*

Proof. That the dual of $\otimes_A$ is $\mathfrak{L}_A$ is immediate:

$$D \otimes_A (B) = [\otimes_A, \otimes_B] = \mathfrak{L}(A, B) = \mathfrak{L}_A(B).$$

The dual of $\mathfrak{L}_A$ is calculated, with the help of Lemma 7.1, as:

$$D\mathfrak{L}_A(B) = [\mathfrak{L}_A, \otimes_B] = \otimes_B(A) = \otimes_A(B).$$

□

Passing now to the non-linear context, let $\mathfrak{Z}_X$ denote the Banach functor that associates to a Banach space Y the Banach space $Z(X, Y)$ and to an operator $T : Y \to W$ the operator $T_* : Z(X,Y) \to Z(X, W)$ of left composition with T. In [11] it was proved that the spaces $[Z(X, Y), Z(\cdot)]$ and $\mathfrak{L}(co_z(X), Y)$ are isometric; hence the functors $\mathfrak{Z}_X$ and $\mathfrak{L}_{co_z(X)}$ are naturally isometric and can be identified. One has:

Proposition 7.3. *$D\mathfrak{Z}_X = \otimes_{co_z(X)}$ and $D\otimes_{co_z(X)} = \mathfrak{Z}_X$*

Proof. Let us make explicit the identification:

1. We assign to each element $p \in co_z(X)\widehat{\otimes}_\pi A$ the natural transformation $\eta_p \in [\mathfrak{Z}_X, \otimes_A]$ that associates to a space Y the operator $\eta_{pY} : Z(X, Y) \longrightarrow Y\widehat{\otimes}_\pi A$:

$$\eta_{pY}(F) = (\phi_F \otimes 1_A)(p)) .$$

2. We assign to each natural transformation $\eta \in [\mathfrak{Z}_X, \otimes_A]$ the element

$$\eta_{co_z(X)}(\Omega_X) \in co_z(X)\widehat{\otimes}_\pi A.$$

□

8 Open problems and their applications

8.1 The remaining bilinear form

There is a third type of bilinear form:

$$\mathcal{B} : co_z(X) \times co_z(Y) \longrightarrow \mathbb{R}.$$

Observe that X and Y play now perfectly symmetric roles. Since $co_z(Y)^* = Y^z$, standard facts yield that such bilinear form can be understood as a linear continuous operator $co_z(X) \to Y^z$. Therefore, after some routine arguments, one gets

Lemma 8.1. *The spaces* $Z(X, Y^z)$ *and* $Z(Y, X^z)$ *are isometric.*

With some care to check that the equivalence relations do nothing irreversible, one gets

Proposition 8.2. *The spaces* $\mathrm{Ext}(X, Y^z)$ *and* $\mathrm{Ext}(Y, X^z)$ *are isometric.*

The first application is a concrete version of the well-known fact that the kernel of a quotient map between two $\mathcal{L}_1$-spaces is itself an $\mathcal{L}_1$-space.

Corollary 8.3. *If* X *is an* $\mathcal{L}_1$*-space then* $co_z(X)$ *is an* $\mathcal{L}_1$*-space.*

Proof. Let X be an $\mathcal{L}_1$-space. Then for every Y one has

$$\mathrm{Ext}(X, Y^z) = 0 = \mathrm{Ext}(Y, X^z).$$

Thus, X^z is injective and its predual $co_z(X)$ is an $\mathcal{L}_1$-space. □

The second identifies a special type of Banach spaces. Recall that a Banach space X is an $\mathcal{L}_1$-space if and only if $\mathrm{Ext}(X, Y^*) = 0$ for all spaces Y. Let us say that X is an $\mathcal{L}_1^z$-space if and only if $\mathrm{Ext}(X, Y^z) = 0$ for all spaces Y.

Corollary 8.4. $co_z(X)$ *is an* $\mathcal{L}_1$*-space if and only if If* X *is an* $\mathcal{L}_1^z$*-space.*

Bourgain [2] obtained an uncomplemented copy Y_1 of l_1 inside l_1. The space $Bo = l_1/Y_1$ is an $\mathcal{L}_1^z$-space which is not an $\mathcal{L}_1$-space. It is not of type $\mathcal{L}_1$ because in that case Y_1 would be complemented in l_1. The proof that it is of type $\mathcal{L}_1^z$ goes a bit beyond the limits of this work, but we present it for the sake of completeness.

$$\begin{aligned}\mathrm{Ext}(Bo, Y^z) &= \mathrm{Ext}(Bo, co_z(Y)^*) = \mathrm{Ext}(co_z(Y), Bo^*)\\ &= \mathrm{Ext}^2(Y, Bo^*) = \mathrm{Ext}^2(Bo, Y^*) = 0\end{aligned}$$

(the proof that $\mathrm{Ext}^2(A, B^*) = \mathrm{Ext}^2(B, A^*)$ is analogous to that of Proposition 5.5; or else, using the results in [11], $\mathrm{Ext}^2(A, B^*) = \mathrm{Ext}(co_z(A), B^*) = \mathrm{Ext}(B, A^z) = \mathrm{Ext}(A, B^z)$ and all the way back.)

8.2 Adjointness, inverse limits and Sobczyk-like results

Recall that a functor $\mathcal{F} : \mathbf{A} \to \mathbf{B}$ is said to be the left adjoint of a functor $\mathcal{G} : \mathbf{B} \to \mathbf{A}$ (and, consequently, G is called a right adjoint for F), something written as $\mathcal{F} \dashv \mathcal{G}$, if for every object A of $\mathbf{A}$ and B of $\mathbf{B}$ there is a natural isomorphism $\mathbf{B}(\mathcal{F}(A), B) = \mathbf{A}(A, \mathcal{G}(B))$. The basic (in fact the only in a Banach space context) adjointness relation is $\otimes_A \dashv \mathfrak{L}_A$.

An immediate consequence of the identification $\mathfrak{Z}_X = \mathfrak{L}_{co_z(X)}$ is that the functor $\otimes_{co_z(X)}$ is the left adjoint of $\mathfrak{Z}_X$, namely

$$\otimes_{co_z(X)} \dashv \mathfrak{Z}_X.$$

An adjointness relationship is always important; and in practical terms, implies a nice behaviour with respect to limits: left adjoints preserve direct limits while right adjoints preserve inverse limits (see [17]). Therefore, one has:

Proposition 8.5. *The functor $\mathfrak{Z}_X$ preserves inverse limits. Namely,*

$$Z(X, \lim_{\leftarrow} X_\alpha) = \lim_{\leftarrow} Z(X, X_\alpha).$$

It would be extremely interesting to know to what extent this property is preserved when passing to Ext; precisely,

Problem. *Under which conditions* $\operatorname{Ext}(X, \lim_{\leftarrow} X_\alpha) = \lim_{\leftarrow} \operatorname{Ext}(X, X_\alpha)$ *?*

This is an important question, let us explain why. Probably the simplest inverse limit construction in Banach space theory is the Banach product (i.e., the l_∞-amalgam). Under minimal necessary restrictions it can be seen ([11]) that in the semi-Banach setting the previous formula is valid, namely,

$$\operatorname{Ext}(X, l_\infty(X_\alpha)) = l_\infty\left(\operatorname{Ext}(X, X_\alpha)\right).$$

Another quite natural inverse limit construction, which only exists in a Banach ambient, is the c_0-amalgam. The paper [11] shows that the identity

$$\operatorname{Ext}(X, c_0(X_\alpha)) = c_0(\operatorname{Ext}(X, X_\alpha))$$

holds for separable Banach spaces with the Bounded Approximation Property, and fails otherwise (which provides in passing the most general possible version of a vector valued Sobczyk theorem). For arbitrary inverse limits, even that fails:
If $\operatorname{Ext}(X, \lim_{\leftarrow} X_\alpha) = \lim_{\leftarrow} \operatorname{Ext}(X, X_\alpha)$ were true for separable Banach spaces with the BAP, then inverse limits of $\mathcal{L}_\infty$ spaces will be of the type $\mathcal{L}_\infty$. But even the kernel of a quotient between two $\mathcal{L}_\infty$ space fails to be an $\mathcal{L}_\infty$-space. Hence, it seems necessary to gain a deep understanding of Banach space structure to provide meaningful solutions to the problem.

8.3 Duality for Ext

Of course that knowing the dual functor of a Banach functor $\mathcal{F}$ is not the same as knowing the dual space of $\mathcal{F}(C)$; a line to explore is to what extent one is obtaining local duals in the sense of [14]. In any case, it will be a remarkable advance to obtain the dual functor of $\operatorname{Ext}(X, \cdot)$.

References

[1] Y. Benyamini and J. Lindenstrauss, *Geometric nonlinear functional analysis,* Vol. 1, Amer. Math. Soc., Providence, 1999.

[2] J. Bourgain, *A counterexample to a complementation problem*, Compo. Math. 43 (1981), pp. 133–144.

[3] F. Cabello Sánchez and J. M. F. Castillo, *Duality and twisted sums of Banach spaces,* J. Funct. Anal. 175 (2000), pp. 1–16.

[4] ______, *The long homology sequence for quasi-Banach spaces with applications,*, Positivity 8 (2004), pp. 379–394.

[5] F. Cabello Sánchez, J. M. F. Castillo, and F. Sánchez, *Nonlinear metric projections in twisted light*, Rev. Real. Acad. Cienc. Madrid 94 (2000) pp. 473–483.

[6] J. M. F. Castillo, *Snarked sums of Banach spaces*, Extracta Math. 12 (1997), pp. 117–128.

[7] J. M. F. Castillo and M. González, *Three-space problems in Banach space theory*, Springer Lecture Notes in Math. 1667, 1997.

[8] J. M. F. Castillo and Y. Moreno, *König-Wittstock quasi-norms on quasi-Banach spaces*, Extracta Math. 17 (2002), pp. 273–280.

[9] ______, *On Isomorphically equivalent extensions of quasi-Banach spaces*, in Recent Progress in Functional Analysis, (K.D. Bierstedt, J. Bonet, M. Maestre, J. Schmets (eds.)), North-Holland Math. Studies 187 (2001), pp. 263–272.

[10] ______, *Extensions of spaces of continuous functions*, Proc. Amer Math. Soc. (to appear).

[11] ______, *Sobczyk cohomology for separable Banach spaces with the bounded approximation property*, preprint 2006.

[12] J. Cigler, V. Losert, and P. W. Michor, *Banach modules on categories of Banach spaces*, Lecture Notes in Pure and Applied mathematics 46 (1979) Marcel-Dekker.

[13] D. B. Fuks, *Eckmann-Hilton duality and the theory of functos in the category of topological spaces*, Russian Math. Surveys 21 (1966), pp. 1–33

[14] M. González and A. Martínez Abejón, *Local dual spaces of a Banach space*, Studia Math. 147 (2001), pp. 155–168.

[15] N. J. Kalton, *The three-space problem for locally bounded F-spaces*, Compo. Math. 37 (1978), pp. 243–276.

[16] N. J. Kalton and N. T. Peck, *Twisted sums of sequence spaces and the three space problem*, Trans. Amer. Math. Soc. 255 (1979), pp. 1–30.

[17] E. Hilton and K. Stammbach, *A course in homological algebra*, Graduate Texts in Mathematics 4, Springer-Verlag 1970.

[18] S. Mac Lane, *Homology*, Grundlehren der mathematischen Wissenschaften 114, Springer-Verlag 1975.

[19] ______, *Categories for the working mathematician*, GTM 5, Springer-Verlag 1971.

[20] P. W. Michor, *Functors and categories in Banach spaces*, Lecture Notes in Math. 651 Springer-Verlag 1978.

[21] B. S. Mitjagin and A. S. Shvarts, *Functors in categories of Banach spaces*, Uspehi Mat. Nauk 19 (1964) n2 (116), pp. 65–130.

[22] J. W. Negrepontis, *Duality of functors in the category of Banach spaces* J. Pure and Appl. Algebra 3 (1973), pp. 119–131.

[23] K. L. Pothoven, *Compact functors and their duals in categories of Banahc spaces* Trans. Amer. Math. Soc. 155 (1971), pp. 149–159.

[24] Z. Semadeni, *Banach spaces of continuous functions*, Polish Scientific Publishers, Warszawa 1971.

From Jesus to Nigel: Sometimes I've been striving for results that in the end Nigel will obtain first. A few other times I've envisioned results Nigel had already got. Now, I'm trying to understand results Nigel obtained time ago. Thanks Nigel, wholeheartedly, from your worst apprentice.

Author information

Jesús M. F. Castillo, Departamento de Matemáticas, Universidad de Extremadura, Avda de Elvas s/n, 060671 Badajoz, Spain.
Email: `castillo@unex.es`

Yolanda Moreno, Departamento de Matemáticas, Escuela Politécnica, Universidad de Extremadura, Avda de la Universidad s/n, 10001 Cáceres, Spain.
Email: `ymoreno@unex.es`

Banach Spaces and their Applications in Analysis, 77–82

Orlicz sequence spaces with denumerable sets of symmetric sequences

Stephen J. Dilworth and Bünyamin Sarı

Abstract. We construct examples of Orlicz sequence spaces whose sets of symmetric sequences are, up to equivalence, countably infinite. This answers a question raised in [3].

Key words. Orlicz sequence spaces, symmetric sequences, spreading models.

AMS classification. Primary 46B20; Secondary 46B45.

1 Introduction

A question raised in [3] asks whether there exists an Orlicz sequence space such that the set of symmetric sequences in the space is, up to equivalence, precisely countably infinite. In this note we give an affirmative answer. In fact, for any countable ordinal γ we construct a reflexive Orlicz sequence space whose set of symmetric sequences with the domination order is, up to equivalence, order-isomorphic to the ordinal interval $[0, \gamma]$ in the reverse order. We also show that the converse holds when the set of symmetric sequences is countable and totally ordered in the domination order. In reflexive Orlicz sequence spaces the set of symmetric sequences coincides with the set of spreading models of the space, and the problem considered here is motivated by the study of the set of spreading models. We refer to [3] and [1] and the references therein for more details.

2 Preliminaries

An *Orlicz function* M is a real-valued continuous non-decreasing and convex function defined on $[0, 1]$ such that $M(0) = 0$ and $M(1) = 1$. For a given M, the *Orlicz sequence space* ℓ_M is the space of all sequences of scalars $x = (a_1, a_2, \ldots)$ such that $\sum_{n=1}^{\infty} M(|a_n|/\rho) < \infty$ for some $\rho > 0$, equipped with the norm

$$\|x\| = \inf\{\rho > 0 : \sum_{n=1}^{\infty} M(|a_n|/\rho) \leq 1\}.$$

We will always assume that M satisfies the Δ_2-*condition* at zero (i.e., that there exists $C > 0$ such that $M(2t) \leq CM(t)$ for all $0 \leq t \leq 1/2$). Then the unit vectors form a normalized symmetric basis for ℓ_M. If N also satisfies the Δ_2-condition at zero then M *dominates* N, denoted by $N \leq M$, if there exists a constant $C > 0$ such that $N(t) \leq CM(t)$ for all $0 \leq t \leq 1$. For M and N satisfying the Δ_2-condition at zero $N \leq M$ if and only if the unit vector basis of ℓ_M dominates that of ℓ_N, that is, there exists a constant $C \geq 1$ such that for every finite sequence (a_i) of scalars,

$$\|\sum_i a_i e_i\|_{\ell_N} \leq C \|\sum_i a_i e_i\|_{\ell_M}.$$

We say that two such functions M and N are *equivalent* if $M \leq N$ and $N \leq M$. Thus M is equivalent to N if and only if the unit vector bases of ℓ_M and of ℓ_N are equivalent.

If M satisfies the Δ_2-condition at zero then an Orlicz sequence space ℓ_N is isomorphic to a subspace of ℓ_M if and only if N is equivalent to some function in $C_{M,1}$, where $C_{M,1}$ is the norm-closed convex hull in $C[0,1]$ of the set

$$E_{M,1} = \overline{\left\{ \frac{M(\lambda t)}{M(\lambda)};\ 0 < \lambda < 1 \right\}}. \tag{2.1}$$

See [2, Lemma 4.a.6, remark (p. 141) and Theorem 4.a.8] for this result. Let (x_i) be a normalized symmetric (or even subsymmetric) sequence in ℓ_M. Then (x_i) is equivalent to the unit vector basis of ℓ_N, where $N \in C_{M,1}$ ([2], Proposition 4.a.7 and Theorem 4.a.8). Thus there is a one-to-one correspondence between the set of symmetric sequences in ℓ_M and the Orlicz functions N in $C_{M,1}$. Moreover, these two sets are *order-isomorphic*. That is, if (x_i) and (y_i) are two normalized symmetric sequences in ℓ_M and N_1 and N_2 are the corresponding functions in $C_{M,1}$ respectively, then (y_i) dominates (x_i) if and only if $N_1 \leq N_2$.

We will use the following method of representing Orlicz functions by sequences of zeros and ones, introduced by Lindenstrauss and Tzafriri [2, p. 161].

Fix $0 < \tau < 1$ and $1 < r < p < \infty$. For every sequence of zeros and ones, $\eta = (\eta(n))_{n=1}^{\infty}$ (i.e. $\eta(n) \in \{0,1\}$ for all n), let M_η be the piecewise linear function defined on $[0,1]$ satisfying $M_\eta(0) = 0$, $M_\eta(1) = 1$, and

$$M_\eta(\tau^k) = \tau^{rk + (p-r)\sum_{n=1}^{k} \eta(n)}, \qquad k = 1, 2, \ldots .$$

In our construction we will use the following particular class of such spaces as considered in [1, Lemma 2.3].

Lemma 2.1 ([1]). *Let $1 < r < p < \infty$, and $0 < \tau < 1$. Let $(n_k) \subset \{1, 2, 2^2, \ldots\}$ with $n_1 = 1$ and put*

$$\rho(i) = \begin{cases} 0 & \text{if } i = n_k \\ 1 & \text{otherwise.} \end{cases}$$

Let $M := M_\rho$ be the corresponding Orlicz function. Then ℓ_M is reflexive and every symmetric sequence in ℓ_M is equivalent to either the unit vector basis of ℓ_M or to the unit vector basis of ℓ_p.

3 The construction

Theorem 3.1. *For every countable ordinal $\gamma \geq 0$ there exists a reflexive Orlicz sequence space ℓ_M such that the set of normalized symmetric sequences in ℓ_M is order-isomorphic to $[0, \gamma]$ with the reverse order.*

Proof. The first part of the construction bears similarities to the one in [1, Theorem 3.1]. For every $\alpha < \gamma$ we construct a sequence ρ_α of the form given in Lemma 2.1 so that the collection of corresponding Orlicz functions $(M_{\rho_\alpha})_{\alpha<\gamma}$ (for the same τ, r and p) is order-isomorphic to $[0,\gamma] \setminus \{\gamma\}$ with the reverse order. Then we patch the ρ_α sequences together into one sequence ϕ in such a way that $C_{M_\phi,1}$ of the corresponding Orlicz function M_ϕ (again with the same τ, r and p) will consist, up to equivalence, only of the collection $\{M_\phi\} \cup \{M_{\rho_\alpha}\}_{\alpha<\gamma} \cup \{t^p\}$. Thus ℓ_{M_ϕ} will be the desired Orlicz sequence space.

The ρ_α sequences will be constructed simultaneously on a finite interval at a time using a variation on what is called the *ε-domination procedure* in [1], which we recall now for completeness.

Let (ρ_j) be a collection of sequences to be constructed, and let $A \subset \mathbb{N}$ and $\varepsilon > 0$. Put $\rho_j(1) = 0$ for all j, and suppose that all the sequences are already defined on $[1, N]$.

The (ε, A)-domination procedure extends the definition of the ρ_j's to an initial segment $[1, N_1]$ for some $N_1 > N$.

Suppose that both A and $\mathbb{N} \setminus A$ are non-empty. Choose a sufficiently large (just how large is specified below) integer $m > N$. For all $k \in \mathbb{N} \setminus A$ place 0's on the coordinates from $[N+1, m] \cap \{2^n : n = 0, 1, 2, \ldots\}$ of the ρ_k's (while the rest of the coordinates of the interval are filled with 1's), and for all $j \in A$ place 1's on *all* the coordinates from $[N+1, m]$ of the ρ_j's , where m is chosen so that

$$\sum_{i=1}^{m} \rho_j(i) - \sum_{i=1}^{m} \rho_k(i)$$

is sufficiently large to ensure that

$$\frac{M_{\rho_j}(\tau^m)}{M_{\rho_k}(\tau^m)} < \varepsilon, \text{ for all } j \in A,\ k \in \mathbb{N} \setminus A.$$

We now pass to the main construction. Let γ be a countable ordinal. For simplicity we shall assume that γ is a limit ordinal (the argument in the successor ordinal case is similar). Let $L = [0,\gamma] \setminus \{\gamma\}$ with the reverse order. Note that every subset of L has a *maximum* element in this order. Let $(\alpha_j)_{j=1}^{\infty}$ be an enumeration of L.

For every $j \in \mathbb{N}$ and every $\varepsilon_k = 2^{-k}$, $k = 1, 2, \ldots$, we carry out an (ε_k, A_j)-domination procedure for $A_j = \{i \in \mathbb{N} : \alpha_i < \alpha_j \text{ in } L \text{ order}\}$ as described above. Note that A_j and $\mathbb{N} \setminus A_j$ are both nonempty. We shall call α_j the *dominating node* and we shall say that α_i is *ε-dominated* by α_j if $i \in A_j$. Since there are countably many choices we can enumerate some order in which to carry out all (ε_k, A_j)-dominations.

At the end we obtain sequences ρ_j for each $\alpha_j \in L$ with the following easily checked properties:

(i) Each ρ_j begins with 0;

(ii) The zeros occur only on coordinates (2^k);

(iii) $\alpha_i \geq \alpha_j$ (in the order of L) if and only if $\rho_i(n) \leq \rho_j(n)$ for all n (we say that ρ_j dominates ρ_i *pointwise*). Thus in this case $M_{\rho_j} \leq M_{\rho_i}$;

(iv) L is order-isomorphic to the collection (M_{ρ_j}).

Now let ϕ be the sequence obtained by writing out the initial segments of the ρ_j's with extra 1's so that each initial segment of length n is followed by n 1's before the next initial segment starts:

$$\phi = (\rho[1], 1, \rho_1[2], 1, 1, \rho_2[2], 1, 1, \rho_1[3], 1, 1, 1, \rho_2[3], 1, 1, 1, \rho_3[3], 1, 1, 1, \rho_1[4], \ldots),$$

where $\rho[n]$ denotes the initial segment of ρ of size n.

We will show that $M := M_\phi$ is the desired Orlicz function. Recall that $N \in E_{M_\rho,1}$ if and only if $N = N_\psi$ where ψ is a pointwise limit of $(T^m\phi)_{m\in\mathbb{N}}$ [2, p. 161]. From the form of ϕ it is clear that the M_{ρ_j} functions and t^p, which is represented by the sequence $(1, 1, 1, \ldots)$, belong to $E_{M,1}$, and, moreover, that M_ϕ strictly dominates each M_{ρ_j}.

We show in the next lemma that, together with M_ϕ, these are the only elements of $E_{M,1}$ up to equivalence.

For $n \geq 0$ by $T^n\rho$ denote the sequence obtained by shifting ρ to the left n coordinates, and for $n < 0$ by $T^n\rho$ denote the sequence ρ with n 1's put at the beginning.

One easily verifies from the definition of $\rho_\alpha := \rho_j$ that $M_{\rho_\alpha}(t) \geq M_{T^m(\rho_\alpha)}(t)$ for all $\alpha = \alpha_j \in L$, $m \in \mathbb{Z}$, and $t \in [0, 1]$. This fact will be used in the proof of our main result.

Lemma 3.2. *The pointwise limits of $(T^m\phi)_{m\in\mathbb{N}}$ are of the form*

(a) all sequences $T^m\rho_j$ for $\alpha_j \in L$ and $m \in \mathbb{Z}$, or

(b) all sequences with at most one term equal to zero and all other terms equal to one.

Proof of Lemma 3.2. Let ψ be such a limit. If ψ is not of type (b) then there is a fixed $m_0 \in \mathbb{Z}$ so that ψ is a pointwise limit of a sequence of sequences of the form $(T^{m_0}\rho_{j_k})_{k\geq 1}$. For each $i \in \mathbb{N}$, let $\alpha_{j(i)}$ be the greatest element in L such that $T^{m_0}\rho_{j(i)}$ agrees with ψ on the first i coordinates. (Every subset of L has a maximum, so $\alpha_{j(i)}$ exists.) Clearly, $\alpha_{j(1)} \geq \alpha_{j(2)} \geq \alpha_{j(3)} \geq \ldots$. Let $\alpha = \lim_i \alpha_{j(i)}$. The limit exists in $L \cup \{\gamma\}$ since $L \cup \{\gamma\}$ is an ordinal in the reverse order.

If $\alpha > \gamma$, then we claim that $\psi = T^{m_0}\rho_\alpha$. Indeed, clearly $T^{m_0}\rho_\alpha \geq \psi$ pointwise since $\rho_\alpha \geq \rho_{j(i)}$ pointwise for all i (by construction). But consider a subinterval of $\mathbb{N}$ on which α is ε-dominated by some β with $\beta > \alpha$. Since $\alpha_{j(i)} \searrow \alpha$ we have $\beta > \alpha_{j(i)}$ for all sufficiently large i. Hence $\alpha_{j(i)}$ is also ε-dominated by β for all sufficiently large i. This proves that $\psi \geq T^{m_0}\rho_\alpha$ pointwise.

If $\alpha = \gamma$ then a similar argument shows that ψ agrees with an ε-dominated sequence on every subinterval of $\mathbb{N}$ on which an ε-domination takes place, and hence ψ is of type (b). □

It remains to show that every function in $C_{M,1}$ is equivalent to M_ϕ or to one of the M_{ρ_j}'s or to t^p.

Let $M_\sigma \in C_{M,1}$. Then by Lemma 3.2 M_σ is equivalent either to M_ϕ or to a uniform limit of a sequence (M_{σ_n}) of finite convex combinations of functions in $E_{M,1}$ of the form

$$M_{\sigma_n} = \sum_{m\in\mathbb{Z},\alpha\geq\gamma} \lambda^n_{m,\alpha} M_{T^m\rho_\alpha},$$

where the positive 'mass' coefficients $\lambda^n_{m,\alpha}$ sum to 1 for each n.

Here $\rho_\gamma := (1, 1, 1, \ldots)$ (so M_{ρ_γ} is equivalent to t^p) represents the minimum symmetric sequence in the domination order.

Define the following function ξ on L:

$$\xi(\alpha) = \lim_{m\to\infty} \limsup_{n\to\infty} \sum_{\beta\geq\alpha, |j|\leq m} \lambda^n_{j,\beta}.$$

We now consider two cases.

In the first case, suppose that ξ is not identically zero. Then there exists a greatest element α_0 with $\xi(\alpha_0) > 0$ (since every subset of L has a maximum element). This implies that there exists $m_0 \in \mathbb{N}$ with

$$\limsup_{n\to\infty} \sum_{\beta\geq\alpha_0, |j|\leq m_0} \lambda^n_{j,\beta} := \delta > 0.$$

We claim that in this case M_σ is equivalent to $M_{\rho_{\alpha_0}}$. Clearly, M_σ dominates $M_{\rho_{\alpha_0}}$ since $\delta > 0$ and $M_{\rho_\beta} \geq M_{\rho_\alpha}$ whenever $\beta \geq \alpha$. Consider a fixed initial segment of $\mathbb{N}$ on which α_0 is ε-dominated at least once. Then the collection of dominating nodes which ε-dominate α_0 on this initial segment is finite and non-empty. So there is a least (in the ordering of L) dominating node $\beta_0 > \alpha_0$ from this collection.

Thus $\xi(\beta_0) = 0$ by the definition of α_0. This means that

$$\lim_{m\to\infty} \limsup_{n\to\infty} \sum_{\beta\geq\beta_0, |j|\leq m} \lambda^n_{j,\beta} = 0.$$

Note that if $|m| > C$, where C depends only on the fixed initial segment of $\mathbb{N}$, then for all $\alpha \in L$, $T^m\rho_\alpha$ will contain at most one zero in that initial segment. But the sequences $T^m\rho_\alpha$ for $\alpha \geq \beta_0$ and $m \leq C$ which occur in the definition of M_{σ_n} are assigned a 'total mass' which tends to zero as $n \to \infty$, and hence their contribution vanishes in the limit. On the other hand, if $\alpha < \beta_0$, then on this initial segment α is ε-dominated whenever α_0 is ε-dominated, and hence ρ_{α_0} is dominated pointwise on this initial segment by ρ_α. Recall that $M_\alpha(t) \geq M_{T^m(\alpha)}(t)$ for all $\alpha \in L$, $m \in \mathbb{Z}$, and $t \in [0, 1]$. Taking the limit as $n \to \infty$ it follows that $M_\sigma \preccurlyeq M_{\rho_{\alpha_0}}$.

In the second case, suppose that $\xi(\alpha) = 0$ for all α. This easily implies that M_σ is equivalent to the minimum element $M_{\rho(\gamma)}$ (which is equivalent to t^p).

Thus, we have proved that $\{\ell_{M_\phi}\} \cup \{\ell_{M_{\rho_j}}\}_{j=1}^\infty \cup \{\ell_p\}$ is the set of symmetric sequences in ℓ_{M_ϕ} up to equivalence. In the domination order, ℓ_{M_ϕ} is the maximum element and ℓ_p is the minimum element, and hence the set is order-isomorphic to the ordinal sum $1 + \gamma^+$ with the reverse order, where γ^+ denotes the successor ordinal to γ, which in turn is isomorphic to the ordinal interval $[0, \gamma]$ with the reverse order. □

Corollary 3.3. *Suppose that L is a countable totally ordered set. Then there exists a reflexive Orlicz space ℓ_M such that the collection of symmetric sequences in ℓ_M with the domination order is order-isomorphic to L if and only if L is order-isomorphic to an ordinal interval $[0, \gamma]$ in the reverse order for some countable ordinal γ.*

Proof. Sufficiency follows from our main result. Conversely, if L is the set of symmetric sequences up to equivalence in ℓ_M, where ℓ_M is reflexive, then L coincides with the set of spreading models of ℓ_M up to equivalence [3], and if L is countable and totally ordered, then L is well-ordered in the reverse order [1, Theorem 3.7], and moreover L has a minimum element [3]. Hence L is order-isomorphic to $[0, \gamma]$ in the reverse order for some countable ordinal γ. □

References

[1] S. J. Dilworth, E. Odell, and B. Sarı, *Lattice structures and spreading models*, Israel J. Math, to appear.

[2] J. Lindenstrauss and L. Tzafriri, *Classical Banach Spaces* vol. I: Sequence spaces, Springer-Verlag, New York, 1977.

[3] B. Sarı, *On the structure of the set of symmetric sequences in Orlicz sequence spaces*, Canad. Math. Bull. 50 (2007), pp. 138–148.

Author information

Stephen J. Dilworth, Department of Mathematics, University of South Carolina, Columbia, SC 29208-0001, USA.
Email: dilworth@math.sc.edu

Bünyamin Sarı, Department of Mathematics, University of North Texas, Denton, TX 76203-1430, USA.
Email: bunyamin@unt.edu

Banach Spaces and their Applications in Analysis, 83–110

Complexity and homogeneity in Banach spaces

Valentin Ferenczi and Christian Rosendal

Abstract. We provide an overview of a number of results concerning the complexity of isomorphism between separable Banach spaces. We also include some new results on the lattice structure of the set of spreading models of a Banach space.

Key words. Isomorphism of Banach spaces, complexity of analytic equivalence relations, minimal Banach spaces, spreading models.

AMS classification. Primary 46B03; Secondary 03E15.

To Nigel Kalton on the occasion of his 60th birthday.

1 Complexity of equivalence relations

A topic much in vogue in descriptive set theory for at least the last twenty years is the study of the relative complexity of Borel and analytic equivalence relations on Polish, i.e., completely metrisable separable spaces. The motivation comes from the general mathematical problem of classifying one class of mathematical objects by another, that is, given some class $\mathcal{A}$ of mathematical objects, e.g., separable Banach spaces, and a corresponding notion of isomorphism, one tries to find complete invariants for the objects in $\mathcal{A}$ up to isomorphism. More explicitly, one would like to assign to each object in $\mathcal{A}$ an object in another category $\mathcal{B}$, such that two objects in $\mathcal{A}$ are isomorphic if and only if their assignments in $\mathcal{B}$ are isomorphic. If this can be done, one purports to have *classified* the elements of $\mathcal{A}$ by the elements of $\mathcal{B}$ up to isomorphism. Thus, for example, the discrete spectrum theorem of Halmos–von Neumann is a classification of the ergodic measure preserving transformations with discrete spectrum by their sets of eigenvalues. Similarly, by Stone duality, boolean algebras can by classified by compact spaces up to homeomorphism.

In descriptive set theory there has been an effort to systematize the notion of classification itself and to determine which classes of objects can properly be said to be classifiable in terms of another. This is done by restricting attention to classes of objects that can readily be made into a standard Borel space (i.e., the measurable space of a Polish, or separable completely metrisable space) and the corresponding equivalence relation of isomorphism. For example, the class of separable Banach spaces can be identified with the set of closed linear subspaces of some isometrically universal space, e.g., $C(2^{\mathbb{N}})$, as we shall see later. However, as uncountable standard Borel spaces are all Borel isomorphic, the perspective changes from the set of objects themselves to the equivalence relation that really encodes the complexity of the objects up to isomorphism. The precise definition is as follows.

Second author: Partially supported by NSF grant DMS 0556368.

Definition 1. Let E and F be equivalence relations on standard Borel spaces X and Y respectively. We say that E is *Borel reducible* to F if there is a Borel function $f : X \to Y$ such that

$$xEy \leftrightarrow f(x)Ff(y)$$

for all $x, y \in X$. We denote this by $E \leqslant_B F$ and informally say that E is less complex than F. If both $E \leqslant_B F$ and $F \leqslant_B E$, then E and F are called *Borel bireducible*, written $E \sim_B F$.

For example, if we let $\mathfrak{B}$ be the standard Borel space of separable Banach spaces, then we will see that the relation of isomorphism is analytic as a subset of $\mathfrak{B}^2$, i.e., is the image by a Borel function of a standard Borel space. Since most other natural isomorphism relations are also analytic or even Borel one usually restricts the attention to this subclass. Thus, if we have two classes $\mathcal{A}$ and $\mathcal{B}$ of mathematical objects, that we have identified with standard Borel spaces $\mathcal{A} \sim X$ and $\mathcal{B} \sim Y$, respectively, and we denote by E and F the corresponding isomorphism relations on X and Y, then a reduction $\phi : X \to Y$ of E to F can be viewed as a classification of the objects in $\mathcal{A}$ by the objects in $\mathcal{B}$. To sum up, one can say that the theory of complexity of equivalence relations is the study of which invariants one can use for various mathematical objects.

We should also briefly mention a slightly other way of viewing Borel reducibility. If $\phi : X \to Y$ is a Borel reduction of E to F, then it is easy to see that ϕ induces an injection $\hat{\phi} : X/E \to Y/F$ and thus that the cardinality of X/E is smaller than that of Y/F. In fact, one can use this to define a notion of *effective cardinality* for quotient spaces, a notion that refines Cantor's concept of cardinality. Thus, for example, the effective cardinals are not wellordered or even linearly ordered.

A large part of the general theory on analytic equivalence relations has concerned the structure of $\leqslant_B$, i.e., the hierarchy of analytic equivalence relations under this ordering, and the place of naturally occurring isomorphism relations in the hierarchy. It is now known that $\leqslant_B$ is extremely complex as an ordering, while, on the other hand, most naturally occurring classification problems tend to be bireducible with a fairly small class of equivalence relations for which there is also a nice structure theory. We now proceed to describe some of these.

Apart from equivalence relations with only countable many classes the simplest Borel equivalence relation is the relation $(\mathbb{R}, =)$ of equality of real numbers. In fact, a result of Silver [46] implies that any Borel equivalence relation either has countably many classes, and thus is just a countable partition of the underlying space into Borel sets, or $(\mathbb{R}, =)$ is Borel reducible to it, and thus has continuum many classes. Equivalence relations reducible to $(\mathbb{R}, =)$, called *smooth*, are simply those "isomorphism" relations that admit real numbers as complete invariants. One outstanding example is Ornstein's theorem that entropy is a complete invariant for Bernoulli shifts.

A deep result due to Harrington, Kechris, and Louveau [28] shows that among non-smooth Borel equivalence relations there is a minimum (with respect to $\leqslant_B$) one, which is called E_0. It is the relation of eventual agreement between infinite binary sequences, i.e., for $x, y \in 2^{\mathbb{N}}$,

$$xE_0y \leftrightarrow \exists n\, \forall m \geqslant n\; x_m = y_m.$$

Besides being minimum above $(\mathbb{R}, =)$, E_0 is also characterized as being maximum for *hyperfinite* Borel equivalence relations, i.e., those that can be written as an increasing union of countably many Borel equivalence relations with finite classes. All Borel actions of $\mathbb{Z}$ give rise to such orbit equivalence relations.

Of special interest among analytic equivalence relations are those that appear as the orbit equivalence relation of a continuous action of a Polish group, i.e., a topological group whose topology is Polish. Since any countable or locally compact second countable group is Polish, this class encompasses most of the orbit equivalence relations usually studied in analysis. It turns out that for each Polish group H there is a maximum, with respect to $\leqslant_B$, orbit equivalence relation induced by H, and even among all orbit equivalence relations induced by actions of countable (or what turns out to be the same, locally compact second countable) groups there is a maximum one denoted by E_∞. It is characterized as the maximum Borel equivalence relation all of whose classes are countable. Similarly, there is a maximum orbit equivalence relation among all those induced by Polish groups, which we denote by E_G.

We should also mention two other equivalence relations of complexity between E_∞ and E_G: the relation $=^+$ on infinite sequences of complex numbers enumerating the same sets, i.e., for $(x_n), (y_n) \in \mathbb{C}^\mathbb{N}$,

$$(x_n)_n =^+ (y_n)_n \leftrightarrow \{x_n\}_n = \{y_n\}_n,$$

and the $\leqslant_B$-maximum orbit equivalence E_{S_∞} induced by the infinite symmetric group S_∞. One easily sees that $=^+$ is induced by an action of S_∞, which is Polish, and thus also E_{S_∞} reduces to E_G. And on the other hand, if E is a Borel equivalence relation with countable classes, then for each z in the domain of E, we can enumerate $[z]_E$ in some way depending on z, which will reduce E to $=^+$. So also $E_\infty \leqslant_B =^+$. The discrete spectrum theorem of Halmos and von Neumann says that two ergodic measure preserving automorphisms T and S of Lebesgue spaces with discrete spectrum are isomorphic if and only if they have the same countable set of eigenvalues, $\Lambda(T)$ and $\Lambda(S)$. Thus, if we fix a Borel function ϕ that to each T picks out an enumeration (λ_n) of the set of eigenvalues $\Lambda(T)$, then we see that T and S are isomorphic if and only if $\phi(T) =^+ \phi(S)$ and hence isomorphism reduces to $=^+$. On the other hand, there is no way of constructing ϕ so that it makes the same choice of enumeration of $\Lambda(T)$ and $\Lambda(S)$ provided the two *sets* are the same. This has to do with the fact that the quotient space $\mathbb{C}^N/{=^+}$ is not countably separated, or in our terminology that $=^+$ is non-smooth. And in fact, isomorphism of measure preserving automorphisms with discrete spectrum is Borel bireducible with $=^+$ (see Foreman [21]).

An interesting discovery due to Kechris and Louveau [32] is that there are analytic equivalence relations that are not reducible to orbit equivalence relations, or equivalently, to E_G. In fact, there is one, minimal among Borel equivalence relations having this property, namely E_1, which is the relation of eventual agreement between infinite sequences of real numbers. I.e., for $x, y \in \mathbb{R}^\mathbb{N}$,

$$xE_1y \leftrightarrow \exists n\, \forall m \geqslant n\; x_m = y_m.$$

To this moment, E_1 is the only known obstruction for Borel equivalence relations to being reducible to an orbit equivalence relation. As with E_0, E_1 is not only characterized by its minimality property, but also by the fact that it is maximum among Borel equivalence relations that can be written as a union of a countable increasing chain of smooth equivalence relations, called *hypersmooth*.

Beyond E_1 there is the relation $E_{\boldsymbol{K}_\sigma}$ maximum among all $\boldsymbol{K}_\sigma$ equivalence relations, i.e., those that can be written as an increasing union of compact sets. It is defined on the space $\prod_{n=1}^{\infty}\{1,\ldots,n\}$ by the following formula

$$xE_{\mathbf{K}_\sigma}y \leftrightarrow \exists N\ \forall n\ |x_n - y_n| \leqslant N.$$

Another realization of this important degree is, for example, a growth relation on functions $f : \mathbb{N} \to \mathbb{N}$ [43]. It is not too hard to see that $E_\infty \leqslant_B E_{\mathbf{K}_\sigma}$, while, on the other hand, $=^+ \nleqslant_B E_{\mathbf{K}_\sigma}$.

As a last example we shall mention the most complex of all analytic equivalence relations, namely the *complete analytic equivalence relation* $E_{\boldsymbol{\Sigma}_1^1}$. It is simply characterized by being maximum among all analytic relations. Combinatorial realizations of this relation may be found in [36] and [17].

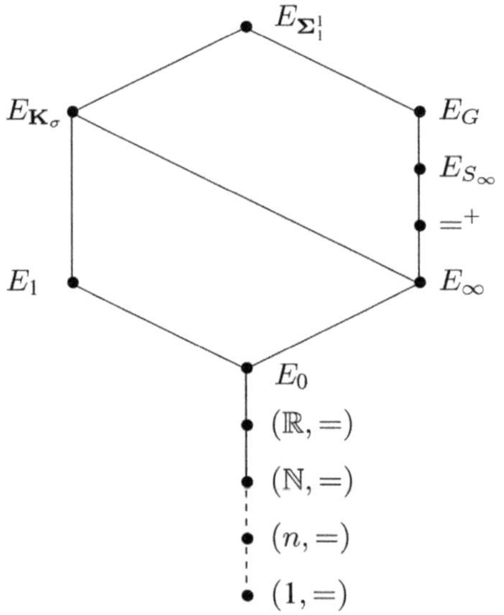

Simplified diagram of complexity of analytic equivalence relations.

2 Standard Borel spaces

For us to consider the class of separable Banach spaces as a standard Borel space, we notice that up to isometry they are all represented as closed subspaces of some metrically universal separable space such as $C(2^{\mathbb{N}})$. We now use a by now standard way to make the closed subspaces of $C(2^{\mathbb{N}})$ into a standard Borel space. First we denote by $F(X)$ the set of all closed subsets of $X = C(2^{\mathbb{N}})$ and equip $F(X)$ with its so called Effros–Borel structure, which is the σ-algebra generated by the sets on the form

$$\{F \in F(X) \mid F \cap U \neq \emptyset\},$$

where U varies over open subsets of X. Equipped with the Effros–Borel structure, $F(X)$ becomes a standard Borel space, i.e., isomorphic as a measure space with $\mathbb{R}$ given its standard Borel algebra. One then easily checks that various relations on $F(X)^2$ and $X \times F(X)$ are Borel, e.g., for $x \in X$ and $F \in F(X)$,

$$x \in F \leftrightarrow \forall n\,(x \in U_n \to F \cap U_n \neq \emptyset),$$

where $\{U_n\}$ is a fixed basis for the topology on X. So the relation '$\in$' is Borel in $X \times F(X)$. We can now also verify that the set $\mathfrak{B}$ of closed *linear* subspaces of X form a Borel set in $F(X)$. To do this, notice that for $Y \in F(X)$, Y is a linear subspace if and only if

$$0 \in Y \;\&\; \forall n, m\; \forall p, q \in \mathbb{Q}$$
$$\big(Y \cap U_n \neq \emptyset \;\&\; Y \cap U_m \neq \emptyset \to Y \cap (p \cdot U_n + q \cdot U_m) \neq \emptyset\big).$$

As all the quantifiers are over countable sets, $\mathfrak{B}$ is a Borel set in $F(X)$ and thus a standard Borel space in its own right and we designate this as *the* standard Borel space of separable Banach spaces. One could of course have constructed this space in other ways, e.g., as the set of norms on a countable set of vectors, but experience shows that any other way of proceeding leads to equivalent results.

Let us now see that, e.g., isometry of separable Banach spaces is an analytic equivalence relation.

$$\begin{aligned} Y \cong_{\mathrm{i}} Z \leftrightarrow & \exists (y_n) \in X^{\mathbb{N}}\; \exists (z_n) \in X^{\mathbb{N}} \\ & \forall m\,\big(Y \cap U_m \neq \emptyset \to \exists n\; y_n \in U_m\big) \\ & \&\; \forall m\,\big(Z \cap U_m \neq \emptyset \to \exists n\; z_n \in U_m\big) \\ & \&\; \forall n\,\big(y_n \in Y\big) \;\&\; \forall n\,\big(z_n \in Z\big) \\ & \&\; \forall n, m\; \|y_n - y_m\| = \|z_n - z_m\|, \end{aligned}$$

which simply expresses that two separable spaces are isometric if and only if they have countable dense subsets which are isometric.

In many cases, we are not interested in all separable Banach spaces, but only in a Borel set of spaces. The most common situation is when we consider only the subspaces of a particular space. But it is not hard to see that if $X \in \mathfrak{B}$, then $\mathfrak{B}(X) = \{Y \in \mathfrak{B} \mid Y \subseteq X\}$ is Borel and we can thus talk about the complexity of the isomorphism relation restricted to this set.

If we are interested in a relation between subsequences of a given basis $(e_n)_n$ of a space X, e.g. isomorphism of the closed linear spans or equivalence of the subsequences, we identify the space $[\mathbb{N}]^{\mathbb{N}}$ of infinite subsets of $\mathbb{N}$ with the set of subsequences of $(e_n)_n$. The associated embedding of $[\mathbb{N}]^{\mathbb{N}}$ into $\mathfrak{B}(X)$ is Borel and therefore we can also compute the complexity of isomorphism etc. between subspaces spanned by subsequences of the basis. The case of block-subspaces is more complicated and will be developed later.

3 Relations between separable Banach spaces

Among the relations of similarity between (infinite dimensional) Banach spaces a few stick out as being of outmost importance, namely, linear isomorphism, (linear) isometry, Lipschitz isomorphism, and uniform homeomorphism. It is well-known that the project of classifying separable Banach spaces up to isomorphism is essentially an impossible task and the tendency nowadays is to settle for something less, namely to find a "basis" for the class of separable Banach spaces, i.e., a list of recognizable spaces such that every space contains a copy of some space in the list.

Another natural question however is also what the complexity of the various classification problems is in the hierarchy of analytic equivalence relations. For example, if one can show that the relation of isomorphism is of high complexity then this lends mathematical sense to the feeling that this relation remains intractable.

Concerning these various relations, Gao and Kechris [22] have shown that isometry between separable complete metric spaces is bireducible with the most complex orbit equivalence relation, E_G, and thus the relation of isometry between separable Banach spaces is at most of this complexity. However, recently, Melleray [38] has been able to show that isometry on $\mathfrak{B}$ is itself Borel bireducible with E_G. This should be contrasted with the result in [36] saying that the relation of (linear) isometric biembeddability between separable Banach spaces is a complete analytic equivalence relation.

Concerning the complexity of isomorphism, lower bounds were successively obtained by Bossard (E_0, [7]), Rosendal (E_1, [41]), and Ferenczi–Galego (the product $E_{\boldsymbol{K}_\sigma} \otimes =^+$, [15]). And the complexity was finally determined by a very recent and yet unpublished result of Ferenczi, Louveau, and Rosendal [17].

Theorem 2 (Ferenczi–Louveau–Rosendal [17]). *The relations of isomorphism, Lipschitz isomorphism, (complemented) biembeddability, and Lipschitz biembeddability between separable Banach spaces are analytic complete, i.e., are maximum among analytic equivalence relations in the Borel reducibility ordering $\leqslant_B$. The same holds for the relation of permutative equivalence of unconditional basic sequences.*

This result thus has the surprising consequence that it is possible to assign in a Borel manner to each separable Banach space X an unconditional basic sequence (e_i^X) such that two spaces X and Y are isomorphic if and only if (e_i^X) and (e_i^Y) are permutatively equivalent. This seems to contradict the feeling that it is somehow easier to check permutative equivalence rather than isomorphism. However, the proof of this result gives no hint as to how this assignment could be computed and obviously the basis (e_i^X) does not itself have to be related to the space X. It would certainly be very interesting to find an "explicit" such assignment which could potentially be of use in applications.

We shall not go into the proof of this result, but only mention that it relies on an elaborate construction due to Argyros and Dodos [4] that allows one to construct spaces essentially containing any specified analytic set of ℓ_p's as its minimal subspaces.

In the fundamental paper [24] Gowers proved his now famous dichotomy theorem stating that any infinite-dimensional Banach space contains either an unconditional basic sequence or an HI subspace. This result, in combination with another result of Komorowski and Tomczak-Jaegermann [33], also solved the homogeneous space problem: ℓ_2 is the only (infinite-dimensional) Banach space which is isomorphic to all its infinite-dimensional closed subspaces.

Given a separable Banach space X which is not isomorphic to ℓ_2, the question remains as to what the possible complexity of isomorphism (and also biembeddability etc.) is on the set of subspaces of X. This is in line with the general question of Gowers concerning the structure of the set of subspaces of a separable Banach space under the quasiorder of embeddability. Even for classical spaces these questions remain unsolved, only lower bounds are obtained.

We shall say that a separable Banach space is *analytic complete* if isomorphism between its subspaces is analytic complete. It is said to be *ergodic* [20] if isomorphism between its subspaces reduces the relation E_0. Analytic complete spaces are those spaces on which isomorphism between subspaces reflects the complexity of isomorphism between all separable Banach spaces (equivalently, of the most complex analytic equivalence relation). Ergodic spaces are such that isomorphism between their subspaces is not smooth though it is not clear whether this is also sufficient for a space to be ergodic.

Question 3. Let X be a separable Banach space which is not isomorphic to ℓ_2. Is X ergodic? Is X analytic complete?

By the proof of [17], the universal unconditional basis of Pełczyński spans an analytic complete space. The spaces c_0 and ℓ_p, $1 \leqslant p < 2$ are ergodic [15], and, in fact, isomorphism between their subspaces has complexity at least $E_{\boldsymbol{K}_\sigma}$; isomorphism between subspaces of L_p, $1 \leqslant p < 2$, has complexity at least $E_{\boldsymbol{K}_\sigma} \otimes =^+$. Concerning $\ell_p, p > 2$, it is only known that there are uncountably many non-isomorphic subspaces [34]. Isomorphism between subspaces of Tsirelson's space T, as well as of its dual T^*, has complexity at least E_1 [41].

When the space X is equipped with a Schauder basis, it is natural to restrict the question of complexity of isomorphism to the class of block-subspaces of X. A natural topological setting for this is the space $bb(X)$ of normalised block-sequences of X, seen as a subspace of $X^{\mathbb{N}}$ where X is equipped with the norm topology. The relation of isomorphism induces a relation denoted $\simeq$ on $bb(X)$, and the canonical map from $bb(X)$ into $\mathfrak{B}(X)$ is Borel.

In this setting, the spaces c_0 and ℓ_p, $1 \leqslant p < +\infty$, with their canonical bases, are the natural homogeneous examples; their bases are *block-homogeneous*, meaning that all their block-subspaces are isomorphic.

Concerning T with its canonical basis, it follows from [41] and [43] that

$$E_1 \leqslant_B (bb(T), \simeq) \leqslant_B E_{K_\sigma},$$

but the exact complexity of $(bb(T), \simeq)$ remains unknown. Note that the basis of T is strongly asymptotic ℓ_1, where a basis (e_i) is *strongly asymptotic* ℓ_p if for some $C < \infty$ and some increasing function $f : \mathbb{N} \to \mathbb{N}$, every normalized sequence $(x_i)_{i=1}^n$ $(n \in \mathbb{N})$ of disjointly supported vectors from $[(e_i)_{i=f(n)}^\infty]$ is C-equivalent to the unit vector basis of ℓ_p^n. For example, the convexification T_p of Tsirelson's space, $1 < p < +\infty$, has a strongly asymptotic ℓ_p basis, and T^* has a strongly asymptotic ℓ_∞ basis. It is more generally proved in [9] that for any space X with a strongly asymptotic ℓ_p basis ($1 \leqslant p \leqslant +\infty$), that is not equivalent to the canonical basis of c_0 or ℓ_p, E_0 is Borel reducible to isomorphism between block-subspaces of X (even to isomorphism between spaces spanned by subsequences of the basis).

It is unknown whether c_0 and ℓ_p, $1 \leqslant p < +\infty$, are the only spaces with a block-homogeneous basis. This would be a natural generalization of Zippin's theorem about perfectly homogeneous bases [47]. Note that a positive answer would imply the homogeneous Banach space theorem of Gowers and Komorowski–Tomczak-Jaegermann, via the fact that every Banach space contains a basic sequence and the non-trivial fact that the spaces c_0 and ℓ_p, $p \neq 2$, are not homogeneous. The following question could be easier to solve than the first part of Question 3.

Question 4. Let X be a separable Banach space with a Schauder basis which is not isomorphic to c_0 or ℓ_p, $1 \leqslant p < +\infty$. Does it follow that $E_0 \leqslant_B (bb(X), \simeq)$?

4 Isomorphism and spaces with an unconditional basis

The general idea of this section is that spaces with an unconditional basis which are not ergodic must satisfy some algebraic properties (isomorphism with their square, their hyperplanes etc.). They therefore resemble Hilbert space more than a generic separable Banach space. Quite similar ideas in the context of the Schröder–Bernstein property for Banach spaces were first developed by Kalton [30].

4.1 Subspaces spanned by subsequences of the basis

Given a space X with a Schauder basis (e_i), we first look at the space $ss(e_i)$ of subsequences of the basis, which we identify with the space $[\mathbb{N}]^{\mathbb{N}}$ of infinite subsets of $\mathbb{N}$ considered as a subset of $2^{\mathbb{N}}$ with its induced topology. The relation induced by isomorphism of the corresponding linear spans will be denoted $\simeq$, and for $K \geqslant 1$, $\simeq^K$ denotes the relation induced by isomorphism with constant at most K.

Theorem 5. *Let X be a Banach space with an unconditional basis (e_i). Then $E_0 \leqslant_B (ss(e_i), \simeq)$, or X is uniformly isomorphic to $X \oplus Y$, for all Y generated by a finite or infinite subsequence of the basis – and therefore isomorphic to its square and to its hyperplanes – and, moreover, uniformly isomorphic to its tail subspaces and isomorphic to an infinite direct sum of uniformly isomorphic copies of itself.*

Proof. The proof uses the following lemma. We need to define the relation E_0' between infinite subsets of $\mathbb{N}$: two sets $A, B \subseteq \mathbb{N}$ are E_0' equivalent if $|(A \cup B) \setminus A| = |(A \cup B) \setminus B| < \infty$, i.e., if A and B have the same finite co-cardinality in $A \cup B$. If (e_i) is a basic sequence and we identify a subset $A \subseteq \mathbb{N}$ with the subsequence it generates, then one easily sees that the relation of equivalence between subsequences of (e_i) is E_0'-invariant.

Lemma 6 (cf. [42]). *Let E be an analytic equivalence relation on $[\mathbb{N}]^{\mathbb{N}}$ that is E_0'-invariant. Then either $E_0 \leqslant_B E$ or E has a comeagre class. In the latter case there are $A, B \in [\mathbb{N}]^{\mathbb{N}}$ such that $AE(\sim A)$ and $BE(B \setminus \min B)$.*

Proof. If E is meagre as a subset of $[\mathbb{N}]^{\mathbb{N}} \times [\mathbb{N}]^{\mathbb{N}}$, then it is not difficult to build finite successive subsets a_n^0 and a_n^1 of $\mathbb{N}$ such that $|a_n^0| = |a_n^1|$ and such that the map $T : 2^{\mathbb{N}} \to [\mathbb{N}]^{\mathbb{N}}$ defined by $T(\alpha) = \bigcup_{k \in \mathbb{N}} a_k^{\alpha(k)}$ Borel reduces E_0 to $\simeq$. If on the other hand E is non-meagre, then by a classical result of Kuratowski–Ulam, [31] Theorem 8.41, some E-class $\mathbb{A}$ is non-meagre. Therefore $\mathbb{A} \cap U$ is comeagre in U for some basic open set U. By E_0'-invariance it follows that $\mathbb{A}$ is comeagre.

Now the map $c : 2^{\mathbb{N}} \to 2^{\mathbb{N}}$ defined by $c(A) = \sim A$ is a homeomorphism of $2^{\mathbb{N}}$ and as $[\mathbb{N}]^{\mathbb{N}}$ is comeagre in $2^{\mathbb{N}}$, also $\mathbb{A}$ is comeagre in $2^{\mathbb{N}}$. But then also $c(\mathbb{A})$ is comeagre in $2^{\mathbb{N}}$ and hence $\mathbb{A} \cap c(\mathbb{A})$ is comeagre too. So pick some $A \in \mathbb{A} \cap c(\mathbb{A})$ which is co-infinite. Then $AE(\sim A)$.

Similarly, for each $n \in \mathbb{N}$, the map $C \in [\mathbb{N}]^{\mathbb{N}} \mapsto C\triangle\{n\} \in [\mathbb{N}]^{\mathbb{N}}$ is an involution homeomorphism of $[\mathbb{N}]^{\mathbb{N}}$ and thus there must be some $B \in \mathbb{A}$ such that for all n, $B\triangle\{n\} \in \mathbb{A}$. In particular, $BE(B \setminus \min B)$. □

Going back to the proof of Theorem 5, assuming E_0 is not Borel reducible to $\simeq$ on $[\mathbb{N}]^{\mathbb{N}}$, we apply Lemma 6 and obtain a comeagre $\simeq$-class $\mathbb{A}$. Classical Banach space arguments about perturbation by finite-dimensional subspaces then ensure that the set $\mathbb{A}_K = \{A \in [\mathbb{N}]^{\mathbb{N}} : A \simeq^K A_0\}$ is comeagre for some fixed $A_0 \in \mathbb{A}$ and some $K \geqslant 1$.

If a and b are non empty subsets of $\mathbb{N}$ and a is finite, recall that a and b are successive, $a < b$, when $\max(a) < \min(b)$. If b is a finite subset of $\mathbb{N}$, say that $A \in [\mathbb{N}]^{\mathbb{N}}$ *passes through* b if $A = a \cup b \cup C$ for some finite subset a of $\mathbb{N}$ and some $C \in [\mathbb{N}]^{\mathbb{N}}$ such that $a < b < C$. We now use the following classical characterization of comeagre subsets of $[\mathbb{N}]^{\mathbb{N}}$, see, e.g., [19] and [42].

Lemma 7. *Let $\mathbb{A}$ be a subset of $[\mathbb{N}]^{\mathbb{N}}$. Then $\mathbb{A}$ is comeagre if and only if there exists a countable family $\{a_k, k \in \mathbb{N}\}$ of finite successive subsets of $\mathbb{N}$ such that whenever $A \in [\mathbb{N}]^{\mathbb{N}}$ passes through a_k for infinitely many k's, then A belongs to $\mathbb{A}$.*

Choose such a family $\{a_k, k \in \mathbb{N}\}$ for our set $\mathbb{A}_K$. To conclude the proof of Theorem 5, we fix $A_0 \in \mathbb{A}_K$ such that $(\sim A_0) \in \mathbb{A}_K$, and we obtain by unconditionality of the basis $(e_n)_{n\in\mathbb{N}}$ of X,

$$X \simeq [e_n, n \in A_0] \oplus [e_n, n \notin A_0] \simeq [e_n, n \in A_0]^2,$$

and

$$[e_n, n \in A_0] \simeq [e_n, n \in a_k, k \in \mathbb{N}] = [e_n, n \in a_{2k}, k \in \mathbb{N}] \oplus [e_n, n \in a_{2k+1}, k \in \mathbb{N}],$$

therefore $[e_n, n \in A_0] \simeq [e_n, n \in A_0]^2 \simeq X$.

Fix $k_0 \in \mathbb{N}$, then $\{1, \ldots, k_0\} \cup \bigcup_{k>k_0} a_k$ and $\bigcup_{k>k_0} a_k$ belong to $\mathbb{A}_K$. Therefore

$$[e_n, n \in \bigcup_{k>k_0} a_k] \simeq [e_1, \ldots, e_{k_0}] \oplus [e_n, n \in \bigcup_{k>k_0} a_k],$$

and by taking a direct sum with the appropriate subspace, we obtain

$$X \simeq [e_n, n > k_0].$$

Whenever I is a subset of $\mathbb{N}$ and $Y = [e_n, n \in I]$, we may find a partition $\{I_1, I_2\}$ of I and two infinite disjoint subsets N_1 and N_2 of $\mathbb{N}$ such that

$$I_1 \cap \Big(\bigcup_{k \in N_2} [\min a_k, \max a_k]\Big) = \emptyset \text{ and } I_2 \cap \Big(\bigcup_{k \in N_1} [\min a_k, \max a_k]\Big) = \emptyset.$$

It follows that $I_1 \cup (\bigcup_{k \in N_2} a_k)$ and $I_2 \cup \bigcup_{k \in N_2} a_k$ belong to $\mathbb{A}_K$ and therefore

$$X \simeq X \oplus [e_n, n \in I_1]$$

and

$$X \simeq X \oplus [e_n, n \in I_2].$$

Thus

$$X \simeq X \oplus [e_n, n \in I_1] \oplus [e_n, n \in I_2] \simeq X \oplus Y.$$

Finally, let $\{B_i, i \in \mathbb{N}\}$ be a partition of $\mathbb{N}$ in infinite subsets and let for $i \in \mathbb{N}$, $C_i = \bigcup_{k \in B_i} a_k$. Then $C_i \in \mathbb{A}_K$ for all i, while also $\bigcup_{i \in \mathbb{N}} C_i \in \mathbb{A}_K$. Therefore

$$X \simeq [e_n, n \in \bigcup_{i \in \mathbb{N}} C_i] \simeq \oplus_{i \in \mathbb{N}} [e_n, n \in C_i],$$

with $[e_n, n \in C_i] \simeq X$, for all $i \in \mathbb{N}$.

All the isomorphisms are obtained with uniform constants depending only on K and the unconditional constant of the basis. □

Note that these results may be generalized to spaces with unconditional decompositions in the same spirit as in [30], see [16].

4.2 Subspaces spanned by block-sequences of the basis

We consider here a space X with a Schauder basis (e_i) and the class of subspaces generated by block-bases, i.e., sequences of vectors with successive supports (called successive vectors). Rather than $bb(X)$, the correct setting for the main theorem here (Theorem 8) seems to be the space denoted $bb_{\mathbf{Q}}(X)$ defined as follows. Let first $\mathbf{Q}$ be a countable subfield of $\mathbb{R}$ such that any finite $\mathbf{Q}$-linear combination of the basis vectors has norm in $\mathbf{Q}$ and let $\mathbb{D}$ be the set of non-zero blocks with coefficients in $\mathbf{Q}$, $\mathbb{D}_1$ be the set of norm 1 vectors in $\mathbb{D}$. The assumption on $\mathbf{Q}$ allows us to normalize while staying in $\mathbf{Q}$. The set $bb_{\mathbf{Q}}(X)$ is then the set of block-bases of vectors in $\mathbb{D}_1$, equipped with the product topology of the discrete topology on $\mathbb{D}_1$. The relation of isomorphism induces a relation $\simeq$ on $bb_{\mathbf{Q}}(X)$. As $\mathbb{D}_1$ is countable, this topology is Polish and epsilon matters may be forgotten until the applications. When we deal with isomorphism classes, they are not relevant since a small enough perturbation preserves the class. Note that the canonical map of $bb_{\mathbf{Q}}(X)$ into $\mathfrak{B}(X)$ is Borel, and this allows us to forget about the

Effros–Borel structure when computing the complexity on block-subspaces. Note also that the reduction of E_0 to $\simeq$ on $bb_{\mathbf{Q}}(X)$ in Theorem 8 will provide a reduction to $\simeq$ on $bb(X)$ as well.

The set $fbb_{\mathbf{Q}}(X)$ denotes the set of finite successive sequences of blocks in $\mathbb{D}_1$. The support $\mathrm{supp}(a)$ of such a sequence a is the union of the supports of the blocks composing the sequence.

Theorem 8 (Ferenczi–Rosendal [20]). *Let X be a Banach space with an unconditional basis. Then E_0 is Borel reducible to $(bb_{\mathbf{Q}}(X), \simeq)$ or there exists a block-subspace X_0 of X which is uniformly isomorphic to $X_0 \oplus Y$ for all block-subspaces Y of X.*

Proof. The argument being similar to the case of subsequences, we shall just sketch the proof. If E_0 is not Borel reducible to $(bb_{\mathbf{Q}}(X), \simeq)$, then there exists a comeagre $\simeq$-class $\mathbb{A} \subset bb_{\mathbf{Q}}(X)$. Now comeagre subsets of $bb_{\mathbf{Q}}(X)$ may be characterized up to small perturbations by the next lemma from [20]. For a a finite block-sequence and b a finite or infinite block-sequence such that $\mathrm{supp}(a) < \mathrm{supp}(b)$, $a^\frown b$ denotes the finite or infinite block-sequence which is the concatenation of a and b. If $b \in fbb_{\mathbf{Q}}(X)$, and $A \in bb_{\mathbf{Q}}(X)$, then A *passes through* b if $A = a^\frown b^\frown C$ for some $a \in fbb_{\mathbf{Q}}(X)$ and some $C \in bb_{\mathbf{Q}}(X)$. If $\mathbb{A}$ is a subset of $bb_{\mathbf{Q}}(X)$ and $\Delta = (\delta_n)_{n\in\mathbb{N}}$ is a sequence of strictly positive real numbers, written $\Delta > 0$, we denote by $\mathbb{A}_\Delta$ the Δ-expansion of $\mathbb{A}$ in $bb_{\mathbf{Q}}(X)$, that is $x = (x_n) \in \mathbb{A}_\Delta$ if and only if there exists $y = (y_n) \in \mathbb{A}$ such that $\|y_n - x_n\| < \delta_n, \forall n \in \mathbb{N}$.

Lemma 9. *Let $\mathbb{A}$ be comeagre in $bb_{\mathbf{Q}}(X)$. Then for all $\Delta > 0$, there exist successive finite block-sequences $a_n, n \in \mathbb{N}$ in $fbb_{\mathbf{Q}}(X)$ such that any element of $bb_{\mathbf{Q}}(X)$ passing trough infinitely many of the a_n's is in $\mathbb{A}_\Delta$.*

By classical perturbation arguments, $\mathbb{A} = \mathbb{A}_\Delta$ for some Δ small enough. Let $a_n = (a_n^1, \dots, a_n^{m_n})$, $n \in \mathbb{N}$ be given by Lemma 9 and let $X_0 = [A_0]$, for some $A_0 \in \mathbb{A}$.

Fix $(y_n)_{n\in\mathbb{N}}$ in $bb_{\mathbf{Q}}(X)$ and $Y = [y_n]_{n\in\mathbb{N}}$. We may find a partition $(I_k)_{k\in\mathbb{N}}$ of $\mathbb{N}$ in successive intervals and an increasing sequence (n_k) of integers such that for all k in $\mathbb{N}$,

$$\mathrm{supp}(y_n, n \in I_k) < \mathrm{supp}(a_{n_{k+1}}) < \mathrm{supp}(y_n, n \in I_{k+2}).$$

Therefore the block sequence $A = (z_n)_{n\in\mathbb{N}}$ defined by

$$\{z_n, n \in \mathbb{N}\} = \bigcup_{k\in\mathbb{N}} \{y_n, n \in I_{2k+1}\} \bigcup \bigcup_{k\in\mathbb{N}} \{a_{n_{2k}}^1, \dots, a_{n_{2k}}^{m_{2k}}\},$$

belongs to $\mathbb{A}$. It follows that

$$X_0 \simeq X_0 \oplus [y_n, n \in \bigcup_{k\in\mathbb{N}} I_{2k+1}],$$

and likewise

$$X_0 \simeq X_0 \oplus [y_n, n \in \bigcup_{k\in\mathbb{N}} I_{2k}],$$

whence finally

$$X_0 \simeq X_0 \oplus Y.$$

Additional care in the proof guarantees uniformity. □

5 Embeddability, biembeddability, and isomorphism

In the paper [24] Gowers proved his dichotomy theorem stating that any infinite-dimensional Banach space contains either an unconditional basic sequence or an HI subspace. Actually Gowers proved more refined structure results that set the stage for a detailed list of inevitable classes of subspaces. To simplify notation in the following we shall use $X \sqsubseteq Y$ to denote that a space X embeds isomorphically into Y. Moreover, we shall assume all spaces considered are infinite-dimensional.

A space is said to be *minimal* if it is $\sqsubseteq$-minimal among its subspaces, *quasi-minimal* if any two subspaces have a common $\sqsubseteq$-minorant, and *strictly quasi-minimal* if it is quasi-minimal but does not contain a minimal subspace. On the other hand, we shall say that a space with a Schauder basis has the *strong Casazza property* (in reference to a property defined by Casazza, see [23]) if no two disjointly supported block-subspaces are isomorphic. Two spaces are said to be *incomparable* in case neither of them embeds into the other, and *totally incomparable* if no space embeds into both of them.

Theorem 10 (Gowers [24]). *Let X be an infinite dimensional Banach space. Then X contains a subspace Y with one of the following properties, which are all possible and mutually exclusive.*

1. *Y is hereditarily indecomposable,*
2. *Y has an unconditional basis with the strong Casazza property,*
3. *Y has an unconditional basis and is strictly quasi-minimal,*
4. *Y has an unconditional basis and is minimal.*

Type (1) spaces were discovered by Gowers and Maurey [25] in 1991, and a type (2) space was constructed by Gowers in [23] and further analyzed in [26]. Tsirelson's space T, the precursor of Banach spaces with "exotic" properties such as Gowers and Maurey's examples, is a typical example of a type (3) space. The spaces c_0, ℓ_p for $1 \leqslant p < +\infty$, the dual T^* of Tsirelson's space, and Schlumprecht's space S [2] are the main known examples of spaces of type (4).

In each case one can ask what the structure of the relations of embeddability, $\sqsubseteq$, biembeddability, $\equiv$, and isomorphism, $\cong$, is on the subspace in question. In his paper [24] Gowers had asked the question of what quasiorders could be realized as the set of subspaces of a separable Banach space ordered by $\sqsubseteq$.

Theorem 11. *Let X be a separable infinite-dimensional Banach space belonging to one of the four types given by Gowers' Theorem 10. Then for each of the relations $\sqsubseteq$, $\equiv$, and $\cong$, we have lower bounds on the complexity as given in the following diagram.*

Type	$\sqsubseteq$	$\equiv$	$\cong$
(1)	$\mathbb{R}$, ω_1, *and* ω_1^**-chains, uncountable Borel antichain*	E_0	E_0
(2)	$\mathbb{R}$, ω_1, *and* ω_1^**-chains,* *uncountable Borel set of totally incomparable spaces*	E_0	E_0
(3)	ω_1^**-chain, uncountable Borel antichain*	E_0	E_0
(4)	*trivial*	*trivial*	*none*

We should mention that by classical results, all uncountable Borel sets are of the size of the continuum, so we get, for example, continuum size antichains in the case of strictly quasiminimal spaces.

Proof. In order to prove this result, we will show how to get it from, in each case, stronger and more refined results that also have somewhat larger scopes.

We may first deduce the results for $\equiv$ and $\cong$ for type (1) and (2) from Lemma 6. For if we are given a Schauder basis (e_i), we can for $\equiv$ and $\cong$ consider the corresponding analytic equivalence relation on $[\mathbb{N}]^{\mathbb{N}}$ obtained by identifying $A \in [\mathbb{N}]^{\mathbb{N}}$ with the subspace $[e_i]_{i\in A}$. As an HI space is non-isomorphic with all of its proper subspaces, we thus see that for $[e_i]$ of type (1) there is no $B \in [\mathbb{N}]^{\mathbb{N}}$ such that $B \equiv B \setminus \min B$ or $B \cong B \setminus \min B$, whence E_0 must reduce to both $\equiv$ and $\cong$. Similarly, in case of type (2), there is no A such that $A \equiv \sim A$ or $A \cong \sim A$, and thus again E_0 reduces to both $\equiv$ and $\cong$.

In the case of type (2), there is also an explicit reduction by the map $\alpha \in 2^{\mathbb{N}} \mapsto [e_{2n+\alpha(n)}, n \in \mathbb{N}]$.

Consider now the chains in the case of type (1) and type (2). Fix first a basic sequence either spanning an HI space or such that any two disjointly supported spaces are totally incomparable.

Assume now that for infinite sets $A, B \subseteq \mathbb{N}$ we have $|A \setminus B| < \infty$, but $|B \setminus A| = \infty$, which we denote by $A \subsetneq^* B$. Then we can find some $B' E_0' B$ such that $A \subseteq B'$ and $|B' \setminus A| = \infty$, whence $[e_i]_{i\in A} \sqsubseteq [e_i]_{i\in B'} \cong [e_i]_{i\in B}$. On the other hand, in the case of HI spaces, $[e_i]_{i\in B} \cong [e_i]_{i\in B'} \not\sqsubseteq [e_i]_{i\in A}$, and in case of type (2), $[e_i]_{i\in A}$ and $[e_i]_{i\in B'\setminus A}$ are totally incomparable and hence $[e_i]_{i\in B} \cong [e_i]_{i\in B'} \not\sqsubseteq [e_i]_{i\in A}$ again. In any case,

$$A \subsetneq^* B \Rightarrow [e_i]_{i\in A} \sqsubseteq [e_i]_{i\in B} \;\&\; [e_i]_{i\in B} \not\sqsubseteq [e_i]_{i\in A}.$$

By simple diagonalisation it is now easy to construct a sequence $(A_\xi)_{\xi<\omega_1}$ such that if $\xi < \zeta < \omega_1$, then $A_\xi \subsetneq^* A_\zeta$, whence $([e_i]_{i\in A_\xi})_{\xi<\omega_1}$ gives a ω_1-chain in the ordering $\sqsubseteq$. Similarly for an ω_1^*-chain. Now to construct the $\mathbb{R}$-chain, we identify $\mathbb{Q}$ with $\mathbb{N}$ and $\mathbb{R}$ with the left parts of the corresponding Dedekind cuts. Thus, if $r < s$ belong to $\mathbb{R}$, then they correspond to subsets $A_r \subsetneq^* A_s$ of $\mathbb{N}$. Thus, $([e_i]_{i\in A_r})_{r\in\mathbb{R}}$ forms an $\mathbb{R}$-chain in $\sqsubseteq$.

To get an uncountable Borel set of totally incomparable subspaces, i.e. such that no subspace of one embeds into the other, in the case of type (2), we simply pick an uncountable Borel set of almost disjoint subsets of $\mathbb{N}$ and notice that if A and B are almost disjoint, i.e., they are both infinite with finite intersection, then $[e_i]_{i\in A}$ and $[e_i]_{i\in B}$ are totally incomparable.

The last fairly simple part of the picture is the existence of the ω_1^*-chain in the case of type (3). Again this is a direct set theoretical diagonalisation. We use here the well-known fact that the embeddability relation on quasiminimal spaces is downwards σ-directed, i.e., that any countable family of subspaces have a common subspace up to isomorphism. This is easy to see, for if X is quasiminimal and (Y_m) is a countable family of infinite dimensional subspaces, where we suppose Y_0 has a basis (e_i), then by quasiminimality, we can inductively pick block sequences $(x_n^{(m)})$ such that $(x_n^{(m+1)})$ is a block of $(x_n^{(m)})$, $[x_n^{(m)}] \sqsubseteq Y_m$ and then take the diagonal sequence $(x_{2n}^{(n)})_n$. Then $[x_{2n}^{(n)}]_n \sqsubseteq [x_n^{(m)}]_n \sqsubseteq Y_m$ for all m. By diagonalisation and using the fact that no subspace is minimal one now constructs an ω_1^* sequence in the ordering $\sqsubseteq$.

The final four results, namely the existence of uncountable Borel antichains in type (1) and (3) and the reduction of E_0 to $\equiv$ and $\cong$ in type (3), however, seems to necessitate more advanced techniques, in particular, metamathematics and methods of effective descriptive set theory. Moreover, also the determinacy result of Gowers [24] is here used in its full force, i.e., for analytic sets. To our knowledge, this is one of the only known applications of his result other than for closed/open sets.

To begin, let us first state a general Ramsey principle for equivalence relations that is specifically adapted for the geometry of Banach spaces.

Theorem 12 (Rosendal [42]). *Let E be an analytic equivalence relation on $[\mathbb{N}]^{\mathbb{N}}$ that is E_0'-invariant, i.e., $E_0' \subseteq E$. Then either $E_0 \leqslant_B E$ or for some infinite subset $A \subseteq \mathbb{N}$, the set $[A]^{\mathbb{N}}$ is contained in a single E-class.*

For example, if (e_i) is a basis and E the induced relation on $[\mathbb{N}]^{\mathbb{N}}$ of isomorphism between spaces spanned by subsequences of the basis, then E is easily E_0'-invariant. We should mention that the only known proof of Theorem 12 uses metamathematical methods and it would be interesting to find a more topological proof of this. Probably, one would have to find some uniformity that could allow for standard methods. As an application of the statement, we see that if E_0 does not reduce to isomorphism on the set of subspaces of $[e_i]$, then (e_i) has an isomorphically homogeneous subsequence, i.e., a subsequence all of whose further subsequences span isomorphic spaces. Nevertheless, this principle does not in itself appear to be enough to get E_0 in type (3). For that we will need a better result due to Ferenczi, relying on the one hand on Theorem 12 and on the other hand on the methods of Pelczar [40], who herself was inspired by the closed case of Gowers' determinacy result.

Theorem 13 (Ferenczi [14]). *Let X be a separable Banach space saturated with isomorphically homogeneous basic sequences. Then X contains a minimal space. In fact, it is enough to suppose that X is saturated with basic sequences whose closed span embed into the closed span of any of their subsequences.*

Pelczar's original result was quite similar, but had a considerably stronger hypothesis, namely, that X was saturated with subsymmetric basic sequences.

Consider now the case of spaces of type (3). If X is a space of type (3), and E_0 does not reduce to isomorphism between its subspaces, then any basic sequence in X

has an isomorphically homogeneous subsequence and hence X is saturated by such sequences. By Theorem 13, X contains a minimal space, which is impossible. Similarly, if E_0 does not reduce to biembeddability between the subspaces of X, then X is saturated by sequences embeddable into all of the spaces spanned by their subsequences and again X must contain a minimal space, which is a contradiction. Therefore, E_0 reduces in each case.

For the record, we spell out the argument in the following corollary.

Corollary 14. *Let X be a separable infinite-dimensional Banach space. Then either E_0 Borel reduces to isomorphism between its subspaces or X contains a minimal space.*

The result we need for the existence of antichains is quite similar, but differs rather by its proof. Again it does not rely heavily on geometric properties of Banach spaces, but mostly on methods of combinatorics and descriptive set theory, and, in particular, not only Gowers' determinacy theorem, but also on the solution to the distortion problem by Odell and Schlumprecht [39].

Theorem 15 (Rosendal [42]). *Let X be a separable infinite dimensional Banach space. Then either X contains a minimal space or there is an uncountable Borel set of incomparable subspaces.*

This immediately implies the existence of uncountable Borel antichains in type (1) and type (3) spaces and thus finishes the proof. □

The remaining case is that of minimal spaces. The structures of $\sqsubseteq$ and $\equiv$ are trivial on such spaces. Concerning $\cong$, there is of course no general result saying that there are many non-isomorphic subspaces in this case, as the space could be Hilbertian, but if the space is not homogeneous, it seems plausible that there must be many isomorphism classes. Indeed, c_0 and ℓ_p, $1 \leqslant p < 2$, [15], as well as Tsirelson's dual space T^* (by the method of [41]) are ergodic. Note also that any non-reflexive minimal space must contain c_0 or ℓ_1 and therefore be ergodic, and that Theorem 5 and Theorem 8 may provide other classes of minimal ergodic spaces.

Question 16. Does there exist an analytic complete minimal Banach space?

We end this section by sketching the proofs of Theorem 15 and of Theorem 13 in the $\cong$-homogeneous case. In each case we may assume that the space X has a Schauder basis (e_i). All vectors will belong to the set $\mathbb{D}$ of non-zero finite $\mathbf{Q}$-linear combinations of the basis vectors (e_i), where $\mathbf{Q}$ is a countable subfield of $\mathbb{R}$ closed under computing norms. We recall that $\mathbb{D}_1$ denotes the set of normalized blocks in $\mathbb{D}$, that the space of infinite $\mathbb{D}_1$-block bases is denoted by $bb_{\mathbf{Q}}(X)$ and the set of finite $\mathbb{D}_1$-block bases is denoted $fbb_{\mathbf{Q}}(X)$.

Sketch of the proof of Theorem 15: We notice first that $\sqsubseteq$ restricted to the standard Borel space of subspaces of X is an analytic quasiorder. So Theorem 15 amounts to saying that either $\sqsubseteq$ has a minimal element or an uncountable Borel antichain. The idea of the proof is to replace $\sqsubseteq$ by a *Borel* quasiorder R containing $\sqsubseteq$ and sufficiently reflecting the properties of the latter. Then one can employ the analysis by Harrington, Marker, and Shelah [29, 27] of Borel quasiorders to deduce the result. The exact result we need can be deduced from [27].

Theorem 17 (Harrington–Marker–Shelah [27]). *If R is a downwards σ-directed Borel quasiorder on a standard Borel space, then either R has an uncountable Borel anti-chain or a minimal element.*

Here R is downwards σ-directed if any countable family in P has a common minorant. We have the following standard observation.

Lemma 18. *Suppose X is quasi-minimal. Then $\sqsubseteq$ is downwards σ-directed on $bb_{\mathbf{Q}}(X)$.*

For $Y = (y_i)$, $Z = (z_i) \in bb_{\mathbf{Q}}(X)$, let $Y \leqslant Z$ if Y is a blocking of Z and put $Y \leqslant^* Z$ if for some k, $(y_i)_{i \geqslant k} \leqslant Z$. Also, if $\Delta = (\delta_i)$ is an infinite sequence of strictly positive reals, write $d(Y, Z) < \Delta$ if $\forall i \ \|y_i - z_i\| < \delta_i$. Set $Y =^* Z$ if $\exists k \ \forall i \geqslant k \ y_i = z_i$. Evidently, $Y =^* Z$ implies $Y \sim Z$.

For a subset $\mathbb{A} \subseteq bb_{\mathbf{Q}}(X)$ let $\mathbb{A}^* = \{Y \in bb_{\mathbf{Q}}(X) \mid \exists Z \in \mathbb{A} \ Z =^* Y\}$ and $\mathbb{A}_\Delta = \{Y \in bb_{\mathbf{Q}}(X) \mid \exists Z \in \mathbb{A} \ d(Z, Y) < \Delta\}$. Notice that if $\mathbb{A}$ is analytic so are both $\mathbb{A}^*$ and $\mathbb{A}_\Delta$. We also set $[Y] = \{Z \in bb_{\mathbf{Q}}(X) \mid Z \leqslant Y\}$ and notice that $[Y]$ is a Borel subset of $bb_{\mathbf{Q}}(X)$. $\mathbb{A}$ is said to be *large* in $[Y]$ if for any $Z \in [Y]$ we have $[Z] \cap \mathbb{A} \neq \emptyset$.

For $\mathbb{A} \subseteq bb_{\mathbf{Q}}(X)$ and $Y \in bb_{\mathbf{Q}}(X)$, the Bagaria–Gowers–López-Abad game $\Game_Y^{\mathbb{A}}$ is defined as follows: Player I plays in the k'th move of the game a vector $z_k \in \mathbb{D}_1$ such that $z_{k-1} < z_k$. Player II responds by either doing nothing or playing a vector $v \in \mathbb{D}_1$ such that $v \in [z_{l+1}, \dots, z_k]$ where l was the last move where II played a vector. We say that player II wins the game if in the end she has produced a block-basis $V = (v_i) \in \mathbb{A}$. A slight variant of this game is shown in [6] to be equivalent to the game studied by Gowers in [24]. It follows from Gowers' determinacy result in [24] that if $\mathbb{A} \subseteq bb_{\mathbf{Q}}(X)$ is analytic, large in $[Y]$ and Δ is given, then for some $Z \in [Y]$, II has a winning strategy in the game $\Game_Z^{\mathbb{A}_\Delta}$. However, due to the complexity of the sets we are dealing with, we need to have a stronger determinacy result, which holds under stronger set-theoretical assumptions.

Lemma 19. (MA + $\neg$CH) *Suppose $\mathbb{W} \subseteq bb_{\mathbf{Q}}(X)$ is a $\mathbf{\Sigma}_2^1$ set, large in some $[Y]$ and $\Delta > 0$. Then II has a winning strategy in $\Game_Z^{\mathbb{W}_\Delta^*}$ for some $Z \in [Y]$.*

Lemma 20. (MA + $\neg$CH) *Suppose that X does not contain a minimal subspace. Then for any $W \in bb_{\mathbf{Q}}(X)$ there is $Y \in [W]$ and a Borel function $\phi \colon [Y] \to [Y]$ such that for all $Z \in [Y]$,*

$$\phi(Z) \leqslant Z$$

and

$$Z \not\sqsubseteq \phi(Z).$$

Proof. We can assume that $W = X$. Also, as c_0 is minimal, X does not contain c_0 and therefore, by the solution to the distortion problem by Odell and Schlumprecht [39], we can by replacing X by a block-subspace suppose that we have two positively separated sets F_0, F_1 of the unit sphere, such that for any $Y \in bb_{\mathbf{Q}}(X)$ there are $x, y \in \mathbb{D}_1$ such that $x \in F_0$, $y \in F_1$, and $x, y \in Y$. We call such sets *inevitable*. Let now

$$\mathbb{A} = \{Y = (y_i) \in bb_{\mathbf{Q}}(X) \mid \forall i\; y_i \in F_0 \cup F_1\}$$

and for $Y \in \mathbb{A}$ let $\alpha(Y) \in 2^{\mathbb{N}}$ be defined by

$$\alpha(Y)(i) = 0 \Leftrightarrow y_i \in F_0.$$

Then $\alpha\colon \mathbb{A} \to 2^{\mathbb{N}}$ is continuous. Fix an uncountable closed set $P \subseteq 2^{\mathbb{N}}$ of almost disjoint subsets of $\mathbb{N}$ and let

$$\mathbb{B} = \{Y \in \mathbb{A} \mid \alpha(Y) \in P\}.$$

As P is closed, so is $\mathbb{B}$ and by the inevitability of F_0 and F_1, $\mathbb{B}$ is large in every $[Y]$.

By fixing a Borel bijection between P and the set of infinite sequences of finite non-zero $\mathbf{Q}$-tuples, we can see each $p \in P$ as coding a way to construct an infinite $\mathbf{Q}$-block sequence of any $Y \in bb_{\mathbf{Q}}(X)$. Denote this infinite block sequence by Y^p and notice that Y^p is a blocking of Y though not necessarily normalized.

Consider now the set

$$\mathbb{W} = \{Y = (y_i) \in bb_{\mathbf{Q}}(X) \mid (y_{2i}) \in \mathbb{B} \;\wedge\; [y_{2i+1}] \not\sqsubseteq [(y_{2i+1})^{\alpha(y_{2i})}]\}.$$

So $\mathbb{W}$ consists of the blocks $(y_i) \in bb_{\mathbf{Q}}(X)$ such that (y_{2i}) codes a subspace of $[y_{2i+1}]$ into which $[y_{2i+1}]$ does not embed. First of all, $\mathbb{W}$ is clearly coanalytic, and again, using the inevitability of F_0 and F_1 and the fact that X contains no minimal subspace, one can verify that $\mathbb{W}$ is large in $bb_{\mathbf{Q}}(X)$.

Take now some $\Delta = (\delta_i)$ depending on the basis constant such that $d(Y, Y') < \Delta$ implies $Y \sim Y'$, and, moreover, such that $\delta_i < \frac{1}{2}d(F_0, F_1)$. By Lemma 19 we can find a $Y \in bb_{\mathbf{Q}}(X)$ such that II has a winning strategy σ in the game $\Game_Y^{\mathbb{W}^*_\Delta}$.

We shall now show how the function $\phi\colon [Y] \to [Y]$ is defined. For this, let $Z \in [Y]$ be given and suppose I plays the sequence $(z_i) = Z$ in the game $\Game_Y^{\mathbb{W}^*_\Delta}$. Then using the strategy σ, II will respond to Z by playing some $V = (v_i) \leqslant Z$, $V \in \mathbb{W}^*_\Delta$. There is therefore some $W = (w_i) \in \mathbb{W}$ such that $V \in \{W\}^*_\Delta$. This W might not, however, be Borel in V. Nevertheless, we can in a Borel manner compute $\alpha(w_{2i}) \in P$, because for almost all i

$$d(v_{2i}, w_{2i}) < \delta_{2i} < \frac{1}{2}d(F_0, F_1)$$

and $w_{2i} \in F_0 \cup F_1$. So by letting $p(i) = 0$ if $d(v_{2i}, F_0) \leqslant d(v_{2i}, F_1)$ and $p(i) = 1$ otherwise, we see that p and $\alpha(w_{2i})$ differ in finitely many coordinates. Moreover, as different elements of P differ in infinitely many coordinates, $\alpha(w_{2i})$ is the unique element of P that differs from p in finitely many coordinates and hence $\alpha(w_{2i})$ is Borel in V. Also by the assumption on $\Delta > 0$,

$$(v_{2i+1}) \sim (w_{2i+1}),$$

and thus

$$(v_{2i+1})^{\alpha(w_{2i})} \sim (w_{2i+1})^{\alpha(w_{2i})}.$$

Now $W \in \mathbb{W}$, so

$$[w_{2i+1}] \not\sqsubseteq [(w_{2i+1})^{\alpha(w_{2i})}],$$

whence also

$$[v_{2i+1}] \not\sqsubseteq [(v_{2i+1})^{\alpha(w_{2i})}].$$

Renormalizing the blocking $(v_{2i+1})^{\alpha(w_{2i})}$, we finally find a $U = (u_i) \leqslant V$ such that $[v_i] \not\sqsubseteq [u_i]$. Thus, $\phi(V) = U$ works. □

The idea of coding with inevitable sets was originally used by J. López-Abad to give a new proof of Gowers' determinacy Theorem [35]. The import of its use here is to impose a relationship between Z and $\phi(Z)$. One is tempted to just apply Gowers' determinacy result directly to get $\phi(Z)$ from Z, but Gowers' Theorem only allows us to force $\phi(Z)$ to belong to a certain set, not such that $\phi(Z)$ stands in a certain relation to Z.

Now to finish the proof of the theorem, we can suppose that X is quasiminimal, but does not contain a minimal subspace. In that case, $(bb_{\mathbf{Q}}(X), \sqsubseteq)$ is a downwards σ-directed analytic quasiorder without a minimal element. Moreover, by replacing X with a subspace, we can suppose that this latter fact is testified by a Borel function ϕ that to each $Y \in bb_{\mathbf{Q}}(X)$ picks out a subspace of Y into which Y does not embed.

The fact that non-minimality is witnessed by a Borel function allows us now to reflect this property to a Borel quasiorder R on $bb_{\mathbf{Q}}(X)$ such that $Y \sqsubseteq Z \Rightarrow YRZ$. But, as R has no minimal element, it must have an uncountable Borel antichain, which thus also is an uncountable Borel antichain for $\sqsubseteq$.

This proves the result under (MA+¬CH), but additional work, again using Gowers' determinacy result and coding with inevitable sets, allows us to show that the property of having a minimal subspace is actually Σ^1_2 and not just its face value Σ^1_3. Similarly, having a continuum of incomparable subspaces is easily Σ^1_2, and thus the statement of the theorem is itself Σ^1_2. By Shoenfield's absoluteness theorem, the additional set-theoretical assumptions can thus be eliminated from the proof. □

Before we prove Theorem 13, we introduce some other notation. We denote by $\mathcal{G}_{\mathbf{Q}}(X)$ the set of subspaces of X spanned by elements of $bb_{\mathbf{Q}}(X)$ and by $Fin_{\mathbf{Q}}(X)$ the set of subspaces spanned by elements of $fbb_{\mathbf{Q}}(X)$. Standard notation will be used concerning successive vectors (respectively finite dimensional subspaces) on (e_i). For $L, M \in \mathcal{G}_{\mathbf{Q}}(X)$, $L \subset^* M$ means that $L = [l_i, i \in \mathbb{N}]$, where $l_i \in M$ for all but finitely many i's.

Sketch of the proof of Theorem 13: Let X be a space which is saturated with isomorphically homogeneous sequences.

By a standard use of Ramsey's Theorem and a diagonalisation, we may assume that there exists $K \geqslant 1$ such that every block-sequence in $bb_{\mathbf{Q}}(X)$ has a further block-sequence in $bb_{\mathbf{Q}}(X)$ in which is K-isomorphically homogeneous. We fix some $C > K^2$.

For L, M two block-subspaces in $\mathcal{G}_{\mathbf{Q}}(X)$, define the infinite game $G_{L,M}$ between two players as follows; for each $k \in \mathbb{N}$, m_k, n_k are integers, x_k is a vector in $\mathbb{D}_1$, y_k a vector in $\mathbb{D}$, and F_k belongs to $Fin_{\mathbf{Q}}(X)$.

1 :		$n_1 < x_1 \in L,$ m_1		$n_2 < x_2 \in L,$ m_2	$\dots$
2 :	n_1		$m_1 < F_1 \subseteq M,$ $y_1 \in F_1, n_2$		$m_2 < F_2 \subseteq M,$ $y_2 \in F_1 + F_2, n_3$ $\dots$

Player 2 wins the game $G_{L,M}$ if $(y_n)_{n\in\mathbb{N}}$ is C-equivalent to $(x_n)_{n\in\mathbb{N}}$.

We shall now provide a stabilizing subspace on which Player 2 has "sufficiently many" winning strategies in games $G_{L,M}$. The reader may look for more about this type of proof in the survey by Todorcevic [5] where it is called "combinatorial forcing". A crucial point in the definition of $G_{L,M}$ is that the moves of the players are asymptotic, in the sense that one player can force the other to play "far enough" along the basis, and this will allow the stabilisation; on the other hand, the use of finite-dimensional spaces F_k leaves enough room for Player 2 to pick vectors y_k which are not necessarily successive on the basis. There are indeed some spaces with a basis, such as T^*, where minimality cannot be proved by finding copies of the basis as successive vectors, and therefore the apparent technicality of the definition is necessary (at least in the case of the weaker hypothesis in Theorem 13).

A *state* is a couple (a, b) with $a \in fbb_{\mathbf{Q}}(X)$ and $b \in (Fin_{\mathbf{Q}}(X) \times \mathbb{D})^{<\omega}$ such that $|a| = |b|$ or $|a| = |b| + 1$. The set S of states is countable, and corresponds to the possible states of a game $G_{L,M}$ after a finite number of moves were made, restricted to elements which do affect the outcome of the game from that state (i.e m_k's and n_k's are forgotten). Thus for $s \in S$, we may define $G_{L,M}(s)$ as the game $G_{L,M}$ starting from the state s. For example, if $s = (a, b)$ with $|a| = 2, |b| = 1$, the game $G_{L,M}(s)$ will start with 1 playing some integer m_2, then 2 playing (F_2, y_2, n_3), etc.

We require a classical "stabilization lemma" used by Maurey in [37].

Lemma 21. *Let N be a countable set and let $\mu : \mathcal{G}_{\mathbf{Q}}(X) \to 2^N$ be a $(\subset^*, \subset)$-monotone map. Then there exists a stabilizing subspace $M_0 \in \mathcal{G}_{\mathbf{Q}}(X)$, i.e., such that $\mu(M) = \mu(M_0)$ for any $M \subset^* M_0$.*

Let now $\tau : \mathcal{G}_{\mathbf{Q}}(X) \to 2^S$ be defined by $s \in \tau(M)$ if there exists $L \subset M$ such that Player 2 has a winning strategy for the game $G_{L,M}(s)$. By the asymptotic nature of the game, τ is $(\subset^*, \subset)$-increasing, and therefore there exists M_0 which is stabilizing for τ. We then define a map $\rho : \mathcal{G}_{\mathbf{Q}}(X) \to 2^S$ by setting $s \in \rho(L)$ if Player 2 has a winning strategy for the game $G_{L,M_0}(s)$. Then ρ is $(\subset^*, \subset)$-decreasing, and therefore there exists a block-subspace $L_0 \in \mathcal{G}_{\mathbf{Q}}(X)$ of M_0 which is stabilizing for ρ. Finally, we check that $\rho(L_0) = \tau(L_0) = \tau(M_0)$. We may assume that $L_0 = [f_n, n \in \mathbb{N}]$, with (f_n) K-isomorphically homogeneous.

We prove that L_0 is minimal. Fix M a block subspace of L_0. We use induction to construct a subsequence $(f_{n_k})_k$ of (f_n), and a sequence $(F_k, y_k)_k$ such that for all k, $F_k \subset M, y_k \in F_1 + \cdots + F_k$,

$$s_k = ((f_{n_1}, \ldots, f_{n_k}), (y_1, \ldots, y_k, F_1, \ldots, F_k)) \in \rho(L_0).$$

Then we are done, since $(f_{n_1}, \ldots, f_{n_k}) \sim^C (y_1, \ldots, y_k)$ for all k, so $(f_{n_k})_k$ is C-equivalent to $(y_k)_k$, and M contains a CK-isomorphic copy of L_0.

Given $s_k = ((f_{n_1}, \ldots, f_{n_k}), (y_1, \ldots, y_k, F_1, \ldots, F_k)) \in \rho(L_0)$, Player 2 has a winning strategy for $G_{L_0,M_0}(s_k)$, therefore if we pick some $n_{k+1} > n_k$ large enough, the state

$$s'_k = ((f_{n_1}, \ldots, f_{n_{k+1}}), (y_1, \ldots, y_k, F_1, \ldots, F_k))$$

belongs to $\rho(L_0) = \tau(M)$, and 2 has a winning strategy for $G_{L,M}(s'_k)$ for some $L \subset M$. Therefore there exist $F_{k+1} \subset M$ and $y_{k+1} \in F_1 + \cdots + F_{k+1} \subset M$, such that

$$s_{k+1} = ((f_{n_1}, \ldots, f_{n_{k+1}}), (y_1, \ldots, y_{k+1}, F_1, \ldots, F_{k+1}))$$

belongs to $\tau(M) = \rho(L_0)$.

It therefore only remains to initiate the induction, i.e. prove that the empty state $(\emptyset, \emptyset)$ belongs to $\rho(L_0)$. To obtain this result, one refines the notion of $\cong$-homogeneity in order to imitate the notion of subsymmetry of basic sequences.

Definition 22. A block-sequence $(x_n)_{n\in\mathbb{N}}$ in X is *C-continuously isomorphically homogeneous* if there exists a continuous map $\phi : [\mathbb{N}]^{\mathbb{N}} \to \mathbb{D}^{\mathbb{N}}$ such for all $A \in [\mathbb{N}]^{\mathbb{N}}$, $\phi(A)$ is a sequence of vectors spanning $[x_n]_{n\in A}$ and is C-equivalent to $(x_n)_{n\in\mathbb{N}}$.

In this definition, the set $\mathbb{D}^{\mathbb{N}}$ is equipped with the product of the discrete topology on $\mathbb{D}$, which turns it into a Polish space. The following result is a consequence of Ellentuck's and Louveau's proofs of the infinite-dimensional Ramsey theorem.

Lemma 23. *Let $(x_n)_{n\in\mathbb{N}} \in bb_{\mathbf{Q}}(X)$ be a block-sequence which is K-isomorphically homogeneous, and let ϵ be positive. Then some subsequence of $(x_n)_{n\in\mathbb{N}}$ is $K^2 + \epsilon$-continuously isomorphically homogeneous.*

The proof of Theorem 13 ends with the final observation:

Lemma 24. *Assume $(l_n)_{n\in\mathbb{N}}$ is a block-sequence in $bb_{\mathbf{Q}}(X)$, which is C-continuously isomorphically homogeneous, and let $L = [l_n, n \in \mathbb{N}]$. Then Player 2 has a winning strategy in the game $G_{L,L}$, therefore $(\emptyset, \emptyset) \in \tau(L)$.*

Therefore some subsequence of (f_n) is $K^2 + \epsilon$-continuously isomorphically homogeneous, and if $K^2 + \epsilon < C$, spans a block-subspace L_{00} such that $(\emptyset, \emptyset) \in \tau(L_{00}) = \rho(L_0)$. □

6 On the Komorowski–Tomczak-Jaegermann side

Anisca [3] developed the techniques of Komorowski–Tomczak-Jaegermann [33] to define finite dimensional decomposition versions of the notion of local unconditional structure and extracted the following consequence.

Theorem 25 (Anisca [3]). *Let X be a separable Banach space with finite cotype and non-isomorphic to ℓ_2. Then for each $k \in \mathbb{N}$, there exists a subspace of $\ell_2(X)$ which has a $k+1$-uniform FDD but not a k-uniform FDD.*

Corollary 26. *Let X be a separable Banach space non-isomorphic to ℓ_2. Then $\ell_2(X)$ has infinitely many non-isomorphic subspaces.*

Anisca actually obtains this corollary assuming X has finite cotype. If X doesn't have finite cotype, then $\ell_2(X)$ contains, e.g., copies of the spaces $(\oplus_{n\in\mathbb{N}} l_p^n)_2$, $1 \leqslant p \leqslant +\infty$, which are easily seen to be mutually non-isomorphic for $p > 2$ (see [9], Corollary 18), and therefore the result holds as well.

Using the techniques of [19, 42] in the case of spaces with an unconditional finite dimensional decomposition, Ferenczi and Galego obtain:

Theorem 27 (Ferenczi–Galego [15]). *Let $1 \leqslant p < +\infty$. Let $X = (\oplus_{n\in\mathbb{N}} F_n)_p$, where the F_n's are finite dimensional. Then X is ergodic or $X \simeq \ell_p(X)$. The similar result holds for c_0-sums.*

The following consequence was observed in [9]. A strongly asymptotic ℓ_p FDD is the obvious generalization of a strongly asymptotic ℓ_p basis; examples are ℓ_p-sums or c_0-sums, as well as Tsirelson sums, of finite dimensional spaces.

Corollary 28 (Dilworth–Ferenczi–Kutzarova–Odell [9]). *Let $1 \leqslant p \leqslant +\infty$. Let X be a Banach space with a strongly asymptotic ℓ_p FDD. Then X is isomorphic to ℓ_2 or X contains infinitely many non-isomorphic subspaces.*

Question 29. What is the exact complexity of isomorphism between subspaces of c_0 or ℓ_p? Is ℓ_p, $p > 2$, ergodic?

Question 30. What is the exact complexity of isomorphism between subspaces of Tsirelson's space T? Between block-subspaces of T?

Recall that $E_1 \leqslant_B (bb(T), \simeq) \leqslant_B E_{K_\sigma}$; computing the exact complexity of $\simeq$ on $bb(T)$ may not be out of reach.

Question 31. What is the exact complexity of isomorphism between subspaces of Schlumprecht's space S? Between block-subspaces of S? Is S ergodic?

Schlumprecht's space is a relevant example by its minimality and the fact that E_0 is Borel reducible to permutative equivalence between its normalized block-sequences [13].

Question 32. Does there exist a space such that the complexity of isomorphism between its subspaces is exactly E_0? Is there a space with a Schauder basis such that the complexity of isomorphism between its block-subspaces is exactly E_0?

Note that the complexity of isomorphism is exactly E_0 between subspaces spanned by subsequences of an unconditional basis with the strong Casazza property (i.e., a basis of space of type (2) in Gowers' theorem).

7 Homogeneity questions

From the solution by Gowers and Komorowski–Tomczak-Jaegermann to the homogeneous Banach space problem it is easy to deduce the slightly stronger statement that a space with a Schauder basis which is isomorphic to all its subspaces spanned by Schauder bases must be isomorphic to ℓ_2. Several questions remain open in that direction:

Question 33. If a Banach space has an unconditional basis and is isomorphic to all its subspaces with an unconditional basis, must it be isomorphic to ℓ_2?

The next question was already mentioned in the introduction and concerns a stronger statement, modulo the fact that c_0 and ℓ_p, $p \neq 2$, contain a subspace with an unconditional basis which is not isomorphic to the whole space [34]:

Question 34. If a Banach space has an unconditional block-homogeneous basis, must it be isomorphic to c_0 or ℓ_p?

Recall that a theorem of Zippin states that a basis which is perfectly homogeneous, i.e., equivalent to all its normalized block-sequences, must be equivalent to the canonical basis of c_0 or ℓ_p, $1 \leqslant p < +\infty$, [47]. Bourgain, Casazza, Lindenstrauss, and Tzafriri extended this result to permutative equivalence [8]. Ferenczi and Rosendal proved that if a normalized Schauder basis is not equivalent to the canonical basis of c_0 or ℓ_p, $1 \leqslant p < +\infty$, then E_0 reduces to equivalence between its normalized block-sequences [19]. Ferenczi [13] obtained that if X has an unconditional basis, then E_0 is Borel reducible to permutative equivalence on $bb(X)$ or every normalized block-sequence has a subsequence equivalent to the unit vector basis of some fixed ℓ_p or c_0.

Some apparently weaker properties turn out to be equivalent to block homogeneity.

Theorem 35 (Ferenczi [12]). *Let Y be a Banach space and let X be a space with an unconditional basis such that every sequence of successive finite block-sequences has a subsequence whose concatenation spans a space isomorphic to Y. Then X is block-homogeneous.*

The techniques used for this result are similar to the one used for Theorem 8. They are based on Lemma 9 and the fact that isomorphism classes in $bb_{\mathbf{Q}}(X)$ verify a topological 0-1 law, i.e., they are either meagre or comeagre in $bb_{\mathbf{Q}}(X)$.

Theorem 36 (Rosendal [44], Assuming Projective Determinacy). *Let X be a Banach space which is not ℓ_1-saturated and such that every weakly null tree has a branch which spans a subspace isomorphic to X. Then X has a block-homogeneous basis.*

Here a *weakly-null tree* in X is a sequence of normalized vectors $(x_s)_{s\in\mathbb{N}^{<\mathbb{N}}}$ indexed by $\mathbb{N}^{<\mathbb{N}}$ such that for all $s \in \mathbb{N}^{<\mathbb{N}}$, the sequence $(x_{s^\frown n})_{n\in\mathbb{N}}$ is weakly-null.

Uniformity results might be necessary to answer Questions 33 or 34. This is in line with the question by Gowers whether there exists a direct proof that a homogeneous Banach space must be uniformly homogeneous.

Question 37. Let X be a space with a block-homogeneous basis. Must it be uniformly block-homogeneous?

An extremely partial answer was obtained in [12].

Theorem 38 (Ferenczi [12]). *If X has a block-homogeneous unconditional basis (e_i) and X is isomorphic to either c_0 or ℓ_p, $1 \leqslant p < +\infty$, then (e_i) is uniformly block-homogeneous.*

8 Spreading models

A fundamental notion in the geometry of Banach spaces is that of a spreading model. We recall that a normalized basic sequence (x_i) is said to generate a spreading model if for all $r_1, \ldots, r_k \in \mathbb{R}$ there is $t \in \mathbb{R}$ such that for any $\epsilon > 0$ there exists N with the following property: for any $N < l_1 < \ldots < l_k$, we have

$$\Big| \, \|r_1 x_{l_1} + \ldots + r_k x_{l_k}\| - t \, \Big| < \epsilon.$$

A more intuitive way of expressing this is by saying that

$$\lim_{l_1<\ldots<l_k,\ l_1\to\infty} \|r_1 x_{l_1} + \ldots + r_k x_{l_k}\|$$

exists for all $r_1, \ldots, r_k$. In this case, we can define a 1-subsymmetric basic sequence $(\tilde{x}_i)$ by the formula

$$\|r_1\tilde{x}_1 + \ldots + r_k\tilde{x}_k\| = \lim_{l_1<\ldots<l_k,\ l_1\to\infty} \|r_1 x_{l_1} + \ldots + r_k x_{l_k}\|,$$

and say that (x_i) generates the *spreading model* $(\tilde{x}_i)$. Though the basic sequence $(\tilde{x}_i)$ is very closely related to the space $[x_i]$ it does not necessarily have to be present there, and indeed this is one of the reasons for its interest.

Presumably, the right notion of isomorphism for spreading models is equivalence and the most natural ordering is majoration. Here a basic sequence (e_i) *majorises* a basic sequence (f_i) if there is a constant K such that for all $r_1, \ldots, r_n$

$$\|r_1 e_1 + \ldots + r_n e_n\| \leqslant K\|r_1 f_1 + \ldots + r_n f_n\|.$$

Some of the major problems about spreading models concern the possible sets of spreading models generated by basic sequences of a given space and the structure of this set of spreading models under the quasiorder of majoration. One particular question that has motivated some research, in particular [1], is the following question of Argyros.

Question 39 (Argyros [1]). Let X be a Banach space such that all spreading models in X are equivalent. Must these spreading models be equivalent to the unit vector basis of c_0 or ℓ_p for some $p \geqslant 1$?

This question is of course motivated by the fact that the spaces ℓ_p have a unique spreading model up to equivalence. For if X is a reflexive space, then any spreading model is generated by a weakly null sequence, and hence in the case of ℓ_p, $1 < p < \infty$, by a normalized block that is itself equivalent with ℓ_p. And, in ℓ_1, any spreading model is, by Rosenthal's ℓ_1-theorem, generated by either an ℓ_1 sequence or a weakly Cauchy sequence. In the case of a weakly Cauchy sequence (x_i), the difference sequence $(x_{2i+1} - x_{2i})$ is weakly null and thus generates ℓ_1 again. This argument, however, does not generalize to c_0 unless one restricts the attention to spreading models generated by weakly null sequences.

Another version of this question had previously been formulated by Rosenthal in another disguise, namely, as a question concerning characterizations of the standard basis of ℓ_p.

Question 40 (Rosenthal). Suppose (e_i) is a basic sequence such that any normalized block-sequence has a subsequence equivalent to (e_i). Is (e_i) then equivalent to the unit vector basis of some ℓ_p or c_0?

It is a fact, shown in [18], that a positive question to Argyros's question leads also to a positive answer to Rosenthal's question. The question of Rosenthal may also open a direction to answer Question 34. For example:

Question 41. If a normalized unconditional basis is block-homogeneous, does it have a block-sequence, or even a subsequence, with the property defined by Rosenthal?

Going back to the structure theory of the set of spreading models under the relation of majoration, we shall here show that under the supposition that there is no uncountable Borel antichain, one can prove quite strong structural results at least when X^* is separable. In the following, it will be assumed that all spaces in question are infinite-dimensional.

Definition 42. Let X be a separable infinite-dimensional Banach space and let $\mathcal{S}_w$ be the set of weakly-null normalized basic sequences (x_i) generating a spreading model $(\tilde{x}_i)$. For (x_i) and (y_i) in $\mathcal{S}_w$, we set $(x_i) \preccurlyeq (y_i)$ if $(\tilde{x}_i)$ is majorised by $(\tilde{y}_i)$. Similarly, we let $(x_i) \approx (y_i)$ if both $(x_i) \preccurlyeq (y_i)$ and $(y_i) \preccurlyeq (x_i)$, i.e, if $(\tilde{x}_i)$ and $(\tilde{y}_i)$ are equivalent.

We notice that $\preccurlyeq$ is a quasiorder on $\mathcal{S}_w$. It seems perhaps more natural to work directly with the set of spreading models, or even the set of spreading models up to equivalence, instead of the set of sequences generating the spreading models. However, the latter is a standard Borel space when X^* is separable, which is not necessarily the case for the former. Thus, $\mathcal{S}_w$ lends itself to the methods of descriptive set theory.

In the fundamental paper [1] by Androulakis, Odell, Schlumprecht, and Tomzcak-Jaegermann it was proved that $(\mathcal{S}_w, \preccurlyeq)$ is an upper semi-lattice, i.e., any two elements have a common least upper bound. Moreover, it was proved that any countable family has an upper bound, though not necessarily a least upper bound. This line of research was continued by B. Sari, who proved the following result.

Theorem 43 (Sari [45]). *Suppose X is a separable Banach space such that $\mathcal{S}_w$ has an infinite increasing sequence with respect to $\preccurlyeq$. Then it has an increasing ω_1-chain.*

This result puts us in a position where we are able to pursue the analysis lying behind Theorem 15, except that, as the relation of majorisation between bases is Borel and not only analytic, the analysis is completely straightforward.

Theorem 44. *Let X be a Banach space with separable dual. Then*

- *either (i) $\mathcal{S}_w/\approx$ is countable or (ii) there is a continuum size antichain in $(\mathcal{S}_w, \preccurlyeq)$,*
- *either (iii) there is a continuum size antichain and also an increasing ω_1-chain in $(\mathcal{S}_w, \preccurlyeq)$ or (iv) $(\mathcal{S}_w, \preccurlyeq)$ is inversely well-founded with a maximal element and for some ordinal $\alpha < \omega_1$ there are no decreasing α-chains.*

Proof. There are two cases. By Sari's result, either $(\mathcal{S}_w, \preccurlyeq)$ is inversely well-founded or has an increasing ω_1-chain.

In the first case, we have by the Kunen–Martin theorem that there is some countable bound on the length of decreasing sequences in $(\mathcal{S}_w, \preccurlyeq)$ and hence for some $\alpha < \omega_1$ there are no decreasing α-chains in $(\mathcal{S}_w, \preccurlyeq)$. Moreover, by the results of [1], $(\mathcal{S}_w, \preccurlyeq)$ is an upper semi-lattice and hence if inversely well-founded it must have a maximal element or otherwise one could construct an infinite increasing sequence.

In the second case, we can apply the results of Harrington, Marker, and Shelah [29, 27] on Borel quasiorderings as follows. They prove that if a Borel quasiorder on a standard Borel space has an ω_1-chain, then it also has a perfect antichain. This thus shows the dichotomy between (iii) and (iv).

Now to see the dichotomy between (i) and (ii), suppose that $(\mathcal{S}_w, \preccurlyeq)$ does not admit an uncountable Borel set of pairwise incomparable elements. Then $(\mathcal{S}_w, \preccurlyeq)$ is inversely well-founded and for some $\alpha < \omega_1$ there are no decreasing α-chains. Moreover, we again have by the results of Harrington, Marker, and Shelah that there is a partition of the space into countably many Borel sets X_n each of which is linearly ordered by $\preccurlyeq$. Therefore, each $(X_n, \preccurlyeq)$ is an inverse prewellordering of countable length and hence each $X_n/\approx$ is countable, whence also $\mathcal{S}_w/\approx$ is countable. □

We notice that P. Dodos [11] has independently arrived at the same result by essentially the same argument, though his setup is slightly different. He also notices that one can discard of the hypothesis that X^* is separable by applying a result of Rosenthal [5]. The following is a reformulation of Dodos's argument in our language:

Let (x_n) be a sequence in a Banach space X. We say that (x_n) is a *Brunel–Sucheston* sequence if

- (x_n) is Cesaro summable,
- (x_n) is a normalized basic sequence,
- for all k and $k \leqslant n_1 < \ldots < n_k$, $k \leqslant m_1 < \ldots < m_k$, we have

$$(x_{n_1}, \ldots, x_{n_k}) \sim_{1+1/k} (x_{m_1}, \ldots, x_{m_k}).$$

Dodos then notices that, by the result of Rosenthal, if X is a Banach space, then, apart from possibly ℓ_1, the spreading models generated by normalized weakly-null sequences of X are exactly those generated by Brunel–Sucheston sequences in X. Moreover, as the set of Brunel–Sucheston sequences is clearly Borel, one can just work with this set instead of $\mathcal{S}_w$.

We should mention that Theorem 44 is in response to a question of Dilworth, Odell, and Sari from [10].

References

[1] G. Androulakis, E. Odell, T. Schlumprecht, and N. Tomczak-Jaegermann, *On the structure of the spreading models of a Banach space*, Canad. J. Math. 57 (2005), pp. 673–707.

[2] G. Androulakis and T. Schlumprecht, *The Banach space S is complementably minimal and subsequentially prime*, Studia Math. 156 (2003), pp. 227–242.

[3] R. Anisca, *Unconditional decompositions in subspaces of $\ell_2(X)$*, Positivity 8 (2004), pp. 423–441.

[4] S. A. Argyros and P. Dodos, *Genericity and amalgamation of classes of Banach spaces*, Adv. Math. 209 (2007), pp. 666–748.

[5] S. A. Argyros and S. Todorcevic, *Ramsey methods in analysis*, Advanced Courses in Mathematics, CRM Barcelona. Birkhäuser Verlag, Basel, (2005).

[6] J. Bagaria and J. López-Abad, *Weakly Ramsey sets in Banach spaces*, Adv. Math. 160 (2001), pp. 133–174.

[7] B. Bossard, *A coding of separable Banach spaces. Analytic and coanalytic families of Banach spaces*, Fund. Math. 172 (2002), pp. 117–152.

[8] J. Bourgain, P. Casazza, J. Lindenstrauss, and L. Tzafriri, *Banach spaces with a unique unconditional basis, up to permutation*, Mem. Amer. Math. Soc. 54 (1985).

[9] S. Dilworth, V. Ferenczi, D. Kutzarova, and E. Odell, *On strongly asymptotic ℓ_p-spaces and minimality*, J. London Math. Soc., 2007; doi: 10.1112/jlms/jdm003.

[10] S. Dilworth, E. Odell, and B. Sari, *Lattice structures and spreading models*, Israel J. Math., to appear.

[11] P. Dodos, *On antichains of spreading models of Banach spaces*, preprint.

[12] V. Ferenczi, *Topological 0-1 laws for subspaces of a Banach space with a Schauder basis*, Illinois J. Math. 49 (2005), pp. 839–856.

[13] ______, *On the number of permutatively inequivalent basic sequences in a Banach space*, J. Funct. Anal. 238 (2006), pp. 353–373.

[14] ______, *Minimal subspaces and isomorphically homogeneous sequences in a Banach space*, Israel J. Math. 156 (2006), pp. 125–140.

[15] V. Ferenczi and E. M. Galego, *Some equivalence relations which are Borel reducible to isomorphism between separable Banach spaces*, Israel J. Math. 152 (2006), pp. 61–82.

[16] ______, *Some results about the Schroeder-Bernstein Property for separable Banach spaces*, Canad. J. Math 591 (2007), pp. 63–84.

[17] V. Ferenczi, A. Louveau, and C. Rosendal, *The complexity of classifying Banach spaces up to isomorphism*, preprint.

[18] V. Ferenczi, A. M. Pelczar, and C. Rosendal, *On a question of Haskell P. Rosenthal concerning a characterization of c_0 and l_p*, Bull. London Math. Soc. 36 (2004), pp. 396–406.

[19] V. Ferenczi and C. Rosendal, *On the number of non-isomorphic subspaces of a Banach space*, Studia Math. 168 (2005), pp. 203–216.

[20] ______, *Ergodic Banach spaces*, Adv. Math. 195 (2005), pp. 259–282.

[21] M. Foreman, *A descriptive view of ergodic theory*, Descriptive set theory and dynamical systems (Marseille-Luminy, 1996), pp. 87–171, London Math. Soc. Lecture Note Ser., 277, Cambridge Univ. Press, Cambridge, 2000.

[22] S. Gao and A. S. Kechris, *On the classification of Polish metric spaces up to isometry*, Mem. Amer. Math. Soc. 161, (2003).

[23] W.T. Gowers, *A solution to Banach's hyperplane problem*, Bul. London Math. Soc. 26 (1994), pp. 523–530.

[24] ______, *An infinite Ramsey theorem and some Banach-space dichotomies*, Ann. of Math. (2) 156 (2002), pp. 797–833.

[25] W. T. Gowers and B. Maurey, *The unconditional basic sequence problem*, J. Amer. Math. Soc. 6 (1993), pp. 851–874.

[26] ______, *Banach spaces with small spaces of operators*, Math. Ann. 307 (1997), pp. 543–568.

[27] L. Harrington, D. Marker, and S. Shelah, *Borel orderings*, Trans. Amer. Math. Soc. 310 (1988), pp. 293–302.

[28] L. A. Harrington, A. S. Kechris, and A. Louveau, *A Glimm-Effros dichotomy for Borel equivalence relations*, J. Amer. Math. Soc. 3 (1990), pp. 903–928.

[29] L. Harrington and S. Shelah, *Counting equivalence classes for co κ Souslin equivalence relations*, Logic Colloquium '80 (Prague, 1980), pp. 147–152, Stud. Logic Foundations Math., 108, North-Holland, Amsterdam New York, 1982.

[30] N. J. Kalton, *A remark on Banach spaces isomorphic to their squares*, Contemp. Math., 232, (1999), pp. 211–217.

[31] A. S. Kechris, *Classical descriptive set theory*, Graduate Texts in Mathematics, 156. Springer-Verlag, New York, 1995. xviii+402 pp.

[32] A. S. Kechris and A. Louveau, *The classification of hypersmooth Borel equivalence relations*, J. Amer. Math. Soc. 10 (1997), pp. 215–242.

[33] R. A. Komorowski and N. Tomczak-Jaegermann, *Banach spaces without local unconditional structure*, Israel J. Math. 89 (1995), pp. 205–226. *Erratum to "Banach spaces without local unconditional structure"*, Israel J. Math. 105 (1998), pp. 85–92.

[34] J. Lindenstrauss and L. Tzafriri, *Classical Banach spaces*, Springer-Verlag New York Heidelberg Berlin. 1979.

[35] J. Lopez-Abad, *Coding into Ramsey sets*, Math. Ann. 332 (2005), pp. 775–794.

[36] A. Louveau and C. Rosendal, *Complete analytic equivalence relations*, Trans. Amer. Math. Soc. 357 (2005), pp. 4839–4866.

[37] B. Maurey, *A note on Gowers' dichotomy theorem*, Convex Geometric Analysis, vol. 34, Cambridge University Press, Cambridge 1998, pp. 149–157.

[38] J. Melleray, *Computing the complexity of the relation of isometry between separable Banach spaces*, to appear in Mathematical Logic Quarterly.

[39] E. Odell and T. Schlumprecht, *The distortion problem*, Acta Math. 173 (1994), pp. 259–281.

[40] A. M. Pelczar, *Subsymmetric sequences and minimal spaces*, Proc. Amer. Math. Soc. 131 (2003), pp. 765–771.

[41] C. Rosendal, *Etude descriptive de l'isomorphisme dans la classe des espaces de Banach*, Thèse de Doctorat de l'Université Paris 6 (2003).

[42] ———, *Incomparable, non-isomorphic and minimal Banach spaces*, Fund. Math. 183 (2004), pp. 253–274.

[43] ———, *Cofinal families of Borel equivalence relations and quasiorders*, J. Symbolic Logic 70 (2005), pp. 1325–1340.

[44] ———, *Infinite asymptotic games*, preprint.

[45] B. Sari, *On Banach spaces with few spreading models*, Proc. Amer. Math. Soc. 134 (2006), pp. 1339–1345.

[46] J. H. Silver, *Counting the number of equivalence classes of Borel and coanalytic equivalence relations*, Ann. Math. Logic 18 (1980), pp. 1–28.

[47] M. Zippin, *On perfectly homogeneous bases in Banach spaces*, Israel J. Math. 4 (1966), pp. 265–272.

Author information

Valentin Ferenczi, Institut de Mathématiques de Jussieu, Projet Analyse Fonctionnelle, Université Pierre et Marie Curie – Paris 6, Boîte 186, 4, Place Jussieu, 75252, Paris Cedex 05, France.
Email: ferenczi@ccr.jussieu.fr

Christian Rosendal, Department of Mathematics, University of Illinois at Urbana-Champaign, 273 Altgeld Hall, MC 382, 1409 W. Green Street, Urbana, IL 61801, USA.
Email: rosendal@math.uiuc.edu

Banach Spaces and their Applications in Analysis, 111–123

Fréchet differentiability of Lipschitz maps and porous sets in Banach spaces

Joram Lindenstrauss, David Preiss, and Jaroslav Tišer

Abstract. This paper is an outline of the main results obtained by the authors on this subject in recent years. The proofs of many of those results are very technical and involved. The complete proofs will be presented in a research monograph the authors are preparing at present. The content of this article is closely related to talks presented by the first named author at conferences: Banach spaces and their applications in analysis, Oxford, Ohio, 2006 and New trends in Banach space theory, Caceres, Spain, 2006.

Key words. Fréchet differentiability, ϵ-Fréchet differentiability, Porous sets, mean value theorem, Γ-null sets, Γ_n-null sets.

AMS classification. 46G05, 46T20, 58C20.

1 Preliminaries

We start with the definition of porous sets and related notions.

Definition 1.1. A set E in a metric space (X, d) is said to be c-porous, $0 < c < 1$, if for every $x \in E$ and $0 < \varepsilon < c$ there is a $z \in X \setminus E$ such that $d(x, z) \leq \varepsilon$ and $B(z, cd(x, z)) \cap E = \emptyset$. A set which is c-porous for some $0 < c < 1$ is called porous.

If X is a Banach space and Y a subspace of X we can talk of a porous set in the direction of Y. A set $E \subset X$ is c-porous in the direction of Y if for every $x \in E$ and $0 < \varepsilon < c$ there is a $y \in Y$, $0 < \|y\| \leq \varepsilon$, such that $B(x + y, c\|y\|) \cap E = \emptyset$. A set which is porous in the direction of a 1-dimensional subspace of X will be called directionally porous.

A countable union of porous sets is called σ-porous. Similarly we define the notion of σ-directionally porous sets in a Banach space.

We will use in the sequel several notions of null sets in separable Banach spaces X. The most simple notion is that of a Haar null set. A Borel set $E \subset X$ is said to be *Haar null* if there is a Borel probability measure μ on X such that $\mu(x + E) = 0$ for all $x \in X$. If $\dim X < \infty$ a set is Haar null iff it is Borel and its Lebesgue measure is 0. This fact will be true also for the other notions of null sets we shall encounter in the future except for the class $\widetilde{\mathcal{A}}$ mentioned below, where it is still open at present. For more informations on Haar null sets as well as most of the notions which appear in this section, and for references to original papers we refer to the book [1].

First author: The research of J. Lindenstrauss was partly supported by a grant from the Israel Science Foundation.
Third author: The research of J. Tišer was supported by a grant GAČR 201/07/0394 and MSM 6840770010.

A map f defined on an open set in a Banach space X into a Banach space Y is said to be *Gâteaux differentiable* at a point x_0 if

$$\lim_{t\to 0} \frac{f(x_0+tu)-f(x_0)}{t} = Tu \tag{1.1}$$

exists for every $u \in X$ and T is a bounded linear operator from X to Y. The operator T is called the Gâteaux derivative and is denoted by $Df(x_0)$.

If for some fixed $u \in X$ the limit

$$f'(x_0;u) := \lim_{t\to 0} \frac{f(x_0+tu)-f(x_0)}{t}$$

exists we call it the directional derivative of f at x_0 in the direction u. Thus f is Gâteaux differentiable at x_0 if and only if $f'(x_0;u)$ exists for every u and forms a bounded linear operator as a function of u. In this case we have

$$f'(x_0;u) = Df(x_0)u.$$

A Banach space is said to have the *Radon Nikodým property*, *RNP* for short, if every Lipschitz map $f\colon \mathbb{R} \longrightarrow X$ has a derivative almost everywhere. This is known to be equivalent to the requirement that every such f has at least one point of differentiability.

It is well known that every Lipschitz map f from a separable space X into a space Y with *RNP* has many points of Gâteaux differentiability. More precisely,

Theorem 1.2. *Every Lipschitz map $f\colon X \longrightarrow Y$ with X separable and Y having the RNP is Gâteaux differentiable almost everywhere.*

This theorem has many versions which depend only on the nature of the notion "almost everywhere", i.e. on the nature of the exceptional sets. One can take as the exceptional sets Haar null sets. A stronger result is obtained by taking a smaller class of exceptional sets, namely Gauss null sets. A Borel set E in a separable Banach space X is said to be *Gauss null* if for every non degenerate (i.e. not supported on a closed hyperplane of X) Gaussian measure μ on X we have $\mu(E) = 0$. At present the strongest known result of Theorem 1.2 (i.e. the smallest class of exceptional sets) is due to Preiss and Zajíček [10]. They define a class $\widetilde{\mathcal{A}}$ of exceptional sets. The definition of this class is somewhat involved and since we are not going to use it we do not reproduce the definition here.

We now pass to the definition of the main notion considered in this paper.

Definition 1.3. A function $f\colon G \longrightarrow Y$ defined on an open set G in a separable Banach space X into a Banach space Y is said to be Fréchet differentiable at a point $x_0 \in G$ if there is a bounded linear operator $T\colon X \longrightarrow Y$ so that

$$f(x_0+u) = f(x_0) + Tu + o(\|u\|),\ u \to 0.$$

Of course, if f is Fréchet differentiable at x_0 then f is Gâteaux differentiable at x_0 and the operator T above is necessarily $Df(x_0)$. Stated otherwise: f is Fréchet differentiable at x_0 if and only if it is Gâteaux differentiable there and the limit in (1.1) exists uniformly on the unit sphere $\{u \mid \|u\| = 1\}$. The notion of Fréchet derivative is the most natural and useful notion of a derivative. If $\dim X < \infty$ and f is Lipschitz then the notions of Fréchet derivative and Gâteaux derivative coincide. This fails if $\dim X = \infty$.

To get a feeling of the difference between these two notions consider the simple example of the function $f\colon L^2[0,1] \longrightarrow L^2[0,1]$ defined by $f(u) = \sin u$. This function is everywhere Gâteaux differentiable with $Df(u) = \cos u$ (i.e. $Df(u)x = x\cos u$), but as it can be easily checked nowhere Fréchet differentiable.

It is an unfortunate fact that existence theorems for Fréchet derivatives are quite rare and usually quite hard to prove.

In the context of Fréchet differentiability we have to assume in existence results that the dual space X^* is separable, i.e. X is what is called an Asplund space. It is known, for example, that if X is separable with X^* non separable, there is an equivalent norm $\|\cdot\|$ on X so that the convex continuous function $f\colon X \longrightarrow \mathbb{R}$ defined by $f(u) = \|u\|$ is nowhere Fréchet differentiable. On the other hand if X is Asplund every convex $f\colon X \to \mathbb{R}$ is Fréchet differentiable on a dense G_δ subset of X, actually even outside a σ-porous set. We should point out in this connection that there is a continuous convex $f\colon \ell^2 \longrightarrow \mathbb{R}$ which is Fréchet differentiable only on a Gauss null set. (For these results and references to original papers we refer as before to [1].)

In view of the difficulty in proving existence results for Fréchet derivatives it is of interest to use some related weaker notions where existence proofs are more accessible (but often also not simple).

Definition 1.4. The function $f\colon X \longrightarrow Y$ is said to be ε-Fréchet differentiable at x_0, $\varepsilon > 0$, if there is a $\delta > 0$ and a bounded linear operator $T\colon X \longrightarrow Y$ so that

$$\|f(x_0 + u) - f(x_0) - Tu\| \le \varepsilon\|u\|$$

if $\|u\| \le \delta$.

If f is ε-Fréchet differentiable at some x_0 for every $\varepsilon > 0$ then clearly f is Fréchet differentiable at x_0. However, if we know only that for every $\varepsilon > 0$ there is a point x_ε at which f is ε-Fréchet differentiable then it is not clear that f is Fréchet differentiable at some point because x_ε depend on ε and in general there is no way to control how x_ε changes with ε.

Another notion which we find convenient to introduce in this paper is *asymptotic Fréchet differentiability*.

Definition 1.5. A function $f\colon X \longrightarrow Y$ is said to be asymptotically Fréchet differentiable at x_0 if there is a bounded linear operator $T\colon X \longrightarrow Y$ so that for every $\varepsilon > 0$ there is $\delta > 0$ and a subspace $Z \subset X$ of finite codimension so that

$$\|f(x_0 + z) - f(x_0) - Tz\| \le \varepsilon\|z\|$$

whenever $z \in Z$ and $\|z\| < \delta$.

We would like to point out a simple connection between the notions of differentiability and the notions of porous set introduced in the beginning of this section. If $E \subset X$ is a porous set then the Lipschitz function $f\colon X \longrightarrow \mathbb{R}$ defined by

$$f(x) := \operatorname{dist}(x, E)$$

is nowhere Fréchet differentiable on E. Indeed, since f attains its minimum on E the only possible derivative of f at a point $x \in E$ is 0. However, by the definition of c-porosity there are points z in X arbitrarily close to x with

$$\|f(z) - f(x)\| = \operatorname{dist}(z, E) \geq c\|z - x\|.$$

Similarly if E is directionally porous then $\operatorname{dist}(x, E)$ is not Gâteaux differentiable at any point of E and thus E is negligible in any sense in which Theorem 1.2 is valid.

Our results below depend on some smoothness assumptions. Let us recall here the basic notions. A Banach space X is uniformly smooth if

$$\rho_X(t) := \sup_{\|x\|=1} \sup_{\|y\|\leq t} \big(\|x + y\| + \|x - y\| - 2\big) = o(t),\ t \searrow 0.$$

For example, let $X = L^p$ and $t \in [0, 1]$. If $1 < p \leq 2$ then $\rho_{L^p}(t) \leq C_p t^p$. If $p = 2$ then $\rho_{L^2}(t) = 2t^2$ and for $2 < p < \infty$ we have $\rho_{L^p}(t) \leq C_p t^2$ and none of these estimates can be improved.

A variant of this notion which was introduced in [3] is asymptotic uniform smoothness. The modulus of asymptotic uniform smoothness of a space X is defined by

$$\bar{\rho}_X(t) := \sup_{\|x\|=1} \inf_{\dim X/Y<\infty} \sup_{\substack{y\in Y\\ \|y\|\leq t}} \big(\|x + y\| - 1\big).$$

Clearly a uniformly smooth space is asymptotically uniform smooth but the converse is false since e.g. for $X = c_0$ the modulus $\bar{\rho}_X(t) \equiv 0$. There are duals to these notions (uniform convexity and asymptotic uniform convexity). We shall not need these notions so we do not define them here. On the other hand we shall need the extension of these notions for general convex functions $\Theta\colon X \longrightarrow \mathbb{R}$. The modulus of smoothness of Θ at the point $x \in X$ is defined by

$$\rho_\Theta(x, t) := \sup_{\|y\|\leq t} \big(\Theta(x + y) + \Theta(x - y) - 2\Theta(x)\big).$$

The modulus of asymptotic smoothness of Θ at the point x is defined by

$$\bar{\rho}_\Theta(x, t) := \inf_{\substack{Y\subset X\\ \dim X/Y<\infty}} \sup_{\substack{y\in Y\\ \|y\|\leq t}} \big(\Theta(x + y) + \Theta(x - y) - 2\Theta(x)\big).$$

Up to a constant factor this extends the previous notions defined for norms, for example $\bar{\rho}_X(t) \leq \sup_{\|x\|=1} \bar{\rho}_{\|\cdot\|}(x, t) \leq 2\bar{\rho}_X(t)$.

An important existence theorem for points of Fréchet differentiability is known for scalar valued Lipschitz functions.

Theorem 1.6. *Let X be an Asplund space and $f\colon X \longrightarrow \mathbb{R}$ a Lipschitz function. Then f is Fréchet differentiable on a dense set in X.*

This theorem is proved in [8]. A somewhat simplified proof (but not simple either) is presented in [6]. This is not an almost everywhere result. The proof in both papers is achieved by a (complicated) iterative construction of a sequence of points x_n which are shown to converge to a point x of Fréchet differentiability. Because the nature of the set of points of Fréchet differentiability is not known we do not know for example if two Lipschitz functions from X to $\mathbb{R}$ have a common point of Fréchet differentiability. Stated otherwise, it is not known if a Lipschitz map $f\colon X \longrightarrow \mathbb{R}^2$ must have a point of Fréchet differentiability. This is the source of difficultly for the arguments in Section 5 where the difficulty is overcome in the most important case, namely for X being a Hilbert space.

Both proofs of Theorem 1.6 yield also a mean value theorem for Fréchet derivatives.

Theorem 1.7. *Let f be a Lipschitz function on an open set G in an Asplund space X into the real line. Then for every pair of points $u, v \in G$ such that the line segment connecting them is in G we have*

$$\begin{aligned}\inf\{&Df(x)(v-u) \mid x \in G,\ f \text{ is Fréchet differentiable at } x\}\\ &\le f(v) - f(u)\\ &\le \sup\{Df(x)(v-u) \mid x \in G,\ f \text{ is Fréchet differentiable at } x\}.\end{aligned}$$

The mean value theorem will have an important role in the coming sections.

In the paper [5] an investigation was carried out on the existence of Fréchet derivatives for functions with a more general range space. In this connection a new class of null sets in Banach spaces was defined as a notion combining measure and category.

Let $T = [0,1]^{\mathbb{N}}$ be endowed with the product topology and the product Lebesgue measure μ. Let $\Gamma(X)$ be the space of all continuous mappings $\gamma\colon T \longrightarrow X$ having continuous partial derivatives $\frac{\partial\gamma}{\partial t_i}$ with respect to all the variables (with one-sided derivatives at the points where the i-th coordinate is 0 or 1). The elements of $\Gamma(X)$ will be called (∞-dimensional) surfaces in X. We equip $\Gamma(X)$ with the topology generated by the semi-norms

$$\|\gamma\|_0 = \sup_{t\in T} \|\gamma(t)\|,\ \|\gamma\|_i = \sup_{t\in T} \left\|\frac{\partial\gamma}{\partial t_i}(t)\right\|,\ 1 \le i < \infty.$$

The space $\Gamma(X)$ with this topology is a Fréchet space (in particular it is a complete separable metric space).

A Borel set E in X will be called Γ-*null* if

$$\mu\{t \in T \mid \gamma(t) \in E\} = 0$$

for residually many $\gamma \in \Gamma(X)$.

This class can be easily shown to have the requirements we put on null sets. In particular, Theorem 1.2 is valid Γ almost everywhere (actually any set belonging to the class $\widetilde{\mathcal{A}}$ in sense of [10] is Γ null).

For the statement of the main result of [5] we need also the notion of a regular point of a function. Let f be a map from an open set in X to a Banach space Y. We say that $x \in X$ is a *regular point* of f if for every $v \in X$ for which $f'(x;v)$ exists

$$\lim_{t \to 0} \frac{f(x+tu+tv) - f(x+tu)}{t} = f'(x;v)$$

uniformly for $\|u\| \leq 1$.

We can now state the main result of [5].

Theorem 1.8. *Suppose that G is an open set in a separable Banach space X. Let $\mathcal{L}$ be a norm separable subspace in the space of all bounded linear operators $L(X,Y)$. Let $f\colon G \longrightarrow Y$ be a Lipschitz function. Then f is Fréchet differentiable at Γ almost every point $x \in X$ at which it is regular, Gâteaux differentiable and $Df(x) \in \mathcal{L}$.*

It is not hard to show that for convex continuous $f\colon X \longrightarrow \mathbb{R}$ every point x is regular. For a general Lipschitz function $f\colon X \longrightarrow Y$ the set of irregular points of f is σ-porous. As a consequence we have

Corollary 1.9. *If X is an Asplund space then every convex continuous $f\colon X \longrightarrow \mathbb{R}$ is Γ almost everywhere Fréchet differentiable.*

Comparing this result with the example of a convex function $f\colon \ell^2 \longrightarrow \mathbb{R}$ which was mentioned above (being Fréchet differentiable only on a Gauss null set) we get that Γ null sets may be "orthogonal" to Gauss null sets in the sense that the subset E of ℓ^2 on which f is Fréchet differentiable is Gauss null while $\ell^2 \setminus E$ is Γ null.

Corollary 1.10. *Let X be an Asplund space. Then every Lipschitz function from X to $\mathbb{R}$ is Γ almost everywhere Fréchet differentiable if and only if every porous set in X is Γ null.*

The "only if" part follows from the observation made above concerning the function $f(x) = \operatorname{dist}(x,E)$ with E porous. It was proved in [5] that the class of spaces X in which every porous set is Γ null includes $C(K)$ spaces with K scattered, subspaces of c_0 and even some reflexive infinite dimensional Banach spaces (but not ℓ^p for $1 < p < \infty$, in particular not the Hilbert space). For spaces satisfying the condition in Corollary 1.10 we can restate Theorem 1.8 as follows

Theorem 1.11. *Let G be an open set in a separable Banach space X in which every porous set is Γ null, $\mathcal{L}$ a norm separable subspace of $L(X,Y)$ and $f\colon G \longrightarrow Y$ a Lipschitz map. Then f is Fréchet differentiable at Γ almost every point in which f is Gâteaux differentiable and $Df(x) \in \mathcal{L}$.*

There is also a formulation of Theorem 1.8 connected to the mean value theorem. This formulation involves slices of the set of Gâteaux derivatives. Let Υ be a subset of $L(X,Y)$. By a *slice* S of Υ we mean a set of the form

$$S(\Upsilon, v_1, \dots, v_m, y_1^*, \dots, y_m^*, \delta) = \Big\{ T \in \Upsilon \;\Big|\; \sum_{i=1}^{m} y_i^*(Tv_i) > \alpha - \delta \Big\},$$

where $m \in \mathbb{N}$, $v_1, \ldots, v_m \in X$, $y_1^*, \ldots y_m^* \in Y^*$, $\delta > 0$ and $\alpha = \sup_{T \in \Upsilon} \sum_{i=1}^m y_i^*(Tv_i)$. The following is proved in [5].

Theorem 1.12. *Let $f: G \longrightarrow Y$ be a Lipschitz map which is Fréchet differentiable at Γ almost every point of an open set G of X. Then for every slice S of the set of Gâteaux derivatives of f the set of points x at which f is Fréchet differentiable and $Df(x) \in S$ is not Γ null.*

To clarify that this is a mean value theorem note that if the range is 1-dimensional then slices are the sets of the form $\{T \in \Upsilon \mid Tx > \alpha - \delta\}$ where $\alpha = \sup\{Tx \mid T \in \Upsilon\}$.

As already pointed out, porous sets are not necessarily Γ-null in spaces such as ℓ_p, $1 < p < \infty$. In such spaces, before the work announced here, there were no results asserting Fréchet differentiability of Lipschitz maps into even two dimensional spaces. However, existence of points of ε-Fréchet differentiability of Lipschitz maps from super-reflexive spaces into finite dimensional spaces was shown in [4] and this result was extended to asymptotically uniformly smooth spaces in [3].

Theorem 1.13. *Let $f: X \longrightarrow Y$ be a Lipschitz map, where X is asymptotically uniformly smooth and Y is finite dimensional. Then for every $\varepsilon > 0$ there is $x \in X$ at which f is both Gâteaux differentiable and ε-Fréchet differentiable.*

In the coming sections we shall announce some of the main results of our recent investigation.

2 Γ_n-null sets and a variational principle

We first introduce the notions of Γ_n-null sets, $n \geq 1$. These are similar to the notion of Γ-null sets from [5] mentioned above in which the infinite dimensional surfaces in X are replaced by the n-dimensional ones.

Let $T_n = [0,1]^n$ be endowed with the product topology and product Lebesgue measure μ_n. We denote by $\Gamma_n(X)$ the space of n-dimensional surfaces in X, i.e., functions $\gamma : [0,1]^n \to X$ which are continuous and continuously differentiable (except for the boundary points where we require only the appropriate one sided derivatives). We norm $\Gamma_n(X)$ by the norm

$$\sup_{t \in T_n} |\gamma(t)| + \sum_{j=1}^n \sup_{t \in T_n} \left| \frac{\partial \gamma}{\partial t_j}(t) \right|.$$

With this norm $\Gamma_n(X)$ becomes a Banach space. A Borel set E in X will be called Γ_n-null if

$$\mu_n\{t \in T_n \mid \gamma(t) \in E\} = 0$$

for residually many $\gamma \in \Gamma_n(X)$.

One dimensional surfaces are called, as usual, curves.

There is a natural connection between Γ_n-null sets and Γ-null sets as the following simple proposition shows.

Proposition 2.1. *If a G_δ set E in X is Γ_n-null for infinitely many values of n then it is Γ-null.*

Our aim is to show that under suitable assumptions, given a porous set $E \subset X$ we can modify a surface $\gamma \in \Gamma_n(X)$ a little so that it cuts E only by a set of measure zero. A natural way of doing this is to construct a sequence $(\gamma_j)_{j=1}^\infty$ so that the measure of $\{t \in T_n \mid \gamma_j(t) \in E\}$ decreases to zero while keeping control of some parameter ("energy") associated with γ_j (e.g., in the case of curves we may use their length) and in the limit to get a surface $\tilde{\gamma}$ which intersect E in a set of measure zero and, thanks to the control of the "energy," is close to γ.

Instead of working with sequences, we found it useful to work with a variational principle. The main difference between this principle and the previous ones is that we need to work in an incomplete metric space (the set of surfaces that meet a given G_δ set in a set of large measure) and we need to minimize a function f that is not lower semi-continuous. Fortunately, this set S of surfaces is a G_δ subset of $\Gamma_n(X)$ even if $\Gamma_n(X)$ is equipped just with the norm $\sup_{t\in T_n} |\gamma(t)|$. So in S, Cauchy sequences satisfying a mild additional assumption converge. A similar observation handles the problem of semi-continuity of f.

Proposition 2.2. *Suppose that X is a set equipped with two metrics $d_0 \leq d$, X is d-complete, $S \subset X$ is G_δ in (X, d_0) and $f : S \to \mathbb{R}$ is d_0-upper semi-continuous. Then there are functions $\delta_j(x_1, \dots, x_j) : S^j \to (0, \infty)$ such that every d-Cauchy sequence $(x_j)_{j=1}^\infty$ in S with*

$$d_0(x_j, x_{j+1}) \leq \delta_j(x_1, \dots, x_j)$$

converges in the metric d to some $x \in S$ and $f(x) \leq \liminf_{j\to\infty} f(x_j)$.

This leads us to a variational principle for spaces equipped with two metrics, which contains the variational principle from [2] as a special case (when $d_0 = 0$).

Theorem 2.3. *Let S be a set equipped with two metrics $d_0 \leq d$ and let $f : S \to \mathbb{R}$ be a function bounded from below. Suppose further that there are functions $\delta_j(x_1, \dots, x_j) : S^j \to (0, \infty)$ with the following property: every d-Cauchy sequence $(x_j)_{j=1}^\infty$ in S such that*

$$d_0(x_j, x_{j+1}) \leq \delta_j(x_1, \dots, x_j),$$

converges in the d metric to some $x \in S$ and $f(x) \leq \liminf_{j\to\infty} f(x_j)$.

Let $F : S \times S \to [0, \infty]$ be d lower semi-continuous in the second variable with $F(x, x) = 0$ for all $x \in S$ and $\inf_{d(x,y)>s} \max(F(x, y), d_0(x, y)) > 0$ for each $s > 0$.

Then for every sequence of positive numbers $(\lambda_j)_{j=1}^\infty$ there is a sequence $(x_j)_{j=1}^\infty$ in S such that for some d_0 continuous $\varphi : S \to [0, \infty)$, the function

$$h(x) := f(x) + \varphi(x) + \sum_{j=1}^\infty \lambda_j F(x_j, x)$$

attains its minimum on S.

Using a somewhat more detailed version of this principle we get the following results.

Theorem 2.4. *Suppose that X admits a convex function Θ which is smooth with modulus of smoothness $o(t^n \log^{n-1}(1/t))$ (as $t \searrow 0$) at every point and satisfies $\Theta(0) = 0$ and $\inf_{\|x\|>s} \Theta(x) > 0$ for every $s > 0$. Then every porous set E in X is Γ_n-null.*

Theorem 2.5. *In an Asplund space every porous set is Γ_1-null.*

Theorem 2.6. *Every σ-directionally porous subset of any Banach space is Γ_1-null as well as Γ_2-null.*

Theorem 2.7. *Let X be a separable Banach space with*

$$\bar{\rho}_X(t) = o(t^n \log^{n-1}(1/t)) \text{ as } t \to 0.$$

Then every porous set in X is contained in a union of a σ-directionally porous set and a Γ_n-null G_δ set.

By combining these results with Proposition 2.1 we get

Theorem 2.8. *Every porous set in a separable Banach space X with $\bar{\rho}_X(t) = o(t^n)$ for each n is Γ-null.*

This gives a somewhat more general class of examples of spaces for which Theorem 1.11 holds.

3 Criteria of ε differentiability

To state the main result of this section, we use the following notation for slices of the set of Gâteaux derivatives of $\mathbb{R}^n$ valued Lipschitz maps. For a Lipschitz map $f = (f_1, \ldots, f_n)$ of a non-empty open set $G \subset X$ of a separable Banach space X into $\mathbb{R}^n$ we denote by $D(f)$ the set of points in G at which f is Gâteaux differentiable. Then slices of the set of Gâteaux derivatives of f are sets of the form

$$S(f, (u_i)_{i=1}^n, \delta) = \Big\{ Df(x) \;\Big|\; x \in D(f),\ \sum_{i=1}^n f_i'(x; u_i) > \alpha - \delta \Big\},$$

where $(u_i)_{i=1}^n$ are points in X and

$$\alpha = \sup_{x \in D(f)} \sum_{i=1}^n f_i'(x; u_i).$$

To understand the connection between smallness of porous sets and differentiability, we recall that Corollary 1.10 says that once we know that porous sets in X are Γ-null,

Fréchet differentiability of Lipschitz maps into finite dimensional spaces follows. It is therefore natural to hope that in spaces in which porous sets are Γ_n-null, Fréchet differentiability of Lipschitz maps into at least spaces of small finite dimension can be shown. In view of the mean value estimates, we may even hope that in these spaces, all slices of the set of Gâteaux derivatives of any Lipschitz map into a space with dimension not exceeding n contain Fréchet derivatives. (See next section for examples showing that in this context the dimension of the target space cannot exceed n.)

These statements remain open, but we were able to prove them after replacing Fréchet derivatives with ε-Fréchet derivatives.

Theorem 3.1. *Suppose that the Asplund space X has the property that every porous set in X can be covered by a union of a Haar null set and a Γ_n-null G_δ set. Let f be a Lipschitz map from a non-empty open set $G \subset X$ to a Banach space of dimension not exceeding n. Then for every $\varepsilon, \delta > 0$ and $(u_i)_{i=1}^n \in X$, the slice $S(f; u_1, \dots, u_n; \delta)$ contains a point of ε-Fréchet differentiability of f.*

4 Examples of big porous sets and not too often Fréchet differentiable maps

In this section we announce the existence of big porous sets and Lipschitz maps into finite dimensional spaces which have no ε-Fréchet derivatives in some slices of their sets of Gâteaux derivatives. This shows that the results of Sections 2 and 3 are quite sharp. We use here the word "quite" to indicate that we need to assume more than the optimal assumption that our space fails to be asymptotically uniformly smooth with modulus of smoothness $o(t^n \log^{n-1}(1/t))$. Firstly, we naturally strengthen this assumption to the requirement that the space be asymptotically uniformly convex with modulus $t^n \log^{n-1}(1/t)$. Secondly, we replace the asymptotic uniform convexity of the space with the asymptotic uniform smoothness of its dual with the dual modulus $t^{n/(n-1)}/\log(1/t)$ (which may be a stronger requirement when the space is not reflexive). Finally, we require that the dual be asymptotically uniformly smooth with a slightly better modulus, $o(t^{n/(n-1)}/\log^{\beta}(1/t))$, $\beta > 1$ for examples concerning porosity and $o(t^q)$, $q > n/(n-1)$ for examples concerning ε-differentiability.

The starting point of these examples is the following observation.

Example 4.1. In ℓ_1 there is a σ-porous set whose complement meets every curve in a set of 1-dimensional Hausdorff measure zero. In particular, this set has Γ_1-null complement and is not Haar null.

Its construction is based on a connection between size of porous sets on curves and existence of functions that are close to being non-differentiable at points where some directional derivative is positive.

Example 4.2. For every $\varepsilon > 0$ there is a porous set E in ℓ_1 and a function $f : \ell_1 \to [0, \varepsilon]$ with $\mathrm{Lip}(f) \le 1$ such that f is increasing in the first coordinate direction e_1 and $f'(x; e_1) = 1$ whenever $x \notin E$.

It follows that for suitable $\eta > 0$, E has small measure on every curve γ with $\|\gamma'(t) - e_1\| < \eta$, from which it is easy to finish the argument showing Example 4.1. Also, it would be easy to use Example 4.2 to find a Lipschitz function $f : \ell_1 \to \mathbb{R}$ such that f is increasing but not constant in the direction e_1 but, however small $\varepsilon > 0$ may be, there is no point of ε-Fréchet differentiability with $f'(x; e_1) > \varepsilon$. Of course, since ℓ_1 is not Asplund, such an example may be found in an easier way by starting from a nowhere ε-Fréchet differentiable function. However, an analogous argument for maps into $\mathbb{R}^n$ gives similar, but considerably more interesting, examples in, e.g., ℓ_p spaces for $1 < p < \infty$. (See Theorem 4.5.)

Returning to Example 4.1 we remark that the method indicated above can be used to show that any space containing ℓ_1 has a σ-porous subset with Γ_1-null complement. By a very different approach, namely by a surprising use of the theory of Lipschitz quotients, this result was proved in [7]. In spite of these results, the question whether the converse to Theorem 2.5 holds, i.e, whether in any non-Asplund space there is a σ-porous subset with Γ_1-null complement, remains open.

Let w be an Orlicz function and use in the Orlicz space associated to w the norm

$$\|\{c_i\}_{i=1}^{\infty}\|_w = \inf\{\lambda > 0 \mid \sum_{i=1}^{\infty} w(|c_i|/\lambda) \le 1\}.$$

A sequence of elements $\{x_i\}_{i=1}^{\infty}$ in a Banach space X is said to satisfy the upper w estimate if there is a constant A so that for every sequence $\{c_i\}_{i=1}^{\infty}$ of real numbers and $k \in N$

$$\|\sum_{i=1}^{k} c_i x_i\| \le A\|\{c_i\}_{i=1}^{k}\|_w.$$

In case $w(t) \approx t^q$ as $t \to 0$ we speak about the upper q estimate.

The following is a rather simple proposition.

Proposition 4.3. *A Banach space which is asymptotically uniformly smooth with modulus w contains a normalized sequence $\{x_i\}_{i=1}^{\infty}$ which satisfies the upper w estimate.*

Note that the converse to Proposition 4.1 is false since $C[0,1]$ contains for any w a normalized sequence satisfying the upper w estimate while it is not isomorphic to an asymptotically smooth space (e.g. since it is not an Asplund space).

The following are the main results in this section.

Theorem 4.4. *Let X be a separable Banach space and $n > 1$. Suppose that for some $\beta > 1$, X^* contains a normalized sequence satisfying the upper $t^{n/(n-1)}/\log^{\beta}(1/t)$ estimate. Then X contains a σ-porous set whose complement meets every n-dimensional surface in a set of n-dimensional Hausdorff measure zero.*

In particular, the σ-porous set from this Theorem has Γ_n-null complement and is neither Haar null nor Γ_n null.

The construction of such a σ-porous set resembles a construction in [9]. In the same paper there is also a much simpler construction in every separable infinite dimensional space of a σ-porous set whose complement has measure zero on every line (see also [1], Chapter 6). Such a set obviously cannot be Haar null.

Theorem 4.5. *Let X be a separable Banach space and let $n > 1$. Suppose that for some $q > n/(n-1)$, X^* contains a normalized sequence satisfying the upper q estimate. Then there is a Lipschitz map $f : X \to \mathbb{R}^n$ and a continuous linear functional y^* on $L(X, \mathbb{R}^n)$ so that $y^*(Df(x)) \geq 0$ whenever f is Gâteaux differentiable at x and for every small $\varepsilon > 0$ the set*

$$\{Df(x) \mid f \text{ is Gâteaux differentiable at } x \text{ with } y^*(Df(x)) > \varepsilon\}$$

is a nonempty slice of the set of all Gâteaux derivatives of f which contains no ε-Fréchet derivative of f.

5 Asymptotic Fréchet differentiability for Lipschitz maps into $\mathbb{R}^n$

The main result in this section actually is concerned with proper Fréchet differentiability.

Theorem 5.1. *Assume that the space X admits a continuous convex function Θ such that for some $n \geq 1$*

(i) $\inf\{\Theta(x); \|x\| > s\} > \Theta(0)$ *for all $s > 0$,*

(ii) $\sup_{x \in X} \rho_\Theta(x,t) = O(t^n \log^{n-1}(1/t))$ *as $t \searrow 0$,*

(iii) for every $x \in X$, $\rho_\Theta(x,t) = o(t^n \log^{n-1}(1/t))$ as $t \searrow 0$.

Let f be a Lipschitz map from a nonempty open set G of X to a space of dimension not exceeding n. Then every slice of the set of Gâteaux derivatives of f contains a Fréchet derivative.

For $n = 1$, this theorem is just a restatement of Theorems 1.7. Condition (iii) above can only hold for $n \leq 2$ since affine maps are the only convex functions which satisfy $\rho_\Theta(x,t) = o(t^2)$ at every $x \in X$ even if $X = \mathbb{R}$ (clearly affine maps fail to satisfy (i)). The proof of Theorem 5.1 resembles the proof in [8]. It involves a delicate iterative construction of a sequence $\{x_n\}_{n=1}^\infty$ in X which is shown to converge to a point x at which f is Fréchet differentiable.

The main special case of Theorem 5.1 is

Corollary 5.2. *Every pair (f,g) of real valued Lipschitz functions defined on an open set G of a Hilbert space possesses a common point of Fréchet differentiability. Moreover, for any $u, v \in X$ and $c \in \mathbb{R}$ the existence of a common point x of Gâteaux differentiability of f and g such that $f'(x,u) + g'(x,v) > c$ implies the existence of such a point of Fréchet differentiability.*

Using a similar argument to the proof of Theorem 5.1 we obtain a result which is meaningful for every finite n.

Theorem 5.3. *Assume that the space X admits a continuous convex function Θ such that for some $n \geq 1$*

(i) $\inf\{\Theta(x); \|x\| > s\} > \Theta(0)$ *for all $s > 0$,*

(ii) $\sup_{x \in X} \overline{\rho}_\Theta(x,t) = O(t^n \log^n(1/t))$ *as $t \searrow 0$,*

(iii) For every $x \in X$, $\overline{\rho}_\Theta(x,t) = o(t^n \log^{n-1}(1/t))$ as $t \searrow 0$.

Then every Lipschitz map f from a nonempty open set G in X to a space of dimension not exceeding n is asymptotically Fréchet differentiable at some points in G.

References

[1] Y. Benyamini and J. Lindenstrauss, *Geometric nonlinear functional analysis*, Vol. 1, Colloquium Publications Vol. 48, Amer. Math. Soc., 2000.

[2] J. M. Borwein and D. Preiss, *A smooth variational principle with applications to subdifferentiability and to differentiability of convex functions*, Trans. Amer. Math. Soc. 303 (1987), pp. 517–528.

[3] W. B. Johnson, J. Lindenstrauss, D. Preiss, and G. Schechtman, *Almost Fréchet differentiability of Lipschitz mappings between infinite dimensional Banach spaces*, Proc. London Math. Soc. (3) 84 (2002), pp. 711–746.

[4] J. Lindenstrauss and D. Preiss, *Almost Fréchet differentiability of finitely many Lipschitz functions*, Mathematika 86 (1996), pp. 393–412.

[5] ———, *On Fréchet differentiability of Lipschitz maps between Banach spaces*, Ann. of Math. 157 (2003), pp. 257–288.

[6] ———, *A new proof of Fréchet differentiability of Lipschitz functions*, J. Eur. Math. Soc. 2 (2000), pp. 199–216.

[7] O. Maleva, *Unavoidable sigma-porous sets*, J. London Math. Soc., to appear.

[8] D. Preiss, *Fréchet differentiability of Lipschitz functions*, J. Funct. Anal. 91 (1990), pp. 312–345.

[9] D. Preiss and J. Tišer, *Two unexpected examples concerning differentiability of Lipschitz functions on Banach spaces*, GAFA Israel Seminar 92-94, Birkhäuser 1995, pp. 219–238.

[10] D. Preiss and L. Zajíček, *Directional derivation of Lipschitz functions*, Israel J. Math. 125 (2001), pp. 1–27.

Author information

Joram Lindenstrauss, Hebrew University, Jerusalem, 91904, Israel.
Email: `joram@math.huji.ac.il`

David Preiss, University of Warwick, Coventry CV4 7AL, UK.
Email: `D.Preiss@warwick.ac.uk`

Jaroslav Tišer, Czech Technical University, 166 27 Prague, Czech Republic.
Email: `tiser@math.feld.cvut.cz`

Banach Spaces and their Applications in Analysis, 125–146

Conditions for stochastic integrability in UMD Banach spaces

Jan van Neerven, Mark Veraar, and Lutz Weis

Abstract. A detailed theory of stochastic integration in UMD Banach spaces has been developed recently by the authors. The present paper is aimed at giving various sufficient conditions for stochastic integrability.

Key words. Stochastic integration in UMD Banach spaces, cylindrical Brownian motion, approximation with elementary processes, γ-radonifying operators, vector-valued Besov spaces.

AMS classification. Primary 60H05; Secondary 46B09.

1 Introduction

In the paper [14] we developed a detailed theory of stochastic integration in UMD Banach spaces and a number of necessary and sufficient conditions for stochastic integrability of processes with values in a UMD space were obtained. The purpose of the present paper is to complement these results by giving further conditions for stochastic integrability.

In Section 2, we prove a result announced in [14] on the strong approximation of stochastically integrable processes by elementary adapted processes. In Section 3 we prove two domination results. In Section 4 we state a criterium for stochastic integrability in terms of the smoothness of the trajectories of the process. This criterium is based on a recent embedding result due to Kalton and the authors [9]. In Section 5 we give an alternative proof of a special case of the embedding result from [9] and we prove a converse result which was left open there. In the final Section 6 we give square function conditions for stochastic integrability of processes with values in a Banach function spaces. We follow the notations and terminology of the paper [14].

2 Approximation

Throughout this note, $(\Omega, \mathcal{F}, \mathbb{P})$ is a probability space endowed with a filtration $\mathbb{F} = (\mathcal{F}_t)_{t\in[0,T]}$ satisfying the usual conditions, H is a separable real Hilbert space with

First author: Supported by a 'VIDI subsidie' (639.032.201) in the 'Vernieuwingsimpuls' programme of the Netherlands Organization for Scientific Research (NWO). Also supported by the Research Training Network "Evolution Equations for Deterministic and Stochastic Systems" (HPRN-CT-2002-00281).
Second author: See first named author. Also supported by the Marie Curie Fellowship Program for a stay at the TU Karlsruhe and supported by the Research Training Network "Phenomena in high dimensions" (MRTN-CT-2004-511953).
Third author: Supported by grants from the Volkswagenstiftung (I/78593) and the Deutsche Forschungsgemeinschaft (We 2847/1-2).

inner product $[\cdot,\cdot]_H$, and E is a real Banach space with norm $\|\cdot\|$.

We call an operator-valued stochastic process $\Phi : [0,T]\times\Omega \to \mathcal{L}(H,E)$ *elementary adapted* with respect to the filtration $\mathbb{F}$ if it is of the form

$$\Phi = \sum_{n=1}^{N}\sum_{m=1}^{M} \mathbf{1}_{(t_{n-1},t_n]\times A_{mn}} \sum_{k=1}^{K} h_k \otimes x_{kmn},$$

where $0 \le t_0 < \cdots < t_N \le T$ with the convention that $(t_{-1},t_0] = \{0\}$, the sets $A_{1n},\dots,A_{Mn} \in \mathcal{F}_{t_{n-1}}$ are disjoint for all $n = 1,\dots,N$, and the vectors $h_1,\dots,h_K \in H$ are orthonormal.

Let $W_H = (W_H(t))_{t\in[0,T]}$ be a *H-cylindrical Brownian motion*, i.e., each $W_H(t)$ is a bounded linear operator from H to $L^2(\Omega)$, for all $h \in H$ the process $W_H h = (W_H(t)h)_{t\in[0,T]}$ is a Brownian motion, and for all $t_1,t_2 \in [0,T]$ and $h_1,h_2 \in H$ we have

$$\mathbb{E}(W_H(t_1)h_1\, W_H(t_2)h_2) = (t_1 \wedge t_2)[h_1,h_2]_H.$$

We will always assume that W_H is adapted to $\mathbb{F}$, i.e., each Brownian motion $W_H h$ is adapted to $\mathbb{F}$. The stochastic integral of an elementary adapted process Φ of the above form with respect to W_H is defined in the obvious way as

$$\int_0^t \Phi\, dW_H = \sum_{n=1}^{N}\sum_{m=1}^{M} \mathbf{1}_{A_{mn}} \sum_{k=1}^{K} \big(W_H(t_n\wedge t)h_k - W_H(t_{n-1}\wedge t)h_k\big) \otimes x_{kmn}.$$

A process $\Phi : [0,T]\times\Omega \to \mathcal{L}(H,E)$ is called *H-strongly measurable* if for all $h\in H$, Φh is strongly measurable. Similarly, Φ is *H-strongly adapted* if for all $h \in H$, Φh is strongly adapted.

A H strongly measurable and adapted process $\Phi : [0,T]\times\Omega \to \mathcal{L}(H,E)$ is called *stochastically integrable* with respect to W_H if there exists a sequence of elementary adapted processes $\Phi_n : [0,T]\times\Omega \to \mathcal{L}(H,E)$ and a $\zeta : \Omega \to C([0,T];E)$ such that

(i) $\lim_{n\to\infty} \langle \Phi_n h, x^*\rangle = \langle \Phi h, x^*\rangle$ in measure for all $h\in H$ and $x^* \in E^*$;

(ii) $\lim_{n\to\infty} \int_0^{\cdot} \Phi_n\, dW_H = \zeta$ in measure in $C([0,T];E)$.

The process ζ is uniquely determined almost surely. We call ζ the *stochastic integral* of Φ, notation:

$$\zeta =: \int_0^{\cdot} \Phi\, dW_H.$$

It is an easy consequence of (i), (ii), and [8, Proposition 17.6] that if Φ is stochastically integrable, then for all $x^* \in E^*$ we have

$$\lim_{n\to\infty} \Phi_n^* x^* = \Phi^* x^* \ \text{ in } L^2(0,T;H) \text{ almost surely.}$$

For UMD spaces E we show that in the definition of stochastic integrability it is possible to strengthen the convergence of the processes $\Phi_n h$ in (i) to strong convergence in measure. The main result of this section was announced without proof in [14] and is closely related to a question raised by McConnell [13, p. 290].

Theorem 2.1. *Let E be a UMD space. If the process $\Phi : [0,T] \times \Omega \to \mathcal{L}(H,E)$ is H-strongly measurable, adapted, and stochastically integrable with respect to W_H, then there exists a sequence of elementary adapted processes $\Phi_n : [0,T] \times \Omega \to \mathcal{L}(H,E)$ such that*

(i)′ $\lim_{n\to\infty} \Phi_n h = \Phi h$ in measure for all $h \in H$;

(ii) $\lim_{n\to\infty} \int_0^{\cdot} \Phi_n \, dW_H = \int_0^{\cdot} \Phi \, dW_H$ in measure in $C([0,T];E)$.

For the definition of the class of UMD Banach spaces and some of its applications in analysis we refer to Burkholder's survey article [4].

Let $\mathcal{H}$ be a separable real Hilbert space and let $(g_n)_{n\geq 1}$ be a sequence of independent standard Gaussian random variables on a probability space $(\Omega', \mathcal{F}', \mathbb{P}')$. A linear operator $R : \mathcal{H} \to E$ is said to be *γ-radonifying* if for some (every) orthonormal basis $(h_n)_{n\geq 1}$ of $\mathcal{H}$ the Gaussian sum $\sum_{n\geq 1} g_n Rh_n$ converges in $L^2(\Omega'; E)$. The linear space of all γ-radonifying operators from $\mathcal{H}$ to E is denoted by $\gamma(\mathcal{H}, E)$. This space is a Banach space when endowed with the norm

$$\|R\|_{\gamma(\mathcal{H},E)} := \Big(\mathbb{E}'\Big\|\sum_{n\geq 1} g_n Rh_n\Big\|^2\Big)^{\frac{1}{2}}.$$

For more information we refer to [3, 5, 10]. The importance of spaces of γ-radonifying operators in the theory of stochastic integration in infinite dimensions is well established; see [14, 15] and the references given therein.

A H-strongly measurable function $\Phi : [0,T] \to \mathcal{L}(H,E)$ is said to *represent* an element $R \in \gamma(L^2(0,T;H),E)$ if for all $x^* \in E^*$ we have $\Phi^* x^* \in L^2(0,T;H)$ and, for all $f \in L^2(0,T;H)$,

$$\langle Rf, x^*\rangle = \int_0^T [f(t), \Phi^*(t)x^*] \, dt.$$

Extending the above definition, we say that an H-strongly measurable process $\Phi : [0,T] \times \Omega \to \mathcal{L}(H,E)$ *represents* a random variable $X : \Omega \to \gamma(L^2(0,T;H),E)$ if for all $x^* \in E^*$ almost surely we have $\Phi^* x^* \in L^2(0,T;H)$ and, for all $f \in L^2(0,T;H)$,

$$\langle Xf, x^*\rangle = \int_0^T [f(t), \Phi^*(t)x^*]_H \, dt \ \text{ almost surely.}$$

For H-strongly measurable process we have the following result [14, Lemma 2.7].

Lemma 2.2. *Let $\Phi : [0,T] \times \Omega \to \mathcal{L}(H,E)$ be a H-strongly measurable process and $X : \Omega \to \gamma(L^2(0,T;H),E)$ be strongly measurable. The following assertions are equivalent:*

1. *Φ represents X.*
2. *$\Phi(\cdot,\omega)$ represents $X(\omega)$ for almost all $\omega \in \Omega$.*

For a Banach space F we denote by $L^0(\Omega; F)$ the vector space of all F-valued random variables on Ω, identifying random variables if they agree almost surely. Endowed with the topology of convergence in measure, $L^0(\Omega; F)$ is a complete metric space. The following result is obtained in [14].

Proposition 2.3. *Let E be a UMD space. For a H-strongly measurable and adapted process $\Phi : [0, T] \times \Omega \to \mathcal{L}(H, E)$ the following assertions are equivalent:*

1. Φ is stochastically integrable with respect to W_H;

2. Φ represents a random variable $X : \Omega \to \gamma(L^2(0, T; H), E)$.

In this case Φh is stochastically integrable with respect to $W_H h$ for all $h \in H$, and for every orthonormal basis $(h_n)_{n\geq 1}$ of H we have

$$\int_0^{\cdot} \Phi \, dW_H = \sum_{n\geq 1} \int_0^{\cdot} \Phi h_n \, dW_H h_n,$$

with almost sure unconditional convergence of the series expansion in $C([0, T]; E)$. Moreover, for all $p \in (1, \infty)$

$$\mathbb{E} \sup_{t\in[0,T]} \Big\| \int_0^t \Phi(s) \, dW_H(s) \Big\|^p \eqsim_{p,E} \mathbb{E}\|X\|^p_{\gamma(L^2(0,T;H),E)}.$$

Furthermore, the mapping $X \mapsto \int_0^{\cdot} \Phi \, dW_H$ has a unique extension to a continuous mapping from $L^0(\Omega; \gamma(L^2(0, T; H), E))$ into $L^0(\Omega; C([0, T]; E))$.

As we will show in a moment, the series expansion in Proposition 2.3 implies that in order to prove Theorem 2.1 it suffices to prove the following weaker version:

Theorem 2.4. *Let E be a UMD space. If the process $\phi : [0, T] \times \Omega \to E$ is strongly measurable, adapted, and stochastically integrable with respect to a Brownian motion W, then there exists a sequence of elementary adapted processes $\phi_n : [0, T] \times \Omega \to E$ such that*

(i)′ $\lim_{n\to\infty} \phi_n = \phi$ in measure;

(ii) $\lim_{n\to\infty} \int_0^{\cdot} \phi_n \, dW = \int_0^{\cdot} \phi \, dW$ in measure in $C([0, T]; E)$.

This theorem may actually be viewed as the special case of Theorem 2.1 corresponding to $H = \mathbb{R}$, by identifying $\mathcal{L}(\mathbb{R}, E)$ with E and identifying $\mathbb{R}$-cylindrical Brownian motions with real-valued Brownian motions.

To see that Theorem 2.1 follows from Theorem 2.4 we argue as follows. Choose an orthonormal basis $(h_n)_{n\geq 1}$ of H and define the processes $\Psi_n : [0, T] \times \Omega \to \mathcal{L}(H, E)$ by

$$\Psi_n h := \sum_{j=1}^{n} [h, h_j]_H \, \Phi h_j.$$

Clearly, $\lim_{n\to\infty}\Psi_n h = \Phi h$ pointwise, hence in measure, for all $h \in H$. In view of the identity $\int_0^{\cdot} \Psi_n \, dW_H = \sum_{j=1}^{n} \int_0^{\cdot} \Phi h_j \, dW_H h_j$ and the series expansion in Proposition 2.3, we also have

$$\lim_{n\to\infty} \int_0^{\cdot} \Psi_n \, dW_H = \int_0^{\cdot} \Phi \, dW_H \quad \text{in measure in } C([0,T];E).$$

With Theorem 2.4, for each $n \geq 1$ we choose a sequence of elementary adapted processes $\phi_{j,n} : [0,T] \times \Omega \to E$ such that $\lim_{j\to\infty} \phi_{j,n} = \Phi h_n$ in measure and

$$\lim_{j\to\infty} \int_0^{\cdot} \phi_{j,n} \, dW_H h_n = \int_0^{\cdot} \Phi h_n \, dW_H h_n \quad \text{in measure in } C([0,T];E).$$

Given $k \geq 1$, choose $N_k \geq 1$ so large that

$$\mathbb{P}\Big\{\Big\| \int_0^{\cdot} \Phi - \Psi_{N_k} \, dW_H \Big\|_\infty > \frac{1}{k}\Big\} < \frac{1}{k}.$$

Let λ be the Lebesgue measure on $[0,T]$. For each $n = 1, \dots, N_k$ choose $j_{k,n} \geq 1$ so large that

$$\lambda \otimes \mathbb{P}\Big\{\|\Phi h_n - \phi_{j_{k,n},n}\| > \frac{1}{kN_k}\Big\} < \frac{1}{kN_k}$$

and

$$\mathbb{P}\Big\{\Big\| \int_0^{\cdot} \Phi h_n - \phi_{j_{k,n},n} \, dW_H h_n \Big\|_\infty > \frac{1}{k}\Big\} < \frac{1}{kN_k}.$$

Define $\Phi_k : [0,T] \times \Omega \to \mathcal{L}(H,E)$ by

$$\Phi_k h := \sum_{n=1}^{N_k} [h, h_n]_H \, \phi_{j_{k,n},n}, \qquad h \in H.$$

Each Φ_k is elementary adapted. For all $h \in H$ with $\|h\|_H = 1$ and all $\delta > 0$ we have, for all $k \geq 1/\delta$,

$$\begin{aligned}
&|\{\|\Phi h - \Phi_k h\| > 2\delta\}| \\
&\qquad \leq \lambda \otimes \mathbb{P}\{\|\Phi h - \Psi_{N_k} h\| > \delta\} + \lambda \otimes \mathbb{P}\{\|\Psi_{N_k} h - \Phi_k h\| > \delta\} \\
&\qquad < \lambda \otimes \mathbb{P}\{\|\Phi h - \Psi_{N_k} h\| > \delta\} + \frac{1}{k}.
\end{aligned}$$

Hence, $\lim_{k\to\infty} \Phi_k h = \Phi h$ in measure for all $h \in H$. Also,

$$\begin{aligned}
&\mathbb{P}\Big\{\Big\| \int_0^T \Phi - \Phi_k \, dW_H \Big\|_\infty > \frac{2}{k}\Big\} \\
&\quad \leq \mathbb{P}\Big\{\Big\| \int_0^T \Phi - \Psi_{N_k} \, dW_H \Big\|_\infty > \frac{1}{k}\Big\} + \mathbb{P}\Big\{\Big\| \int_0^T \Psi_{N_k} - \Phi_k \, dW_H \Big\|_\infty > \frac{1}{k}\Big\} \\
&\quad < \frac{1}{k} + \frac{1}{k} = \frac{2}{k},
\end{aligned}$$

and therefore $\lim_{k\to\infty} \int_0^{\cdot} \Phi_k \, dW_H = \int_0^{\cdot} \Phi \, dW_H$ in measure in $C([0,T];E)$. Thus the processes Φ_k have the desired properties.

This matter having been settled, the remainder of the section is aimed at proving Theorem 2.4. The following argument will show that it suffices to prove Theorem 2.4 for *uniformly bounded* processes ϕ. To see why, for $n \geq 1$ define

$$\phi_n := \mathbf{1}_{\{\|\phi\| \leq n\}} \phi.$$

The processes ϕ_n are uniformly bounded, strongly measurable and adapted, and we have $\lim_{n\to\infty} \phi_n = \phi$ pointwise, hence also in measure.

We claim that each ϕ_n is stochastically integrable with respect to W and

$$\lim_{n\to\infty} \int_0^{\cdot} \phi_n \, dW = \int_0^{\cdot} \phi \, dW \ \text{ in measure in } C([0,T];E).$$

To see this, let $X : \Omega \to \gamma(L^2(0,T), E)$ be the element represented by ϕ. Put

$$X_n(\omega)f := X(\omega)(\mathbf{1}_{\{\|\phi(\cdot,\omega)\| \leq n\}} f), \qquad f \in L^2(0,T).$$

Then by [14, Proposition 2.4], $\lim_{n\to\infty} X_n = X$ almost surely in $\gamma(L^2(0,T), E)$. It is easily checked that ϕ_n represents X_n, and therefore ϕ_n is stochastically integrable by Proposition 2.3. The convergence in measure of the stochastic integrals now follows from the continuity assertion in Proposition 2.3. This completes the proof of the claim. A more general result in this spirit will be proved in Section 3.

It remains to prove Theorem 2.4 for uniformly bounded processes Φ. Let $\mathcal{D}_n$ denote the finite σ-field generated by the n-th dyadic equipartition of the interval $[0,T]$ and let $\mathcal{G}_n = \mathcal{D}_n \otimes \mathcal{F}$ be the product σ-field in $[0,T] \times \Omega$. Then $\mathbb{G} = \{\mathcal{G}_n\}_{n\geq 1}$ is a filtration in $[0,T] \times \Omega$ with $\bigvee_{n\geq 1} \mathcal{G}_n = \mathcal{B} \otimes \mathcal{F}$, where $\mathcal{B}$ is the Borel σ-algebra of $[0,T]$. In what follows with think of $[0,T] \times \Omega$ as probability space with respect to the product measure $\frac{dt}{T} \otimes \mathbb{P}$. Note that for all $f \in L^2([0,T] \times \Omega; E)$, for almost all $\omega \in \Omega$ we have

$$\mathbb{E}(f|\mathcal{G}_n)(\cdot,\omega) = \mathbb{E}(f(\cdot,\omega)|\mathcal{D}_n) \ \text{ in } L^2(0,T;E).$$

Define the operators G_n on $L^2([0,T] \times \Omega; E)$ by

$$G_n f := \tau_n \mathbb{E}(f|\mathcal{G}_n),$$

where τ_n denotes the right translation operator over $2^{-n}T$ in $L^2([0,T] \times \Omega; E)$, i.e., $\tau_n f(t,\omega) = \mathbf{1}_{[2^{-n}T,T]} f(t - 2^{-n}T, \omega)$.

Lemma 2.5. *Let $\phi : [0,T] \times \Omega \to E$ be strongly measurable, adapted, uniformly bounded, and stochastically integrable with respect to W. Then the processes $\phi_n : [0,T] \times \Omega \to E$ defined by $\phi_n := G_n \phi$ are strongly measurable, adapted, uniformly bounded, and stochastically integrable with respect to W. Moreover, $\lim_{n\to\infty} \phi_n = \phi$ in measure and*

$$\lim_{n\to\infty} \int_0^{\cdot} \phi_n \, dW = \int_0^{\cdot} \phi \, dW \ \textit{ in measure in } C([0,T];E). \tag{2.1}$$

Proof. First note that each process ϕ_n is strongly measurable, uniformly bounded, strongly measurable and adapted. By the vector-valued martingale convergence theorem and the strong continuity of translations we have

$$\begin{aligned}
&\lim_{n\to\infty} \|\phi - \phi_n\|_{L^2([0,T]\times\Omega;E)} \\
&\quad\le \lim_{n\to\infty} \|\phi - \tau_n\phi\|_{L^2([0,T]\times\Omega;E)} + \lim_{n\to\infty} \|\tau_n\phi - \tau_n\mathbb{E}(\phi|\mathcal{G}_n)\|_{L^2([0,T]\times\Omega;E)} \\
&\quad\le \lim_{n\to\infty} \|\phi - \tau_n\phi\|_{L^2([0,T]\times\Omega;E)} + \lim_{n\to\infty} \|\phi - \mathbb{E}(\phi|\mathcal{G}_n)\|_{L^2([0,T]\times\Omega;E)} = 0.
\end{aligned}$$

It follows that $\lim_{n\to\infty}\phi_n = \phi$ in $L^2([0,T]\times\Omega;E)$, and therefore also in measure.

Let $X : \Omega \to \gamma(L^2(0,T),E)$ be the random variable represented by ϕ. For all $n \ge 1$ let the random variable $X_n : \Omega \to \gamma(L^2(0,T),E)$ defined by

$$X_n(\omega) := X(\omega)\circ\tau_n^*\circ\mathbb{E}(\,\cdot\,|\mathcal{D}_n)$$

where $\tau_n^* \in \mathcal{L}(L^2(0,T))$ denotes the left translation operator. It is easily seen that for all $n \ge 1$, X_n is represented by ϕ_n, and therefore ϕ_n is stochastically integrable with respect to W by Proposition 2.3. By [14, Proposition 2.4] we obtain $\lim_{j\to\infty} X_n = X$ almost surely in $\gamma(L^2(0,T),E)$. Hence, $\lim_{n\to\infty} X_n = X$ in measure in $\gamma(L^2(0,T),E)$, and (2.1) follows from the continuity assertion in Proposition 2.3. □

Now we can complete the proof of Theorem 2.4 for uniformly bounded processes ϕ. The processes ϕ_n in Lemma 2.5 can be represented as

$$\phi_n = \sum_{j=1}^{2^n} \mathbf{1}_{I_j}\phi_{j,n},$$

where the I_j is the j-th interval in the n-th dyadic partition of $[0,T]$ and the random variable $\phi_{j,n} : \Omega \to E$ is uniformly bounded and $\mathcal{F}_j$-measurable, where $\mathcal{F}_j = \mathcal{F}_{2^{-n}(j-1)T}$. The proof is completed by approximating the $\phi_{j,n}$ in $L^0(\Omega,\mathcal{F}_j;E)$ with simple random variables.

Let E_1 and E_2 be real Banach spaces. Theorem 2.1 can be strengthened for processes with values in $\mathcal{L}(E_1,E_2)$, integrable with respect to an E_1-valued Brownian motion.

Let μ be a centered Gaussian Radon measure on E_1 and let W_μ be an E_1-valued Brownian motion with distribution μ, i.e., for all $t \ge 0$ and $x^* \in E_1^*$ we have

$$\mathbb{E}\langle W_\mu(t), x^*\rangle^2 = t\int_{E_1} \langle x, x^*\rangle^2\, d\mu(x).$$

Let H_μ denote the reproducing kernel Hilbert space associated with μ and let $i_\mu : H_\mu \hookrightarrow E_1$ be the inclusion operator. We can associate an H_μ-cylindrical Brownian motion W_{H_μ} with W_μ by the formula

$$W_{H_\mu}(t)i_\mu^* x^* := \langle W_\mu(t), x^*\rangle.$$

We say that $\Phi : [0,T] \times \Omega \to \mathcal{L}(E_1, E_2)$ is *E_1-strongly measurable and adapted* if for all $x \in E_1$, Φx is strongly measurable and adapted. An E_1-strongly measurable and adapted process $\Phi : [0,T] \times \Omega \to \mathcal{L}(E_1, E_2)$ is called *stochastically integrable* with respect to the E_1-valued Brownian motion W_μ if the process $\Phi \circ i_\mu : [0,T] \times \Omega \to \mathcal{L}(H_\mu, E_2)$ is stochastically integrable with respect to W_{H_μ}. In this case we write

$$\int_0^\cdot \Phi \, dW_\mu := \int_0^\cdot \Phi \circ i_\mu \, dW_{H_\mu}.$$

By the Pettis measurability theorem and the separability of H_μ, the E_1-strong measurability of Φ implies the H_μ-strong measurability of $\Phi \circ i_\mu$. We call Φ an *elementary adapted* process if $\Phi \circ i_\mu$ is elementary adapted.

Theorem 2.6. *Let E_1 be a Banach space, E_2 be a UMD space and $p \in (1,\infty)$. Let W_μ be an E_1-valued Brownian motion with distribution μ. If the process $\Phi : [0,T] \times \Omega \to \mathcal{L}(E_1, E_2)$ is E_1-strongly measurable and adapted and stochastically integrable with respect to W_μ, then there exists a sequence of elementary adapted processes $\Phi_n : [0,T] \times \Omega \to \mathcal{L}(E_1, E_2)$ such that*

(i)″ $\lim_{n\to\infty} \Phi_n x = \Phi x$ in measure for μ-almost all $x \in E_1$;

(ii) $\displaystyle\lim_{n\to\infty} \int_0^\cdot \Phi_n \, dW_\mu = \int_0^\cdot \Phi \, dW_\mu$ in measure in $C([0,T]; E_2)$.

The proof depends on some well-known facts about measurable linear extensions. We refer to [3, 6] for more details. If μ is a centered Gaussian Radon measure on E_1 with reproducing kernel Hilbert space H_μ and $(h_n)_{n\geq 1}$ is an orthonormal basis for H_μ, then the coordinate functionals $h \mapsto [h, h_n]_{H_\mu}$ can be extended to μ-measurable linear mappings $\overline{h_n}$ from E_1 to $\mathbb{R}$. Moreover, these extensions are μ-essentially unique in the sense that every two such extensions agree μ-almost everywhere. Putting

$$\overline{P_n} x := \sum_{j=1}^n \overline{h_j} x \, h_j, \qquad x \in E_1,$$

we obtain a μ-measurable linear extension of the orthogonal projection P_n in H_μ onto the span of the vectors $h_1, \ldots, h_n$. Again this extension is μ-essentially unique, and we have

$$\lim_{n\to\infty} i_\mu \overline{P_n} x = \sum_{n\geq 1} \overline{h_j} x \, i_\mu h_j = x \ \text{ for } \mu\text{-almost all } x \in E_1. \tag{2.2}$$

Proof. [Proof of Theorem 2.6] We will reduce the theorem to Theorem 2.4. Choose an orthonormal basis $(h_n)_{n\geq 1}$ of the reproducing kernel Hilbert space H_μ and define the processes $\Psi_n : [0,T] \times \Omega \to \mathcal{L}(E_1, E_2)$ by

$$\Psi_n x := \Phi i_\mu \overline{P_n} x, \qquad x \in E_1.$$

By (2.2), $\lim_{n\to\infty} \Psi_n x = \Phi x$ in measure for μ-almost all $x \in E_1$. Also,

$$\lim_{n\to\infty} \int_0^{\cdot} \Psi_n \, dW_\mu = \lim_{n\to\infty} \int_0^{\cdot} \Psi_n \circ i_\mu \, dW_{H_\mu}$$
$$\overset{(*)}{=} \int_0^{\cdot} \Phi \circ i_\mu \, dW_{H_\mu} = \int_0^{\cdot} \Phi \, dW_\mu \quad \text{in measure in } C([0,T]; E_2),$$

where the identity $(*)$ follows by series representation as in the argument following the statement of Theorem 2.4. The proof may now be completed along the lines of this argument; for Φ_k we take

$$\Phi_k x := \sum_{n=1}^{N_k} \overline{h_{N_k} x} \, \phi_{j_{k,n},n}, \qquad x \in E_1,$$

where the elementary adapted processes $\phi_{j,n}$ approximate $\Phi i_\mu h_n$ and the indices N_k are chosen as before. □

As a final comment we note that L^p-versions of the results of this section hold as well; for these one has to replace almost sure convergence by L^p-convergence in the proofs.

3 Domination

In this section we present two domination results which were implicit in the arguments so far, and indeed some simple special cases of them have already been used.

The first comparison result extends [15, Corollary 4.4], where the case of functions was considered.

Theorem 3.1 (Domination). *Let E be a UMD space. Let $\Phi, \Psi : [0,T]\times\Omega \to \mathcal{L}(H,E)$ be H-strongly measurable and adapted processes and assume that Ψ is stochastically integrable with respect to W_H. If for all $x^* \in E^*$ we have*

$$\int_0^T \|\Phi^*(t)x^*\|_H^2 \, dt \le \int_0^T \|\Psi^*(t)x^*\|_H^2 \, dt \ \text{ almost surely},$$

then Φ is stochastically integrable and for all $p \in (1,\infty)$,

$$\mathbb{E} \sup_{t\in[0,T]} \Big\| \int_0^t \Phi(s) \, dW_H(s) \Big\|^p \lesssim_{p,E} \mathbb{E} \sup_{t\in[0,T]} \Big\| \int_0^t \Psi(s) \, dW_H(s) \Big\|^p,$$

whenever the right hand side is finite.

Proof. Since Φ and Ψ are H-strongly measurable and adapted, without loss of generality we may assume that E is separable. By Proposition 2.3, Ψ represents a random variable $Y : \Omega \to \gamma(L^2(0,T;H),E)$. In particular, for all $x^* \in E^*$ we have

$\Psi^* x^* \in L^2(0,T;H)$ almost surely. We claim that almost surely,

$$\int_0^T \|\Phi^*(t)x^*\|_H^2\,dt \le \int_0^T \|\Psi^*(t)x^*\|_H^2\,dt \ \text{ for all } x^* \in E^*.$$

Indeed, by the reflexivity and separability of E we may choose a countable, norm dense, $\mathbb{Q}$-linear subspace F of E^*. Let N_1 be a null set such that

$$\int_0^T \|\Phi^*(t,\omega)x^*\|_H^2\,dt \le \int_0^T \|\Psi^*(t,\omega)x^*\|_H^2\,dt \tag{3.1}$$

for all $\omega \in \complement N_1$ and all $x^* \in F$. By Lemma 2.2 there exists a null set N_2 such that $\Psi(\cdot,\omega)$ represents $Y(\omega)$ for all $\omega \in \complement N_2$. Fix $y^* \in E^*$ arbitrary and choose a sequence $(y_n^*)_{n\ge1}$ in F such that $\lim_{n\to\infty} y_n^* = y^*$ in E^* strongly. Fix an arbitrary $\omega \in \complement(N_1 \cup N_2)$. We will prove the claim by showing that

$$\int_0^T \|\Phi^*(t,\omega)y^*\|_H^2\,dt \le \int_0^T \|\Psi^*(t,\omega)y^*\|_H^2\,dt. \tag{3.2}$$

By the closed graph theorem there exists a constant C_ω such that

$$\|\Psi^*(\cdot,\omega)x^*\|_{L^2(0,T;H)} \le C_\omega \|x^*\| \ \text{ for all } x^* \in E^*.$$

Hence, $\Psi^*(\cdot,\omega)y^* = \lim_{n\to\infty} \Psi^*(\cdot,\omega)y_n^*$ in $L^2(0,T;H)$, by the strong convergence of the y_n^*'s to y^*. It follows from (3.1), applied to the functionals $y_n^* - y_m^* \in F$, that $(\Phi^* y_n^*)_{n\ge1}$ is a Cauchy sequence in $L^2(0,T;H)$. Identification of the limit shows that $\Phi^*(\cdot,\omega)y^* = \lim_{n\to\infty} \Phi^*(\cdot,\omega)y_n^*$ in $L^2(0,T;H)$. Now (3.2) follows from the corresponding inequality for y_n^* by letting $n \to \infty$.

By the claim and [15, Theorem 4.2 and Corollary 4.4], almost every function $\Phi(\cdot,\omega)$ represents an element $X(\omega) \in \gamma(L^2(0,T;H),E)$ for which we have

$$\|X(\omega)\|_{\gamma(L^2(0,T;H),E)} \le \|Y(\omega)\|_{\gamma(L^2(0,T;H),E)}.$$

By [14, Remark 2.8] X is strongly measurable as a $\gamma(L^2(0,T;H),E)$-valued random variable. Since Φ represents X, Φ is stochastically integrable by Proposition 2.3. Moreover, from Proposition 2.3 we deduce that

$$\mathbb{E} \sup_{t\in[0,T]} \Big\| \int_0^t \Phi(s)\,dW_H(s) \Big\|^p \eqsim_{p,E} \mathbb{E}\|X\|^p_{\gamma(L^2(0,T;H),E)} \le \mathbb{E}\|Y\|^p_{\gamma(L^2(0,T;H),E)}$$

$$\eqsim_{p,E} \mathbb{E} \sup_{t\in[0,T]} \Big\| \int_0^t \Psi(s)\,dW_H(s) \Big\|^p.$$

□

The next result extends [15, Theorem 6.2].

Corollary 3.2 (Dominated convergence). *Let E be a UMD space and fix $p \in (1, \infty)$. For $n \geq 1$, let $\Phi_n : [0,T] \times \Omega \to \mathcal{L}(H,E)$ be H-strongly measurable and adapted and stochastically integrable processes and assume that there exists a H-strongly measurable and adapted process $\Phi : [0,T] \times \Omega \to \mathcal{L}(H,E)$ such that for all $x^* \in E^*$,*

$$\lim_{n\to\infty} \Phi_n^* x^* = \Phi^* x^* \text{ almost surely in } L^2(0,T;H). \tag{3.3}$$

Assume further that there exists a H-strongly measurable and adapted process Ψ : $[0,T] \times \Omega \to \mathcal{L}(H,E)$ that is stochastically integrable and for all n and all $x^* \in E^*$,

$$\int_0^T \|\Phi_n^*(t)x^*\|_H^2 \, dt \leq \int_0^T \|\Psi^*(t)x^*\|_H^2 \, dt \text{ almost surely}. \tag{3.4}$$

Then Φ is stochastically integrable and

$$\lim_{n\to\infty} \int_0^{\cdot} \Phi_n - \Phi \, dW_H = 0 \text{ in measure in } C([0,T];E).$$

Proof. The assumptions (3.3) and (3.4) imply that for all n and $x^* \in E^*$,

$$\int_0^T \|\Phi_n^*(t)x^*\|_H^2 \, dt \leq \int_0^T \|\Psi^*(t)x^*\|_H^2 \, dt \text{ almost surely}. \tag{3.5}$$

Theorem 3.1 therefore implies that each Φ_n is stochastically integrable, and by passing to the limit $n \to \infty$ in (3.5) we see that the same is true for Φ. Let $Z_n : \Omega \to \gamma(L^2(0,T;H),E)$ be the element represented by $\Phi_n - \Phi$. By Proposition 2.3 it suffices to prove that

$$\lim_{n\to\infty} Z_n = 0 \text{ in measure in } \gamma(L^2(0,T;H),E). \tag{3.6}$$

As in the proof of Theorem 3.1, (3.4) implies that for almost all $\omega \in \Omega$,

$$\int_0^T \|\Phi_n^*(t,\omega)x^*\|_H^2 \, dt \leq \int_0^T \|\Psi^*(t,\omega)x^*\|_H^2 \, dt \text{ for all } n \geq 1 \text{ and } x^* \in E^*, \tag{3.7}$$

and

$$\int_0^T \|\Phi^*(t,\omega)x^*\|_H^2 \, dt \leq \int_0^T \|\Psi^*(t,\omega)x^*\|_H^2 \, dt \text{ for all } n \geq 1 \text{ and } x^* \in E^*. \tag{3.8}$$

Denoting by $Y : \Omega \to \gamma(L^2(0,T;H),E)$ the element represented by Ψ, we obtain that, for almost all $\omega \in \Omega$, for all $x^* \in E^*$,

$$\|Z_n^*(\omega)x^*\|_{L^2(0,T;H)} \leq 2\|Y^*(\omega)x^*\|_{L^2(0,T;H)}. \tag{3.9}$$

Let N_1 be a null set such that (3.7) and (3.8) hold for all $\omega \in \complement N_1$. Then for all $\omega \in \complement N_1$ there is a constant $C(\omega)$ such that for all $x^* \in E^*$ and all $n \geq 1$,

$$\int_0^T \|\Phi^*(t,\omega) - \Phi_n^*(t,\omega)x^*\|_H^2 \, dt \leq C^2(\omega)\|x^*\|^2. \tag{3.10}$$

Let $(x_j^*)_{j\geq 1}$ be a dense sequence in E^*. By (3.3) we can find a null set N_2 such that for all $\omega \in \complement N_2$ and all $j \geq 1$ we have

$$\lim_{n\to\infty} \Phi_n^*(\cdot,\omega)x_j^* = \Phi^*(\cdot,\omega)x_j^* \text{ in } L^2(0,T;H). \tag{3.11}$$

Clearly, (3.10) and (3.11) imply that for all $\omega \in \complement(N_1 \cup N_2)$ we have

$$\lim_{n\to\infty} \Phi_n^*(\cdot,\omega)x^* = \Phi^*(\cdot,\omega)x^* \text{ in } L^2(0,T;H) \text{ for all } x^* \in E^*,$$

hence for almost all $\omega \in \Omega$, for all $x^* \in E^*$,

$$\lim_{n\to\infty} Z_n^*(\omega)x^* = 0 \text{ in } L^2(0,T;H). \tag{3.12}$$

By (3.9), (3.12), and a standard tightness argument as in [15, Theorem 6.2] we obtain that for almost all $\omega \in \Omega$, $\lim_{n\to\infty} Z_n(\omega) = 0$ in $\gamma(L^2(0,T;H),E)$. This gives (3.6). □

Again we leave it to the reader to formulate the L^p-version of these results.

4 Smoothness – I

Extending a result of Rosiński and Suchanecki (who considered the case $H = \mathbb{R}$), it was shown in [15] (for arbitrary Banach spaces E and functions Φ) and [14] (for UMD Banach spaces and processes Φ) that if E is a Banach space with type 2, then every H-strongly measurable and adapted process $\Phi : [0,T]\times\Omega \to \mathcal{L}(H,E)$ with trajectories in $L^2(0,T;\gamma(H,E))$ is stochastically integrable with respect to a H-cylindrical Brownian motion W_H. Moreover, for $H = \mathbb{R}$ this property characterizes the spaces E with type 2. Below (Theorem 4.2) we shall obtain an extension of this result for processes in UMD spaces with type $p \in [1,2)$.

The results will be formulated in terms of vector valued Besov spaces. We briefly recall the definition. We follow the approach of Peetre; see [1, 7, 18, 19]. The Fourier transform of a function $f \in L^1(\mathbb{R}^d;E)$ will be normalized as

$$\widehat{f}(\xi) = \frac{1}{(2\pi)^{d/2}} \int_{\mathbb{R}^d} f(x)e^{-ix\cdot\xi}\,dx, \quad \xi \in \mathbb{R}^d.$$

Let $\phi \in \mathcal{S}(\mathbb{R}^d)$ be a fixed Schwartz function whose Fourier transform $\widehat{\phi}$ is nonnegative and has support in $\{\xi \in \mathbb{R}^d : \frac{1}{2} \leq |\xi| \leq 2\}$ and which satisfies

$$\sum_{k\in\mathbb{Z}} \widehat{\phi}(2^{-k}\xi) = 1 \quad \text{for } \xi \in \mathbb{R}^d \setminus \{0\}.$$

Define the sequence $(\varphi_k)_{k\geq 0}$ in $\mathcal{S}(\mathbb{R}^d)$ by

$$\widehat{\varphi_k}(\xi) = \widehat{\phi}(2^{-k}\xi) \quad \text{for } k = 1,2,\ldots \quad \text{and } \widehat{\varphi_0}(\xi) = 1 - \sum_{k\geq 1} \widehat{\varphi_k}(\xi), \quad \xi \in \mathbb{R}^d.$$

For $1 \le p, q \le \infty$ and $s \in \mathbb{R}$ the *Besov space* $B^s_{p,q}(\mathbb{R}^d; E)$ is defined as the space of all E-valued tempered distributions $f \in \mathcal{S}'(\mathbb{R}^d; E)$ for which

$$\|f\|_{B^s_{p,q}(\mathbb{R}^d;E)} := \Big\| \big(2^{ks} \varphi_k * f\big)_{k \ge 0} \Big\|_{l^q(L^p(\mathbb{R}^d;E))}$$

is finite. Endowed with this norm, $B^s_{p,q}(\mathbb{R}^d; E)$ is a Banach space, and up to an equivalent norm, it is independent of the choice of the initial function ϕ. The sequence $(\varphi_k * f)_{k \ge 0}$ is called the *Littlewood–Paley decomposition* of f associated with the function ϕ.

Next we define the Besov space for domains. Let D be a nonempty bounded open domain in $\mathbb{R}^d$. For $1 \le p, q \le \infty$ and $s \in \mathbb{R}$ we define

$$B^s_{p,q}(D; E) = \{f|_D : \ f \in B^s_{p,q}(\mathbb{R}^d; E)\}.$$

This space is a Banach space endowed with the norm

$$\|g\|_{B^s_{p,q}(D;E)} = \inf_{f|_D = g} \|f\|_{B^s_{p,q}(\mathbb{R}^d;E)}.$$

See [20, Section 3.2.2] (where the scalar case is considered) and [2].

We have the following embedding result, which is a straightforward extension of [9, Theorems 1.1 and 3.2] where the case $H = \mathbb{R}$ was considered:

Proposition 4.1. *Let E be a Banach space and H be a non-zero separable Hilbert space. Let $D \subseteq \mathbb{R}^d$ be an open domain and let $p \in [1, 2]$. Then E has type p if and only if we have a continuous embedding*

$$B^{\frac{d}{p} - \frac{d}{2}}_{p,p}(D; \gamma(H, E)) \hookrightarrow \gamma(L^2(D; H); E).$$

If we combine this result with Proposition 2.3 we obtain the following condition for stochastic integrability of processes.

Theorem 4.2. *Let H be a separable Hilbert space and E be a UMD Banach space with type $p \in [1, 2]$. If $\Phi : [0, T] \times \Omega \to \mathcal{L}(H, E)$ is an H-strongly measurable process and adapted process with trajectories in $B^{\frac{1}{p} - \frac{1}{2}}_{p,p}(0, T; \gamma(H, E))$ almost surely, then Φ is stochastically integrable with respect W_H. Moreover, for all $q \in (1, \infty)$,*

$$\mathbb{E} \sup_{t \in [0,T]} \Big\| \int_0^t \Phi(s)\, dW_H(s) \Big\|^q \eqsim_{p,E} \mathbb{E}\|\Phi\|^q_{B^{\frac{1}{p} - \frac{1}{2}}_{p,p}(0,T;\gamma(H,E))}.$$

A similar result can be given for processes with Hölder continuous trajectories. In particular, invoking [9, Corollary 3.4] we see that Theorem 4.2 may be applied to functions in $C^\alpha([0, 1]; \gamma(H, E))$ and, if E is a UMD space, to processes with paths almost surely in $C^\alpha([0, 1]; \gamma(H, E))$, where $\alpha > 1/p - 1/2$. Since UMD spaces always have non-trivial type, there exists an $\varepsilon > 0$ such that every H-strongly measurable and adapted process with paths in $C^{\frac{1}{2} - \varepsilon}([0, 1]; \gamma(H, E))$ is stochastically integrable with respect to W_H. In the converse direction, [9, Theorem 3.5] implies that if E is a Banach space failing type $p \in (1,2)$, then for any $0 < \alpha < 1/p - 1/2$ there exist examples of functions in $C^\alpha([0,1];E)$ which fail to be stochastically integrable with respect to scalar Brownian motions.

5 Smoothness – II

In this section we give an alternative proof of Proposition 4.1 in the case D is a finite interval. The argument uses the definition of the Besov space from [11] instead of the Fourier analytic definition of Peetre.

For $s \in (0,1)$ and $p,q \in [1,\infty]$ we will recall the definition of the Besov space $\Lambda^s_{p,q}(0,T;E)$ from [11]. Since it is not obvious that this space is equal to the Besov space of Section 4 we use the notation $\Lambda^s_{p,q}(0,T;E)$ instead of $B^s_{p,q}(0,T;E)$.

Let $I = (0,T)$. For $h \in \mathbb{R}$ and a function $\phi : I \to E$ we define the function $T(h)\phi : I \to E$ as the translate of ϕ by h, i.e.

$$(T(h)\phi)(t) := \begin{cases} \phi(t+h) & \text{if } t+h \in I, \\ 0 & \text{otherwise.} \end{cases}$$

For $h \in \mathbb{R}$ put $I[h] := \{r \in I : \ r+h \in I\}$. For a strongly measurable function $\phi \in L^p(I;E)$ and $t > 0$ let

$$\varrho_p(\phi,t) := \sup_{|h|\le t} \Big(\int_{I[h]} \|T(h)\phi(r) - \phi(r)\|^p \, dr \Big)^{\frac{1}{p}}.$$

We use the obvious modification if $p = \infty$. Now define

$$\Lambda^s_{p,q}(I;\mathcal{L}(E,F)) := \{\phi \in L^p(I;E) : \|\phi\|_{\Lambda^s_{p,q}(I;E)} < \infty\},$$

where

$$\|\phi\|_{\Lambda^s_{p,q}(I;E)} = \Big(\int_0^T \|\phi(t)\|^p \, dt \Big)^{\frac{1}{p}} + \Big(\int_0^1 (t^{-s} \varrho_p(\phi,t))^q \, \frac{dt}{t} \Big)^{\frac{1}{q}} \tag{5.1}$$

with the obvious modification for $q = \infty$. Endowed with the norm $\|\cdot\|_{\Lambda^s_{p,q}(I;E)}$, $\Lambda^s_{p,q}(I;E)$ is a Banach space. The continuous inclusions $\Lambda^s_{p,q_1}(I;E) \hookrightarrow \Lambda^s_{p,q_2}(I;E)$, $\Lambda^{s_1}_{p,q}(I;E) \hookrightarrow \Lambda^{s_2}_{p,q}(I;E)$, and $\Lambda^s_{p_1,q}(I;E) \hookrightarrow \Lambda^s_{p_2,q}(I;E)$ hold for all $s, s_1, s_2 \in (0,1)$, $p, p_1, p_2, q, q_1, q_2 \in [1,\infty]$ with $1 \le p_2 \le p_1 \le \infty$, $q_1 \le q_2$, $s_2 \le s_1$. For all $p \in [1,\infty)$ we have

$$\Lambda^s_{p,q}(I;E) = B^s_{p,q}(I;E)$$

with equivalent norms. Here $B^s_{p,q}(I;E)$ is the space defined in Section 4. Since we could not find a reference for this, we include the short argument. If $I = \mathbb{R}$ this follows from [16, Proposition 3.1] (also see [18, Theorem 4.3.3]). Therefore, for general I the inclusion "$\supseteq$" follows from the definitions. For the other inclusion notice that by [11, Theorem 3.b.7] one has

$$\Lambda^s_{p,q}(I;E) = (L^p(I;E), W^{1,p}(I;E))_{s,q}.$$

It is well known that there is a common extension operator from the spaces $L^p(I;E)$ and $W^{1,p}(I;E)$ into $L^p(\mathbb{R};E)$ and $W^{1,p}(\mathbb{R};E)$ for all $p \in [1,\infty]$. Therefore, by interpolation we obtain an extension operator from the space $(L^p(I;E), W^{1,p}(I;E))_{s,q}$ into $(L^p(\mathbb{R};E), W^{1,p}(\mathbb{R};E))_{s,q}$. Now the latter is again equal to $B^s_{p,q}(\mathbb{R};E)$ and therefore "$\subseteq$" holds as well.

We put, for $t > 0$, $\varphi^s_p(\phi,t) := t^{-s}\varrho_p(\phi,t)$ and observe for later use the easy fact that there is a constant $c_{q,s} > 0$ such that for all $\phi \in \Lambda^s_{p,q}(I;E)$ we have

$$c_{q,s}^{-1}\|\varphi^s_p(\phi,\cdot)\|_{L^q(0,1;\frac{dt}{t})} \le \big\|\big(\varphi^s_p(\phi,2^{-n})\big)_{n\ge 0}\big\|_{l^q} \le c_{q,s}\|\varphi^s_p(\phi,\cdot)\|_{L^q(0,1;\frac{dt}{t})}. \tag{5.2}$$

Theorem 5.1. *Let H be a separable Hilbert space, E be a Banach space, and $p \in [1,2)$. Then E has type p if and only if $\Lambda^{\frac{1}{p}-\frac{1}{2}}_{p,p}(0,T;\gamma(H,E)) \hookrightarrow \gamma(L^2(0,T;H),E)$ continuously.*

Proof. For the proof that E has type p if the inclusion holds we refer to [9, Theorem 3.3]. To prove the converse we may assume $T = 1$. Let $(g_{00}, g_{nk} : n \ge 0, k = 1,\dots,2^n)$ be the L^2-normalized Haar system on $[0,1]$, i.e. $g_{00} \equiv 1$ and for all other n and k let

$$g_{nk} = \begin{cases} 2^{\frac{n}{2}} & \text{on } [(k-1)2^{-n}, (k-\frac{1}{2})2^{-n}) \\ -2^{\frac{n}{2}} & \text{on } [(k-\frac{1}{2})2^{-n}, k2^{-n}) \\ 0 & \text{otherwise.} \end{cases}$$

Let $(h_i)_{i\ge 1}$ be an orthonormal basis for H. Then $(g_{nk}\otimes h_i)_{m,k,i}$ is an orthonormal basis for $L^2(0,1;H)$. Let $(\gamma_i)_i$, $(\gamma_{nki})_{n,k,i}$ be Gaussian sequences and let $(r_{nk})_{n,k}$ be an independent Rademacher sequence. Let $\Phi \in \Lambda^{\frac{1}{p}-\frac{1}{2}}_{p,p}(0,T;\gamma(H,E))$ be arbitrary. Since E has type p, $L^2(\Omega;E)$ has type p with $T_p(L^2(\Omega;E)) = T_p(E)$ ([5]) and we have

$$\begin{aligned}
&\Big(\mathbb{E}\Big\|\sum_{i\ge 1}\gamma_{nki}I_\Phi g_{00}\otimes h_i + \sum_{n\ge 0}\sum_{k=1}^{2^n}\sum_{i\ge 1}\gamma_{nki}I_\Phi g_{nk}\otimes h_i\Big\|^2\Big)^{\frac{1}{2}} \\
&= \Big(\mathbb{E}_r\mathbb{E}\Big\|\sum_{i\ge 1}\gamma_{nki}I_\Phi g_{00}\otimes h_i + \sum_{n\ge 0}\sum_{k=1}^{2^n}\sum_{i\ge 1}r_{nk}\gamma_{nki}I_\Phi g_{nk}\otimes h_i\Big\|^2\Big)^{\frac{1}{2}} \\
&\le \Big\|\sum_{i\ge 1}\gamma_i I_\Phi g_{00}\otimes h_i\Big\|_{L^2(\Omega;E)} + T_p(E)\Big(\sum_{n\ge 0}\sum_{k=1}^{2^n}\Big\|\sum_{i\ge 1}\gamma_i I_\Phi g_{nk}\otimes h_i\Big\|^p_{L^2(\Omega;E)}\Big)^{\frac{1}{p}}.
\end{aligned}$$

Now one easily checks that

$$\Big\|\sum_{i\ge 1}\gamma_i I_\Phi g_{00}\otimes h_i\Big\|_{L^2(\Omega;E)} \le \|\Phi\|_{L^p(0,1;\gamma(H,E))}.$$

For the other term note that

$$I_\Phi g_{nk}\otimes h_i = 2^{\frac{n}{2}}\int_{(k-1)2^{-n}}^{(k-\frac{1}{2})2^{-n}} (\Phi(s) - \Phi(s+2^{-n-1}))h_i\, ds.$$

Therefore,

$$\begin{aligned}
&\sum_{k=1}^{2^n}\Big\|\sum_{i\geq 1}\gamma_i I_\Phi g_{nk}\otimes h_i\Big\|^p_{L^2(\Omega;E)}\\
&= 2^{\frac{np}{2}}\sum_{k=1}^{2^n}\Big\|\int_{(k-1)2^{-n}}^{(k-\frac12)2^{-n}}\Phi(s)-\Phi(s+2^{-n-1})\,ds\Big\|^p_{\gamma(H,E)}\\
&\leq 2^{\frac{np}{2}}2^{(n+1)(1-p)}\sum_{k=1}^{2^n}\int_{(k-1)2^{-n}}^{(k-\frac12)2^{-n}}\|\Phi(s)-\Phi(s+2^{-n-1})\|^p_{\gamma(H,E)}\,ds\\
&\leq 2^{-p+1}2^{n(1-\frac{p}{2})}\int_0^{1-2^{-n-1}}\|\Phi(s)-\Phi(s+2^{-n-1})\|^p_{\gamma(H,E)}\,ds.
\end{aligned}$$

We conclude that

$$\begin{aligned}
&\Big(\sum_{n\geq 0}\sum_{k=1}^{2^n}\mathbb{E}\Big\|\sum_{i\geq 1}\gamma_i I_\Phi g_{nk}\otimes h_i\Big\|^p\Big)^{\frac1p}\\
&\leq 2^{-1+\frac1p}\Big(\sum_{n\geq 0}2^{n(1-\frac{p}{2})}\int_0^{1-2^{-n-1}}\|\Phi(s)-\Phi(s+2^{-n-1})\|^p_{\gamma(H,E)}\,ds\Big)^{\frac1p}\\
&\lesssim_p \|\Phi\|_{\Lambda^{\frac1p-\frac12}_{p,p}(0,T;\gamma(H,E))},
\end{aligned}$$

where the last inequality follows from (5.2). □

If $(0,T)$ is replaced with $\mathbb{R}$, one can use the Haar basis on each interval $(j,j+1)$ to obtain the analogous embedding result for $\mathbb{R}$. More generally, the proof can be adjusted to the case of finite or infinite rectangles $D\subseteq\mathbb{R}^d$. Furthermore, using extension operators one can extend the embedding result to bounded regular domains.

As a consequence of Theorem 5.1 we recover a Hölder space embedding result from [9]. Using [15, Theorem 2.3] this can be reformulated as follows.

Proposition 5.2. *Let E be a Banach space and $p\in[1,2)$. If E is of type p, then for all $\alpha>1/p-1/2$, $\phi\in C^\alpha([0,1];E)$ implies that ϕ is stochastically integrable with respect to W. Moreover, there exists a constant C only depending on the type p constant of E such that*

$$\mathbb{E}\Big\|\int_0^1\phi\,dW\Big\|^2\leq C^2\|\phi\|^2_{C^\alpha([0,1];E)}.$$

In [9] a converse to this result is obtained as well: if all functions in $C^\alpha([0,1];E)$ are stochastically integrable, then E has type p for all $p\in[1,2)$ satisfying $\alpha<1/p-1/2$. However, the case that $\alpha=1/p-1/2$ is left open there and will be considered in the following theorem. For the definition of stable type p we refer to [12].

Theorem 5.3. *Let E be a Banach space, let $\alpha \in (0, 1/2]$ and let $p \in [1, 2)$ be such that $\alpha = 1/p - 1/2$. If every function in $C^\alpha([0,1]; E)$ is stochastically integrable with respect to W, then E has stable type p.*

Since l^p spaces for $p \in [1, 2)$ do not have stable type p, it follows from Theorem 5.3 that there exists a $(1/p - 1/2)$-Hölder continuous function $\phi : [0,1] \to l^p$ that is not stochastically integrable with respect to W. An explicit example can be obtained from the construction below. This extends certain examples in [17, 21].

Proof. *Step 1:* Fix an integer $N \geq 1$. First we construct an certain function with values in l^p_N. Let $\varphi_{00}, \varphi_{nk}$ for $n \geq 0, k = 1, \ldots, 2^n$ be the Schauder functions on $[0,1]$, i.e., $\varphi_{nk}(x) = \int_0^x g_{nk}(t)\,dt$ where g_{nk} are the L^2-normalized Haar functions. Let $(e_n)_{n=1}^N$ be the standard basis in l^p_N. Let $\psi : [0,1] \to l^p$ be defined as

$$\psi(t) = \sum_{n=0}^{N} \sum_{k=1}^{2^n} 2^{\frac{(p-1)n}{p}} \varphi_{nk}(t) e_{2^n+k}.$$

Then ψ is stochastically integrable and

$$\begin{aligned}
\mathbb{E}\Big\| \int_0^1 \psi\,dW \Big\|^p &= \mathbb{E}\Big\| \sum_{n=0}^{N} \sum_{k=1}^{2^n} 2^{\frac{(p-1)n}{p}} \int_0^1 \varphi_{nk}\,dW e_{2^n+k} \Big\|^p \\
&= \sum_{n=0}^{N} \sum_{k=1}^{2^n} 2^{(p-1)n} \mathbb{E}\Big| \int_0^1 \varphi_{nk}\,dW \Big|^p \\
&= m_p^p \sum_{n=0}^{N} \sum_{k=1}^{2^n} 2^{(p-1)n} \Big(\mathbb{E}\Big| \int_0^1 \varphi_{nk}\,dW \Big|^2 \Big)^{\frac{p}{2}} \\
&= m_p^p \sum_{n=0}^{N} \sum_{k=1}^{2^n} 2^{(p-1)n} \Big(\int_0^1 \varphi_{nk}^2(t)\,dt \Big)^{\frac{p}{2}} \\
&= m_p^p \sum_{n=0}^{N} \sum_{k=1}^{2^n} 2^{(p-1)n} \Big(\frac{2^{-2n-2}}{3} \Big)^{\frac{p}{2}} = m_p^p \frac{N}{12^{\frac{p}{2}}},
\end{aligned}$$

where $m_p = (\mathbb{E}|W(1)|^p)^{\frac{1}{p}}$. Therefore,

$$\Big(\mathbb{E}\Big\| \int_0^1 \psi\,dW \Big\|^2 \Big)^{\frac{1}{2}} \geq \Big(\mathbb{E}\Big\| \int_0^1 \psi\,dW \Big\|^p \Big)^{\frac{1}{p}} = K_p N^{\frac{1}{p}}, \tag{5.3}$$

with $K_p = m_p/\sqrt{12}$. On the other hand ψ is α-Hölder continuous with

$$\|\psi\|_{C^\alpha([0,1];E)} = \sup_{t\in[0,1]} \|\psi(t)\| + \sup_{0\leq s<t\leq 1} \frac{\|\psi(t) - \psi(s)\|}{(t-s)^\alpha} \leq C_p, \tag{5.4}$$

where C_p is a constant only depending on p. Indeed, for each $t \in [0,1]$, we have

$$\|\psi(t)\| = \Big(\sum_{n=0}^{N}\sum_{k=1}^{2^n} 2^{(p-1)n}|\varphi_{nk}(t)|^p\Big)^{\frac{1}{p}} \le \Big(\sum_{n=0}^{N} 2^{(p-1)n}2^{-\frac{(n+2)p}{2}}\Big)^{\frac{1}{p}}$$
$$= \Big(\sum_{n=0}^{N} 2^{-(1-p/2)n-p}\Big)^{\frac{1}{p}} \le \frac{1}{2}\Big(\frac{2^{1-p/2}}{2^{1-p/2}-1}\Big)^{\frac{1}{p}}.$$

Now fix $0 \le s < t \le 1$. Let n_0 be the largest integer such that there exists an integer k with the property that $s,t \in [(k-1)2^{-n_0}, k2^{-n_0}]$. Now for each $0 \le n \le n_0$ let k_n be the unique integer such that $s,t \in [(k_n-1)2^{-n}, k_n 2^{-n}]$. Then it follows that

$$|\varphi_{nk_n}(t) - \varphi_{nk_n}(s)| \le 2^{\frac{n}{2}}(t-s) \le 2^{\frac{n}{2}}2^{-n_0(1-\alpha)}(t-s)^{\alpha}.$$

For $n_0 < n \le N$, let $k_n > \ell_n$ be the unique integers such that $t \in [(k_n-1)2^{-n}, k_n 2^{-n}]$ and $s \in [(\ell_n-1)2^{-n}, \ell_n 2^{-n}]$. Then

$$|\varphi_{nk_n}(t) - \varphi_{nk_n}(s)| = |\varphi_{nk_n}(t)| \le 2^{-\frac{n}{2}-1},$$
$$|\varphi_{n\ell_n}(t) - \varphi_{n\ell_n}(s)| = |\varphi_{n\ell_n}(s)| \le 2^{-\frac{n}{2}-1}.$$

We conclude that

$$\|\psi(t)-\psi(s)\|^p$$
$$\le \sum_{n=0}^{n_0}\sum_{k=1}^{2^n} 2^{(p-1)n}|\varphi_{nk}(t)-\varphi_{nk}(s)|^p + \sum_{n=n_0+1}^{N}\sum_{k=1}^{2^n} 2^{(p-1)n}|\varphi_{nk}(t)-\varphi_{nk}(s)|^p$$
$$\le \sum_{n=0}^{n_0} 2^{(p-1)n}2^{\frac{np}{2}}2^{-n_0(1-\alpha)p}(t-s)^{\alpha p} + \sum_{n=n_0+1}^{N} 2^{(p-1)n}2^{-\frac{np}{2}}$$
$$\le \frac{1}{2^{\frac{3}{2}p-1}-1}(t-s)^{\alpha p} + \frac{2^{-(n_0+1)(1-\frac{p}{2})}}{1-2^{-(1-\frac{p}{2})}}.$$

Noting that $2^{-(n_0+1)} \le (t-s)$ and $(1-\frac{p}{2}) = \alpha p$, it follows that

$$\|\psi(t)-\psi(s)\| \le \Big(\frac{1}{2^{\frac{3}{2}p-1}-1} + \frac{1}{1-2^{-(1-\frac{p}{2})}}\Big)^{\frac{1}{p}}(t-s)^{\alpha}.$$

Therefore, (5.4) follows.

Step 2: Assume that every function in $C^{\alpha}([0,1];E)$ is stochastically integrable. It follows from the closed graph theorem that there exists a constant C such that for all $\phi \in C^{\alpha}([0,1];E)$ we have

$$\Big(\mathbb{E}\Big\|\int_0^1 \phi\, dW\Big\|^2\Big)^{\frac{1}{2}} \le C\|\phi\|_{C^{\alpha}([0,1];E)}. \tag{5.5}$$

Now assume that E does not have stable type p. By the Maurey–Pisier theorem [12, Theorem 9.6] it follows that l^p is finitely representable in E. In particular it follows that for each integer N there exists an operator $T_N : l^p_N \to E$ such that $\|x\| \le \|T_N x\| \le 2\|x\|$ for all $x \in l^p_N$. Now let $\phi : [0,1] \to E$ be defined as $\phi(t) = T_N \psi_N(t)$, where $\psi_N : [0,1] \to l^p_N$ is the function constructed in Step 1. Then it follows from (5.3), (5.4), and (5.5) that

$$K_p N^{\frac{1}{p}} \le \Big(\mathbb{E}\Big\| \int_0^1 \psi\, dW \Big\|^2\Big)^{\frac{1}{2}} \le \Big(\mathbb{E}\Big\| \int_0^1 \phi\, dW \Big\|^2\Big)^{\frac{1}{2}}$$
$$\le C\|\phi\|_{C^\alpha([0,1];E)} \le 2C\|\psi\|_{C^\alpha([0,1];l^p_N)} \le 2CC_p.$$

This cannot hold for N large and therefore E has stable type p. □

As a corollary we obtain the result that the set of all $\alpha \in (0, 1/2]$ such that every $f \in C^\alpha([0,1]; E)$ is stochastically integrable is relatively open.

Corollary 5.4. *Let E be a Banach space, $\alpha \in (0, 1/2]$, and let $p \in [1, 2)$ be such that $\alpha = 1/p - 1/2$. If every function in $C^\alpha([0,1]; E)$ is stochastically integrable with respect to W, then E has (stable) type p_1 for some $p_1 > p$. In particular, there exists an $\varepsilon \in (0,\alpha)$ such that every function in $C^{\alpha-\varepsilon}([0,1]; E)$ is stochastically integrable.*

Proof. The first part follows from Theorem 5.3 and [12, Corollary 9.7, Proposition 9.12]. The last statement is a consequence of this and Proposition 5.2, where $\varepsilon > 0$ may be taken such that $\alpha - \varepsilon = 1/p_1 - 1/2$. □

6 Banach function spaces

In this section we prove a criterium (Theorem 6.2) for stochastic integrability of a process in the case E is a UMD Banach function space which was stated without proof in [14]. It applies to the spaces $E = L^p(S)$, where $p \in (1,\infty)$ and (S, Σ, μ) is a σ-finite measure space.

We start with the case where Φ is a function with values in $\mathcal{L}(H, E)$. The following proposition extends [15, Corollary 2.10], where the case $H = \mathbb{R}$ was considered.

Proposition 6.1. *Let E be Banach function space with finite cotype over a σ-finite measure space (S, Σ, μ). Let $\Phi : [0,T] \to \mathcal{L}(H, E)$ be a H-strongly measurable function and assume that there exists a strongly measurable function $\phi : [0,T] \times S \to H$ such that for all $h \in H$ and $t \in [0,T]$,*

$$(\Phi(t)h)(\cdot) = [\phi(t,\cdot), h]_H \quad \text{in } E.$$

Then Φ is stochastically integrable if and only if

$$\Big\| \Big(\int_0^T \|\phi(t,\cdot)\|_H^2\, dt\Big)^{\frac{1}{2}} \Big\|_E < \infty. \tag{6.1}$$

In this case we have

$$\Big(\mathbb{E}\Big\|\int_0^T \Phi\,dW_H\Big\|_E^2\Big)^{\frac12} \eqsim_E \Big\|\Big(\int_0^T \|\phi(t,\cdot)\|_H^2\,dt\Big)^{\frac12}\Big\|_E.$$

Proof. First assume that Φ is stochastically integrable. Let $\mathcal{N} = \{n \in \mathbb{N} : 1 \le n < \dim(H)+1\}$, and $(e_m)_{m\in\mathcal{N}}$ be the standard unit basis for $L^2(\mathcal{N},\tau)$, where τ denotes the counting measure on $\mathcal{N}$. Choose orthonormal bases $(f_n)_{n\ge1}$ for $L^2(0,T)$ and $(h_n)_{n\in\mathcal{N}}$ for H. Define $\Psi : [0,T]\times\mathcal{N} \to E$ by $\Psi(t,n) := \Phi(t)h_n$ and define the integral operator $I_\Psi : L^2([0,T]\times\mathcal{N},\, dt\times\tau) \to E$ by

$$I_\Psi f := \int_{\mathcal{N}}\int_{[0,T]} f(t,n)\Psi(t,n)\,dt\,d\tau(n) = \sum_{n\in\mathcal{N}}\int_0^T f(t,n)\Phi(t)h_n\,dt.$$

Note that the integral on the right-hand side is well defined as a Pettis integral. Let $I_\Phi \in \gamma(L^2(0,T;H),E)$ be the operator representing Φ as in Proposition 2.3 (the special case for functions). Then $I_\Psi \in \gamma(L^2([0,T]\times\mathcal{N},\, dt\times\tau),E)$ and

$$\Big(\mathbb{E}\Big\|\int_0^T \Phi\,dW_H\Big\|_E^2\Big)^{\frac12} = \|I_\Phi\|_{\gamma(L^2(0,T;H),E)} = \|I_\Psi\|_{\gamma(L^2([0,T]\times\mathcal{N},\,dt\times\tau),E)}.$$

On the other hand, by a similar calculation as in [15, Corollary 2.10] one obtains, with (r_{mn}) denoting a doubly indexed sequence of Rademacher variables on a probability space $(\Omega',\mathcal{F}',\mathbb{P}')$,

$$\begin{aligned}
\|I_\Psi\|_{\gamma(L^2([0,T]\times\mathcal{N},\,dt\times\tau),E)} &\eqsim_E \Big(\mathbb{E}'\Big\|\sum_{m,n} r_{mn}\int_0^T \sum_k \Psi(t,k)e_m(k)f_n(t)\,dt\Big\|_E^2\Big)^{\frac12}\\
&\eqsim_E \Big\|\Big(\int_0^T \sum_k |\Psi(t,k)(\cdot)|^2\,dt\Big)^{\frac12}\Big\|_E\\
&= \Big\|\Big(\int_0^T \|\phi(t,\cdot)\|_H^2\,dt\Big)^{\frac12}\Big\|_E.
\end{aligned}$$

For the converse one can read all estimates backwards, but we have to show that Φ belongs to $L^2(0,T;H)$ scalarly if (6.1) holds. For all $x^* \in E^*$ we have

$$\begin{aligned}
\|\Phi^*x^*\|_{L^2(0,T;H)}^2 &= \Big(\sum_{m,n}\Big(\int_0^T [\Phi^*(t)x^*, h_m]_H f_n(t)\,dt\Big)^2\Big)^{\frac12}\\
&= \Big(\sum_{n,m}\Big(\int_0^T \sum_k \langle\Psi(t,k),x^*\rangle e_m(k)f_n(t)\,dt\Big)^2\Big)^{\frac12}\\
&\le \Big(\mathbb{E}'\Big\|\sum_{n,m} r_{mn}\int_0^T \sum_k \Psi(t,k)e_m(k)f_n(t)\,dt\Big\|_E^2\Big)^{\frac12}\|x^*\|.
\end{aligned}$$

□

Combining this proposition with Proposition 2.3 and recalling the fact that UMD spaces have finite cotype, we obtain:

Theorem 6.2. *Let E be UMD Banach function space over a σ-finite measure space (S, Σ, μ) and let $p \in (1, \infty)$. Let $\Phi : [0, T] \times \Omega \to \mathcal{L}(H, E)$ be a H-strongly measurable and adapted process and assume that there exists a strongly measurable function ϕ : $[0, T] \times \Omega \times S \to H$ such that for all $h \in H$ and $t \in [0, T]$,*

$$(\Phi(t)h)(\cdot) = [\phi(t, \cdot), h]_H \quad in\ E.$$

Then Φ is stochastically integrable if and only if

$$\Big\| \Big(\int_0^T \|\phi(t, \cdot)\|_H^2 \, dt \Big)^{\frac{1}{2}} \Big\|_E < \infty \quad almost\ surely.$$

In this case for all $p \in (1, \infty)$ we have

$$\mathbb{E} \sup_{t \in [0,T]} \Big\| \int_0^t \Phi(t) \, dW_H(t) \Big\|^p \eqsim_{p,E} \mathbb{E} \Big\| \Big(\int_0^T \|\phi(t, \cdot)\|_H^2 \, dt \Big)^{\frac{1}{2}} \Big\|_E^p.$$

Acknowledgments. The second named author thanks S. Kwapień for the helpful discussion that led to Theorem 5.3.

References

[1] H. Amann, *Operator-valued Fourier multipliers, vector-valued Besov spaces, and applications*, Math. Nachr. 186 (1997), pp. 5–56.

[2] ———, *Compact embeddings of vector-valued Sobolev and Besov spaces*, Glas. Mat. Ser. III 35 (55) (2000), pp. 161–177.

[3] V.I. Bogachev, *Gaussian measures*, Mathematical Surveys and Monographs, vol. 62. American Mathematical Society, Providence, RI, 1998.

[4] D. L. Burkholder, *Martingales and singular integrals in Banach spaces*, Handbook of the geometry of Banach spaces, vol. 1, North-Holland, Amsterdam, 2001, pp. 233–269.

[5] J. Diestel, H. Jarchow, and A. Tonge, *Absolutely summing operators*, Cambridge Studies in Advanced Mathematics, vol. 43. Cambridge University Press, Cambridge, 1995.

[6] D. Feyel and A. de La Pradelle, *Opérateurs linéaires gaussiens*, Potential Anal. 3 (1994), pp. 89–105.

[7] M. Girardi and L. Weis, *Operator-valued Fourier multiplier theorems on Besov spaces*, Math. Nachr. 251 (2003), pp. 34–51.

[8] O. Kallenberg, *Foundations of modern probability*, second ed, Probability and its Applications. Springer-Verlag, New York, 2002.

[9] N. J. Kalton, J. M. A. M. van Neerven, M. C. Veraar, and L. Weis, *Embedding vector-valued Besov spaces into spaces of γ-radonifying operators*, to appear in Math. Nachr.

[10] N. J. Kalton and L. Weis, *The H^∞-functional calculus and square function estimates*, in preparation.

[11] H. König, *Eigenvalue distribution of compact operators*, Operator Theory: Advances and Applications, vol. 16. Birkhäuser Verlag, Basel, 1986.

[12] M. Ledoux and M. Talagrand, *Probability in Banach spaces. Isoperimetry and processes*, Ergebnisse der Mathematik und ihrer Grenzgebiete, vol. 23. Springer-Verlag, Berlin, 1991.

[13] T. R. McConnell, *Decoupling and stochastic integration in UMD Banach spaces*, Probab. Math. Statist. 10 (1989), pp. 283–295.

[14] J. M. A. M. van Neerven, M. C. Veraar, and L. Weis, *Stochastic integration in UMD Banach spaces*, to appear in Ann. Prob.

[15] J. M. A. M. van Neerven and L. Weis, *Stochastic integration of functions with values in a Banach space*, Studia Math. 166 (2005), pp. 131–170.

[16] A. Pelczyński and M. Wojciechowski, *Molecular decompositions and embedding theorems for vector-valued Sobolev spaces with gradient norm*, Studia Math. 107 (1993), pp. 61–100.

[17] J. Rosiński and Z. Suchanecki, *On the space of vector-valued functions integrable with respect to the white noise*, Colloq. Math. 43 (1980), pp. 183–201.

[18] H.-J. Schmeisser, *Vector-valued Sobolev and Besov spaces*, Seminar analysis of the Karl-Weierstraß-Institute of Mathematics 1985/86 (Berlin, 1985/86), Teubner Texte Math. vol. 96, Teubner, Leipzig, 1987, pp. 4–44.

[19] H. Triebel, *Interpolation theory, function spaces, differential operators*, North-Holland Mathematical Library, vol. 18. North-Holland Publishing Co., Amsterdam, 1978.

[20] ———, *Theory of function spaces*, Monographs in Mathematics, vol. 78. Birkhäuser Verlag, Basel, 1983.

[21] M. Yor, *Sur les intégrales stochastiques à valeurs dans un espace de Banach*, Ann. Inst. H. Poincaré Sect. B (N.S.) 10 (1974), pp. 31–36.

Author information

Jan van Neerven, Delft Institute of Applied Mathematics, Delft University of Technology, P.O. Box 5031, 2600 GA Delft, The Netherlands.
Email: J.M.A.M.vanNeerven@tudelft.nl

Mark Veraar, Delft Institute of Applied Mathematics, Delft University of Technology, P.O. Box 5031, 2600 GA Delft, The Netherlands.
Email: M.C.Veraar@tudelft.nl,mark@profsonline.nl

Lutz Weis, Mathematisches Institut I, Technische Universität Karlsruhe, D-76128 Karlsruhe, Germany.
Email: Lutz.Weis@math.uni-karlsruhe.de

Banach Spaces and their Applications in Analysis, 147–182

A new infinite game in Banach spaces with applications

Edward Odell, Thomas Schlumprecht, and András Zsák

Abstract. We consider the following two-player game played on a separable, infinite-dimensional Banach space X. Player S chooses $k_1 \in \mathbb{N}$ and $X_1 \in \mathrm{cof}(X)$, the set of finite-codimensional subspaces of X. Then player P chooses $x_1 \in S_{X_1}$, the unit sphere of X_1. Moves alternate thusly, forever. We study this game in the following setting. Certain normalized, 1-unconditional sequences (u_i) and (v_i) are fixed so that S has a winning strategy to force P to select x_i's so that if the moves are $(k_1, X_1, x_1, k_2, X_2, x_2, \dots)$, then (x_i) is dominated by (u_{k_i}) and/or (x_i) dominates (v_{k_i}). In particular, we show that for suitable (u_i) and (v_i) if X is reflexive and S can win both of the games above, then X embeds into a reflexive space Z with an FDD which also satisfies analogous block upper (u_i) and lower (v_i) estimates. Certain universal space consequences ensue.

Key words. Two-player game, Banach space, finite-dimensional decomposition, universal space.

AMS classification. 46B10, 46B20, 05C05.

Dedicated to Nigel J. Kalton on the occasion of his 60th *birthday.*

1 Introduction

Let X be a separable infinite-dimensional Banach space, and let $\mathcal{A}$ be a set of normalized sequences in X. We can consider a two-player game in X each move of which consists of player S (subspace chooser) selecting some element Y from the set $\mathrm{cof}(X)$ of finite-codimensional subspaces of X, and P (point chooser) responding by selecting a vector y from the unit sphere S_Y of Y. The game, which we shall refer to as the $\mathcal{A}$-game, consists of an infinite sequence of such moves generating a sequence $X_1, x_1, X_2, x_2, \dots$, where $X_i \in \mathrm{cof}(X)$ and $x_i \in S_{X_i}$ for all $i \in \mathbb{N}$. S wins the $\mathcal{A}$-game if $(x_i)_{i=1}^{\infty} \in \mathcal{A}$.

Of course this game, which has its roots in the game described by Gowers [5] and in the notion of asymptotic structure [9], has certain limitations. Unlike the theory of asymptotic structure (where, for each $n \in \mathbb{N}$, a game is considered that consists of n moves, where each move is the same as above), there is generally no unique smallest class $\mathcal{A}$ (depending on X) for which S has a winning strategy. However, one can hypothesize certain specific classes $\mathcal{A}$ for which S has a winning strategy for a given X and deduce certain structural consequences. For example, if for some $K > 0$ we let $\mathcal{A}$ be the class of sequences K-equivalent to the unit vector basis of ℓ_p $(1 < p < \infty)$, then any reflexive space X in which S has a winning strategy for the $\mathcal{A}$-game in X, embeds into an ℓ_p-sum of finite-dimensional spaces [10]. In fact, it was the problem of classifying subspaces of ℓ_p-sums of finite-dimensional spaces that motivated the study of this game.

First author: Research supported by the National Science Foundation.
Second author: Research supported by the National Science Foundation.

The general theme here is to take a coordinate-free property of a space X, recast it in terms of S having a winning strategy in the $\mathcal{A}$-game for a suitable class $\mathcal{A}$, and then to show that X embeds into a space with an FDD (finite-dimensional decomposition) which has the "coordinatized" version of the property we started with. In addition to the ℓ_p result in [10] cited above this general theme was followed in [11] and [13]. In [11] reflexive spaces X were studied for which S has a winning strategy for both games corresponding to the classes $\mathcal{A}_p$ of normalized basic sequences with an ℓ_p-lower estimate and $\mathcal{A}^q$ of normalized basic sequences with an ℓ_q-upper estimate $(1 < q \leq p < \infty)$. The end result was that X embeds into a reflexive space with an FDD such that every block sequence satisfies ℓ_p-lower and ℓ_q-upper estimates. A consequence of this is that one can construct a separable, reflexive space universal for the class of separable, uniformly convex spaces or, more generally, for the class $C_\omega = \{X \,:\, X \text{ is separable, reflexive}\,, \mathrm{Sz}(X) \leq \omega,\ \mathrm{Sz}(X^*) \leq \omega\}$, where $\mathrm{Sz}(Y)$ denotes the Szlenk index of a separable Banach space Y. Recently an alternative proof of the universal result was given [3] using powerful set-theoretical notions (although, the FDD structural results cannot be obtained in this way). We should also note that a set-theoretical study of $\mathcal{A}$-games was given by Rosendal [14].

The motivation behind this paper arose from a problem posed to us by Pełczyński. Given $\alpha < \omega_1$, does there exist a separable, reflexive space universal for the class C_α (defined as above with ω replaced by α). Thus far the authors of [3] have been unable to extend their techniques to this problem. In researching Pełczyński's problem we discovered that it was necessary to consider a new game and solve the corresponding embedding problem in this context.

The game is played as follows. In each move of the game S (subspace chooser) selects $k \in \mathbb{N}$ and $Y \in \mathrm{cof}(X)$, and then P (point chooser) responds by choosing $y \in S_Y$. The game then consists of an infinite sequence of moves generating a sequence $k_1, X_1, x_1, k_2, X_2, x_2, \ldots$, where $k_i \in \mathbb{N}$, $X_i \in \mathrm{cof}(X)$ and $x_i \in S_{X_i}$ for all $i \in \mathbb{N}$. Given a normalized, 1-unconditional sequence (v_i), S is declared winner of this game if (v_{k_i}) is dominated by (x_i). We also consider, for given normalized 1-unconditional sequence (u_i), the version where S wins the game if (x_i) is dominated by (u_{k_i}). In the case where (v_i) and (u_i) are the unit vector bases of ℓ_p and ℓ_q, respectively, this conforms to the games considered in [11], but Pełczyński's problem requires us to consider sequences (v_i) and (u_i) that are not subsymmetric. In [13] in order to solve the problem of embedding an asymptotic ℓ_p space into one with an asymptotic ℓ_p FDD it was necessary to extend the results of [11] concerning ℓ_p-lower and ℓ_q-upper estimates to more general (v_i)-lower and (u_i)-upper estimates, but the game played did not change.

The main results of this paper are given in Section 4, Theorems 4.1 and 4.5, and Corollary 4.3. In brief these theorems say the following. Suppose we are given normalized, 1-unconditional bases (v_i) and (u_i) with certain properties, and a reflexive space X. Assume that S wins the subsequential (v_i)-lower and the subsequential (u_i)-upper games described above. Then X embeds into a space Z with an FDD such that every block sequence satisfies subsequential (v_i)-lower and (u_i)-upper estimates. (Precise definitions of these estimates will be given below.)

One application of these theorems is a new proof of the results of [11]. The application to the Pełczyński problem will appear in [12], where further machinery is necessary to exploit the results obtained here.

In Section 5 we derive some universal space consequences of our embedding theorems. Section 2 introduces our terminology, in particular we give precise definitions of various lower and upper norm estimates. Section 2 also contains some straightforward duality results concerning such norm estimates, and a combinatorial result (Proposition 2.18) that is key to embedding spaces satisfying the coordinate-free version of a certain property into a space with an FDD satisfying the "coordinatized" version of the same property. In Section 3 we define the space $Z^V(E)$, where Z is a Banach space with an FDD $E=(E_i)$ and V is the closed linear span of a normalized, 1-unconditional sequence (v_i). We develop the properties of $Z^V(E)$, in particular proving that, under appropriate hypotheses, $Z^V(E)$ is a reflexive space admitting subsequential V-lower estimates.

2 Definitions and preliminary results

We begin with fixing some terminology. Let Z be a Banach space with an FDD $E=(E_n)$. For $n\in\mathbb{N}$ we denote by P_n^E the n-th *coordinate projection, i.e.,* $P_n^E: Z\to E_n$ is the map defined by $\sum_i z_i \mapsto z_n$, where $z_i\in E_i$ for all $i\in\mathbb{N}$. For a finite set $A\subset\mathbb{N}$ we put $P_A^E=\sum_{n\in A} P_n^E$. The *projection constant* $K(E,Z)$ *of* (E_n) *(in* Z*)* is defined by

$$K=K(E,Z)=\sup_{m<n}\|P^E_{[m,n]}\|,$$

where $[m,n]$ denotes the interval $\{m,m+1,\dots,n\}$ in $\mathbb{N}$. Recall that K is always finite and, as in the case of bases, we say that (E_n) *is bimonotone (in* Z*)* if $K=1$. By passing to the equivalent norm

$$|||\cdot|||: Z\to\mathbb{R}\,, \qquad z\mapsto \sup_{m<n}\|P^E_{[m,n]}(z)\|,$$

we can always renorm Z so that $K=1$.

For a sequence (E_i) of finite-dimensional spaces we define the vector space

$$c_{00}(\oplus_{i=1}^\infty E_i)=\big\{(z_i):\, z_i\in E_i \text{ for all } i\in\mathbb{N}, \text{ and } \{i\in\mathbb{N}:\, z_i\neq 0\} \text{ is finite}\big\},$$

which is dense in each Banach space for which (E_i) is an FDD. For a set $A\subset\mathbb{N}$ we denote by $\bigoplus_{i\in A} E_i$ the linear subspace of $c_{00}(\oplus E_i)$ generated by the elements of $\bigcup_{i\in A} E_i$. As usual we denote the vector space of sequences in $\mathbb{R}$ which are eventually zero by c_{00}. We sometimes will consider for the same sequence (E_i) of finite-dimensional spaces different norms on $c_{00}(\oplus E_i)$. In order to avoid confusion we will therefore often index the norm by the Banach space whose norm we are using, *i.e.,* $\|\cdot\|_Z$ denotes the norm of the Banach space Z.

If Z has an FDD (E_i), the vector space $c_{00}(\oplus_{i=1}^\infty E_i^*)$, where E_i^* is the dual space of E_i for each $i\in\mathbb{N}$, can be identified in a natural way with a w^*-dense subspace of Z^*. Note however that the embedding $E_i^*\hookrightarrow Z^*$ is, in general, not isometric unless $K=1$. We will always consider E_i^* with the norm it inherits from Z^* instead of the norm it has as the dual space of E_i. We denote the norm closure of $c_{00}(\oplus_{i=1}^{\infty} E_i^*)$ in Z^* by $Z^{(*)}$. Note that $Z^{(*)}$ is w^*-dense in Z^*, the unit ball $B_{Z^{(*)}}$ norms Z, and (E_i^*) is an FDD of $Z^{(*)}$ having a projection constant not exceeding $K(E,Z)$. If $K(E,Z)=1$, then $B_{Z^{(*)}}$ is 1-norming for Z and $Z^{(*)(*)}=Z$.

For $z\in c_{00}(\oplus E_i)$ we define *the support* $\text{supp}_E(z)$ *of* z *with respect to* (E_i) by

$$\text{supp}_E(z) = \{i\in\mathbb{N} :\ P_i^E(z)\neq 0\},$$

and we define the *range* $\text{ran}_E(z)$ *of* Z *with respect to* (E_i) to be the smallest interval in $\mathbb{N}$ containing $\text{supp}_E(z)$. A sequence (z_i) (finite or infinite) of non-zero vectors in $c_{00}(\oplus E_i)$ is called *a block sequence of* (E_i) if

$$\max \text{supp}_E(z_n) < \min \text{supp}_E(z_{n+1}) \quad \text{whenever } n\in\mathbb{N} \text{ (or } n<\text{length}(z_i)).$$

A block sequence (z_i) of (E_i) is called *normalized (in Z)* if $\|z_i\|_Z=1$ for all $i\in\mathbb{N}$.

Let $\bar\delta=(\delta_i)\subset(0,1)$ with $\delta_i\downarrow 0$. A (finite or infinite) sequence (z_i) in S_Z is called *a* $\bar\delta$*-skipped block sequence of* (E_i) if there exists a sequence $1\le k_0<k_1<k_2<\dots$ in $\mathbb{N}$ such that

$$\|z_n - P^E_{(k_{n-1},k_n)}(z_n)\| < \delta_n \quad \text{for all } n\in\mathbb{N} \text{ (or } n\le\text{length}(z_i)).$$

Remark 2.1. A sequence (F_i) of finite-dimensional spaces is called *a blocking of* (E_i) if for some sequence $m_1<m_2<\dots$ in $\mathbb{N}$ we have $F_n=\bigoplus_{j=m_{n-1}+1}^{m_n} E_j$ for all $n\in\mathbb{N}$ ($m_0=0$). If (F_i) is a blocking of (E_i), and if (x_i) is a $\bar\delta$-skipped block sequence of (F_i), then (x_i) is not necessarily a $\bar\delta$-skipped block sequence of (E_i) (since in the definition of skipped block sequence we skip exactly one coordinate). Nevertheless it is clear that (x_i) is a $2K\bar\delta$-skipped block sequence of (E_i), where K is the projection constant of (E_i) in Z.

Definition 2.2. Given two sequences (e_i) and (f_i) in some Banach spaces, and given a constant $C>0$, we say that (f_i) *C-dominates* (e_i), or that (e_i) *is C-dominated by* (f_i), if

$$\Big\|\sum a_i e_i\Big\| \le C\Big\|\sum a_i f_i\Big\| \qquad \text{for all } (a_i)\in c_{00}.$$

We say that (f_i) *dominates* (e_i), or that (e_i) *is dominated by* (f_i), if there exists a constant $C>0$ such that (f_i) C-dominates (e_i).

We shall now introduce certain lower and upper norm estimates for FDD's.

Definition 2.3. Let Z be a Banach space with an FDD (E_n), let V be a Banach space with a normalized, 1-unconditional basis (v_i) and let $1\le C<\infty$.

We say that (E_n) *satisfies subsequential C-V-lower estimates (in Z)* if every normalized block sequence (z_i) of (E_n) in Z C-dominates (v_{m_i}) with $m_i=\min\text{supp}_E(z_i)$ for all $i\in\mathbb{N}$, and (E_n) *satisfies subsequential C-V-upper estimates (in Z)* if every normalized block sequence (z_i) of (E_n) in Z is C-dominated by (v_{m_i}), where $m_i=\min\text{supp}_E(z_i)$ for all $i\in\mathbb{N}$.

If U is another space with a normalized and 1-unconditional basis (u_i), we say that (E_n) *satisfies subsequential C-(V,U) estimates (in Z)* if it satisfies subsequential C-V-lower and C-U-upper estimates in Z.

We say that (E_n) satisfies *subsequential V-lower, U-upper* or *(V,U) estimates (in Z)* if for some $C \geq 1$ it satisfies subsequential C-V-lower, C-U-upper or C-(V,U) estimates in Z, respectively.

Remark 2.4. Assume that (E_n) satisfies subsequential C-V-lower estimates in Z and that (z_i) is a normalized block sequence of (E_n). If

$$\max \operatorname{supp}_E(z_{i-1}) < m_i \leq \min \operatorname{supp}_E(z_i)$$

for all $i \in \mathbb{N}$ (where $\max \operatorname{supp}_E(z_0) = 0$), then (z_i) C-dominates (v_{m_i}).

Another easy fact is that if every normalized block sequence (z_i) of (E_n) in Z dominates (v_{m_i}), where $m_i = \min \operatorname{supp}_E(z_i)$ for all $i \in \mathbb{N}$, then (E_n) satisfies subsequential V-lower estimates in Z.

Analogous statements hold for upper estimates.

We shall need a coordinate-free version of subsequential lower and upper estimates. One way of defining this is reminiscent of the notion of asymptotic structure. Let V be a Banach space with a normalized and 1-unconditional basis (v_i), and let $C \in [1, \infty)$. Assume that we are given an infinite-dimensional Banach space X. We say that X *satisfies subsequential C-V-lower estimates* (respectively, *subsequential C-V-upper estimates*) if

$$\begin{array}{lll} \exists\, k_1 \in \mathbb{N} & \exists\, X_1 \in \operatorname{cof}(X) & \forall\, x_1 \in S_{X_1} \\ \exists\, k_2 \in \mathbb{N} & \exists\, X_2 \in \operatorname{cof}(X) & \forall\, x_2 \in S_{X_2} \\ \exists\, k_3 \in \mathbb{N} & \exists\, X_3 \in \operatorname{cof}(X) & \forall\, x_3 \in S_{X_3} \\ \vdots & \vdots & \vdots \end{array}$$

such that $k_1 < k_2 < \dots$ and (v_{k_i}) is C-dominated by (respectively, C-dominates) (x_i). If U is another Banach space with a normalized, 1-unconditional basis (u_i), then we say that X *satisfies subsequential C-(V,U) estimates* if it satisfies subsequential C-V-lower and C-U-upper estimates. Finally, we say that *the Banach space X satisfies subsequential V-lower, U-upper or (V,U) estimates* if for some constant C it satisfies subsequential C-V-lower, C-U-upper or C-(V,U) estimates, respectively.

The above definitions are given more formally in the language of games. Let us recall from the introduction that in our games each move consists of S (subspace chooser) selecting $k \in \mathbb{N}$ and $Y \in \operatorname{cof}(X)$, and then P (point chooser) responding by choosing $y \in S_Y$. The game then consists of an infinite sequence of moves generating a sequence $k_1, X_1, x_1, k_2, X_2, x_2, \dots$, where $k_i \in \mathbb{N}$, $X_i \in \operatorname{cof}(X)$ and $x_i \in S_{X_i}$ for all $i \in \mathbb{N}$. Player S wins the game if (v_{k_i}) is C-dominated by (x_i), otherwise P is declared the winner. The space X then satisfies subsequential C-V-lower estimates if and only if S has a winning strategy, *i.e.*, there is a function ϕ such that given sequences (k_i) in $\mathbb{N}$, (X_i) in $\operatorname{cof}(X)$, and (x_i) in X such that $x_i \in S_{X_i}$ *and* $(k_i, X_i) = \phi(x_1, x_2, \dots, x_{i-1})$ for all

$i \in \mathbb{N}$, then $k_1 < k_2 < \dots$ and (v_{k_i}) is C-dominated by (x_i). The notions of subsequential C-V-upper estimates, subsequential C-(V,U) estimates, etc. can be formalized in a similar way.

Another way of expressing subsequential lower and upper estimates in a coordinate-free way uses infinite, countably branching trees (see Proposition 2.9 below). This is not surprising since a winning strategy in the game described above corresponds naturally to such a tree. We define for $\ell \in \mathbb{N}$

$$T_\ell = \big\{(n_1, n_2, \dots, n_\ell) : n_1 < n_2 < \dots < n_\ell \text{ are in } \mathbb{N}\big\}$$

and

$$T_\infty = \bigcup_{\ell=1}^{\infty} T_\ell\,, \qquad T_\infty^{\text{even}} = \bigcup_{\ell=1}^{\infty} T_{2\ell}.$$

If $\alpha = (m_1, \dots, m_\ell) \in T_\ell$, we call ℓ the *length of* α and denote it by $|\alpha|$, and $\beta = (n_1, \dots, n_k) \in T_\infty$ is called an *extension of* α, or α is called *a restriction of* β, if $\ell < k$ and $m_i = n_i$ for $i = 1, \dots, \ell$. We then write $\alpha < \beta$ and with this order both T_∞ and T_∞^{even} are trees.

An *even tree* in a Banach space X is a family $(x_\alpha)_{\alpha \in T_\infty^{\text{even}}}$ in X. Sequences of the form $\big(x_{(\alpha,n)}\big)_{n > n_{2\ell-1}}$, where $\ell \in \mathbb{N}$ and $\alpha = (n_1, n_2, \dots, n_{2\ell-1}) \in T_\infty$, are called *nodes* of the tree. For a sequence $n_1 < n_2 < \dots$ of positive integers the sequence $\big(x_{(n_1,n_2,\dots,n_{2\ell})}\big)_{\ell=1}^{\infty}$ is called a *branch* of the tree.

If $(x_\alpha)_{\alpha \in T_\infty^{\text{even}}}$ is an even tree in a Banach space X and if $T' \subset T_\infty^{\text{even}}$ is closed under taking restrictions so that for each $\alpha \in T' \cup \{\emptyset\}$ and for each $m \in \mathbb{N}$ the set $\{n \in \mathbb{N} : (\alpha, m, n) \in T'\}$ is either empty or has infinite size, and moreover the latter occurs for infinitely many values of m, then we call $(x_\alpha)_{\alpha \in T'}$ a *full subtree of* $(x_\alpha)_{\alpha \in T_\infty^{\text{even}}}$. Note that $(x_\alpha)_{\alpha \in T'}$ could then be relabeled to a family indexed by T_∞^{even}, and note that the branches of $(x_\alpha)_{\alpha \in T'}$ are branches of $(x_\alpha)_{\alpha \in T_\infty^{\text{even}}}$ and that the nodes of $(x_\alpha)_{\alpha \in T'}$ are subsequences of certain nodes of $(x_\alpha)_{\alpha \in T_\infty^{\text{even}}}$.

An even tree $(x_\alpha)_{\alpha \in T_\infty^{\text{even}}}$ in a Banach space X is called *normalized* if $\|x_\alpha\| = 1$ for all $\alpha \in T_\infty^{\text{even}}$, and is called *weakly null* if every node is a weakly null sequence. If X has an FDD (E_n), then $(x_\alpha)_{\alpha \in T_\infty^{\text{even}}}$ is called a *block even tree of* (E_n) if every node is a block sequence of (E_n).

Definition 2.5. Let V be a Banach space with a normalized and 1-unconditional basis (v_i), and let $C \in [1, \infty)$. Assume that we are given an infinite-dimensional Banach space X. We say that X *satisfies subsequential* C-V-*lower tree estimates* if every normalized, weakly null even tree $(x_\alpha)_{\alpha \in T_\infty^{\text{even}}}$ in X has a branch $\big(x_{(n_1,n_2,\dots,n_{2i})}\big)$ which C-dominates $(v_{n_{2i-1}})$.

We say that X *satisfies subsequential* C-V-*upper tree estimates* if every normalized, weakly null even tree $(x_\alpha)_{\alpha \in T_\infty^{\text{even}}}$ in X has a branch $\big(x_{(n_1,n_2,\dots,n_{2i})}\big)$ which is C-dominated by $(v_{n_{2i-1}})$.

If U is a second space with a 1-unconditional and normalized basis (u_i), we say that X *satisfies subsequential* C-(V,U) *tree estimates* if it satisfies subsequential C-V-lower and C-U-upper tree estimates.

We say that X *satisfies subsequential V-lower, U-upper* or (V,U) *tree estimates* if for some $1 \leq C < \infty$ X satisfies subsequential C-V-lower, C-U-upper or C-(V,U) tree estimates, respectively.

Remark 2.6. As in the FDD case, we do not need to fix a constant C in the above definitions: if every normalized, weakly null even tree $(x_\alpha)_{\alpha \in T_\infty^{\text{even}}}$ in X has a branch $(x_{(n_1,n_2,\dots,n_{2i})})$ which dominates $(v_{n_{2i-1}})$, then there exists a constant $C \geq 1$ such that X satisfies subsequential C-V-lower tree estimates. The analogous statement for upper estimates also holds. (See [13, Proposition 1.2].)

Proposition 2.9 below shows that, under some mild hypotheses, the two coordinate-free versions of lower and upper estimates given above are essentially the same. Before stating this result we need a certain property of basic sequences defined in [1].

Definition 2.7. Let V be a Banach space with a normalized, 1-unconditional basis (v_i) and let $1 \leq C < \infty$.

We say that (v_i) is *C-right-dominant* (respectively, *C-left-dominant*) if for all sequences $m_1 < m_2 < \dots$ and $n_1 < n_2 < \dots$ of positive integers with $m_i \leq n_i$ for all $i \in \mathbb{N}$ we have that (v_{m_i}) is C-dominated by (respectively, C-dominates) (v_{n_i}). We say that (v_i) is *right-dominant* or *left-dominant* if for some $C \geq 1$ it is C-right-dominant or C-left-dominant, respectively.

Remark 2.8. For (v_i) to be right-dominant (respectively, left-dominant) it is enough to have the property that (v_{m_i}) is dominated by (respectively, dominates) (v_{n_i}) for all sequences $m_1 < m_2 < \dots$ and $n_1 < n_2 < \dots$ of positive integers with $m_i \leq n_i$ for all $i \in \mathbb{N}$. Also, (v_i) is C-right-dominant (respectively, C-left-dominant) if and only if the sequence (v_i^*) of biorthogonal functionals in $V^{(*)}$ is C-left-dominant (respectively, C-right-dominant).

Proposition 2.9. *Let V be a Banach space with a normalized and 1-unconditional basis (v_i) and let $C, D \in [1, \infty)$. Let X be an infinite-dimensional Banach space.*

(a) Assume that (v_i) is D-left-dominant. If X satisfies subsequential C-V-lower estimates, then for all $\varepsilon > 0$ X satisfies subsequential $(CD+\varepsilon)$-V-lower tree estimates.

(b) Assume that X^* is separable. If X satisfies subsequential C-V-lower tree estimates, then it also satisfies subsequential C-V-lower estimates.

Remark 2.10. Analogous results hold for upper estimates. For that in (a) we need to assume that (v_i) is D-right-dominant.

Proof. (a) Assume that for some $\varepsilon > 0$ there is a normalized, weakly null even tree $(x_\alpha)_{\alpha \in T_\infty^{\text{even}}}$ in X such that for any sequence $n_1 < n_2 < \dots$ of positive integers the branch $(x_{(n_1,n_2,\dots,n_{2i})})$ does not $(CD+\varepsilon)$-dominate $(v_{n_{2i-1}})$. We show that in this case X does not satisfy subsequential C-V-lower estimates by exhibiting a winning strategy for the point chooser P. Fix a sequence $(\delta_i)\subset(0,1)$ with $\Delta=\sum_i \delta_i$ satisfying

$$C<\frac{CD+\varepsilon}{D}\Big(1+\Delta\frac{CD+\varepsilon}{D}\Big)^{-1}.$$

Suppose the game starts with S picking $k_1\in\mathbb{N}$ and $X_1\in\mathrm{cof}(X)$. Since the nodes of $(x_\alpha)_{\alpha\in T_\infty^{\mathrm{even}}}$ are weakly null, there exist $n_1,n_2\in\mathbb{N}$ such that $k_1\leq n_1<n_2$ and $d\big(x_{(n_1,n_2)},X_1\big)<\delta_1$. P's response will be a $y_1\in S_{X_1}$ with $\big\|x_{(n_1,n_2)}-y_1\big\|<\delta_1$. In the second move S picks $k_2\in\mathbb{N}$ and $X_2\in\mathrm{cof}(X)$. Then there exist $n_3,n_4\in\mathbb{N}$ such that $n_2<n_3<n_4$, $k_2\leq n_3$ and $d\big(x_{(n_1,n_2,n_3,n_4)},X_2\big)<\delta_2$. P's response will be some $y_2\in S_{X_2}$ with $\big\|x_{(n_1,n_2,n_3,n_4)}-y_2\big\|<\delta_2$. In general, on the jth move of the game ($j\geq 2$), S picks $k_j\in\mathbb{N}$, $X_j\in\mathrm{cof}(X)$. Then one can find $n_{2j-1},n_{2j}\in\mathbb{N}$ such that $n_{2j-2}<n_{2j-1}<n_{2j}$, $k_j\leq n_{2j-1}$ and $d\big(x_{(n_1,n_2,\dots,n_{2j})},X_j\big)<\delta_j$. P's j^{th} move will be some $y_j\in S_{X_j}$ such that $\big\|x_{(n_1,n_2,\dots,n_{2j})}-y_j\big\|<\delta_j$.

Since the branch $\big(x_{(n_1,n_2,\dots,n_{2i})}\big)$ does not $(CD+\varepsilon)$-dominate $(v_{n_{2i-1}})$, there exists $(a_i)\in c_{00}$ such that

$$\Big\|\sum_i a_i v_{n_{2i-1}}\Big\|>(CD+\varepsilon)\Big\|\sum_i a_i x_{(n_1,n_2,\dots,n_{2i})}\Big\|.$$

We may assume without loss of generality that $\big\|\sum a_i v_{k_i}\big\|=1$. Using the D-left-dominant property of (v_i) and that $k_i\leq n_{2i-1}$ for all $i\in\mathbb{N}$, together with the choice of Δ, an easy computation now gives

$$\Big\|\sum_i a_i v_{k_i}\Big\|>C\Big\|\sum_i a_i y_i\Big\|.$$

Thus P wins the game.

(b) Assume that X does not satisfy subsequential C-V-lower estimates. This means that S does not have a winning strategy, which in turn implies that there is a winning strategy ϕ for the point chooser (this follows from the fact that closed games [4] or, more generally, Borel games [8] are determined). Thus given sequences (k_i) in $\mathbb{N}$, (X_i) in $\mathrm{cof}(X)$, and (x_i) in X such that $k_1<k_2<\dots$ *and* $x_n=\phi(k_1,X_1,k_2,X_2,\dots,k_n,X_n)$ for all $n\in\mathbb{N}$, then $x_i\in S_{X_i}$ for all $i\in\mathbb{N}$ and (v_{k_i}) is not C-dominated by (x_i). Fix a sequence (X_i) of finite-codimensional subspaces of X such that every bounded sequence (x_i) with $x_i\in X_i$ for all $i\in\mathbb{N}$ is weakly null. This exists by the assumption that X has separable dual.

We now construct a normalized, weakly null even tree in X by recursion to show that X does not satisfy subsequential C-V-lower tree estimates. For $\ell\in\mathbb{N}$ and $\alpha=(n_1,n_2,\dots,n_{2\ell})\in T_{2\ell}$ we set $x_\alpha=\phi(n_1,X_{n_2},n_3,X_{n_4},\dots,n_{2\ell-1},X_{n_{2\ell}})$. It is easy to verify that $(x_\alpha)_{\alpha\in T_\infty^{\mathrm{even}}}$ is a normalized, weakly null even tree in X, and that for any sequence $n_1<n_2<\dots$ of positive integers the branch $\big(x_{(n_1,n_2,\dots,n_{2i})}\big)$ does not C-dominate $(v_{n_{2i-1}})$. □

If V is a Banach space with a normalized, 1-unconditional basis, and if N is an infinite subset of $\mathbb{N}$, we write V_N for the closed linear span of $\{v_i\,:\,i\in N\}$. When we talk about subsequential V_N-lower estimates, etc., it will be with respect to the normalized, 1-unconditional basis $(v_i)_{i\in N}$ of V_N. We shall also write $\mathbb{N}^{(\omega)}$ for the set of all infinite subsets of $\mathbb{N}$.

Note that if V is a Banach space with a normalized, 1-unconditional, left-dominant basis (v_i), then for any space Z with an FDD (E_i) satisfying subsequential V-lower estimates in Z, the FDD (E_i) will also satisfy subsequential V_N-lower estimates for any $N\in\mathbb{N}^{(\omega)}$. Later on we shall need a result that allows us to pass from subsequential V_N-lower estimates for some $N\in\mathbb{N}^{(\omega)}$ to subsequential V-lower estimates. Before stating this result we need a definition.

Definition 2.11. Let V be a Banach space with a normalized, 1-unconditional basis (v_i) and let $1\leq C<\infty$. We say that (v_i) is *C-block-stable* if any two normalized block bases (x_i) and (y_i) with

$$\max\big(\mathrm{supp}(x_i)\cup\mathrm{supp}(y_i)\big) < \min\big(\mathrm{supp}(x_{i+1})\cup\mathrm{supp}(y_{i+1})\big) \quad \text{for all } i\in\mathbb{N}$$

are C-equivalent. We say that (v_i) is *block-stable* if it is C-block-stable for some constant C.

Remark 2.12. It is routine to check that (v_i) is C-block-stable if and only if the sequence (v_i^*) of biorthogonal functionals in $V^{(*)}$ is C-block-stable.

A block-stable basis is a special case of a *block-norm-determined FDD* introduced by H. P. Rosenthal, who has initiated an exhaustive study of such FDDs [15].

Lemma 2.13. *Let V and U be Banach spaces with normalized, 1-unconditional block-stable bases (v_i) and (u_i), respectively, and assume that (v_i) is dominated by (u_i). Let $M\in\mathbb{N}^{(\omega)}$ and let Z be a Banach space with an FDD $E=(E_i)$ satisfying subsequential (V_M,U_M) estimates in Z. Then $W=Z\oplus_{\ell_\infty}V_{\mathbb{N}\setminus M}$ has an FDD $F=(F_i)$ satisfying subsequential (V,U) estimates in W.*

Proof. Choose constants $B,C,D\in[1,\infty)$ such that (v_i) and (u_i) are B-block-stable, (v_i) is D-dominated by (u_i), and (E_i) satisfies subsequential C-(V_M,U_M) estimates in Z. For each $n\in\mathbb{N}$ define

$$F_n=\begin{cases} E_i & \text{if } n=m_i \text{ for some } i\in\mathbb{N}\,,\\ \mathbb{R}\cdot v_n & \text{if } n\notin M\,.\end{cases}$$

Then $F=(F_n)$ is an FDD for W with projection constant $K(F,W)\leq K(E,Z)$. We now show that (F_i) satisfies subsequential $\overline{C}$-(V,U) estimates in W, where $\overline{C}=B\cdot\max\{2C,D\}$.

Let (z_i) be a normalized block sequence of (F_n) in W. For each $i\in\mathbb{N}$ let $k_i=\min\mathrm{supp}_F(z_i)$ and write

$$z_i=z_i^{(1)}+z_i^{(2)}, \quad \text{where } z_i^{(1)}\in Z\,,\ z_i^{(2)}\in V_{\mathbb{N}\setminus M}.$$

Fix $(a_i)\in c_{00}$. We have

$$\Big\|\sum_i a_i z_i^{(1)}\Big\|_Z \geq \frac{1}{C}\Big\|\sum_i a_i\cdot\|z_i^{(1)}\|_Z\cdot v_{\min\operatorname{supp}_F(z_i^{(1)})}\Big\|_V \geq \frac{1}{BC}\Big\|\sum_i a_i\cdot\|z_i^{(1)}\|_Z\cdot v_{k_i}\Big\|_V$$

and

$$\Big\|\sum_i a_i z_i^{(2)}\Big\|_V \geq \frac{1}{B}\Big\|\sum_i a_i\cdot\|z_i^{(2)}\|_Z\cdot v_{k_i}\Big\|_V .$$

It follows that

$$\begin{aligned}\Big\|\sum_i a_i z_i\Big\|_W &\geq \frac{1}{BC}\max\Big\{\Big\|\sum_i a_i\cdot\|z_i^{(1)}\|_Z\cdot v_{k_i}\Big\|_V, \Big\|\sum_i a_i\cdot\|z_i^{(2)}\|_Z\cdot v_{k_i}\Big\|_V\Big\}\\ &\geq \frac{1}{2BC}\Big\|\sum_i a_i v_{k_i}\Big\|_V .\end{aligned}$$

Similarly, we have

$$\Big\|\sum_i a_i z_i^{(1)}\Big\|_Z \leq C\Big\|\sum_i a_i\cdot\|z_i^{(1)}\|_Z\cdot u_{\min\operatorname{supp}_F(z_i^{(1)})}\Big\|_U \leq BC\Big\|\sum_i a_i\cdot\|z_i^{(1)}\|_Z\cdot u_{k_i}\Big\|_U$$

and

$$\Big\|\sum_i a_i z_i^{(2)}\Big\|_V \leq B\Big\|\sum_i a_i\cdot\|z_i^{(2)}\|_Z\cdot v_{k_i}\Big\|_V \leq BD\Big\|\sum_i a_i\cdot\|z_i^{(2)}\|_Z\cdot u_{k_i}\Big\|_U .$$

It follows that

$$\begin{aligned}\Big\|\sum_i a_i z_i\Big\|_W &\leq \max\Big\{BC\Big\|\sum_i a_i\cdot\|z_i^{(1)}\|_Z\cdot u_{k_i}\Big\|_U, BD\Big\|\sum_i a_i\cdot\|z_i^{(2)}\|_Z\cdot u_{k_i}\Big\|_U\Big\}\\ &\leq B\max\{C,D\}\Big\|\sum_i a_i u_{k_i}\Big\|_U .\end{aligned}$$

The proof of the lemma is complete. □

The next two results show how norm estimates in a space and in its dual are related.

Proposition 2.14. *Assume that Z has an FDD (E_i), and let V be a space with a normalized and 1-unconditional basis (v_i). The following statements are equivalent:*

(a) (E_i) satisfies subsequential V-lower estimates in Z.
(b) (E_i^*) satisfies subsequential $V^{(*)}$-upper estimates in $Z^{(*)}$.

(Here subsequential $V^{(*)}$-upper estimates are with respect to (v_i^*), the sequence of biorthogonal functionals to (v_i)).

Moreover, if (E_i) is bimonotone in Z, then the equivalence holds true if one replaces, for some $C\geq 1$, V-lower estimates by C-V-lower estimates in (a) and $V^{(*)}$-upper estimates by C-$V^{(*)}$-upper estimates in (b).

Remark 2.15. By duality, Proposition 2.14 holds if we interchange the words *lower* and *upper* in (a) and (b).

Proof. Without loss of generality we may assume that (E_i) is bimonotone in Z.
"(a)⇒(b)" Let (z_i^*) be a normalized block sequence of $E^* = (E_n^*)$ in $Z^{(*)}$, and for each $i \in \mathbb{N}$ let $m_i = \min \mathrm{supp}_{E^*}(z_i^*)$. Given $(a_i) \in c_{00}$, choose $z \in S_Z$ with finite support with respect to (E_n) such that $\|\sum a_i z_i^*\| = \sum a_i z_i^*(z)$. For each $i \in \mathbb{N}$ write

$$P^E_{[\min \mathrm{supp}_{E^*}(z_i^*), \min \mathrm{supp}_{E^*}(z_{i+1}^*))}(z) = b_i z_i ,$$

where $z_i \in S_Z$ and $|b_i| \leq 1$. Since (E_n) satisfies subsequential C-V-lower estimates in Z, we have $\|\sum b_i v_{m_i}\| \leq C \|\sum b_i z_i\| \leq C$. Hence

$$\begin{aligned}\Big\|\sum a_i z_i^*\Big\| &= \sum a_i b_i z_i^*(z_i) \leq \sum |a_i||b_i| \\ &\leq \Big\|\sum a_i v_{m_i}^*\Big\| \cdot \Big\|\sum b_i v_{m_i}\Big\| \leq C \Big\|\sum a_i v_{m_i}^*\Big\|,\end{aligned}$$

as required.
"(b)⇒(a)" Let (z_i) be a normalized block sequence of (E_n) in Z, and for each $i \in \mathbb{N}$ let $m_i = \min \mathrm{supp}_E(z_i)$. Given $(a_i) \in c_{00}$, choose $(b_i) \in c_{00}$ such that $\|\sum b_i v_{m_i}^*\| = 1$ and $\|\sum a_i v_{m_i}\| = \sum a_i b_i$. For each $i \in \mathbb{N}$ there exists $z_i^* \in S_{Z^{(*)}}$ such that $z_i^*(z_i) = 1$ and $\mathrm{ran}_{E^*}(z_i^*) \subset \mathrm{ran}_E(z_i)$. Since (E_n^*) satisfies subsequential C-$V^{(*)}$-upper estimates in $Z^{(*)}$, we have $\|\sum b_i z_i^*\| \leq C$, and hence

$$\Big\|\sum a_i z_i\Big\| \geq \frac{1}{C} \sum a_i b_i = \frac{1}{C} \Big\|\sum a_i v_{m_i}\Big\|.$$

This completes the proof. □

Proposition 2.16. *Assume that U is a space with a normalized, 1-unconditional basis (u_i) which is D-right-dominant for some $D \geq 1$, and that X is a reflexive space which satisfies subsequential C-U-upper tree estimates for some $C \geq 1$.*

Then, for any $\varepsilon > 0$, X^* satisfies subsequential $(2CD+\varepsilon)$-$U^{(*)}$-lower tree estimates.

Remark 2.17. One might ask whether or not the converse of Proposition 2.16 is true, *i.e.*, similar to the FDD case, whether X satisfies subsequential U-upper tree estimates if X^* satisfies subsequential $U^{(*)}$-lower tree estimates.

The answer is affirmative under certain conditions on U, but we do not give a direct proof for that fact. Instead, we shall deduce it from one of our main embedding theorems (see Corollary 4.4 in Section 4).

Proof. We start with a simple observation. Let (x_i^*) be a normalized, weakly null sequence in X^*. For each $n \in \mathbb{N}$ pick $x_n \in S_X$ with $x_n^*(x_n) = 1$. There exist $y \in X$ and $k_1 < k_2 < \ldots$ in $\mathbb{N}$ such that $x_{k_n} \xrightarrow{w} y$. Given $\eta \in (0,1)$, there exists $n_0 \in \mathbb{N}$ such that $|x_{k_n}^*(y)| < \eta$ for all $n \geq n_0$. Set

$$y_n^* = x_{k_{n_0+n}}^* \quad \text{and} \quad y_n = \frac{x_{k_{n_0+n}} - y}{\|x_{k_{n_0+n}} - y\|}, \qquad n \in \mathbb{N}.$$

We have found, for given $\eta \in (0,1)$, a subsequence (y_i^*) of (x_i^*) and a normalized, weakly null sequence (y_i) in X satisfying $y_n^*(y_n) > (1-\eta)/2$ for all $n \in \mathbb{N}$.

Now let $(x_\alpha^*)_{\alpha \in T_\infty^{\text{even}}}$ be a normalized, weakly null even tree in X^*. By the above observation we can find a normalized, weakly null even tree $(y_\alpha)_{\alpha \in T_\infty^{\text{even}}}$ in X and a full subtree $(y_\alpha^*)_{\alpha \in T_\infty^{\text{even}}}$ of $(x_\alpha^*)_{\alpha \in T_\infty^{\text{even}}}$ such that $y_\alpha^*(y_\alpha) > (1-\eta)/2$ for all $\alpha \in T_\infty^{\text{even}}$. By a further pruning of these trees, we can also assume that $|y_\alpha^*(y_\beta)| < \eta/2^{\max\{|\alpha|,|\beta|\}}$ whenever $\alpha < \beta$ or $\beta < \alpha$.

By assumption, there exist $m_1 < m_2 < \dots$ in $\mathbb{N}$ such that $\big(y_{(m_1,m_2,\dots,m_{2i})}\big)$ is C-dominated by $(u_{m_{2i-1}})$. Let $(a_i) \in c_{00}$. There exists $(b_i) \in c_{00}$ such that $\big\| \sum b_i u_{m_{2i-1}} \big\| = 1$ and $\sum a_i b_i = \big\| \sum a_i u_{m_{2i-1}}^* \big\|$. So $\big\| \sum b_i y_{(m_1,m_2,\dots,m_{2i})} \big\| \le C$, and hence

$$\Big\| \sum a_i y^*_{(m_1,m_2,\dots,m_{2i})} \Big\| \ge \frac{1}{C} \Big(\sum a_i b_i \frac{1-\eta}{2} - \sum_{i \ne j} |a_i||b_j| \eta / 2^{\max\{i,j\}} \Big)$$

$$> \frac{1}{2C + \varepsilon/D} \Big\| \sum a_i u^*_{m_{2i-1}} \Big\|$$

provided η is sufficiently small. Now the branch $\big(y^*_{(m_1,m_2,\dots,m_{2i})}\big)$ of $\big(y_\alpha^*\big)_{\alpha \in T_\infty^{\text{even}}}$ corresponds to a branch $\big(x^*_{(n_1,n_2,\dots,n_{2i})}\big)$ of $\big(x_\alpha^*\big)_{\alpha \in T_\infty^{\text{even}}}$, where $n_1 < n_2 < \dots$ and $m_i \le n_i$ for all $i \in \mathbb{N}$. Since (u_i) is D-right-dominant, it follows that (u_i^*) is D-left-dominant, and hence the above inequality shows that $\big(x^*_{(n_1,n_2,\dots,n_{2i})}\big)$ $(2CD+\varepsilon)$-dominates $\big(u^*_{n_{2i-1}}\big)$.□

We conclude this section with a key combinatorial result. We need to fix some terminology first.

Given a Banach space X, we let $(\mathbb{N} \times S_X)^\omega$ denote the set of all sequences (k_i, x_i), where $k_1 < k_2 < \dots$ are positive integers, and (x_i) is a sequence in S_X. We equip the set $(\mathbb{N} \times S_X)^\omega$ with the product topology of the discrete topologies of $\mathbb{N}$ and S_X. Given $\mathcal{A} \subset (\mathbb{N} \times S_X)^\omega$ and $\varepsilon > 0$, we let

$$\mathcal{A}_\varepsilon = \Big\{ (\ell_i, y_i) \in (\mathbb{N} \times S_X)^\omega : \exists\, (k_i, x_i) \in \mathcal{A} \quad k_i \le \ell_i\,,\ \|x_i - y_i\| < \varepsilon \cdot 2^{-i}\ \forall\, i \in \mathbb{N} \Big\},$$

and we let $\overline{\mathcal{A}}$ be the closure of $\mathcal{A}$ in $(\mathbb{N} \times S_X)^\omega$.

Given $\mathcal{A} \subset (\mathbb{N} \times S_X)^\omega$, we say that an even tree $(x_\alpha)_{\alpha \in T_\infty^{\text{even}}}$ in X *has a branch in* $\mathcal{A}$ if there exist $n_1 < n_2 < \dots$ in $\mathbb{N}$ such that $\big(n_{2i-1}, x_{(n_1,n_2,\dots,n_{2i})}\big) \in \mathcal{A}$.

Proposition 2.18. *Let X be an infinite-dimensional (closed) subspace of a reflexive space Z with an FDD (E_i). Let $\mathcal{A} \subset (\mathbb{N} \times S_X)^\omega$. Then the following are equivalent.*

(a) *For all $\varepsilon > 0$ every normalized, weakly null even tree in X has a branch in $\overline{\mathcal{A}_\varepsilon}$.*

(b) *For all $\varepsilon > 0$ there exist $(K_i) \subset \mathbb{N}$ with $K_1 < K_2 < \dots$, $\bar\delta = (\delta_i) \subset (0,1)$ with $\delta_i \downarrow 0$, and a blocking $F = (F_i)$ of (E_i) such that if $(x_i) \subset S_X$ is a $\bar\delta$-skipped block sequence of (F_n) in Z with $\|x_i - P^F_{(r_{i-1},r_i)} x_i\| < \delta_i$ for all $i \in \mathbb{N}$, where $1 \le r_0 < r_1 < r_2 < \dots$, then $(K_{r_{i-1}}, x_i) \in \overline{\mathcal{A}_\varepsilon}$.*

Proof. For each $m\in\mathbb{N}$ we set $Z_m=\overline{\bigoplus_{i>m} E_i}$. Given $\varepsilon>0$, we consider the following game between players S (subspace chooser) and P (point chooser). The game has an infinite sequence of moves; on the nth move ($n\in\mathbb{N}$) S picks $k_n, m_n\in\mathbb{N}$ and P responds by picking $x_n\in S_X$ with $d(x_n, Z_{m_n})<\varepsilon'\cdot 2^{-n}$, where $\varepsilon'=\min\{\varepsilon, 1\}$. S wins the game if the sequence (k_i, x_i) the players generate ends up in $\overline{\mathcal{A}_{5\varepsilon}}$, otherwise P is declared the winner. We will refer to this as the $(\mathcal{A},\varepsilon)$-game and show that statements (a) and (b) above are equivalent to:

(c) For all $\varepsilon>0$ S has a winning strategy for the $(\mathcal{A},\varepsilon)$-game.

Note that statement (b) yields a particular winning strategy for S, so the implication (b)⇒(c) is clear. This is also included in the sequence of implications (a)⇒(c)⇒(b) and (b)⇒(a) which are what we are about to demonstrate.

"(a)⇒(c)" Assume that for some $\varepsilon>0$ S does not have a winning strategy for the $(\mathcal{A},\varepsilon)$-game. Then there is a winning strategy ϕ for the point chooser P. Thus ϕ is a function taking values in S_X such that for all sequences $(k_i), (m_i)$ in $\mathbb{N}$ if $x_n=\phi(k_1, m_1, k_2, m_2, \dots, k_n, m_n)$ for all $n\in\mathbb{N}$, then $d(x_i, Z_{m_i})<\varepsilon'\cdot 2^{-i}$ for all $i\in\mathbb{N}$ and $(k_i, x_i)\notin\overline{\mathcal{A}_{5\varepsilon}}$. We will now construct a normalized, weakly null even tree $(x_\alpha)_{\alpha\in T_\infty^{\text{even}}}$ in X to show that (a) fails. This will be a recursive construction which also builds auxiliary trees $(y_\alpha)_{\alpha\in T_\infty^{\text{even}}}$ in X and $(m_\alpha)_{\alpha\in T_\infty^{\text{even}}}$ in $\mathbb{N}$.

Fix positive integers ℓ, $n_1<n_2<\dots<n_{2\ell-1}$. Let $\alpha=(n_1, n_2, \dots, n_{2\ell-1})\in T_\infty$, and for $1\le j\le\ell$ set $k_j=n_{2j-1}$. Assume that for $1\le j<\ell$ we have already defined $x_j=x_{(n_1,n_2,\dots,n_{2j})}$, $y_j=y_{(n_1,n_2,\dots,n_{2j})}$ and $m_j=m_{(n_1,n_2,\dots,n_{2j})}$ such that $y_j=\phi(k_1, m_1, k_2, m_2, \dots, k_j, m_j)$. We will now construct the nodes $\big(x_{(\alpha,n)}\big)$, $\big(y_{(\alpha,n)}\big)$ and $\big(m_{(\alpha,n)}\big)$. Set

$$z_i=\phi(k_1, m_1, \dots, k_{\ell-1}, m_{\ell-1}, k_\ell, i)\,,\quad i\in\mathbb{N}.$$

Note that $z_i\in S_X$ and $d(z_i, Z_i)<\varepsilon'\cdot 2^{-\ell}$ for all $i\in\mathbb{N}$. We now pass to a weakly convergent subsequence: there exist $i_1<i_2<\dots$ in $\mathbb{N}$ and $z\in X$ such that $z_{i_n}\xrightarrow{w} z$ as $n\to\infty$. Note that $\|z\|\le\varepsilon'\cdot 2^{-\ell}$. For each $n\in\mathbb{N}$ set

$$w_n=\frac{z_{i_n}-z}{\|z_{i_n}-z\|}.$$

Note that (w_n) is a normalized, weakly null sequence in X, and

$$\|z_{i_n}-w_n\|\le\frac{2\varepsilon' 2^{-\ell}}{1-\varepsilon' 2^{-\ell}}\le 4\varepsilon 2^{-\ell}$$

for all $n\in\mathbb{N}$. We now set

$$x_{(n_1,n_2,\dots,n_{2\ell-1},n)}=w_n\,,\quad y_{(n_1,n_2,\dots,n_{2\ell-1},n)}=z_{i_n}\quad\text{and}\quad m_{(n_1,n_2,\dots,n_{2\ell-1},n)}=i_n$$

for all $n\in\mathbb{N}$ with $n>n_{2\ell-1}$. This completes the recursive construction.

It follows by induction that $(x_\alpha)_{\alpha\in T_\infty^{\text{even}}}$ is a normalized, weakly null even tree in X and $(y_\alpha)_{\alpha\in T_\infty^{\text{even}}}$ is a normalized even tree in X such that for all $\alpha\in T_\infty^{\text{even}}$ we have $\|x_\alpha-y_\alpha\|\le 4\varepsilon\cdot 2^{-|\alpha|/2}$. Moreover, given a sequence $n_1<n_2<\dots$ in $\mathbb{N}$, setting $k_j=n_{2j-1}$, $m_j=m_{(n_1,n_2,\dots,n_{2j})}$ and $y_j=y_{(n_1,n_2,\dots,n_{2j})}$ for all $j\in\mathbb{N}$, we have

$$y_n=\phi(k_1, m_1, k_2, m_2, \dots, k_n, m_n)\quad\text{for all } n\in\mathbb{N}.$$

Hence no branch of $(y_\alpha)_{\alpha\in T_\infty^{\text{even}}}$ is in $\overline{\mathcal{A}_{5\varepsilon}}$, and no branch of $(x_\alpha)_{\alpha\in T_\infty^{\text{even}}}$ is in $\overline{\mathcal{A}_\varepsilon}$.

"(c)⇒(b)" Let (ϕ,ψ) be a winning strategy for S in the $(\mathcal{A},\varepsilon)$-game. Thus ϕ and ψ are functions taking values in $\mathbb{N}$ such that for all sequences (k_i), (m_i) in $\mathbb{N}$ and (x_i) in S_X if $d(x_n, Z_{m_n}) < \varepsilon'\cdot 2^{-n}$, $k_n = \phi(x_1, x_2, \dots, x_{n-1})$ and $m_n \geq \psi(x_1, x_2, \dots, x_{n-1})$ for all $n \in \mathbb{N}$, then $(k_i, x_i) \in \overline{\mathcal{A}_{5\varepsilon}}$. For each interval $I \subset \mathbb{N}$ and $\delta > 0$ fix a finite set $S_{I,\delta} \subset S_X$ such that for all $x \in S_{I,\delta}$ we have $\|x - P_I^E x\| < \delta$ and for all $y \in S_X$ if $\|y - P_I^E y\| < \delta$, then there exists $x \in S_{I,\delta}$ such that $\|x - y\| < 3\delta$.

We now construct a blocking (F_i) of (E_i) by recursion. Let $m_1 = \psi()$ and $F_1 = \bigoplus_{i=1}^{m_1} E_i$. Choose any $m_2 > m_1$ and set $F_2 = \bigoplus_{i=m_1+1}^{m_2} E_i$. Assume that for some $n \in \mathbb{N}$, $n \geq 3$, we have already chosen $m_1 < \ldots < m_{n-1}$ and we have set $F_j = \bigoplus_{i=m_{j-1}+1}^{m_j} E_i$ for $1 \leq j < n$ ($m_0 = 0$). We now choose $m_n > m_{n-1}$ such that if $\ell \in \mathbb{N}$, $1 \leq r_0 < r_1 < \ldots < r_\ell \leq n$ and

$$x_j \in S_{[m_{r_{j-1}}+1, m_{r_j}-1], \varepsilon' 2^{-j}} \quad \text{for } 1 \leq j \leq \ell,$$

then $m_n \geq \psi(x_1, x_2, \dots, x_\ell)$. Finally, we set $F_n = \bigoplus_{i=m_{n-1}+1}^{m_n} E_i$. This completes the recursive construction.

For each $n \in \mathbb{N}$ let $\delta_n = \varepsilon' \cdot 2^{-n}$, and let K_n be chosen so that $K_n \geq \phi()$, and if $\ell \in \mathbb{N}$, $1 \leq r_0 < r_1 < \ldots < r_\ell \leq n$ and $x_j \in S_{[m_{r_{j-1}}+1, m_{r_j}-1], \varepsilon'\cdot 2^{-j}}$ for $1 \leq j \leq \ell$, then $K_n \geq \phi(x_1, \dots, x_\ell)$. We can of course also ensure that the sequence (K_i) is strictly increasing. Let $\bar\delta = (\delta_i)$. We will now verify that (b) holds. Let (y_i) be a $\bar\delta$-skipped block sequence of (F_n): there exist $1 \leq r_0 < r_1 < r_2 < \ldots$ such that

$$\|y_i - P^F_{[r_{i-1}+1, r_i-1]} y_i\| < \varepsilon' 2^{-i} \quad \text{for all } i \in \mathbb{N},$$

that is to say,

$$\|y_i - P^E_{[m_{r_{i-1}}+1, m_{r_i}-1]} y_i\| < \varepsilon' 2^{-i} \quad \text{for all } i \in \mathbb{N}.$$

For each $i \in \mathbb{N}$ there exists $x_i \in S_{[m_{r_{i-1}}+1, m_{r_i}-1], \varepsilon' 2^{-i}}$ such that $\|x_i - y_i\| < 3\varepsilon' \cdot 2^{-i}$. Set

$$k_i = \phi(x_1, \dots, x_{i-1}) \quad \text{for each } i \in \mathbb{N}.$$

Consider the sequence $k_1, m_{r_0}, x_1, k_2, m_{r_1}, x_2, \dots$. We have $x_i \in S_X$ and

$$d(x_i, Z_{m_{r_{i-1}}}) \leq \|x_i - P^E_{[m_{r_{i-1}}+1, m_{r_i}-1]} x_i\| < \varepsilon' 2^{-i}$$

for all $i \in \mathbb{N}$. Moreover $m_{r_0} \geq m_1 \geq \psi()$, $K_{r_0} \geq K_1 \geq \phi() = k_1$, and given $\ell \in \mathbb{N}$, setting $n = r_\ell$, we have $1 \leq r_0 < r_1 < \ldots < r_\ell \leq n$ and $x_i \in S_{[m_{r_{i-1}}+1, m_{r_i}-1], \varepsilon' 2^{-i}}$ for $1 \leq i \leq \ell$. It follows that ($n \geq 3$ and) $m_{r_\ell} = m_n \geq \psi(x_1, \dots, x_\ell)$ and $k_{\ell+1} \leq K_n = K_{r_\ell}$. So $(k_i, x_i) \in \overline{\mathcal{A}_{5\varepsilon}}$, and hence $(K_{r_{i-1}}, y_i) \in \overline{\mathcal{A}_{8\varepsilon}}$.

"(b)⇒(a)" Given $\varepsilon > 0$, let (K_i), $\bar\delta = (\delta_i)$ and (F_i) be as in statement (b). First note that if (x_i) is a normalized, weakly null sequence in X, then

$$\forall \eta > 0\, \forall p \in \mathbb{N} \quad \exists n \in \mathbb{N}\, \exists q > p \quad \text{such that} \quad \|x_n - P^F_{(p,q)} x_n\| < \eta\,.$$

Indeed, the sequence $\big(P^F_{[1,p]}x_i\big)$ is weakly null, and hence norm-null, so there exists $n\in\mathbb{N}$ such that $\|P^F_{[1,p]}x_n\|<\eta/2$. One can then choose $q>p$ such that $\|P^F_{[q,\infty)}x_n\|<\eta/2$. The claim now follows by triangle-inequality.

Now let $(x_\alpha)_{\alpha\in T_\infty^{\text{even}}}$ be a normalized, weakly null even tree in X. We choose positive integers $n_1<n_2<\dots$ and $1=r_0<r_1<r_2<\dots$ by recursion. For $\ell\in\mathbb{N}$ we first choose $n_{2\ell-1}>K_{r_{\ell-1}}$ such that $n_{2\ell-1}>n_{2\ell-2}$ ($n_0=0$), and then choose $n_{2\ell}>n_{2\ell-1}$ and $r_\ell>r_{\ell-1}$ such that

$$\big\|x_{(n_1,n_2,\dots,n_{2\ell})}-P^F_{(r_{\ell-1},r_\ell)}x_{(n_1,n_2,\dots,n_{2\ell})}\big\|<\varepsilon 2^{-\ell}.$$

By assumption (b) we have $\big(n_{2i-1},x_{(n_1,n_2,\dots,n_{2i})}\big)\in\overline{\mathcal{A}_\varepsilon}$. □

3 The space $\mathbf{Z^V(E)}$

Let Z be a space with an FDD $E=(E_i)$, and let V be a space with a 1-unconditional and normalized basis (v_i). The space $Z^V=Z^V(E)$ is defined to be the completion of $c_{00}(\oplus E_i)$ with respect to the following norm $\|\cdot\|_{Z^V}$.

$$\|z\|_{Z^V}=\max_{\substack{k\in\mathbb{N}\\ 1\le n_0<n_1<n_2<\dots<n_k}}\Big\|\sum_{j=1}^{k}\|P^E_{[n_{j-1},n_j)}(z)\|_Z\cdot v_{n_{j-1}}\Big\|_V\quad\text{for } z\in c_{00}(\oplus E_i).$$

Note that if (v_i) is C-block-stable and D-right-dominant, then the projection constant $K(E,Z^V)$ of (E_i) in Z^V satisfies

$$K(E,Z^V)\le\min\{K(E,Z),C,D,2\}.$$

Here we allow $C=\infty$ or $D=\infty$ if (v_i) is not block-stable or not right-dominant, respectively. Note also that if $\|\cdot\|$ and $\|\cdot\|'$ are equivalent norms on Z, then the corresponding norms $\|\cdot\|_{Z^V}$ and $\|\cdot\|'_{Z^V}$ are equivalent on $c_{00}(\oplus E_i)$. This often allows us, when examining the space Z^V, to assume that (E_i) is bimonotone in Z.

Our first set of results culminating in Corollary 3.4 determine when the space $Z^V(E)$ is reflexive.

Lemma 3.1. *Every normalized block sequence (z_i) of (E_n) in Z^V 1-dominates some block sequence (b_i) of (v_n) that satisfies*

$$1/2\le\|b_i\|_V\le 1\quad\textit{and}\quad \operatorname{ran}(b_i)\subset\operatorname{ran}_E(z_i)\quad\textit{for all } i\in\mathbb{N}.$$

(Here the range, $\operatorname{ran}(x)$, *of $x=\sum a_iv_i\in V$ is the set $\{i\in\mathbb{N}:\ a_i\neq 0\}$.)*

Proof. Let $z\in S_{Z^V}$ have finite support with respect to (E_i). Choose k, $1\le n_0<n_1<\dots<n_k$ in $\mathbb{N}$ such that

$$\|z\|_{Z^V}=\Big\|\sum_{j=1}^{k}\|P^E_{[n_{j-1},n_j)}(z)\|_Z\cdot v_{n_{j-1}}\Big\|_V.$$

Without loss of generality we can assume that $n_0 \le \min \mathrm{supp}_E(z) \le n_1 - 1$ and that $n_{k-1} \le \max \mathrm{supp}_E(z) = n_k - 1$. Set $m_0 = \min \mathrm{supp}_E(z)$, $m_j = n_j$ for $1 \le j < k$, and let $m_k > \max \mathrm{supp}_E(z)$. By the triangle-inequality we have

$$\begin{aligned}\|z\|_{Z^V} &\le \Big\| \|P^E_{[n_0,n_1)}(z)\|_Z \cdot v_{n_0} \Big\|_V + \Big\| \sum_{j=2}^{k} \|P^E_{[n_{j-1},n_j)}(z)\|_Z \cdot v_{n_{j-1}} \Big\|_V \\ &\le 2 \Big\| \sum_{j=1}^{k} \|P^E_{[m_{j-1},m_j)}(z)\|_Z \cdot v_{m_{j-1}} \Big\|_V .\end{aligned}$$

Now let (z_i) be a normalized block sequence of (E_n) in Z^V. It follows from the above that there exist positive integers $1 \le n_0 < n_1 < \dots$ and $1 = k_1 < k_2 < \dots$ such that $n_{k_\ell - 1} = \min \mathrm{supp}_E(z_\ell)$, $n_{k_{\ell+1}-2} \le \max \mathrm{supp}_E(z_\ell)$, and

$$\|z_\ell\|_{Z^V} \le 2 \Big\| \sum_{j=k_\ell}^{k_{\ell+1}-1} \|P^E_{[n_{j-1},n_j)}(z_\ell)\|_Z \cdot v_{n_{j-1}} \Big\|_V \quad \text{for all } \ell \in \mathbb{N}.$$

It follows that

$$b_\ell = \sum_{j=k_\ell}^{k_{\ell+1}-1} \|P^E_{[n_{j-1},n_j)}(z_\ell)\|_Z \cdot v_{n_{j-1}}$$

satisfies $1/2 \le \|b_\ell\|_V \le 1$ and $\mathrm{ran}(b_\ell) \subset \mathrm{ran}_E(z_\ell)$ for all $\ell \in \mathbb{N}$. Moreover, given $(a_i) \in c_{00}$, setting $z = \sum a_i z_i$ we have

$$\begin{aligned}\|z\|_{Z^V} &\ge \Big\| \sum_{j=1}^{\infty} \|P^E_{[n_{j-1},n_j)}(z)\|_Z \cdot v_{n_{j-1}} \Big\|_V \\ &= \Big\| \sum_{\ell=1}^{\infty} \sum_{j=k_\ell}^{k_{\ell+1}-1} \|P^E_{[n_{j-1},n_j)}(z)\|_Z \cdot v_{n_{j-1}} \Big\|_V \\ &= \Big\| \sum_{\ell=1}^{\infty} a_\ell b_\ell \Big\|_V .\end{aligned}$$

□

Corollary 3.2. *Let V be a Banach space with a normalized and 1-unconditional basis (v_i), and let Z be a space with an FDD $E = (E_i)$.*

If the basis (v_i) is boundedly complete, then (E_i) is a boundedly complete FDD for $Z^V(E)$.

Proof. Let (z_i) be a normalized block sequence of (E_n) in Z^V. Let (b_i) be a block sequence of (v_n) given by Lemma 3.1. Given $\varepsilon > 0$, let (a_i) be a scalar sequence with $|a_i| > \varepsilon$ for all $i \in \mathbb{N}$. Since (v_i) is boundedly complete, and since (z_i) dominates (b_i) it follows that

$$\sup_n \Big\| \sum_{i=1}^{n} a_i z_i \Big\|_{Z^V} = \infty .$$

Hence (E_i) is a boundedly complete FDD for $Z^V(E)$. □

Lemma 3.3. *Let V be a Banach space with a normalized and 1-unconditional basis (v_i), and assume that the space Z has an FDD $E=(E_i)$.*

If the basis (v_i) is shrinking and if (E_i) is a shrinking FDD for Z then (E_i) is also a shrinking FDD for $Z^V(E)$.

Proof. Without loss of generality we may assume that (E_i) is bimonotone in Z. We first note that, given positive integers $1\le n_0<n_1<\dots$ and vectors $z_j^*\in\bigoplus_{i\in[n_{j-1},n_j)}E_i^*$ with $\|z_j^*\|_{Z^*}\le 1$ for each $j\in\mathbb{N}$, if $v^*=\sum_{i=1}^\infty a_i v_{n_{i-1}}^*$ converges in V^* with $\|v^*\|\le 1$, then the series $z^*=\sum_{i=1}^\infty a_i z_i^*$ converges in $(Z^V)^*$ and $\|z^*\|_{(Z^V)^*}\le 1$. Indeed, for $p\le q$ in $\mathbb{N}$ there exists $z\in S_{Z^V}$ with $\mathrm{supp}_E(z)\subset[n_{p-1},n_q)$ such that

$$\begin{aligned}\Big\|\sum_{j=p}^q a_j z_j^*\Big\|_{(Z^V)^*}&=\sum_{j=p}^q a_j z_j^*(z)\le\sum_{j=p}^q|a_j|\cdot\|P^E_{[n_{j-1},n_j)}(z)\|_Z\\&\le\Big\|\sum_{j=p}^q a_j v_{n_{j-1}}^*\Big\|_{V^*}\cdot\Big\|\sum_{j=p}^q\|P^E_{[n_{j-1},n_j)}(z)\|_Z\cdot v_{n_{j-1}}\Big\|_V\qquad(3.1)\\&\le\Big\|\sum_{j=p}^q a_j v_{n_{j-1}}^*\Big\|_{V^*}\cdot\|z\|_{Z^V}\le\Big\|\sum_{j=p}^q a_j v_{n_{j-1}}^*\Big\|_{V^*},\end{aligned}$$

which implies the claim. Next define K to be the union of the following two sets K_1 and K_2:

$$K_1=\Big\{\sum_{i=1}^\infty a_i z_i^*:1\le n_0<n_1<\dots,\quad z_j^*\in\bigoplus_{i\in[n_{j-1},n_j)}E_i^*\text{ and }\|z_j^*\|_{Z^*}\le 1\ \forall j\in\mathbb{N},\ \Big\|\sum_{i=1}^\infty a_i v_{n_{i-1}}^*\Big\|_{V^*}\le 1\Big\},$$

$$K_2=\Big\{\sum_{i=1}^\ell a_i z_i^*:\ell\in\mathbb{N},\ 1\le n_0<n_1<\dots<n_\ell\le\infty,\quad z_j^*\in\bigoplus_{i\in[n_{j-1},n_j)}E_i^*\text{ and }\|z_j^*\|_{Z^*}\le 1\text{ for }1\le j\le\ell,\ \Big\|\sum_{i=1}^\ell a_i v_{n_{i-1}}^*\Big\|_{V^*}\le 1\Big\}.$$

An element $\sum_{i=1}^\ell a_i z_i^*$ of K_2 will also be written as an infinite sum $\sum_{i=1}^\infty a_i z_i^*$ by setting $a_i=0$ and $z_i^*=0$ for all $i>\ell$. Clearly, K is a Z^V-norming subset (isometrically) of $B_{(Z^V)^*}$. We claim that K is w^*-compact. Indeed, for each $k\in\mathbb{N}$ let $y_k^*=\sum_{i=1}^\infty a_i^{(k)}z_{(k,i)}^*\in K$, where for some (finite or infinite) sequence $1\le n_0^{(k)}<n_1^{(k)}<n_2^{(k)}<\dots$ in $\mathbb{N}\cup\{\infty\}$ we have $z_{(k,j)}^*\in\bigoplus_{i\in[n_{j-1}^{(k)},n_j^{(k)})}E_i^*$ and $\|z_{(k,j)}^*\|_{Z^*}\le 1$ for all j, and $\left\| \sum_i a_i^{(k)} v^*_{n^{(k)}_{i-1}} \right\|_{V^*} \leq 1$. After passing to a subsequence we can assume that

$$a_i = \lim_{k\to\infty} a_i^{(k)} \quad \text{exists for all } i \in \mathbb{N},$$

and there exists $\ell \in \mathbb{N} \cup \{0, \infty\}$ such that:

- $\lim_{k\to\infty} n_j^{(k)} = n_j$ exists for $0 \leq j < \ell$,
- $\lim_{k\to\infty} n_\ell^{(k)} = \infty$ if $\ell \in \mathbb{N} \cup \{0\}$,
- $\lim_{k\to\infty} z^*_{(k,j)} = z^*_j$ exists (in norm) for $0 \leq j < \ell$,
- $w^* - \lim_{k\to\infty} z^*_{(k,\ell)} = z^*_\ell$ exists if $\ell \in \mathbb{N} \cup \{0\}$,
- $w^* - \lim_{k\to\infty} \sum_{i\in\mathbb{N},\, i>\ell} a_i^{(k)} z^*_{(k,i)} = 0$ if $\ell \in \mathbb{N} \cup \{0\}$.

Consider the case when $\ell = \infty$. We have $1 \leq n_0 < n_1 < \dots$, $z_j^* \in \bigoplus_{[n_{j-1}, n_j)} E_i^*$ and $\|z_j^*\|_{Z^*} \leq 1$ for all $j \in \mathbb{N}$. Moreover, since (v_i^*) is a boundedly complete basis of V^*, the series $\sum_{i=1}^\infty a_i v^*_{n_{i-1}}$ converges and $\left\| \sum_{i=1}^\infty a_i v^*_{n_{i-1}} \right\|_{V^*} \leq 1$. Hence $z^* = \sum_{i=1}^\infty a_i z_i^*$ belongs to K. Finally, given $z \in S_Z$ with finite support with respect to (E_i), for sufficiently large $r \in \mathbb{N}$ we have

$$\lim_{k\to\infty} y_k^*(z) = \lim_{k\to\infty} \sum_{i=1}^r a_i^{(k)} z^*_{(k,i)}(z) = \sum_{i=1}^r a_i z_i^*(z) = z^*(z).$$

It follows that $y_k^* \xrightarrow{w^*} z^*$ as $k \to \infty$. The case $\ell \in \mathbb{N} \cup \{0\}$ is similar. We have $1 \leq n_0 < n_1 < \dots < n_\ell \leq \infty$, $z_j^* \in \bigoplus_{[n_{j-1}, n_j)} E_i^*$ and $\|z_j^*\|_{Z^*} \leq 1$ for $1 \leq j \leq \ell$, and $\sum_{i=1}^\ell a_i v^*_{n_{i-1}} \in B_{V^*}$. So $z^* = \sum_{i=1}^\ell a_i z_i^* \in K$ and $y_k^* \xrightarrow{w^*} z^*$ as $k \to \infty$. This completes the proof that K is w^*-closed.

We deduce that Z^V is embedded in $C(K)$, the space of continuous functions on K. Let (z_i) be a bounded block sequence of (E_n) in Z^V, and let $z^* \in K$. Using the notation as in the definition of K, if $z^* \in K_1$, then computing as in (3.1)

$$z^*(z_i) = \sum_{j,\ n_j \geq \min \mathrm{supp}_E(z_i)} a_j z_j^*(z_i) \leq \|z_i\|_{Z^V} \cdot \Big\| \sum_{j, n_j \geq \min \mathrm{supp}_E(z_i)} a_j v^*_{n_{j-1}} \Big\|_{V^*},$$

which converges to zero as $i \to \infty$; and if $z^* \in K_2$, then for all sufficiently large values of i

$$z^*(z_i) = \sum_{j=1}^{\ell} a_j z_j^*(z_i) = a_\ell z_\ell^*(z_i),$$

which converges to zero as $i \to \infty$, since (E_i) is assumed a shrinking FDD for Z.

It follows that (z_i) is weakly null in $C(K)$, and thus in Z^V. Since (z_i) was an arbitrary bounded block sequence in Z^V, this finishes the proof that (E_i) is shrinking in Z^V. □

From Lemma 3.3 and Corollary 3.2 we obtain the following result.

Corollary 3.4. *Assume that V is a reflexive Banach space with a normalized and 1-unconditional basis (v_i) and that Z is a space with a shrinking FDD $E=(E_i)$. Then $Z^V(E)$ is reflexive.*

The idea of the norm $\|\cdot\|_{Z^V}$ is, of course, to introduce a subsequential V-lower-estimate. The next lemma determines when this is the case.

Lemma 3.5. *Let V be a Banach space with a normalized and 1-unconditional basis (v_i), and let Z be a Banach space with an FDD $E=(E_i)$.*

If, for some $C\geq 1$, (v_i) is C-block stable, then (E_i) satisfies subsequential $2C$-V-lower estimates in $Z^V(E)$.

Proof. Let (z_i) be a normalized block sequence in $Z^V(E)$, and for each $i\in\mathbb{N}$ let $m_i=\min \operatorname{supp}_E(z_i)$. By Lemma 3.1, there exists a block sequence (b_i) of (v_n) with $1/2\leq\|b_i\|_V\leq 1$ and $\operatorname{ran}(b_i)\subset\operatorname{ran}_E(z_i)$ for all $i\in\mathbb{N}$, which is 1-dominated by (z_i). Since (v_i) is 1-unconditional and C-block-stable, it follows that (b_i) $2C$-dominates (v_{m_i}), which proves the lemma. □

The final result in this section shows when subsequential U-upper estimates are preserved under $Z\mapsto Z^V$.

Lemma 3.6. *Let V and U be Banach spaces with normalized, 1-unconditional and block-stable bases (v_i) and (u_i), respectively, and assume that (v_i) is dominated by (u_i). Let Z be a Banach space with an FDD (E_i).*

If (E_i) satisfies subsequential U-upper estimates in Z, then (E_i) also satisfies subsequential U-upper estimates in Z^V.

Proof. Choose constants B_V, B_U, D and C in $[1,\infty)$ such that (v_i) is B_V-block-stable, (u_i) is B_U-block-stable, (v_i) is D-dominated by (u_i), and (E_i) satisfies subsequential C-U-upper estimates in Z. Let K be the projection constant of (E_i) in Z, and set $\overline{C}=B_VD+B_UCD+2B_VDK$. We show that for any finite block sequence $(z_i)_{i=1}^{\ell}$ of (E_n), and for any k and $n_0<n_1<\ldots<n_k$ in $\mathbb{N}$ we have (putting $z=\sum_{i=1}^{\ell} z_i$ and $m_j=\min\operatorname{supp}_E(z_j)$ for $1\leq j\leq\ell$)

$$\Big\|\sum_{j=1}^{k}\|P^E_{[n_{j-1},n_j)}(z)\|_Z v_{n_{j-1}}\Big\|_V\leq\overline{C}\cdot\Big\|\sum_{i=1}^{\ell}\|z_i\|_{Z^V}\cdot u_{m_i}\Big\|_U. \tag{3.2}$$

Taking then the supremum of the left side of (3.2) over all choices of k and $n_0<n_1<\ldots<n_k$ in $\mathbb{N}$, we obtain

$$\Big\|\sum_{i=1}^{\ell} z_i\Big\|_{Z^V}\leq\overline{C}\Big\|\sum_{i=1}^{\ell}\|z_i\|_{Z^V}\cdot u_{m_i}\Big\|_U,$$

and thus that (E_i) satisfies subsequential $\overline{C}$-U-upper estimates in Z^V. Note that in proving (3.2) we can of course assume that $n_k\leq\max\operatorname{supp}_E(z_\ell)+1$.

For $i=1,2,\dots,\ell$, put

$$J_i = \{j\in\{1,2,\dots,k\} : \min\operatorname{supp}_E(z_i)\le n_{j-1} < n_j \le \min\operatorname{supp}_E(z_{i+1})\}$$

with $\min\operatorname{supp}_E(z_{\ell+1})=\max\operatorname{supp}_E(z_\ell)+1$ and $J_0=\{1,2,\dots,k\}\setminus\bigcup_{i=1}^{\ell} J_i$.

For $j=1,2,\dots,k$ put

$$I_j = \{i\in\{1,2,\dots,\ell\} : n_{j-1}\le\min\operatorname{supp}_E(z_i) \le \max\operatorname{supp}_E(z_i)<n_j\}$$

and $I_0=\{1,2,\dots,\ell\}\setminus\bigcup_{j=1}^{k} I_j$.

Firstly, if $b_i=\sum_{j\in J_i}\|P^E_{[n_{j-1},n_j)}(z_i)\|_Z\cdot v_{n_{j-1}}$ for $1\le i\le\ell$, we have

$$\begin{aligned}\Big\|\sum_{i=1}^{\ell}\sum_{j\in J_i}\|P^E_{[n_{j-1},n_j)}(z_i)\|_Z\cdot v_{n_{j-1}}\Big\|_V &= \Big\|\sum_{i=1}^{\ell} b_i\Big\|_V\\ &\le B_V\Big\|\sum_{i=1}^{\ell}\|b_i\|_V\cdot v_{m_i}\Big\|_V \qquad (3.3)\\ &\le B_V D\Big\|\sum_{i=1}^{\ell}\|z_i\|_{Z^V}\cdot u_{m_i}\Big\|_U.\end{aligned}$$

Secondly, if $b_j=\sum_{i\in I_j}\|z_i\|_Z\cdot u_{m_i}$ for each $j\in J_0$, then

$$\begin{aligned}\Big\|\sum_{j\in J_0}\Big\|\sum_{i\in I_j} z_i\Big\|_Z\cdot v_{n_{j-1}}\Big\|_V &\le C\Big\|\sum_{j\in J_0}\|b_j\|_U\cdot v_{n_{j-1}}\Big\|_V\\ &\le CD\Big\|\sum_{j\in J_0}\|b_j\|_U\cdot u_{n_{j-1}}\Big\|_U\\ &\le CDB_U\Big\|\sum_{j\in J_0} b_j\Big\|_U \le CDB_U\Big\|\sum_{i=1}^{\ell}\|z_i\|_{Z^V}\cdot u_{m_i}\Big\|_U.\end{aligned} \qquad (3.4)$$

Thirdly, given $j\in J_0$ and $i\in I_0$ such that $P^E_{[n_{j-1},n_j)}(z_i)\neq 0$, we have either

$$n_{j-1} < \min\operatorname{supp}_E(z_i) < n_j \le \max\operatorname{supp}_E(z_i)\,, \qquad \text{or} \qquad (3.5)$$

$$\min\operatorname{supp}_E(z_i) < n_{j-1} \le \max\operatorname{supp}_E(z_i) < n_j\,. \qquad (3.6)$$

Let $J_{0,1}$ be the set of all $j\in J_0$ for which there exists an $i\in I_0$ such that (3.5) holds and let i_j^1 denote the unique such $i\in I_0$. Similarly, we let $J_{0,2}$ be the set of all $j\in J_0$ for which there exists an $i \in I_0$ such that (3.6) holds and we denote by i_j^2 the unique such $i \in I_0$. We now obtain

$$
\begin{aligned}
&\Big\| \sum_{j\in J_0} \sum_{i\in I_0} \|P^E_{[n_{j-1},n_j)}(z_i)\|_Z \cdot v_{n_{j-1}} \Big\|_V \qquad (3.7)\\
&\qquad \le K \Big\| \sum_{j\in J_{0,1}} \|z_{i_j^1}\|_Z \cdot v_{n_{j-1}} \Big\|_V + K \Big\| \sum_{j\in J_{0,2}} \|z_{i_j^2}\|_Z \cdot v_{n_{j-1}} \Big\|_V \\
&\qquad \le K B_V \Big\| \sum_{j\in J_{0,1}} \|z_{i_j^1}\|_Z \cdot v_{m_{i_j^1}} \Big\|_V + K B_V \Big\| \sum_{j\in J_{0,2}} \|z_{i_j^2}\|_Z \cdot v_{m_{i_j^2}} \Big\|_V \\
&\qquad \le 2 K B_V D \Big\| \sum_{i=1}^{\ell} \|z_i\|_{Z^V} \cdot u_{m_i} \Big\|_U .
\end{aligned}
$$

Finally, we deduce from (3.3), (3.4) and (3.7) that

$$
\begin{aligned}
&\Big\| \sum_{j=1}^{k} \|P^E_{[n_{j-1},n_j)}(z)\|_Z \cdot v_{n_{j-1}} \Big\|_V \\
&\qquad \le \Big\| \sum_{i=1}^{\ell} \sum_{j\in J_i} \|P^E_{[n_{j-1},n_j)}(z_i)\|_Z \cdot v_{n_{j-1}} \Big\|_V \\
&\qquad\qquad + \Big\| \sum_{j\in J_0} \|P^E_{[n_{j-1},n_j)}(z)\|_Z \cdot v_{n_{j-1}} \Big\|_V \\
&\qquad \le \Big\| \sum_{i=1}^{\ell} \sum_{j\in J_i} \|P^E_{[n_{j-1},n_j)}(z_i)\|_Z \cdot v_{n_{j-1}} \Big\|_V \\
&\qquad\qquad + \Big\| \sum_{j\in J_0} \Big\| \sum_{i\in I_j} z_i \Big\|_Z \cdot v_{n_{j-1}} \Big\|_V \\
&\qquad\qquad + \Big\| \sum_{j\in J_0} \sum_{i\in I_0} \|P^E_{[n_{j-1},n_j)}(z_i)\|_Z \cdot v_{n_{j-1}} \Big\|_V \\
&\qquad \le (B_V D + B_U C D + 2 B_V D K) \Big\| \sum_{i=1}^{m} \|z_i\|_{Z^V} \cdot u_{m_i} \Big\|_U ,
\end{aligned}
$$

which finishes the proof of (3.2). □

4 Embedding theorems

In this section we will prove and deduce some consequences of

Theorem 4.1. *Assume that V is a Banach space with a normalized, 1-unconditional and left-dominant basis (v_i). Let X be a separable, infinite-dimensional, reflexive space with subsequential V-lower tree estimates.*

(a) For every reflexive space Z with an FDD $E=(E_i)$ which contains X there is a blocking $H=(H_i)$ of (E_i), and there exists $N\in\mathbb{N}^{(\omega)}$ such that X naturally isomorphically embeds into $Z^{V_N}(H)$.

(b) There is a space $\tilde{Y}$ with a bimonotone, shrinking FDD $\tilde{G}=(\tilde{G}_i)$, and there exists $N\in\mathbb{N}^{(\omega)}$ such that X is a quotient of $\tilde{Y}^{V_N}(\tilde{G})$.

Recall that $\mathbb{N}^{(\omega)}$ denotes the set of all infinite subsets of $\mathbb{N}$, and if V is a Banach space with a normalized, 1-unconditional basis (v_i), and if $N\in\mathbb{N}^{(\omega)}$, then we write V_N for the closed linear span of $\{v_i : i\in N\}$. When we talk about subsequential V_N-lower estimates, etc., it is with respect to the normalized, 1-unconditional basis $(v_i)_{i\in N}$ of V_N.

Remark 4.2. Theorem 4.1 has a quantitative version. Let $C,D\in[1,\infty)$ and assume that (v_i) is D-left-dominant and that X satisfies subsequential C-V-lower tree estimates.

Then for all $K\in[1,\infty)$ there is a constant $M=M(C,D,K)\in[1,\infty)$ such that in part (a) if $K(E,Z)\leq K$, then in the conclusion X M-embeds into $Z^{V_N}(H)$. Indeed, this follows directly from the proof. What is important is that M depends only on the constants C,D and K.

Also, there exists a constant $L=L(C,D)\in[1,\infty)$ such that in the conclusion of part (b) we get an onto map $\tilde{Q}\colon \tilde{Y}^{V_N}(\tilde{G})\to X$ with $\|\tilde{Q}\|=1$ and $\tilde{Q}(L\cdot B_{\tilde{Y}^{V_N}})\supset B_X$. This also follows directly from the proof. However, the proof of part (b) uses [10, Lemma 3.1], which in turn appeals to a theorem of Zippin [18]. The theorem of Zippin we need here states that every separable, reflexive space embeds isometrically into a reflexive space with an FDD. A quantitative version of this result claims the existence of a universal constant K such that every separable, reflexive space embeds isometrically into a reflexive space with an FDD whose projection constant is at most K. Indeed, if this wasn't true, then for all $n\in\mathbb{N}$ there would be a "bad" space X_n, and then the ℓ_2-sum of the sequence (X_n) would contradict Zippin's theorem. The existence of this universal constant K gives a quantitative version of (a special case of) [10, Lemma 3.1]: there is a universal constant $\overline{K}$ such that every separable, reflexive space X embeds isometrically into a reflexive space Z with an FDD $E=(E_i)$ with $K(E,Z)\leq\overline{K}$ such that $c_{00}(\oplus_{i=1}^{\infty}E_i)\cap X$ is dense in X. The proof of part (b) now really does give the quantitative version of (b) stated above.

The consequences of all this are quantitative analogues of Corollaries 4.3 and 4.4, and of Theorem 4.5. We shall state (without proof) the quantitative analogue of Theorem 4.5, and leave the reader to formulate the analogues of Corollaries 4.3 and 4.4. The proofs are straightforward: one simply needs to keep track of the various constants in the proofs of the qualitative statements.

Corollary 4.3. *Assume that V is a reflexive Banach space with a normalized and 1-unconditional basis (v_i), and that (v_i) is left-dominant and block-stable. Let X be a separable, infinite-dimensional, reflexive space with subsequential V-lower tree estimates.*

Then X is a subspace of a reflexive space Z with an FDD satisfying subsequential V-lower estimates and it is a quotient of a reflexive space Y with an FDD satisfying subsequential V-lower estimates.

Proof. By a theorem of Zippin [18] we can embed X into a reflexive space W with an FDD $E=(E_i)$. Using Theorem 4.1 (a) we find a blocking $F=(F_i)$ of (E_i) and $L\in\mathbb{N}^{(\omega)}$ such that X embeds into $Z=W^{V_L}(F)$.

Theorem 4.1 (b) provides a space $\tilde{Y}$ with a shrinking FDD $\tilde{G}=(\tilde{G}_i)$ and $M\in\mathbb{N}^{(\omega)}$ such that X is a quotient of $Y=\tilde{Y}^{V_M}(\tilde{G})$.

By Corollary 3.4 the spaces Z and Y are reflexive. It follows from Lemma 3.5 that (F_i) satisfies subsequential V_L-lower estimates in Z, and that $(\tilde{G}_i)$ satisfies subsequential V_M-lower estimates in Y. The result now follows from Lemma 2.13 (with (u_i) the unit vector basis of $U=\ell_1$). □

From Corollary 4.3 and Proposition 2.14 we deduce in certain instances the inverse implication of Proposition 2.16.

Corollary 4.4. *Let V be a reflexive Banach space with a normalized, 1-unconditional basis (v_i), and assume that (v_i) is left-dominant and block-stable.*

If X is a separable, infinite-dimensional, reflexive space which satisfies subsequential V-lower tree estimates, then X^* satisfies subsequential V^*-upper tree estimates.

Proof. By Corollary 4.3 X is a quotient of a reflexive space with an FDD satisfying subsequential V-lower estimates. Hence, by Proposition 2.14, X^* is the subspace of a reflexive space Z with an FDD (E_i) satisfying subsequential V^*-upper estimates.

Now let $(x_\alpha)_{\alpha\in T_\infty^{\text{even}}}$ be a normalized, weakly null even tree in X^*. One can recursively choose $n_1<n_2<\ldots$ in $\mathbb{N}$ such that

$$\|x_{(n_1,n_2,\ldots,n_{2i})} - P^E_{[n_{2i-1},n_{2i+1})}\big(x_{(n_1,n_2,\ldots,n_{2i})}\big)\| < 2^{-i} \qquad \text{for all } i\in\mathbb{N}\,.$$

Set

$$z_i = \frac{P^E_{[n_{2i-1},n_{2i+1})}\big(x_{(n_1,n_2,\ldots,n_{2i})}\big)}{\|P^E_{[n_{2i-1},n_{2i+1})}\big(x_{(n_1,n_2,\ldots,n_{2i})}\big)\|} \qquad \text{for all } i\in\mathbb{N}\,.$$

Then (z_i) is dominated by $(v^*_{n_{2i-1}})$ since (E_i) satisfies subsequential V^*-upper estimates. It follows that $\big(x_{(n_1,n_2,\ldots,n_{2i})}\big)$ is also dominated by $\big(v^*_{n_{2i-1}}\big)$. □

Theorem 4.5. *Let V and U be reflexive Banach spaces with 1-unconditional, normalized and block-stable bases (v_i) and (u_i), respectively. Further assume that (v_i) is left-dominant, (u_i) is right-dominant, and that (v_i) is dominated by (u_i).*

If X is a separable, infinite-dimensional, reflexive Banach space which satisfies subsequential (V,U)-tree estimates, then X can be embedded into a reflexive Banach space Z with an FDD (G_i) which satisfies subsequential (V,U)-estimates in Z.

Proof. By Proposition 2.16 X^* satisfies subsequential U^*-lower tree estimates, and we can apply Corollary 4.3 to deduce that X^* is the quotient of a reflexive space Y^* with an FDD (E_i^*) (Y^* being the dual of a space Y with an FDD (F_i)) satisfying subsequential U^*-lower estimates in Y^*. Thus X is a subspace of the reflexive space Y having an FDD (E_i) which, by Proposition 2.14, satisfies subsequential U-upper estimates in Y.

Theorem 4.1 part (a) yields a blocking $F=(F_i)$ of (E_i) and an infinite subset M of $\mathbb{N}$ such that X embeds into $Z=Y^{V_M}(F)$.

By Corollary 3.4 the space Z is reflexive, and by Lemma 3.5 (F_i) satisfies subsequential V_M-lower estimates in Z. Since (E_i) satisfies subsequential U-upper estimates in Y, there exists $N\in\mathbb{N}^{(\omega)}$ such that (F_i) satisfies subsequential U_N-upper estimates in Y. Since (u_i) is right-dominant, we may assume after replacing N if necessary that $m_i\leq n_i$ for all $i\in\mathbb{N}$, where m_i and n_i are the ith elements of M and N, respectively. Now $(v_i)_{i\in M}$ is dominated by $(u_i)_{i\in N}$, so by Lemma 3.6 (F_i) also satisfies subsequential U_N-upper estimates in Z. Finally, since (v_i) is left-dominant, (F_i) satisfies subsequential (V_N,U_N) estimates in Z. An application of Lemma 2.13 completes the argument. □

Before proceeding to the proof of Theorem 4.1 we state the quantitative version of Theorem 4.5 as promised earlier.

Theorem 4.6. *For all $B,C,D,L,R\in[1,\infty)$ there exist constants $\overline{C}=\overline{C}(B,D,R)$ and $K=K(C,L,R)$ in $[1,\infty)$ such that the following holds. Let V and U be reflexive Banach spaces with 1-unconditional, normalized and B-block-stable bases (v_i) and (u_i), respectively. Further assume that (v_i) is L-left-dominant, (u_i) is R-right-dominant, and that (v_i) is D-dominated by (u_i).*

If X is a separable, infinite-dimensional, reflexive Banach space which satisfies subsequential C-(V,U)-tree estimates, then X can be K-embedded into a reflexive Banach space Z which has a bimonotone FDD (G_i) satisfying subsequential $\overline{C}$-(V,U)-estimates in Z.

Proof. [Proof of Theorem 4.1 part (a)] Choose constants C and D in $[1,\infty)$ such that X satisfies subsequential C-V-lower tree estimates and (v_i) is D-left-dominant. Let K be the projection constant of (E_i) in Z. Set

$$\mathcal{A}=\left\{(k_i,x_i)\in(\mathbb{N}\times S_X)^\omega : (v_{k_i}) \text{ is } C\text{-dominated by } (x_i)\right\},$$

and choose $\varepsilon>0$ such that

$$\overline{\mathcal{A}_\varepsilon}\subset\left\{(k_i,x_i)\in(\mathbb{N}\times S_X)^\omega : (v_{k_i}) \text{ is } 2CD\text{-dominated by } (x_i)\right\}.$$

By Proposition 2.18 there exist $(K_i)\subset\mathbb{N}$ with $K_1<K_2<\dots$, $\bar{\delta}=(\delta_i)\subset(0,1)$ with $\delta_i\downarrow 0$, and a blocking $F=(F_i)$ of (E_i) such that if $(x_i)\subset S_X$ is a $(2K\bar{\delta})$-skipped block sequence of (F_n) in Z with $\|x_i-P^F_{(r_{i-1},r_i)}x_i\|<2K\delta_i$ for all $i\in\mathbb{N}$, where $1\leq r_0<r_1<r_2<\dots$, then $(v_{K_{r_{i-1}}})$ is $2CD$-dominated by (x_i).

It is easy to see that we can block (F_i) into an FDD $G=(G_i)$ such that there exists $(e_n)\subset S_X$ with

$$\|e_n - P_n^G(e_n)\| < \delta_n/2K \qquad \text{for all } n\in\mathbb{N}\,.$$

Let (v_i'') be a subsequence of (v_i) such that if $(x_i)\subset S_X$ is a $\bar{\delta}$-skipped block sequence of (G_n) in Z with $\|x_i - P^G_{(r_{i-1},r_i)}x_i\| < \delta_i$ for all $i\in\mathbb{N}$, where $1\le r_0<r_1<r_2<\dots$, then $(v''_{r_{i-1}})$ is $2CD$-dominated by (x_i). Note that if $G_j=\bigoplus_{i=m_{j-1}+1}^{m_j}$, $j\in\mathbb{N}$, $0=m_0<m_1<m_2<\dots$, then $(v_i'')=(v_{K_{m_i}})$ will do.

In order to continue we need the following result from [10], which is due (in a different form) to W. B. Johnson [6].

Proposition 4.7. *Let X be a Banach space which is a subspace of a reflexive space Z with an FDD $A=(A_i)$ having projection constant K. Let $\bar{\eta}=(\eta_i)\subset(0,1)$ with $\eta_i\downarrow 0$. Then there exist positive integers $N_1<N_2<\dots$ such that the following holds. Given positive integers $1\le k_0<k_1<\dots$ and $x\in S_X$, there exist $x_i\in X$ and $t_i\in(N_{k_{i-1}-1},N_{k_{i-1}})$ ($i\in\mathbb{N}$, $N_0=0$) such that*

(a) $x=\sum_{i=1}^\infty x_i$, *and for all $i\in\mathbb{N}$ we have (putting $t_0=0$),*
(b) *either* $\|x_i\|<\eta_i$ *or* $\|x_i-P^A_{(t_{i-1},t_i)}x_i\|<\eta_i\|x_i\|$,
(c) $\|x_i-P^A_{(t_{i-1},t_i)}x\|<\eta_i$,
(d) $\|x_i\|<K+1$,
(e) $\|P^A_{t_i}x\|<\eta_i$.

This result is in fact a slight variation of (and follows easily from the proof of) Corollary 4.4 in [10].

We now apply Proposition 4.7 with $A=G$ and $\bar{\eta}=\bar{\delta}$ to obtain an appropriate sequence $N_1<N_2<\dots$ of positive integers. Set $H_j=\bigoplus_{i=N_{j-1}+1}^{N_j}G_i$ for each $j\in\mathbb{N}$ (and with $N_0=0$), and let (v_i') be the subsequence of (v_i) defined by $v_i'=v''_{N_i}$ for all $i\in\mathbb{N}$. Let $N\in\mathbb{N}^{(\omega)}$ be chosen such that $(v_i)_{i\in N}$ is the subsequence (v_i') of (v_i).

Fix $x\in S_X$ and a sequence $1\le n_0<n_1<\dots$ in $\mathbb{N}$. We will show that

$$\Big\|\sum_{i=1}^\infty \|P^H_{[n_{i-1},n_i)}(x)\|_Z\cdot v'_{n_{i-1}}\Big\|_V \le 4KD^2C(K+2\Delta+2)+K(K+1)+3K\Delta\,, \quad (4.1)$$

where $\Delta=\sum_{i=1}^\infty\delta_i$. Taking then the supremum over all choices of (n_i), we obtain that the norms $\|\cdot\|_Z$ and $\|\cdot\|_{Z^{V_N}(H)}$ are equivalent when restricted to X, and hence statement (a) follows.

Set $M_i=N_{n_i-1}$ for $i=0,1,2,\dots$. We thus have to show that

$$\Big\|\sum_{i=1}^\infty \|P^G_{(M_{i-1},M_i]}(x)\|_Z\cdot v'_{n_{i-1}}\Big\|_V \le 4KD^2C(K+2\Delta+2)+K(K+1)+3K\Delta\,.$$

For each $i\in\mathbb{N}$ choose $x_i\in X$ and $t_i\in(M_{i-1},N_{n_{i-1}})$ such that (a)–(e) of Proposition 4.7 hold with $A=G$ and $\bar{\eta}=\bar{\delta}$.

For each $i \in \mathbb{N}$ let $\bar{x}_i = \frac{x_{i+1}}{\|x_{i+1}\|}$ and $\alpha_i = \|x_{i+1}\|$ if $\|x_{i+1}\| \geq \delta_{i+1}$, and let $\bar{x}_i = e_{M_i}$ and $\alpha_i = 0$ if $\|x_{i+1}\| < \delta_{i+1}$. Observe that $\|\bar{x}_i - P^G_{(t_i,t_{i+1})}(\bar{x}_i)\| < \delta_i$ for all $i \in \mathbb{N}$, from which it follows that (v''_{t_i}) is $2CD$-dominated by $(\bar{x}_i)$. Hence

$$\begin{aligned}\Big\|\sum_{i=1}^{\infty} x_i\Big\|_Z &\geq \Big\|\sum_{i=1}^{\infty} \alpha_i \bar{x}_i\Big\|_Z - \|x_1\|_Z - \Delta \\ &\geq \frac{1}{2CD}\Big\|\sum_{i=1}^{\infty} \alpha_i v''_{t_i}\Big\|_V - (K+1) - \Delta \\ &\geq \frac{1}{2CD}\Big\|\sum_{i=1}^{\infty} \|x_{i+1}\|_Z \cdot v''_{t_i}\Big\|_V - \frac{1}{2CD}\Delta - (K+1) - \Delta\,,\end{aligned}$$

and thus

$$\Big\|\sum_{i=1}^{\infty} \|x_{i+1}\|_Z \cdot v''_{t_i}\Big\|_V \leq 2CD(K + 2\Delta + 2)\,. \tag{4.2}$$

For each $i \in \mathbb{N}$ we have (putting $t_0 = 0$)

$$\|P^G_{(M_{i-1},M_i]}(x)\|_Z \leq K\|P^G_{(t_{i-1},t_{i+1})}(x)\|_Z \leq K\big(\|x_i\|_Z + \|x_{i+1}\|_Z + 3\delta_i\big).$$

Since $v'_{n_{i-1}} = v''_{N_{n_{i-1}}}$ and $N_{n_{i-1}} > t_i$ for all $i \in \mathbb{N}$, follows that

$$\begin{aligned}&\Big\|\sum_{i=1}^{\infty} \|P^G_{(M_{i-1},M_i]}(x)\|_Z \cdot v'_{n_{i-1}}\Big\|_V \\ &\qquad \leq K\Big\|\sum_{i=1}^{\infty} \|x_i\|_Z \cdot v'_{n_{i-1}}\Big\|_V + K\Big\|\sum_{i=1}^{\infty} \|x_{i+1}\|_Z \cdot v'_{n_{i-1}}\Big\|_V + 3K\Delta \\ &\qquad \leq K\Big\|\sum_{i=1}^{\infty} \|x_{i+1}\|_Z \cdot v'_{n_i}\Big\|_V + K\Big\|\sum_{i=1}^{\infty} \|x_{i+1}\|_Z \cdot v'_{n_{i-1}}\Big\|_V + K(K+1) + 3K\Delta \\ &\qquad \leq 2KD\Big\|\sum_{i=1}^{\infty} \|x_{i+1}\|_Z \cdot v''_{t_i}\Big\|_V + K(K+1) + 3K\Delta \\ &\qquad \leq 4KD^2C(K + 2\Delta + 2) + K(K+1) + 3K\Delta\,.\end{aligned}$$

□

Before we prove part (b) of Theorem 4.1 we need a blocking result due to Johnson and Zippin.

Proposition 4.8 ([7]). *Let $T\colon Y \to Z$ be a bounded linear operator from a space Y with a shrinking FDD (G_i) into a space Z with an FDD (H_i). Let $\varepsilon_i \downarrow 0$. Then there exist blockings $E = (E_i)$ of (G_i) and $F = (F_i)$ of (H_i) so that for all $m < n$ and $y \in S_{\bigoplus_{i\in(m,n)} E_i}$ we have $\|P^F_{[1,m)} Ty\| < \varepsilon_m$ and $\|P^F_{[n,\infty)} Ty\| < \varepsilon_n$.*

Proof. [Proof of Theorem 4.1 part (b)] By Lemma 3.1 in [10] we can, after renorming X if necessary, regard X^* (isometrically) as a subspace of a reflexive space Y^* (being the dual of a reflexive space Y with bimonotone FDD (E_i)) such that $c_{00}(\oplus_{i=1}^{\infty} E_i^*) \cap X^*$ is dense in X^*. We have a natural quotient map $Q\colon Y \to X$. By a theorem of Zippin [18] we may regard X (isometrically) as a subspace of a reflexive space Z with an FDD (F_i'). Let K be the projection constant of (F_i') in Z, and choose constants C and D in $[1, \infty)$ such that X satisfies subsequential C-V-lower tree estimates and (v_i) is D-left-dominant.

Using Proposition 2.18 as in the proof of part (a), we find sequences $(K_i) \subset \mathbb{N}$ with $K_1 < K_2 < \dots$, $\bar{\delta} = (\delta_i) \subset (0,1)$ with $\delta_i \downarrow 0$, and a blocking (F_i) of (F_i') such that if $(x_i) \subset S_X$ is a $2K\bar{\delta}$-skipped block sequence of (F_n) in Z with $\|x_i - P^F_{(r_{i-1},r_i)} x_i\|_Z < 2K\delta_i$ for all $i \in \mathbb{N}$, where $1 \le r_0 < r_1 < r_2 < \dots$, then $(v_{K_{r_{i-1}}})$ is $2CD$-dominated by (x_i), and moreover, using standard perturbation arguments and making $\bar{\delta}$ smaller if necessary, we can assume that if $(z_i) \subset Z$ satisfies $\|x_i - z_i\|_Z < \delta_i$ for all $i \in \mathbb{N}$, then (z_i) is a basic sequence equivalent to (x_i) with projection constant at most $2K$. We also require that

$$\Delta = \sum_{i=1}^{\infty} \delta_i < \frac{1}{7} \,. \tag{4.3}$$

Choose a sequence $\bar{\varepsilon} = (\varepsilon_i) \subset (0,1)$ with $\varepsilon_i \downarrow 0$ and

$$3K(K+1) \sum_{j=i}^{\infty} \varepsilon_j < \delta_i^2 \qquad \text{for all } i \in \mathbb{N} \,. \tag{4.4}$$

After blocking (F_i) if necessary, we can assume that for any subsequent blocking D of F there is a sequence (e_i) in S_X such that

$$\|e_i - P_i^D(e_i)\|_Z < \varepsilon_i / 2K \qquad \text{for all } i \in \mathbb{N} \,. \tag{4.5}$$

By Proposition 4.8 we may assume, after further blocking our FDDs if necessary, that

$$\text{for all } m<n \text{ and } y \in S_{\oplus_{i\in(m,n)} E_i} \text{ we have} \tag{4.6}$$
$$\|P^F_{[1,m)} \circ Q(y)\| < \varepsilon_m \qquad \text{and} \qquad \|P^F_{[n,\infty)} \circ Q(y)\| < \varepsilon_n \,,$$

and moreover the same holds if one passes to any blocking of (E_i) and the corresponding blocking of (F_i).

For $i \in \mathbb{N}$ let $\tilde{E}_i$ be the quotient space of E_i determined by Q, i.e., if $y \in E_i$, then the norm of $\tilde{y}$, the equivalence class of y in E_i, is given by $|||\tilde{y}||| = \|Q(y)\|$. Passing to a further blocking of (E_i) (and the corresponding blocking of (F_i)), we may assume that $\tilde{E}_i \neq \{0\}$ for all $i \in \mathbb{N}$. Given $y = \sum y_i \in c_{00}(\oplus_{i=1}^{\infty} E_i)$, $y_i \in E_i$ for all $i \in \mathbb{N}$, we set $\tilde{y} = \sum \tilde{y}_i \in c_{00}(\oplus_{i=1}^{\infty} \tilde{E}_i)$ and

$$|||\tilde{y}||| = \max_{m<n} \Big\| \sum_{i=m}^{n} Q(y_i) \Big\| = \max_{m<n} \|Q \circ P^E_{[m,n]}(y)\| .$$

We let $\tilde{Y}$ be the completion of $c_{00}(\oplus_{i=1}^{\infty}\tilde{E}_i)$ with respect to $|||\cdot|||$. Since (E_i) is a bimonotone FDD in Y, we have $|||\tilde{y}|||\leq\|y\|$ for all $y\in c_{00}(\oplus_{i=1}^{\infty}E_i)$, and hence the map $y\mapsto\tilde{y}$ extends to a norm one map from Y to $\tilde{Y}$. By the definition of $|||\cdot|||$ we have $\|Qy\|\leq|||\tilde{y}|||$ for any $y\in c_{00}(\oplus_{i=1}^{\infty}E_i)$. It follows that $\tilde{y}\mapsto Q(y)$ extends to a norm one map $\tilde{Q}\colon\tilde{Y}\to X$ with $\tilde{Q}(\tilde{y})=Q(y)$ for all $y\in Y$.

In order to continue our proof we will need the following proposition from [11].

Proposition 4.9 ([11, Proposition 2.6]).

(a) *$(\tilde{E}_i)$ is a bimonotone, shrinking FDD for $\tilde{Y}$.*

(b) *$\tilde{Q}$ is a quotient map from $\tilde{Y}$ onto X. More precisely if $x\in X$ and $y\in Y$ is such that $Q(y)=x$, $\|y\|=\|x\|$ and $y=\sum y_i$ with $y_i\in E_i$ for all $i\in\mathbb{N}$, then $\tilde{y}=\sum\tilde{y}_i\in\tilde{Y}$, $|||\tilde{y}|||=\|y\|$ and $\tilde{Q}(\tilde{y})=x$.*

(c) *Let $(\tilde{y}_i)$ be a block sequence of $(\tilde{E}_n)$ in $B_{\tilde{Y}}$, and assume that $(\tilde{Q}(\tilde{y}_i))$ is a basic sequence with projection constant $\overline{K}$ and that $a=\inf_i\|\tilde{Q}(\tilde{y}_i)\|>0$. Then for all $(a_i)\in c_{00}$ we have*

$$\Big\|\sum a_i\tilde{Q}(\tilde{y}_i)\Big\|\leq\Big|\Big|\Big|\sum a_i\tilde{y}_i\Big|\Big|\Big|\leq\frac{3\overline{K}}{a}\Big\|\sum a_i\tilde{Q}(\tilde{y}_i)\Big\|\,.$$

To finish the proof of Theorem 4.1 (b) it suffices to find a constant $L<\infty$, a subsequence (v_i') of (v_i), and a blocking $\tilde{G}=(\tilde{G}_i)$ of $(\tilde{E}_i)$ with the following property. For each $x\in S_X$ there exists a $\tilde{y}=\sum\tilde{y}_i\in\tilde{Y}$, $\tilde{y}_i\in\tilde{G}_i$ for all $i\in\mathbb{N}$, such that

$$\|\tilde{Q}(\tilde{y})-x\|<1/2, \tag{4.7}$$

and for any choice of k and $1\leq n_0<n_1<n_2<\dots$ in $\mathbb{N}$,

$$\Big\|\sum_{j=1}^{\infty}|||P^{\tilde{G}}_{[n_{j-1},n_j)}(\tilde{y})|||\cdot v'_{n_{j-1}}\Big\|_V\leq L. \tag{4.8}$$

Once this is accomplished, we consider the space $\tilde{Y}^{V_N}=\tilde{Y}^{V_N}(\tilde{G})$, where $N\in\mathbb{N}^{(\omega)}$ is chosen so that $(v_i)_{i\in N}$ is the subsequence (v_i') of (v_i). Given $x=x_0\in S_X$, the property of $\tilde{G}$ allows us to recursively choose $x_n\in\frac{1}{2^n}B_X$ and $\tilde{y}_n\in\frac{L}{2^{n-1}}B_{\tilde{Y}^{V_N}}$, $n\in\mathbb{N}$, so that $x_n=x_{n-1}-\tilde{Q}(\tilde{y}_n)$ for all $n\in\mathbb{N}$. It follows that $\sum_{n=1}^{\infty}\tilde{y}_n$ converges in $\tilde{Y}^{V_N}$ with $\|\sum_{n=1}^{\infty}\tilde{y}_n\|_{\tilde{Y}^{V_N}}\leq 2L$ and $\tilde{Q}(\sum_{n=1}^{\infty}\tilde{y}_n)=x$. Thus $\tilde{Q}\colon\tilde{Y}^{V_N}\to X$ remains surjective, which finishes the proof.

In order to show the existence of a suitable blocking $\tilde{G}$ of $\tilde{E}$ we need the following result from [11].

Lemma 4.10 ([11, Lemma 2.7]). *Assume that (4.6) holds for our original map $Q\colon Y\to X$. Then there exist integers $0=N_0<N_1<\dots$ so that if for each $i\in\mathbb{N}$ we define:*

- $C_i=\bigoplus_{j=N_{i-1}+1}^{N_i}E_j,\qquad D_i=\bigoplus_{j=N_{i-1}+1}^{N_i}F_j;$

- $L_i = \left\{ j \in \mathbb{N} : N_{i-1} < j \le \dfrac{N_{i-1}+N_i}{2} \right\}$;
- $R_i = \left\{ j \in \mathbb{N} : \dfrac{N_{i-1}+N_i}{2} < j \le N_i \right\}$;
- $C_{i,L} = \bigoplus_{j \in L_i} E_j$ *and* $C_{i,R} = \bigoplus_{j \in R_i} E_j$,

then the following holds. Let $x \in S_X$*,* $0 \le m < n$ *and* $\varepsilon > 0$*, and assume that* $\|x - P^D_{(m,n)}(x)\| < \varepsilon$*. Then there exists* $y \in B_Y$ *with* $y \in C_{m,R} \oplus \left(\bigoplus_{i\in(m,n)} C_i\right) \oplus C_{n,L}$ *(where* $C_{0,R} = \{0\}$*) and* $\|Qy - x\| < K(2\varepsilon + \varepsilon_{m+1})$ *(recall that* K *is the projection constant of* (F'_i) *in* Z*).*

Let (C_i) and (D_i) be the blockings given by Lemma 4.10. Note that the sequence (N_i) in the lemma used to define these blockings will not be needed in the sequel, so we can discard it. We now apply Proposition 4.7 with $(A_i) = (D_i)$ and $\bar{\eta} = \bar{\varepsilon}$ to obtain a sequence $N_1 < N_2 < \dots$ in $\mathbb{N}$ so that the conclusions of the proposition are satisfied. Let (v''_i) be a subsequence of (v_i) such that if $(x_i) \subset S_X$ is a $\bar{\delta}$-skipped block sequence of (D_n) in Z with $\|x_i - P^D_{(r_{i-1},r_i)} x_i\|_Z < \delta_i$ for all $i \in \mathbb{N}$, where $1 \le r_0 < r_1 < r_2 < \dots$, then $(v''_{r_{i-1}})$ is $2CD$-dominated by (x_i), and moreover, if $(z_i) \subset Z$ satisfies $\|x_i - z_i\| < \delta_i$ for all $i \in \mathbb{N}$, then (z_i) is a basic sequence equivalent to (x_i) with projection constant at most $2K$. Let (v'_i) be the subsequence of (v_i) defined by setting $v'_i = v''_{N_i}$ for all $i \in \mathbb{N}$. We now come to our final blockings: for each $i \in \mathbb{N}$ set $G_i = \bigoplus_{j=N_{i-1}+1}^{N_i} C_j$ and let $H_i = \bigoplus_{j=N_{i-1}+1}^{N_i} D_j$ ($N_0 = 0$). Put $G = (G_i)$, let $\tilde{G} = (\tilde{G}_i)$ be the corresponding blocking of $(\tilde{E}_i)$, and set $H = (H_i)$.

Fix a sequence (e_i) in S_X so that (4.5) holds. Let $x \in S_X$. By the choice of $N_1, N_2, \dots$, for each $i \in \mathbb{N}$, there are $x_i \in (K+1)B_X$ and $t_i \in (N_{i-1}, N_i)$ such that $x = \sum_{i=1}^{\infty} x_i$ and for all $i \in \mathbb{N}$ either $\|x_i\| < \varepsilon_i$ or $\|P^D_{(t_{i-1},t_i)} x_i - x_i\| < \varepsilon_i \|x_i\|$ ($t_0 = 0$). For each $i \in \mathbb{N}$ let $\bar{x}_i = x_{i+1}/\|x_{i+1}\|$ and $\alpha_i = \|x_{i+1}\|$ if $\|x_{i+1}\| \ge \varepsilon_{i+1}$, and let $\bar{x}_i = e_{N_i}$ and $\alpha_i = 0$ if $\|x_{i+1}\| < \varepsilon_{i+1}$. Since

$$\|\bar{x}_i - P^D_{(t_i,t_{i+1})}(\bar{x}_i)\| < \varepsilon_{i+1} \quad \text{for all } i \in \mathbb{N}, \tag{4.9}$$

there exists $(y_i) \subset B_Y$ with $y_i \in C_{t_i,R} \oplus \left(\bigoplus_{j\in(t_i,t_{i+1})} C_j\right) \oplus C_{t_{i+1},L}$ and

$$\|Q(y_i) - \bar{x}_i\| < 3K\varepsilon_{i+1}, \quad i \in \mathbb{N}. \tag{4.10}$$

Also, if $\|x_1\| < \varepsilon_1$, then set $y_0 = 0$, and if $\|x_1\| \ge \varepsilon_1$, then choose $y_0 \in (K+1)B_Y$ such that $y_0 \in \left(\bigoplus_{j\in(0,t_1)} C_j\right) \oplus C_{t_1,L} \subset G_1$ and $\|Q(y_0) - x_1\| < 3K(K+1)\varepsilon_1$.

Set $\bar{x} = x_1 + \sum_{i=1}^{\infty} \alpha_i \bar{x}_i$, and note that (this series converges and) by (4.3) and (4.4)

$$\|x - \bar{x}\| \le \sum_{i=2}^{\infty} \varepsilon_i < \frac{1}{4}. \tag{4.11}$$

As a $\bar{\delta}$-skipped block sequence of (D_i) (this follows from (4.9) and (4.4)), $(\bar{x}_i)$ is a basic sequence with projection constant at most $2K$ that $2CD$-dominates (v''_{t_i}). Since, by (4.10), $\|\tilde{Q}(\tilde{y}_i)-\bar{x}_i\|<3K\varepsilon_{i+1}<\delta_i$ for all $i\in\mathbb{N}$, the sequence $\big(\tilde{Q}(\tilde{y}_i)\big)$ is also a basic sequence with projection constant at most $2K$ and is equivalent to $(\bar{x}_i)$. Furthermore, we have $\inf_i\|\tilde{Q}(\tilde{y}_i)\|\geq\inf_i\big(\|\bar{x}_i\|-\delta_i\big)>6/7$, and thus, by Proposition 4.9 (c),

$$\Big\|\sum a_i\tilde{Q}(\tilde{y}_i)\Big\|\leq\Big|\Big|\Big|\sum a_i\tilde{y}_i\Big|\Big|\Big|\leq 7K\Big\|\sum a_i\tilde{Q}(\tilde{y}_i)\Big\|\quad\text{for all } (a_i)\in c_{00}. \tag{4.12}$$

Thus $(\tilde{y}_i)$ is a basic sequence equivalent to $(\bar{x}_i)$ and, in particular, $\sum_{i=1}^{\infty}\alpha_i\tilde{y}_i$ converges. Putting $\tilde{y}=\tilde{y}_0+\sum_{i=1}^{\infty}\alpha_i\tilde{y}_i$ we have

$$\begin{aligned}\|\tilde{Q}\tilde{y}-\bar{x}\| &\leq \|\tilde{Q}\tilde{y}_0-x_1\|+\sum_{i=1}^{\infty}|\alpha_i|\cdot\|\tilde{Q}\tilde{y}_i-\bar{x}_i\|\\ &\leq 3K(K+1)\sum_{i=1}^{\infty}\varepsilon_i<1/4,\end{aligned}$$

and hence, by (4.11), $\|\tilde{Q}\tilde{y}-x\|<1/2$, so we have (4.7).

We now fix integers $1\leq n_0<n_1<n_2<\dots$. We have $\tilde{y}_i\in\tilde{G}_i\oplus\tilde{G}_{i+1}$ for each $i\in\mathbb{N}$, and $\tilde{y}_0\in\tilde{G}_1$. It follows that

$$\begin{aligned}\Big\|\sum_{s=1}^{\infty}|||P^{\tilde{G}}_{[n_{s-1},n_s)}(\tilde{y})|||\cdot v'_{n_{s-1}}\Big\|_V &\leq |||\tilde{y}_0|||+\Big\|\sum_{s=1}^{\infty}\alpha_{n_{s-1}-1}\cdot v'_{n_{s-1}}\Big\|_V\\ &\quad+\Big\|\sum_{s=1}^{\infty}\Big|\Big|\Big|\sum_{i=n_{s-1}}^{n_s-1}\alpha_i\tilde{y}_i\Big|\Big|\Big|\cdot v'_{n_{s-1}}\Big\|_V,\end{aligned} \tag{4.13}$$

where we put $\alpha_0=0$ in case $n_0=1$. We now show how to bound each of the three terms of the right-hand side of the above inequality, and hence obtain (4.8) with $L=126CD^2K^3$.

We already have $|||\tilde{y}_0|||\leq K+1$. Since $(\bar{x}_i)$ $2CD$-dominates (v''_{t_i}) we get

$$\begin{aligned}2CD^2\Big\|\sum_i\alpha_i\bar{x}_i\Big\|_Z &\geq D\Big\|\sum_i\alpha_i\cdot v''_{t_i}\Big\|_V\\ &\geq\Big\|\sum_i\alpha_i\cdot v'_{i+1}\Big\|_V.\end{aligned}$$

Since $v'_{i+1}=v''_{N_{i+1}}$ and $N_{i+1}>t_i$ for all $i\in\mathbb{N}$,

$$2CD^2\Big\|\sum_i\alpha_i\bar{x}_i\Big\|_Z\geq\Big\|\sum_{s=1}^{\infty}\alpha_{n_{s-1}-1}\cdot v'_{n_{s-1}}\Big\|_V.$$

Moreover, it follows from (4.11) that

$$\Big\|\sum_i\alpha_i\bar{x}_i\Big\|_Z=\|\bar{x}-x_1\|_Z\leq K+3. \tag{4.14}$$

This yields the bound of $2CD^2(K+3)$ for the second term of (4.13).

For each $s \in \mathbb{N}$ let $\tilde{w}_s = \sum_{i=n_{s-1}}^{n_s-1} \alpha_i \tilde{y}_i$ and $b_s = \sum_{i=n_{s-1}}^{n_s-1} \alpha_i \bar{x}_i$. Note that by (4.9) and (4.4), we have for all $s \in \mathbb{N}$,

$$\begin{aligned} \|b_s - P^D_{(t_{n_{s-1}},t_{n_s})}(b_s)\| &\leq \sum_{i=n_{s-1}}^{n_s-1} |\alpha_i| \cdot 2K \cdot \|\bar{x}_i - P^D_{(t_i,t_{i+1})}\bar{x}_i\| \\ &< 2K(K+1) \sum_{i=n_{s-1}}^{n_s-1} \varepsilon_{i+1} < \delta_s^2. \end{aligned} \tag{4.15}$$

For each $s \in \mathbb{N}$ set $\bar{b}_s = b_s/\|b_s\|$ and $\beta_s = \|b_s\|$ if $\|b_s\| \geq \delta_s$, and set $\bar{b}_s = \bar{x}_{n_{s-1}}$ and $\beta_s = 0$ if $\|b_s\| < \delta_s$. It follows from (4.15) and (4.9) that $(\bar{b}_s) \subset S_X$ is a $\bar{\delta}$-skipped block sequence of (D_i) in Z with $\|\bar{b}_s - P^D_{(t_{n_{s-1}},t_{n_s})}(\bar{b}_s)\| < \delta_s$ for all $s \in \mathbb{N}$, and hence it is a basic sequence that $2CD$-dominates $(v''_{t_{n_{s-1}}})$.

From (4.10) and (4.12) we have

$$\begin{aligned} \|\tilde{Q}(\tilde{w}_s) - b_s\| &\leq \sum_{i=n_{s-1}}^{n_s-1} |\alpha_i| \cdot \|\tilde{Q}(\tilde{y}_i) - \bar{x}_i\| \\ &< 3K(K+1) \sum_{i=n_{s-1}}^{n_s-1} \varepsilon_{i+1} < \delta_s \end{aligned} \tag{4.16}$$

and for all $s \in \mathbb{N}$,

$$|||\tilde{w}_s||| \leq 7K\|\tilde{Q}(\tilde{w}_s)\|. \tag{4.17}$$

We now obtain the following sequence of inequalities.

$$\begin{aligned} \Big\| \sum_{s=1}^{\infty} |||\tilde{w}_s||| \cdot v'_{n_{s-1}} \Big\|_V &\leq 7K \Big\| \sum_{s=1}^{\infty} \|\tilde{Q}(\tilde{w}_s)\| \cdot v'_{n_{s-1}} \Big\|_V && \text{(from (4.17))} \\ &\leq 7K \Big\| \sum_{s=1}^{\infty} \|b_s\| \cdot v'_{n_{s-1}} \Big\|_V + 7K\Delta && \text{(from (4.16))} \\ &\leq 7K \Big\| \sum_{s=1}^{\infty} \beta_s v'_{n_{s-1}} \Big\|_V + 14K\Delta \\ &\leq 7KD \Big\| \sum_{s=1}^{\infty} \beta_s v''_{t_{n_{s-1}}} \Big\|_V + 14K\Delta. \end{aligned}$$

As (v_i) is D-left-dominant, $(v''_{t_{n_{s-1}}})$,

$$\Big\| \sum_{s=1}^{\infty} |||\tilde{w}_s||| \cdot v'_{n_{s-1}} \Big\|_V \leq 14CD^2K \Big\| \sum_{s=1}^{\infty} \beta_s \bar{b}_s \Big\| + 14K\Delta.$$

Since $(\bar{b}_s)$ $2CD$-dominates $(v''_{t_{n_{s-1}}})$,

$$\Big\|\sum_{s=1}^{\infty}|||\tilde{w}_s|||\cdot v'_{n_{s-1}}\Big\|_V \le 14CD^2K\Big\|\sum_{i=n_0}^{\infty}\alpha_i\bar{x}_i\Big\| + 14CD^2K\Delta + 14K\Delta.$$

Finally, since $(\bar{x}_i)$ is a basic sequence with projection constant at most $2K$, it follows from (4.14) that

$$\Big\|\sum_{i=n_0}^{\infty}\alpha_i\bar{x}_i\Big\|_Z \le 2K\Big\|\sum_{i=1}^{\infty}\alpha_i\bar{x}_i\Big\|_Z \le 2K(K+3).$$

This provides an upper bound of $116CD^2K^3$ for the third term of (4.13), which leads to (4.8) with $L = 126CD^2K^3$, as claimed. This completes the proof of part (b) of Theorem 4.1. □

5 Universal constructions and applications

Let V and U be reflexive spaces with normalized, 1-unconditional, block-stable bases (v_i) and (u_i), respectively, such that (v_i) is left-dominant, (u_i) is right-dominant and (v_i) is dominated by (u_i). For each $C \in [1,\infty)$ let $\mathcal{A}_{V,U}(C)$ denote the class of all separable, infinite-dimensional, reflexive Banach spaces that satisfy subsequential C-(V,U)-tree estimates. We also let

$$\mathcal{A}_{V,U} = \bigcup_{C\in[1,\infty)} \mathcal{A}_{V,U}(C)\ ,$$

which is the class of all separable, infinite-dimensional, reflexive Banach spaces that satisfy subsequential (V,U)-tree estimates.

Theorem 5.1. *The class $\mathcal{A}_{V,U}$ defined above contains an element which is universal for the class.*

More precisely, for all $B,D,L,R\in[1,\infty)$ there exists a constant $\overline{C}=\overline{C}(B,D)\in[1,\infty)$ and for all $C\in[1,\infty)$ there is a constant $K(C)=K_{B,D,L,R}(C)\in[1,\infty)$ such that if (v_i) is B-block-stable and L-left-dominant, if (u_i) is B-block-stable and R-right-dominant, and if (v_i) is D-dominated by (u_i), then there exists $Z\in\mathcal{A}_{V,U}$ such that for all $C\in[1,\infty)$ every $X\in\mathcal{A}_{V,U}(C)$ $K(C)$-embeds into Z, and moreover Z has a bimonotone FDD satisfying subsequential $\overline{C}$-(V,U) estimates in Z.

Proof. By a result of Schechtman [16] there exists a space W with a bimonotone FDD $E=(E_i)$ with the property that any bimonotone FDD is naturally almost isometric to a subsequence $\overline{\bigoplus_{i=1}^{\infty}E_{k_i}}$ which is 1-complemented in W. More precisely, given a Banach space X with a bimonotone FDD (F_i) and given $\varepsilon>0$, there is a subsequence (E_{k_i}) of (E_i) and a $(1+\varepsilon)$-embedding $T\colon X\to W$ such that $T(F_i)=E_{k_i}$ for all $i\in\mathbb{N}$, and $\sum_{i=1}^{\infty}P^E_{k_i}$ is a norm-1 projection of W onto $\overline{\bigoplus_{i=1}^{\infty}E_{k_i}}$.

We shall now modify the norm on W in two stages. We first consider the space $(W^{(*)})^{U^*}(E^*)$. By Corollary 3.2 the sequence (E_i^*) is a boundedly complete (and bimonotone) FDD for this space. It follows that (E_i) is a bimonotone, shrinking FDD for a space Y with $Y^* = (W^{(*)})^{U^*}(E^*)$. By Lemma 3.5 and Proposition 2.14 (E_i) satisfies subsequential $2B$-U-upper estimates in Y.

We now let $Z = Y^V(E)$. By Corollary 3.4 Z is reflexive, by Lemma 3.5 (E_i) satisfies subsequential $2B$-V-lower estimates in Z, and by Lemma 3.6 (E_i) also satisfies subsequential $(3BD{+}2B^2D)$-U-upper estimates in Z. Thus (E_i) is a bimonotone FDD satisfying subsequential $\overline{C}$-(V,U)-estimates in Z, where $\overline{C} = 3BD+2B^2D$. It remains to show that Z is universal for $\mathcal{A}_{V,U}$.

Let $C \in [1,\infty)$ and let $X \in \mathcal{A}_{V,U}(C)$. By Theorem 4.6 there exist constants $K_1 = K_1(B,D,R)$ and $K_2 = K_2(C,L,R)$ in $[1,\infty)$ such that X K_2-embeds into a reflexive space $\tilde{X}$ which has a bimonotone FDD (F_i) satisfying subsequential K_1-(V,U) estimates in $\tilde{X}$. Now we can find a subsequence (E_{k_i}) of (E_i) and a 2-embedding $T\colon \tilde{X} \to W$ such that $T(F_i) = E_{k_i}$ for all $i \in \mathbb{N}$ and $\sum_i P_{k_i}^E$ is a norm-1 projection of W onto $\overline{\bigoplus_i E_{k_i}}$. It follows in particular that (E_{k_i}) satisfies subsequential $2K_1$-(V,U) estimates in W, *i.e.*, if (w_i) is a normalized block sequence of (E_{k_n}) in W with $\min \operatorname{supp}_E(w_i) = k_{m_i}$ for all $i \in \mathbb{N}$, then (w_i) $2K_1$-dominates (v_{m_i}) and is $2K_1$-dominated by (u_{m_i}). Hence by Proposition 2.14 $(E_{k_i}^*)$ satisfies subsequential $2K_1$-(U^*,V^*) estimates in $W^{(*)}$. (Note that the dual of the subspace $\overline{\bigoplus_i E_{k_i}}$ of W is naturally isometrically isomorphic to the subspace $\overline{\bigoplus_i E_{k_i}^*}$ of $W^{(*)}$.) We shall now use this to show that the norms $\|\cdot\|_W$, $\|\cdot\|_Y$ and $\|\cdot\|_Z$ are all equivalent when restricted to $c_{00}(\oplus_i E_{k_i})$, which implies that $\tilde{X}$ and hence also X embed into Z.

Fix $w^* \in c_{00}(\oplus_i E_{k_i}^*)$. Clearly we have $\|w^*\|_{W^{(*)}} \le \|w^*\|_{Y^*}$. Choose $1 \le m_0 < m_1 < \dots$ in $\mathbb{N}$ such that $\|w^*\|_{Y^*} = \Big\| \sum_{i=1}^\infty \|P_{[m_{i-1},m_i)}^{E^*} w^*\|_{W^{(*)}} \cdot u_{m_{i-1}}^* \Big\|_{U^*}$.
We may assume that $m_0 = 1$ and $P_{[m_{i-1},m_i)}^{E^*} w^* \neq 0$ for all $i \in \mathbb{N}$. Then there exist $j_1 < j_2 < \dots$ in $\mathbb{N}$ such that $k_{j_i} = \min \operatorname{supp}_{E^*} P_{[m_{i-1},m_i)}^{E^*} w^*$ for all $i \in \mathbb{N}$. Since (u_i^*) is B-block-stable and R-left-dominant, and since $(E_{k_i}^*)$ satisfies subsequential $2K_1$-U^*-lower estimates in $W^{(*)}$, we have

$$\begin{aligned} \|w^*\|_{Y^*} &\le B \Big\| \sum_{i=1}^\infty \|P_{[m_{i-1},m_i)}^{E^*} w^*\|_{W^{(*)}} \cdot u_{k_{j_i}}^* \Big\|_{U^*} \\ &\le BR \Big\| \sum_{i=1}^\infty \|P_{[m_{i-1},m_i)}^{E^*} w^*\|_{W^{(*)}} \cdot u_{j_i}^* \Big\|_{U^*} \\ &\le 2BRK_1 \|w^*\|_{W^{(*)}} . \end{aligned} \tag{5.1}$$

This shows that $\|\cdot\|_{W^{(*)}}$ and $\|\cdot\|_{Y^*}$ are equivalent on $c_{00}(\oplus_i E_{k_i}^*)$. It is easy to verify that $\sum_i P_{k_i}^{E^*}$, which defines a norm-1 projection of $W^{(*)}$ onto $\overline{\bigoplus_i E_{k_i}^*}$, is also a norm-1 projection of Y^* onto $\overline{\bigoplus_i E_{k_i}^*}$. It follows that $\frac{1}{2BRK_1}\|w\|_W \le \|w\|_Y \le \|w\|_W$ for all $w \in c_{00}(\oplus_i E_{k_i})$.

A very similar argument shows that $\|y\|_Y \le \|y\|_Z \le 2BLK_1\|y\|_Y$ for all $y \in c_{00}(\oplus_i E_{k_i})$. Indeed, the first inequality is clear from the definition of $\|\cdot\|_Z$, whereas the second one is obtained by a computation similar to the one in (5.1).

We have thus shown that the 2-embedding $T\colon \tilde{X} \to W$ becomes a $8B^2LRK_1^2$-embedding viewed as a map $\tilde{X} \to Z$. Hence X K-embeds into Z, where $K = 8B^2LRK_1^2K_2$. □

We conclude this paper with two applications of our embedding theorems. The first one is the observation that our results here give an alternative proof to the main theorem in [11].

Theorem 5.2 ([11]). *Let X be a separable, reflexive Banach space and let $1 \le q \le p \le \infty$. The following are equivalent.*

(a) X satisfies (p,q) tree estimates.
(b) X is isomorphic to a subspace of a reflexive space Z having an FDD which satisfies (p,q) estimates.
(c) X is isomorphic to a quotient of a reflexive space Z having an FDD which satisfies (p,q) estimates.

Here an FDD (E_n) of a Banach space Z is said to satisfy (p,q) estimates if there is a constant $C>0$ such that for every block sequence (x_i) of (E_n) we have

$$C^{-1}\Big(\sum \|x_i\|^p\Big)^{1/p} \le \Big\|\sum x_i\Big\| \le C\Big(\sum \|x_i\|^q\Big)^{1/q},$$

and a Banach space X is said to satisfy (p,q) estimates if every normalized, weakly null tree $(x_\alpha)_{\alpha\in T_\infty}$ in X has a branch that dominates the unit vector basis of ℓ_p and that is dominated by the unit vector basis of ℓ_q. The family $(x_\alpha)_{\alpha\in T_\infty}$ in X is called a normalized, weakly null tree if for all $\alpha \in T_\infty \cup \{\emptyset\}$ the sequence $(x_{(\alpha,n)})$ is normalized and weakly null, and a branch of $(x_\alpha)_{\alpha\in T_\infty}$ is a sequence $(x_{(n_1,n_2,\dots,n_i)})$, where $n_1 < n_2 < \dots$.

The second, and main, application concerns the existence of universal spaces for the classes C_α defined in the Introduction. Recall that for each countable ordinal α the class C_α consists of all separable, reflexive spaces X such that both X and its dual X^* have Szlenk index at most α. Szlenk introduced his index to show that there is no separable, reflexive space that contains isomorphic copies of every separable, reflexive space [17].

The Szlenk index $\mathrm{Sz}(\cdot)$ has the following properties [17]: for a separable space X, $\mathrm{Sz}(X) < \omega_1$ if and only if X^* is separable (so $\bigcup_{\alpha<\omega_1} C_\alpha$ is the class of all separable, reflexive spaces); if Y embeds into X, then $\mathrm{Sz}(Y) \le \mathrm{Sz}(X)$; for all $\alpha < \omega_1$ there exists a separable, reflexive space X such that $\mathrm{Sz}(X) > \alpha$. From these properties it follows immediately that if a separable space Z contains isomorphic copies of every separable, reflexive space, then Z^* is not separable, and so Z cannot be reflexive. (Later J. Bourgain showed that such a space Z must contain $C[0,1]$, and hence all separable Banach spaces [2].)

It seems natural to ask if there is, for each countable ordinal α, a separable, reflexive space that is universal for C_α. This question was indeed raised by Pełczyński motivated by the results of [11], which imply an affirmative answer for $\alpha=\omega$. In [12] we show that Pełczyński's question has an affirmative answer for all $\alpha<\omega_1$.

Theorem 5.3. *For each countable ordinal α there is a separable, reflexive space Z which is universal for the class C_α.*

This is a simplified version of our result which also includes estimates on embedding constants and determines the class C_β in which the universal space Z lives. The proof, which is given in [12], splits into two parts. We first prove that if $X\in C_\alpha$, then there exists $\gamma<\alpha$ such that X satisfies subsequential $((T_{\gamma,\frac{1}{2}})^*, T_{\gamma,\frac{1}{2}})$ tree estimates, where $T_{\gamma,\frac{1}{2}}$ is the Tsirelson space of order γ. The ingredient for the second part of the proof is Theorem 5.1 from this paper. We fix a sequence (α_n) of ordinals with $\alpha=\sup_n(\alpha_n+1)$, and for each $n\in\mathbb{N}$ we let Z_n be a separable, reflexive space which is universal for the class $\mathcal{A}_{((T_{\alpha_n,\frac{1}{2}})^*, T_{\alpha_n,\frac{1}{2}})}$. The ℓ_2-direct sum Z of the sequence (Z_n) is then the required universal space for C_α.

References

[1] S. F. Bellenot, R. Haydon, and E. Odell, *Quasi-reflexive and tree spaces constructed in the spirit of R. C. James*. Banach Space Theory (Iowa City, IA, 1987), Contemp. Math. 85 (1989), pp. 19–43.

[2] J. Bourgain, *On separable Banach spaces, universal for all separable reflexive spaces*, Proc. Amer. Math. Soc. 79 (1980), pp. 241–246.

[3] P. Dodos and V. Ferenczi, *Some strongly bounded classes of Banach spaces*, preprint.

[4] D. Gale and F. M. Stewart, *Infinite games with perfect information*. 28, Contributions to the theory of games, pp. 245–266. Princeton University Press, 1953.

[5] W. T. Gowers, *An infinite Ramsey theorem and some Banach-space dichotomies*, Ann. of Math. (2) 156 (2002), pp. 797–833.

[6] W. B. Johnson, *On quotients of L_p which are quotients of l_p*, Compositio Math. 34 (1977), pp. 69–89.

[7] W. B. Johnson and M. Zippin, *On subspaces of quotients of $(\sum G_n)_{l_p}$ and $(\sum G_n)_{c_0}$*, Israel J. Math. 13 (1972), pp. 311–316. Proceedings of the International Symposium on Partial Differential Equations and the Geometry of Normed Linear Spaces, Jerusalem, 1972.

[8] D. A. Martin, *Borel determinacy*, Ann. of Math. (2) 102 (1975), pp. 363–371.

[9] B. Maurey, V. D. Milman, and N. Tomczak-Jaegermann, *Asymptotic infinite-dimensional theory of Banach spaces*, Oper. Theory: Adv. Appl. 77 (1994), pp. 149–175.

[10] E. Odell and Th. Schlumprecht, *Trees and branches in Banach spaces*, Trans. Amer. Math. Soc. 354 (2002), pp. 4085–4108.

[11] ———, *A universal reflexive space for the class of uniformly convex Banach spaces*, Math. Ann. 335 (2006), pp. 901–916.

[12] E. Odell, Th. Schlumprecht, and A. Zsák, *Banach spaces with bounded Szlenk index*, preprint.

[13] ———, *On the structure of asymptotic ℓ_p spaces*, submitted.

[14] C. Rosendal, *Infinite asymptotic games*, preprint.

[15] H. P. Rosenthal, *Block determined finite dimensional decompositions*, preprint.

[16] G. Schechtman, *On Pełczyński's paper "Universal bases"*, Israel J. Math. 22 (1975), pp. 181–184.

[17] W. Szlenk, *The non-existence of a separable reflexive Banach space universal for all separable reflexive Banach spaces*, Studia Math. 30 (1968), pp. 53–61.

[18] M. Zippin, *Banach spaces with separable duals*, Trans. Amer. Math. Soc. 310 (1988), pp. 371–379.

Author information

Edward Odell, Department of Mathematics, The University of Texas, 1 University Station C1200, Austin, TX 78712, USA.
Email: odell@math.utexas.edu

Thomas Schlumprecht, Department of Mathematics, Texas A&M University, College Station, TX 78712, USA.
Email: schlump@math.tamu.edu

András Zsák, School of Mathematical Sciences, University of Nottingham, University Park, Nottingham NG7 2RD, United Kingdom.
Email: andras.zsak@maths.nottingham.ac.uk

Banach Spaces and their Applications in Analysis, 183–191 © de Gruyter 2007

Extremal configurations for moments of sums of independent positive random variables

Gideon Schechtman

Abstract. We find the extremal configuration for the p-moment of sums of independent positive random variables while constraining the sum of the expectations of the random variables and the sum of their p-moments.

Key words. Rosenthal inequality, extremal moment problems.

AMS classification. 60E15, 60G50.

1 Introduction

If $X_1, X_2, \dots, X_n$ are non-negative independent random variables and $p > 1$ then

$$\max\{\sum_{i=1}^{n} \mathbb{E}X_i, (\sum_{i=1}^{n} \mathbb{E}X_i^p)^{1/p}\} \le (\mathbb{E}(\sum_{i=1}^{n} X_i)^p)^{1/p} \le K_p \max\{\sum_{i=1}^{n} \mathbb{E}X_i, (\sum_{i=1}^{n} \mathbb{E}X_i^p)^{1/p}\} \tag{1.1}$$

where K_p depends only on p. This and a similar and better known inequality for mean zero random variables were proved by Rosenthal in [8]. These inequalities and their variants were quite useful in both Probability and Functional Analysis. The left inequality is easy and the constant one in it is easily seen to be best possible. The constant in the right hand side inequality has been shown in [6] to be of order $p/\log p$. The latest result in this direction is contained in [7] and gives a version of the inequality above (and more importantly of its version for symmetric variables) with constants independent of p. Of course the term $\max\{\sum_{i=1}^{n} \mathbb{E}X_i, \sum_{i=1}^{n} \mathbb{E}X_i^p\}$ is replaced in [7] with another (equivalent) quantity still depending on the individual X_i-s only. The trigger to the present note is the recent paper [1] in which a related extremal problem is treated.

In this paper we find, for each $0 < A, B < \infty$, the supremum of $\mathbb{E}(\sum X_i)^p$ subject to: $\{X_i\}$ are non-negative, independent random variables with $\sum \mathbb{E}X_i = A$ and $\sum \mathbb{E}X_i^p = B$. We also find the random variables which asymptotically achieve this supremum. As a corollary one finds the actual best possible constant K_p in (1.1). The determination of this constant has previously been done in [4] (see also [2] for a more general result).

The similar problem for symmetric random variables was treated and solved in [9] and [3]. The proofs here are based on ideas we developed in collaboration with Figiel, Hitczenko, Johnson, and Zinn in the preliminary stages of [3] (where we treated the symmetric case). The proofs that actually appeared in [3] are quite different from these

Supported by the Israel Science Foundation.

preliminary ideas. Now seem to us to be a good opportunity to expose these ideas which didn't appear in print. We see this as the main purpose of this note. Moreover, dealing with the positive rather than the symmetric case turn out to be substantially simpler. We find the proof quite interesting and we hope these ideas will find more applications.

Our main result is

Theorem 1.1. *The supremum of* $\mathbb{E}(\sum_{i=1}^n X_i)^p$ *over all* n *and all independent non-negative random variables* X_i *satisfying*

$$\sum_{i=1}^n \mathbb{E}X_i = A \text{ and } \sum_{i=1}^n \mathbb{E}X_i^p = B$$

is

$$\mathbb{E}\Big(\Big(\frac{B}{A}\Big)^{1/(p-1)} P_{A^{p/(p-1)}/B^{1/(p-1)}}\Big)^p \text{ for } p \geq 2$$

where P_μ *denotes a Poisson random variable with parameter* μ, *and*

$$A^p + B \text{ for } 1 < p \leq 2.$$

Next we show how to (asymptotically) achieve the supremum. The proof of the theorem will show that this is basically the only way the supremum can be achieved. For $p \geq 2$ and large enough k let X_i, $i = 1, \ldots, k$, be independent identically distributed random variables each taking the values

$$(B/A)^{1/(p-1)} \text{ with probability } A^{p/(p-1)}/kB^{1/(p-1)}$$

and

$$0 \text{ with probability } 1 - A^{p/(p-1)}/kB^{1/(p-1)}.$$

Then

$$\sum_{i=1}^k \mathbb{E}X_i = A, \ \sum_{i=1}^k \mathbb{E}X_i^p = B$$

and

$$\mathbb{E}(\sum_{i=1}^k X_i)^p \overset{k\to\infty}{\longrightarrow} \mathbb{E}\Big(\Big(\frac{B}{A}\Big)^{1/(p-1)} P_{A^{p/(p-1)}/B^{1/(p-1)}}\Big)^p.$$

For $1 < p \leq 2$, let $X_1, \ldots, X_k$ denote independent random variables each taking the value $1/k$ with probability A (assuming as we may that $A \leq 1$) and the value 0 with probability $1 - A$, and let Y be a random variable independent of the X_i-s and taking the value $k^{1/p}$ with probability B/k (assuming $k \geq B$) and the value 0 with probability $1 - B/k$. Then $\sum_{i=1}^k \mathbb{E}X_i + \mathbb{E}Y \overset{k\to\infty}{\longrightarrow} A$, $\sum_{i=1}^k \mathbb{E}X_i^p + \mathbb{E}Y^p \overset{k\to\infty}{\longrightarrow} B$ and $\mathbb{E}(\sum_{i=1}^k X_i + Y)^p \overset{k\to\infty}{\longrightarrow} A^p + B$.

By taking $A^p = B$ in Theorem 1.1 it is easy to get the following corollary, previously obtained in [2],

Corollary 1.2. *The best value of* K_p *is* $(\mathbb{E}(P_1)^p)^{1/p}$ *for* $p \geq 2$ *and* $2^{1/p}$ *for* $1 < p \leq 2$.

2 Proof of the theorem

The proof of the theorem has two main ingredients: In Proposition 2.1 and Corollary 2.2 we reduce the problem to that of a maximization of the p-th moment of linear combination of independent Poisson random variables subject to certain constraints on the coefficients and the parameters involved. This reduces the problem to what looks like a sophisticated calculus problem. After calculating the needed partial derivatives in Proposition 2.3 we solve a related and somewhat stronger maximization problem in the Main Proposition 2.4. The proof involves a certain trick – adding a seemingly superfluous constraint.

Proposition 2.1. *Let $p \geq 1$, let A and B be two disjoint events in a probability space of equal probability and let $\tilde{A}$ and $\tilde{B}$ be two independent events of the same probability as A. Let a, b, c be non-negative numbers then*

$$\mathbb{E}(c + a1_A + b1_B)^p \leq \mathbb{E}(c + a1_{\tilde{A}} + b1_{\tilde{B}})^p.$$

We shall present two proofs.

Proof 1. Put $\alpha = P(A)$. Then,

$$\mathbb{E}(c + a1_A + b1_B)^p = (1 - 2\alpha)c^p + \alpha((a+c)^p + (b+c)^p)$$

and

$$\mathbb{E}(c + a1_{\tilde{A}} + b1_{\tilde{B}})^p = (1-\alpha)^2 c^p + \alpha(1-\alpha)((a+c)^p + (b+c)^p) + \alpha^2(a+b+c)^p.$$

Subtracting, cancelling α and dividing by c (assuming as we may it is positive) we see that it is enough to show that

$$f(a,b) = 1 + (1+a+b)^p - (1+a)^p - (1+b)^p \geq 0$$

for all $a, b \geq 0$. For all $b \geq 0$ $g_b(a) = (1+a+b)^p - (1+a)^p$ is increasing in a so

$$f(a,b) = g_b(a) - g_b(0) \geq 0.$$

□

Proof 2. This proof provides a more general statement: $\tilde{A}$ and $\tilde{B}$ can be *any* two events of that same probability as A, not necessarily independent. Also, one can replace the power p function with any convex function.

By [5, Lemma 4], $a1_A + b1_B$ is in the convex hull of all random variables on the same probability space which have the same distribution as $a1_{\tilde{A}} + b1_{\tilde{B}}$. Clearly this implies that $c + a1_A + b1_B$ is in the convex hull of all random variables on the same probability space which have the same distribution as $c + a1_{\tilde{A}} + b1_{\tilde{B}}$. The result now follows by convexity. □

Corollary 2.2. *For $p \geq 1$ and $A, B > 0$ put $K_p(A,B) = \sup(\mathbb{E}(\sum X_i)^p)^{1/p}$ where the* sup *is taken over all independent random variables $X_i \geq 0$ with $\sum \mathbb{E}(X_i) \leq A$ and $\sum \mathbb{E}(X_i)^p \leq B^p$. Then*

$$K_p(A,B) = \sup(\mathbb{E}(\sum a_i P_{\mu_i})^p)^{1/p}$$

where the sup *is taken over all* $a_i, \mu_i \geq 0$ *with* $\sum a_i\mu_i \leq A$, $\sum a_i^p \mu_i \leq B^p$ *and independent random variables* P_{μ_i} *where* P_{μ_i} *has Poisson distribution with parameter* μ_i.

Proof. Let $X_i = \sum_{j=1}^{m_i} \alpha_{i,j} 1_{A_{i,j}}$, $i = 1, \dots, n$ be a finite sequence of simple non-negative random variables (i.e., for each i, $\{A_{i,j}\}_{j=1}^{m_i}$ are disjoint) satisfying the constraints:

$$\sum_{i=1}^{n}\sum_{j=1}^{m_i} \alpha_{i,j} P(A_{i,j}) = A \text{ and } \sum_{i=1}^{n}\sum_{j=1}^{m_i} \alpha_{i,j}^p P(A_{i,j}) = B.$$

Let $Y_{i,j}$ have the same distribution as $\alpha_{i,j} 1_{A_{i,j}}$ for each i, j while $Y_{i,j}$, $i = 1, \dots, n$, $j = 1, \dots, m_i$ are independent. Then $\sum_{i,j} \mathbb{E}Y_{i,j} = A$, $\sum_{i,j} \mathbb{E}Y_{i,j}^p = B$ and, by Proposition 2.1,

$$\mathbb{E}(\sum_{i,j} \alpha_{i,j} Y_{i,j})^p \geq \mathbb{E}(\sum_i X_i)^p.$$

Using Proposition 2.1 again we can replace each $Y_{i,j}$ with a Poisson random variable $P_{\mu_{i,j}}$, with $\mu_{i,j} = P(A_{i,j})$ while $\mathbb{E}(\sum_{i,j} \alpha_{i,j} P_{\mu_{i,j}})^p \geq \mathbb{E}(\sum_{i,j} \alpha_{i,j} Y_{i,j})^p$. Indeed, divide each $A_{i,j}$ into k disjoint sets of equal measure $A_{i,j,\ell}$, $\ell = 1, \dots, k$. Let $\tilde{A}_{i,j,\ell}$, $i = 1, \dots, n$, $j = 1, \dots, m_i$, $\ell = 1, \dots, k$ be independent with $P(\tilde{A}_{i,j,\ell}) = P(A_{i,j,\ell})$. Then

$$\mathbb{E}(\sum_{i,j} \alpha_{i,j} P_{\mu_{i,j}})^p = \lim_{k\to\infty} \mathbb{E}(\sum_{i,j,\ell} \alpha_{i,j,\ell} 1_{\tilde{A}_{i,j,\ell}})^p \geq \mathbb{E}(\sum_{i,j} \alpha_{i,j} Y_{i,j})^p.$$

This proves that $K_p(A, B) \leq \sup(\mathbb{E}(\sum a_i P_{\mu_i})^p)^{1/p}$. The other inequality follows easily by presenting each P_{μ_i} as a limit of appropriate sum of independent indicator functions. Since we are going to use only the inequality proved above, we leave the details to the reader. □

In the next proposition we compute some partial derivatives needed in the proof of the following Main Proposition.

Proposition 2.3. *Let* X *be a positive random variable which have finite* p*-th moment,* $p > 1$. *Let* P_μ *denote an independent Poisson variable with parameter* μ. *Then, for all* $a > 0$, $\mu > 0$,

$$\frac{\partial}{\partial a}\mathbb{E}(aP_\mu + X)^p = p\mu\mathbb{E}(a + aP_\mu + X)^{p-1} \tag{2.1}$$

and

$$\frac{\partial}{\partial \mu}\mathbb{E}(aP_\mu + X)^p pa\mathbb{E}(aU + aP_\mu + X)^{p-1} \tag{2.2}$$

where U *is a random variable uniformly distributed over* $[0, 1]$ *and independent of* (P_μ, X).

Proof. Assume first that $1 < p < 2$. Recall that for every positive variable Y with p-th moment

$$\mathbb{E}Y^p = c_p \int_0^\infty \frac{\varphi_Y(t) - 1 - it\mathbb{E}Y}{t^{p+1}} dt$$

and

$$\mathbb{E}Y^{p-1} = c_{p-1} \int_0^\infty \frac{1 - \varphi_Y(t)}{t^p} dt$$

where $c_p = (\int_0^\infty \frac{e^{it}-1-it}{t^{p+1}} dt)^{-1}$, $c_{p-1} = (\int_0^\infty \frac{1-e^{it}}{t^p} dt)^{-1}$, and $\varphi_Y(t) = \mathbb{E}e^{itY}$. Note that integration by parts gives $\frac{-ic_p}{c_{p-1}} = p$.

Differentiating formally,

$$\begin{aligned}
\frac{\partial}{\partial a}\mathbb{E}(aP_\mu + X)^p &= c_p \frac{\partial}{\partial a} \int_0^\infty \frac{e^{\mu(e^{ita}-1)}\varphi_X(t) - 1 - it(a\mu + \mathbb{E}X)}{t^{p+1}} dt \\
&= c_p \int_0^\infty \frac{it\mu e^{ita} e^{\mu(e^{ita}-1)}\varphi_X(t) - it\mu}{t^{p+1}} dt \\
&= -i\mu c_p \int_0^\infty \frac{1 - e^{ita} e^{\mu(e^{ita}-1)}\varphi_X(t)}{t^p} dt \\
&= \frac{-i\mu c_p}{c_{p-1}} \mathbb{E}(a + aP_\mu + X)^{p-1}.
\end{aligned} \tag{2.3}$$

It is easy to justify the differentiation under the integral sign. Similarly,

$$\begin{aligned}
\frac{\partial}{\partial \mu}\mathbb{E}(aP_\mu + X)^p &= c_p \frac{\partial}{\partial \mu} \int_0^\infty \frac{e^{\mu(e^{ita}-1)}\varphi_X(t) - 1 - it(a\mu + \mathbb{E}X)}{t^{p+1}} dt \\
&= c_p \int_0^\infty \frac{(e^{ita} - 1)e^{\mu(e^{ita}-1)}\varphi_X(t) - ita}{t^{p+1}} dt \\
&= -iac_p \int_0^\infty \frac{1 - \frac{e^{ita}-1}{ia} e^{\mu(e^{ita}-1)}\varphi_X(t)}{t^p} dt \\
&= \frac{-iac_p}{c_{p-1}} \mathbb{E}(aU + aP_\mu + X)^{p-1}.
\end{aligned} \tag{2.4}$$

This concludes the proof for $1 < p < 2$. For the other values of p one can use a similar proof, using

$$\mathbb{E}Y^p = c_p \int_0^\infty \frac{\varphi_Y(t) - \sum_{j=0}^k \frac{\mathbb{E}(itY)^j}{j!}}{t^{p+1}} dt \quad \text{for } k < p < k+1, \quad k \text{ an integer}$$

and

$$\mathbb{E}Y^k = (-i)^k \frac{\partial^k \mathbb{E}\varphi_Y(t)}{\partial t^k}_{|t=0}, \quad k \text{ an integer.}$$

Alternatively, one can notice first that the four quantities:

$$\mathbb{E}\frac{\partial}{\partial a}(aP_\mu + X)^q, \;\; aq\mathbb{E}(a + aP_\mu + X)^{q-1}$$

and

$$\mathbb{E}\frac{\partial}{\partial \mu}(aP_\mu + X)^q, \;\; \mu q\mathbb{E}(aU + aP_\mu + X)^{q-1}$$

are well defined for all complex q in an open strip containing $1 < \mathrm{Real}(q) \leq p$ and are analytic functions of q there (we omit the justification). Since the first two coincide on the interval $(0,1)$ they coincide in the whole strip. The same holds for the last two quantities. Finally, it is easy to justify changing the order of $\mathbb{E}$ and the differentiation.□

The next proposition is the main one.

Proposition 2.4 (Main Proposition). *Given a positive random variable X with p-th moment $p > 1$, the supremum of $\mathbb{E}(aP_\mu + bP_\nu + X)^p$ subject to $a, b, \mu, \nu > 0$, $a\mu + b\nu = A$, $a^p\mu + b^p\nu = B$ and where P_μ, P_ν are Poisson variables with the indicated parameters and $P_\mu, P_\nu$and X are independent, is given by*

$$\mathbb{E}\Big(\Big(\frac{B}{A}\Big)^{1/(p-1)} P_{A^{p/(p-1)}/B^{1/(p-1)}} + X\Big)^p \ \textit{for}\ p \geq 2$$

and

$$\mathbb{E}(A+X)^p + B \ \textit{for}\ 1 < p \leq 2.$$

Proof. It is tempting to use Proposition 2.3 to find the supremum by the method of Lagrange multipliers. We were not successful trying to do it. Moreover, as we shall see below, in the case $1 < p < 2$ the supremum is not attained at an inner point of the domain. Instead, we first add another constraint, $a\mu = C$ $(0 < C < A)$. Solving for b, μ and ν in terms of a we get:

$$\mu = \frac{C}{a},\ b = \Big(\frac{B - Ca^{p-1}}{A - C}\Big)^{1/(p-1)},\ \nu = \frac{(A-C)^{p/(p-1)}}{(B - Ca^{p-1})^{1/(p-1)}}. \tag{2.5}$$

These quantities are well defined and positive as long as $a < (B/C)^{1/(p-1)}$. We thus need to find the supremum of

$$g(a) = \mathbb{E}(aP_\mu + bP_\nu + X)^p$$

(where μ, b, ν are given by (2.5)) in the range $0 < a < (B/C)^{1/(p-1)}$. For that we compute the derivatives of μ, b, ν with respect to a:

$$\begin{aligned} &\frac{d\mu}{da} = \frac{-C}{a^2},\ \frac{db}{da} \frac{-Ca^{p-2}}{(B - Ca^{p-1})^{(p-2)/(p-1)}(A-C)^{1/(p-1)}}, \\ &\frac{d\nu}{da} = \frac{C(A-C)^{p/(p-1)}a^{p-2}}{(B - Ca^{p-1})^{p/(p-1)}}, \end{aligned} \tag{2.6}$$

and use them and the formulas (2.1) and (2.2) to find $g'(a)$. Denote $Y = aP_\mu + bP_\nu + X$ then, by the chain rule,

$$\begin{aligned} g'(a) = &\frac{pC}{a}\mathbb{E}(a+Y)^{p-1} - \frac{pC}{a}\mathbb{E}(aU+Y)^{p-1} \\ &- \frac{pC\nu a^{p-2}}{(B - Ca^{p-1})^{(p-2)/(p-1)}(A-C)^{1/(p-1)}}\mathbb{E}(b+Y)^{p-1} \\ &+ \frac{pC(A-C)^{p/(p-1)}ba^{p-2}}{(B - Ca^{p-1})^{p/(p-1)}}\mathbb{E}(bU+Y)^{p-1} \end{aligned} \tag{2.7}$$

where b and ν are given by (2.5) and U is a standard uniform random variable independent of Y. Using (2.5) again we can write (2.7) as

$$g'(a) = \frac{pC}{a}(\mathbb{E}(a+Y)^{p-1} - \mathbb{E}(aU+Y)^{p-1}) - \frac{pCa^{p-2}}{b^{p-1}}(\mathbb{E}(b+Y)^{p-1} - \mathbb{E}(bU+Y)^{p-1})$$

or

$$g'(a) = pCa^{p-2}\Big(\mathbb{E}\Big(1+\frac{Y}{a}\Big)^{p-1} - \mathbb{E}\Big(U+\frac{Y}{a}\Big)^{p-1} - \mathbb{E}\Big(1+\frac{Y}{b}\Big)^{p-1} + \mathbb{E}\Big(U+\frac{Y}{b}\Big)^{p-1}\Big). \tag{2.8}$$

We now separate between different ranges of p. First, for $p = 2$ it is easy to see that $\mathbb{E}(aP_\mu + bP_\nu + X)^2 = A^2 + B + \mathbb{E}X^2 + 2A\mathbb{E}X$ for all a, b, μ, ν satisfying the two original constraints. It is also easy to check that this coincides with the (two) conclusion(s) of the proposition for $p = 2$.

If $p > 2$ then we claim that $g'(a) > 0$ if and only if $a < b$ (equivalently, if and only if $a < (B/A)^{1/(p-1)}$). Indeed, by (2.8), it is enough to show that

$$h(x) = (1+x)^{p-1} - \mathbb{E}(U+x)^{p-1} = (1+x)^{p-1} - \int_0^1 (t+x)^{p-1}dt$$

is an increasing function on $(0,\infty)$. Now

$$h'(x) = (p-1)\int_0^1 ((1+x)^{p-2} - (t+x)^{p-2})dt > 0$$

since $p > 2$. (For later use note that, if $1 < p < 2$, h is decreasing on $(0,\infty)$.) This proves that the maximum of g is attained for $a = b = (B/A)^{1/(p-1)}$. The corresponding values of μ and ν are

$$\mu = C(A/B)^{1/(p-1)} \text{ and } \nu = (A-C)(A/B)^{1/(p-1)}.$$

For these values, the distribution of $aP_\mu + bP_\nu$ is the same as that of

$$\Big(\frac{B}{A}\Big)^{1/(p-1)} P_{A^{p/(p-1)}/B^{1/(p-1)}}$$

(using the fact that the sum of two independent Poisson random variables is a Poisson variable whose parameter is the sum of the parameters of the original variables). Thus, the supremum of $\mathbb{E}(aP_\mu + bP_\nu + X)^p$ given the three constraints is

$$\mathbb{E}\Big(\Big(\frac{B}{A}\Big)^{1/(p-1)} P_{A^{p/(p-1)}/B^{1/(p-1)}} + X\Big)^p.$$

Since C does not appear in this last quantity the same holds under the two original constraints.

The case $1 < p < 2$ is a bit more complicated. As was remarked above, h is decreasing on $(0,\infty)$ and thus the supremum of g is given at one of the end points; i.e., as a tends to 0 or to $(B/C)^{1/(p-1)}$. In the first case the distribution of aP_μ tends to that of the constant C, as is easy to see from the behavior of the characteristic function: $\mathbb{E}e^{itaP_\mu} = e^{\mu(e^{ita}-1)} \to e^{itC}$ (since $a\mu = C$ and $a \to 0$). In the second case, since $b \to 0$ we get similarly that bP_ν tends to the constant $A - C$. We thus get that the supremum of $g(a)$ is the maximum between

$$\mathbb{E}\Big(C + \Big(\frac{B}{A-C}\Big)^{1/(p-1)} P_{(A-C)^{p/(p-1)}/B^{1/(p-1)}} + X\Big)^p$$

and

$$\mathbb{E}\Big(A - C + \Big(\frac{B}{C}\Big)^{1/(p-1)} P_{C^{p/(p-1)}/B^{1/(p-1)}} + X\Big)^p.$$

It follows that the supremum of $\mathbb{E}(aP_\mu + bP_\nu + X)^p$ subject to $a, b, \mu, \nu > 0$, $a\mu + b\nu = A$, $a^p\mu + b^p\nu = B$ is

$$\sup_{0<C<A} \mathbb{E}\Big(A - C + \Big(\frac{B}{C}\Big)^{1/(p-1)} P_{C^{p/(p-1)}/B^{1/(p-1)}} + X\Big)^p.$$

To find this supremum denote

$$\varphi(C) = \mathbb{E}\Big(A - C + \Big(\frac{B}{C}\Big)^{1/(p-1)} P_{C^{p/(p-1)}/B^{1/(p-1)}} + X\Big)^p.$$

By Proposition 2.3,

$$\begin{aligned}\varphi'(C) = &-p\mathbb{E}\Big(A - C + \Big(\frac{B}{C}\Big)^{1/(p-1)} P_{C^{p/(p-1)}/B^{1/(p-1)}} + X\Big)^{p-1} \\ &- \frac{p}{p-1}\mathbb{E}\Big(\Big(\frac{B}{C}\Big)^{1/(p-1)} + A - C + \Big(\frac{B}{C}\Big)^{1/(p-1)} P_{C^{p/(p-1)}/B^{1/(p-1)}} + X\Big)^{p-1} \\ &+ \frac{p^2}{p-1}\mathbb{E}\Big(\Big(\frac{B}{C}\Big)^{1/(p-1)} U + A - C + \Big(\frac{B}{C}\Big)^{1/(p-1)} P_{C^{p/(p-1)}/B^{1/(p-1)}} + X\Big)^{p-1}\end{aligned}$$

where U is a standard uniform random variable independent of the other ones. We would like to show that φ is a decreasing function. For that it would be enough to show that for any $s, t > 0$

$$\psi(s,t) = -t^{p-1} - \frac{1}{p-1}(s+t)^{p-1} + \frac{p}{p-1}\mathbb{E}(sU + t)^{p-1} < 0.$$

(Take $s = \big(\frac{B}{C}\big)^{1/(p-1)}$ and $tA - C + \big(\frac{B}{C}\big)^{1/(p-1)} P_{C^{p/(p-1)}/B^{1/(p-1)}} + X$ in the previous equation.) This is simple:

$$\begin{aligned}\psi(s,t) &= -\,t^{p-1} - \frac{1}{p-1}(s+t)^{p-1} + \frac{1}{p-1}\frac{(s+t)^p - t^p}{s} \\ &= -\,t^{p-1} + \frac{t}{p-1}\frac{(s+t)^{p-1} - t^{p-1}}{s}.\end{aligned}$$

The concavity of the function x^{p-1} implies that, for each $t > 0$, ψ in a decreasing function of s. Since $\lim_{s\to 0^+} \psi(s,t) = 0$ we get that ψ is negative. We conclude that

$$\sup \mathbb{E}(aP_\mu + bP_\nu + X)^p = \lim_{C\to 0} \mathbb{E}\Big(A - C + \Big(\frac{B}{C}\Big)^{1/(p-1)} P_{C^{p/(p-1)}/B^{1/(p-1)}} + X\Big)^p.$$

When $C \to 0$ the support of $P_{C^{p/(p-1)}/B^{1/(p-1)}}$ tends to zero. Also, using to the definition of the Poisson distribution, it is easy to see that

$$\lim_{C\to 0} \mathbb{E}\Big(\Big(\frac{B}{C}\Big)^{1/(p-1)} P_{C^{p/(p-1)}/B^{1/(p-1)}}\Big)^p = B.$$

Consequently,

$$\sup \mathbb{E}(aP_\mu + bP_\nu + X)^p = \mathbb{E}(A + X)^p + B.$$

□

The proof of Theorem 1.1 is a simple consequence of Corollary 2.2 and Proposition 2.4.

References

[1] D. Berend and T. Tassa, *Estimates for moments of sums of random variables.*

[2] V. H. De la Pena, R. Ibragimov, and S. Sharakhmetov, *On Extremal distributions and sharp L_p-bounds for sums of multilinear forms*, Ann Probab. 31 (2003), pp. 630–675.

[3] T. Figiel, P. Hitczenko, W. B. Johnson, G. Schechtman, and J. Zinn, *Extremal properties of Rademacher functions with applications to the Khintchine and Rosenthal inequalities*, Trans. Amer. Math. Soc. 349 (1997), pp. 997–1027.

[4] R. Ibragimov and S. Sarakhmetov, *Exact bounds on the moments of symmetric statistics*, Seventh Vilinius Conference on Probability Theory and Mathematical Statistics (1998). Abstracts of Communications, pp. 243–244.

[5] W. B. Johnson and G. Schechtman, *Sums of independent random variables in rearrangement invariant function spaces*, Ann. Probab. 17 (1989), pp. 789–808.

[6] W. B. Johnson, G. Schechtman, and J. Zinn, *Best constants in moment inequalities for linear combinations of independent and exchangeable random variables*, Ann. Probab. 13 (1985), pp. 234–253.

[7] R. Latała, *Estimation of moments of sums of independent real random variables*, Ann. Probab. 25 (1997), pp. 1502–1513.

[8] H. P. Rosenthal, *On the subspaces of L^p $(p > 2)$ spanned by sequences of independent random variables*, Israel J. Math. 8 (1970), pp. 273–303.

[9] S. A. Utev, *Extremal problems in moment inequalities*, (Russian) Limit theorems of probability theory, pp. 56–75, 175, Trudy Inst. Mat., 5, "Nauka" Sibirsk. Otdel., Novosibirsk, 1985.

Author information

Gideon Schechtman, Department of Mathematics, Weizmann Institute of Science, Rehovot, Israel.
Email: gideon.schechtman@weizmann.ac.il

Banach Spaces and their Applications in Analysis, 193–208

Greedy approximation in Banach spaces

Vladimir Temlyakov

Abstract. Recently, a particular kind of nonlinear approximation, namely greedy approximation has attracted a lot of attention in both theoretical and applied settings. Greedy type algorithms have proven to be very useful in various applications such as image compression, signal processing, design of neural networks, and the numerical solution of nonlinear partial differential equations. A theory of greedy approximation is now emerging. Some fundamental convergence results have already been established and many fundamental problems remain unsolved. In this survey we place emphasis on the study of the efficiency of greedy algorithms with regard to redundant systems (dictionaries). We discuss greedy algorithms in a Banach space from the point of view of convergence and rate of convergence.

Key words. Greedy algorithms, Banach spaces, greedy expansion, convergence.

AMS classification. Primary 41A65; Secondary 41A25, 41A46, 46B20.

1 Introduction

This paper is a survey on greedy approximation in Banach spaces. In the paper we concentrate on studying convergence and rate of convergence of greedy algorithms with regard to redundant systems (dictionaries). It is clear that the case of Hilbert spaces is a particular case of the case of Banach spaces. The theory of greedy approximation in Hilbert spaces is much better developed than the corresponding theory in Banach spaces. There are survey papers [4], [19], [23] on greedy approximation in Hilbert spaces. In this paper we will not give a survey on greedy approximation in Hilbert spaces. Greedy approximation in Banach spaces is naturally subdivided into two big areas of research: greedy approximation with regard to bases or minimal systems and greedy approximation with regard to redundant systems. In this survey we do not discuss the first area of research. The reader can find a corresponding discussion in recent surveys [14], [24]. We begin with notations that have become standard in the theory of greedy approximation.

Let X be a Banach space with norm $\|\cdot\|$. We say that a set of elements (functions) $\mathcal{D}$ from X is a dictionary (symmetric dictionary) if each $g \in \mathcal{D}$ has norm bounded by one ($\|g\| \le 1$),

$$g \in \mathcal{D} \quad \text{implies} \quad -g \in \mathcal{D},$$

and $\overline{\operatorname{span}}\mathcal{D} = X$. We denote the closure (in X) of the convex hull of $\mathcal{D}$ by $A_1(\mathcal{D})$. We introduce a new norm, associated with a dictionary $\mathcal{D}$, in the dual space X^* by the formula

$$\|F\|_{\mathcal{D}} := \sup_{g \in \mathcal{D}} F(g), \qquad F \in X^*.$$

We will discuss in this paper greedy algorithms with regard to $\mathcal{D}$. For a nonzero element

$f \in X$ we denote by F_f a norming (peak) functional for f:

$$\|F_f\| = 1, \qquad F_f(f) = \|f\|.$$

The existence of such a functional is guaranteed by Hahn–Banach theorem.

There are two major classes of greedy algorithms in Banach spaces. We begin with a classical Pure Greedy Algorithm (PGA) defined in a Hilbert space and show how we obtain two different classes of greedy algorithms in Banach spaces. We note that the PGA is known under different names in different areas of research. The PGA has been introduced in [9] for a special dictionary $\mathcal{D}$ under the name *projection pursuit*. This name is used in statistics. In signal processing the PGA is known under the name *matching pursuit*.

Pure Greedy Algorithm (PGA).
We define $f_0 := f$. Then for each $m \geq 1$, we inductively define:

1) $\varphi_m \in \mathcal{D}$ is any satisfying (we assume existence)

$$\langle f_{m-1}, \varphi_m \rangle = \sup_{g \in \mathcal{D}} \langle f_{m-1}, g \rangle;$$

2) $f_m := f_{m-1} - \langle f_{m-1}, \varphi_m \rangle \varphi_m;$

3) $G_m(f, \mathcal{D}) := \sum_{j=1}^{m} \langle f_{j-1}, \varphi_j \rangle \varphi_j.$

The greedy step (the first step) of the PGA can be interpreted in two ways. First, we look at the mth iteration for an element $\varphi_m \in \mathcal{D}$ and a number λ_m satisfying

$$\|f_{m-1} - \lambda_m \varphi_m\|_H = \inf_{g \in \mathcal{D}, \lambda} \|f_{m-1} - \lambda g\|_H. \tag{1.1}$$

Second, we look for an element $\varphi_m \in \mathcal{D}$ such that

$$\langle f_{m-1}, \varphi_m \rangle = \sup_{g \in \mathcal{D}} \langle f_{m-1}, g \rangle. \tag{1.2}$$

In a Hilbert space both versions (1.1) and (1.2) result in the same PGA. In a general Banach space the corresponding versions of (1.1) and (1.2) lead to different greedy algorithms. The Banach space version of (1.1) is straightforward: instead of the Hilbert norm $\|\cdot\|_H$ in (1.1) we use the Banach norm $\|\cdot\|_X$. This results in the following greedy algorithm.

X-Greedy Algorithm (XGA).
We define $f_0 := f$, $G_0 := 0$. Then, for each $m \geq 1$, we inductively define

1) $\varphi_m \in \mathcal{D}$, $\lambda_m \in \mathbb{R}$ are such that (we assume existence)

$$\|f_{m-1} - \lambda_m \varphi_m\|_X = \inf_{g \in \mathcal{D}, \lambda} \|f_{m-1} - \lambda g\|_X. \tag{1.3}$$

2) Denote $f_m := f_{m-1} - \lambda_m \varphi_m, \qquad G_m := G_{m-1} + \lambda_m \varphi_m.$

The second version of the PGA in a Banach space is based on the concept of a norming (peak) functional. We note that in a Hilbert space a norming functional F_f acts as follows

$$F_f(g) = \langle f/\|f\|, g \rangle.$$

Therefore, (1.2) can be rewritten in terms of the norming functional $F_{f_{m-1}}$ as

$$F_{f_{m-1}}(\varphi_m) = \sup_{g \in \mathcal{D}} F_{f_{m-1}}(g). \tag{1.4}$$

This observation leads to the class of dual greedy algorithms. We define the Weak Dual Greedy Algorithm with weakness τ (WDGA) (see [19], [6]) that is a generalization of the Weak Greedy Algorithm (see [16]).

Weak Dual Greedy Algorithm (WDGA(τ)).
Let $\tau := \{t_m\}_{m=1}^{\infty}$, $t_m \in [0,1]$, be a weakness sequence. We define $f_0 := f$. Then, for each $m \geq 1$, we inductively define

1) $\varphi_m \in \mathcal{D}$ is any satisfying

$$F_{f_{m-1}}(\varphi_m) \geq t_m \|F_{f_{m-1}}\|_{\mathcal{D}}. \tag{1.5}$$

2) Define a_m as $\|f_{m-1} - a_m \varphi_m\| = \min_{a \in \mathbb{R}} \|f_{m-1} - a\varphi_m\|$.

3) Denote $f_m := f_{m-1} - a_m \varphi_m$.

Let us make a remark that justifies the idea of the dual greedy algorithms in terms of real analysis. We consider here approximation in uniformly smooth Banach spaces. For a Banach space X we define the modulus of smoothness

$$\rho(u) := \sup_{\|x\|=\|y\|=1} \left(\frac{1}{2}(\|x+uy\| + \|x-uy\|) - 1\right).$$

The uniformly smooth Banach space is the one with the property

$$\lim_{u \to 0} \rho(u)/u = 0.$$

It is easy to see that for any Banach space X its modulus of smoothness $\rho(u)$ is an even convex function satisfying the inequalities

$$\max(0, u-1) \leq \rho(u) \leq u, \quad u \in (0, \infty).$$

The following well-known proposition is a simple corollary of the definition of a uniformly smooth Banach space.

Proposition 1.1. *Let X be a uniformly smooth Banach space. Then for any $x \neq 0$ and y we have*

$$F_x(y) = \left(\frac{d}{du}\|x+uy\|\right)(0) = \lim_{u \to 0}(\|x+uy\| - \|x\|)/u. \tag{1.6}$$

Proposition 1.1 shows that in the WDGA we are looking for an element $\varphi_m \in \mathcal{D}$ that provides a big derivative of the quantity $\|f_{m-1} + ug\|$. Thus, we have two classes of greedy algorithms in Banach spaces. The first one is based on a greedy step of the form (1.3). We call this class the class of *X-greedy algorithms*. The second one is based on a greedy step of the form (1.5). We call this class the class of *dual greedy algorithms*. A very important feature of the dual greedy algorithms is that they can be modified into a weak form. The term "weak" in the definition of the WDGA means that at the greedy step (1.5) we do not shoot for the optimal element of the dictionary which realizes the corresponding sup but are satisfied with weaker property than being optimal. The obvious reason for this is that we do not know in general that the optimal one exists. Another, practical reason is that the weaker the assumption the easier to satisfy it and, therefore, easier to realize in practice.

Both algorithms (XGA and WDGA) have two steps at each iteration: 1) a greedy step of choosing φ_m; 2) an approximation step of building an approximant G_m and a residual f_m. The approximation steps of both algorithms coincide. In both cases we look for a best approximation of f_{m-1} by $\lambda\varphi_m$, $\lambda \in \mathbb{R}$. The greedy step of WDGA is different from the greedy step of XGA. By a greedy step, in choosing an mth element $\varphi_m \in \mathcal{D}$ to be used in an m-term approximant, we mean one which maximizes (minimizes) a certain functional determined by information from previous iterations of the algorithm. As it is explained above we obtain two different classes of greedy algorithms by varying the above-mentioned functional. Also, we obtain different types of greedy algorithms by using different ways of constructing (choosing coefficients of the linear combination) the m-term approximation from already found m elements of the dictionary. The XGA and WDGA update the residual f_{m-1} by subtracting $c_m(f)\varphi_m$. Therefore, both algorithms provide for an element $f \in X$ an expansion into a series

$$f \sim \sum_{m=1}^{\infty} c_m(f)\varphi_m.$$

We discuss convergence of the XGA and WDGA in Section 2. We also discuss there some modifications of the XGA and WDGA that provide expansions. In Section 3 we discuss the Weak Chebyshev Greedy Algorithm (WCGA) that belongs to the class of dual greedy algorithms. In the WCGA we build an approximant G_m in a way to maximally use the approximation power of the elements $\varphi_1, \dots, \varphi_m$. In Section 4 we discuss algorithms that utilize the following idea of relaxation. At the mth iteration of an algorithm we build an approximant G_m as a linear combination of the previous approximant G_{m-1} and a new element φ_m. In Section 5 we discuss some further results and give historical remarks.

2 Greedy expansions

From the definition of a dictionary it follows that any element $f \in X$ can be approximated arbitrarily well by finite linear combinations of the dictionary elements. The primary goal of this section is to discuss representations of an element $f \in X$ by a series

$$f \sim \sum_{j=1}^{\infty} c_j(f) g_j(f), \quad g_j(f) \in \mathcal{D}, \quad c_j(f) > 0, \quad j = 1, 2, \dots . \tag{2.1}$$

In building a representation (2.1) we should construct two sequences: $\{g_j(f)\}_{j=1}^{\infty}$ and $\{c_j(f)\}_{j=1}^{\infty}$. In this paper the construction of $\{g_j(f)\}_{j=1}^{\infty}$ will be based on ideas used in greedy-type nonlinear approximation (greedy-type algorithms). This justifies the use of the term *greedy expansion* for (2.1) considered in the paper. We will consider two different classes of greedy expansions that correspond to the above two classes of greedy algorithms. The first class is the class of *X-greedy expansions* obtained by X-greedy algorithms and the second class is the class of *dual greedy expansions* obtained by dual greedy algorithms. The X-Greedy Algorithm provides a X-greedy expansion and the WDGA provides a dual greedy expansion.

There is no results on X-Greedy Algorithms in general uniformly smooth Banach spaces. We formulate the following two open problems in this regard.

Open Problem 1. Does the XGA converge for all dictionaries $\mathcal{D}$ and each element $f \in X$ in uniformly smooth Banach spaces X with modulus of smoothness of fixed power type q, $1 < q \leq 2$, $(\rho(u) \leq \gamma u^q)$?

Open Problem 2. Characterize the class of Banach spaces X such that the X-Greedy Algorithm converges for all dictionaries $\mathcal{D}$ and each element f.

The Open Problem 2 has been formulated in [19, p. 73]. A little more is known about convergence of the WDGA. We formulate some more open problems from [19].

Open Problem 3. Characterize the class of Banach spaces X such that the WDGA(t), $t \in (0, 1]$, converges for all dictionaries $\mathcal{D}$ and each element f.

Open Problem 4. (Conjecture). Prove that the WDGA(t), $t \in (0, 1]$, converges for all dictionaries $\mathcal{D}$ and each element $f \in X$ in uniformly smooth Banach spaces X with modulus of smoothness of fixed power type q, $1 < q \leq 2$, $(\rho(u) \leq \gamma u^q)$.

Open Problem 5. Find necessary and sufficient conditions on a weakness sequence τ to guarantee convergence of the Weak Dual Greedy Algorithm in uniformly smooth Banach spaces X with modulus of smoothness of fixed power type q, $1 < q \leq 2$, $(\rho(u) \leq \gamma u^q)$ for all dictionaries $\mathcal{D}$ and each element $f \in X$.

We note that Problem 3 and Problem 4 were formulated in [19] for the case $t = 1$. We now formulate a convergence result for the WDGA due to Ganichev and Kalton [10] for uniformly smooth Banach spaces satisfying an extra condition called *property* Γ.

Definition 2.1. A uniformly smooth Banach space has *property* Γ if there is a constant $\beta > 0$ such that for any $x, y \in X$, satisfying $F_x(y) = 0$, we have

$$\|x + y\| \geq \|x\| + \beta F_{x+y}(y).$$

Theorem 2.2. *Let X be a uniformly smooth Banach space with property Γ. Then the WDGA(τ) with $\tau = \{t\}$, $t \in (0,1]$, converges for each dictionary and all $f \in X$.*

It is known (see [10]) that L_p spaces with $p \in (1,\infty)$ have property Γ.

Open Problem 6. Let $p \in (1,\infty)$. Find necessary and sufficient conditions on a weakness sequence τ to guarantee convergence of the Weak Dual Greedy Algorithm in the L_p space for all dictionaries $\mathcal{D}$ and each element $f \in L_p$.

We now proceed to a discussion of greedy expansions (2.1). The construction of $\{g_j(f)\}_{j=1}^{\infty}$ is, clearly, the most important and difficult part in building a representation (2.1). It was pointed out in [18] that we have a great flexibility in choosing the coefficients $c_j(f)$ of the expansion (2.1). In [18] this flexibility was used for constructing a dual greedy expansion that converges in any uniformly smooth Banach space. We discuss these results in more detail. The construction from [18] depends on two numerical parameters $t \in (0,1]$ (the weakness parameter) and $b \in (0,1)$ (the tuning parameter of the approximation method). The construction also depends on a majorant μ of the modulus of smoothness of the Banach space X.

Dual Greedy Algorithm with parameters (t, b, μ) (DGA(t, b, μ)).
Let X be a uniformly smooth Banach space with modulus of smoothness $\rho(u)$ and $\mu(u)$ be a continuous majorant of $\rho(u)$: $\rho(u) \le \mu(u)$, $u \in [0,\infty)$. For parameters $t \in (0,1]$, $b \in (0,1]$ we define sequences $\{f_m\}_{m=0}^{\infty}$, $\{\varphi_m\}_{m=1}^{\infty}$, $\{c_m\}_{m=1}^{\infty}$ inductively. Let $f_0 := f$. If for $m \ge 1$ $f_{m-1} = 0$ then we set $f_j = 0$ for $j \ge m$ and stop. If $f_{m-1} \ne 0$ then we conduct the following three steps:

1) take any $\varphi_m \in \mathcal{D}$ such that $F_{f_{m-1}}(\varphi_m) \ge t\|F_{f_{m-1}}\|_{\mathcal{D}}$;

2) choose $c_m > 0$ from the equation $\|f_{m-1}\|\mu(c_m/\|f_{m-1}\|) = \frac{tb}{2}c_m\|F_{f_{m-1}}\|_{\mathcal{D}}$;

3) define $f_m := f_{m-1} - c_m\varphi_m$.

In [18] we proved the following convergence result.

Theorem 2.3. *Let X be a uniformly smooth Banach space with modulus of smoothness $\rho(u)$ and let $\mu(u)$ be a continuous majorant of $\rho(u)$ with the property $\mu(u)/u \downarrow 0$ as $u \to +0$. Then for any $t \in (0,1]$ and $b \in (0,1)$ the DGA(t,b,μ) converges for each dictionary $\mathcal{D}$ and all $f \in X$.*

We proceed to results on Banach spaces with power type modulus of smoothness $\rho(u) \le \gamma u^q$, $q \in (1,2]$. It is well known (see [15]) that power type modulus of smoothness of nontrivial Banach spaces are limited to the case $q \in [1,2]$. The following result from [18] gives the rate of convergence.

Theorem 2.4. *Assume X has modulus of smoothness $\rho(u) \le \gamma u^q$, $q \in (1,2]$. Denote $\mu(u) := \gamma u^q$. Then for any dictionary $\mathcal{D}$ and any $f \in A_1(\mathcal{D})$ the rate of convergence of the DGA(t,b,μ) is given by*

$$\|f_m\| \le C(t,b,\gamma,q)m^{-\frac{t(1-b)}{p(1+t(1-b))}}, \quad p := \frac{q}{q-1}.$$

In the paper [22] we pushed the flexibility in choosing the coefficients $c_j(f)$ in the expansion (2.1) to the extreme. We made these coefficients independent of an element $f \in X$. Surprisingly, for properly chosen coefficients we obtained results for the corresponding dual greedy expansion similar to the above Theorem 2.3 and Theorem 2.4. Even more surprisingly, we obtained similar results for the corresponding X-greedy expansions. We proceed to the formulation of these results. Let $\mathcal{C} := \{c_m\}_{m=1}^{\infty}$ be a fixed sequence of positive numbers. We restrict ourselves to positive numbers because of the symmetry of the dictionary $\mathcal{D}$.

X-Greedy Algorithm with coefficients $\mathcal{C}$ (XGA($\mathcal{C}$)).
We define $f_0 := f$, $G_0 := 0$. Then, for each $m \geq 1$, we inductively define:
1) $\varphi_m \in \mathcal{D}$ is such that (we assume existence)

$$\|f_{m-1} - c_m\varphi_m\|_X = \inf_{g \in \mathcal{D}} \|f_{m-1} - c_m g\|_X.$$

2) Denote $f_m := f_{m-1} - c_m\varphi_m, \qquad G_m := G_{m-1} + c_m\varphi_m.$

Dual Greedy Algorithm with weakness τ and coefficients $\mathcal{C}$ (DGA(τ,$\mathcal{C}$)).
Let $\tau := \{t_m\}_{m=1}^{\infty}$, $t_m \in [0,1]$, be a weakness sequence. We define $f_0 := f$, $G_0 := 0$. Then, for each $m \geq 1$, we inductively define
1) $\varphi_m \in \mathcal{D}$ is any satisfying $F_{f_{m-1}}(\varphi_m) \geq t_m \|F_{f_{m-1}}\|_{\mathcal{D}}$.
2) Define $f_m := f_{m-1} - c_m\varphi_m, \qquad G_m := G_{m-1} + c_m\varphi_m.$

In the case $\tau = \{t\}$, $t \in (0,1]$, we write t instead of τ in the notation. The first result on convergence properties of the DGA($t,\mathcal{C}$) has been obtained in [18]. We formulate it here.

Theorem 2.5. *Let X be a uniformly smooth Banach space with modulus of smoothness $\rho(u)$. Assume $\mathcal{C} = \{c_j\}_{j=1}^{\infty}$ is such that*

$$\sum_{j=1}^{\infty} c_j = \infty, \quad \text{and, for any } y > 0, \quad \sum_{j=1}^{\infty} \rho(yc_j) < \infty.$$

Then for the DGA($t,\mathcal{C}$), we have $\liminf_{m\to\infty} \|f_m\| = 0$.

In [22] we proved an analogue of Theorem 2.5 for the XGA($\mathcal{C}$) and improved upon the convergence in Theorem 2.5 in the case of uniformly smooth Banach spaces with power type modulus of smoothness. Under an extra assumption on $\mathcal{C}$ we replaced $\liminf$ by $\lim$. Here is the corresponding result from [22].

Theorem 2.6. *Let $\mathcal{C} \in \ell_q \setminus \ell_1$ be a monotone sequence. Then the DGA($t,\mathcal{C}$) and the XGA($\mathcal{C}$) converge for each dictionary and all $f \in X$ in any uniformly smooth Banach space X with modulus of smoothness $\rho(u) \leq \gamma u^q$, $q \in (1,2]$.*

In [22] we also addressed a question of rate of approximation for $f \in A_1(\mathcal{D})$. We proved the following theorem.

Theorem 2.7. *Let X be a uniformly smooth Banach space with modulus of smoothness $\rho(u) \leq \gamma u^q$, $q \in (1,2]$. We set $s := (1+1/q)/2$ and $\mathcal{C}_s := \{k^{-s}\}_{k=1}^{\infty}$. Then the DGA($t, \mathcal{C}_s$) and XGA($\mathcal{C}_s$) (for this algorithm $t = 1$) converge for $f \in A_1(\mathcal{D})$ with the following rate: for any $r \in (0, t(1-s))$*

$$\|f_m\| \leq C(r,t,q,\gamma)m^{-r}.$$

For the case $t = 1$, Theorem 2.7 provides rate of convergence of order m^{-r} for $f \in A_1(\mathcal{D})$ with r arbitrarily close to $(1-1/q)/2$. Theorem 2.4 provides similar rate of convergence. It would be interesting to understand if the rate $m^{-(1-1/q)/2}$ is the best that can be achieved in greedy expansions (for each $\mathcal{D}$, any $f \in A_1(\mathcal{D})$, and any X with $\rho(u) \leq \gamma u^q$, $q \in (1,2]$). We note that there are greedy approximation methods that provide error bound of the order $m^{1/q-1}$ for $f \in A_1(\mathcal{D})$ (see surveys [19] and [25] for recent results). However, these approximation methods do not provide an expansion. We will discuss such algorithms in Section 3 and Section 4.

3 Weak Chebyshev Greedy Algorithm

Let $\tau := \{t_k\}_{k=1}^{\infty}$ be a given sequence of nonnegative numbers $t_k \leq 1$, $k = 1, \dots$. We define the Weak Chebyshev Greedy Algorithm (WCGA) (see [17]) that is a generalization for Banach spaces of the Weak Orthogonal Greedy Algorithm defined and studied in [16] (see also [5] for Orthogonal Greedy Algorithm).

Weak Chebyshev Greedy Algorithm (WCGA(τ)).
We define $f_0^c = f_0^{c,\tau} := f$. Then for each $m \geq 1$ we inductively define:

1) $\varphi_m^c := \varphi_m^{c,\tau} \in \mathcal{D}$ is any satisfying $F_{f_{m-1}^c}(\varphi_m^c) \geq t_m \|F_{f_{m-1}^c}\|_{\mathcal{D}}$.

2) Define $\Phi_m := \Phi_m^{\tau} := \operatorname{span}\{\varphi_j^c\}_{j=1}^m$, and define $G_m^c := G_m^{c,\tau}$ to be the best approximant to f from Φ_m.

3) Denote $f_m^c := f_m^{c,\tau} := f - G_m^c$.

We begin with a convergence result. Recall that the modulus of smoothness $\rho(u)$ of a uniformly smooth Banach space is an even convex function such that $\rho(0) = 0$ and $\lim_{u\to 0} \rho(u)/u = 0$. The following function $s(u) := \rho(u)/u$, $s(0) := 0$, associated with $\rho(u)$ is a continuous increasing on $[0,\infty)$ function. Therefore, the inverse function $s^{-1}(\cdot)$ is well defined.

Theorem 3.1. *Let X be a uniformly smooth Banach space with modulus of smoothness $\rho(u)$. Assume that a weakness sequence $\tau := \{t_m\}_{m=1}^{\infty}$ satisfies the condition: for any $\theta > 0$, $\sum_{m=1}^{\infty} t_m s^{-1}(\theta t_m) = \infty$. Then for any $f \in X$ we have for the WCGA(τ), $\lim_{m\to\infty} \|f_m^c\| = 0$.*

Corollary 3.2. *Let a Banach space X have modulus of smoothness $\rho(u)$ of power type $1 < q \le 2$; $(\rho(u) \le \gamma u^q)$. Assume that*

$$\sum_{m=1}^{\infty} t_m^p = \infty, \quad p = \frac{q}{q-1}. \tag{3.1}$$

Then the WCGA(τ) converges for any $f \in X$.

It is proved in [17] that the condition (3.1) in Corollary 3.2 is sharp. It is shown on the example $X = \ell_q$, $1 < q \le 2$, and $\mathcal{D}$ is a standard basis $\{e_j\}_{j=1}^{\infty}$, $e_j := (0,\dots,0,1,0,\dots)$, for ℓ_q. The following convergence result has been proved in [6].

Theorem 3.3. *Let X be a separable reflexive Banach space. Then, X admits an equivalent norm for which the WCGA(t), $t \in (0,1]$, converges for every dictionary $\mathcal{D}$ and for every $f \in X$.*

Theorem 3.3 was obtained in [6] as a corollary to the following theorem:

Theorem 3.4 ([6, Proposition 3]). *Suppose that X is a reflexive Banach space which has both the Kadec–Klee property and a Fréchet differentiable norm. Then, the WCGA(t), $t \in (0,1]$, converges for every dictionary $\mathcal{D}$ and for every $f \in X$.*

We recall that X has the Kadec–Klee property (known also as the Radon–Riesz property or property H) if the following holds: whenever a sequence $\{x_n\}_{n=1}^{\infty}$ in X converges weakly to $x \in X$, and $\|x\| = \lim_{n\to\infty} \|x_n\|$, then, the $\{x_n\}_{n=1}^{\infty}$ converges strongly to x. It is known that every separable Banach space admits an equivalent norm with the Kadec–Klee property (see [3, Theorem 2.6]).

Open Problem 7. Characterize Banach spaces X such that the WCGA(t), $t \in (0,1]$, converges for every dictionary $\mathcal{D}$ and for every $f \in X$.

We proceed to results on the rate of convergence. The following theorem from [17] provides the rate of convergence of the WCGA.

Theorem 3.5. *Let X be a uniformly smooth Banach space with modulus of smoothness $\rho(u) \le \gamma u^q$, $1 < q \le 2$. Then for a sequence $\tau := \{t_k\}_{k=1}^{\infty}$, $t_k \le 1$, $k = 1,2,\dots$, we have for any $f \in A_1(\mathcal{D})$ that*

$$\|f_m^{c,\tau}\| \le C(q,\gamma)\Big(1 + \sum_{k=1}^{m} t_k^p\Big)^{-1/p}, \quad p := \frac{q}{q-1},$$

with a constant $C(q,\gamma)$ which may depend only on q and γ.

This theorem gives the rate of convergence of WCGA for f in $A_1(\mathcal{D})$. It was pointed out in [1] that it is important to have estimates for the rate of approximation of greedy algorithms for more general functions. We will now formulate the corresponding variant of Theorem 3.5 from [25]. Theorem 3.5 was derived in [17] from the following lemma:

Lemma 3.6 ([17, Lemma 2.3]). *Let X be a uniformly smooth Banach space with modulus of smoothness $\rho(u)$. Take a number $\varepsilon \geq 0$ and two elements f, f^ε from X such that $\|f - f^\varepsilon\| \leq \varepsilon$, $f^\varepsilon/A(\varepsilon) \in A_1(\mathcal{D})$, with some number $A(\varepsilon) > 0$. Then we have for $m = 1, 2, \ldots$*

$$\|f_m^{c,\tau}\| \leq \|f_{m-1}^{c,\tau}\| \inf_\lambda \left(1 - \lambda t_m A(\varepsilon)^{-1}\left(1 - \frac{\varepsilon}{\|f_{m-1}^{c,\tau}\|}\right) + 2\rho\left(\frac{\lambda}{\|f_{m-1}^{c,\tau}\|}\right)\right).$$

It is noted in [25] that in the same way as Theorem 3.5 was derived from Lemma 3.6 in [17] one obtains the following result.

Theorem 3.7. *Let X be a uniformly smooth Banach space with modulus of smoothness $\rho(u) \leq \gamma u^q$, $1 < q \leq 2$. Take a number $\varepsilon \geq 0$ and two elements f, f^ε from X such that $\|f - f^\varepsilon\| \leq \varepsilon$, $f^\varepsilon/A(\varepsilon) \in A_1(\mathcal{D})$, with some number $A(\varepsilon) > 0$. Then we have*

$$\|f_m^{c,\tau}\| \leq \max\left(2\varepsilon, C(q,\gamma)(A(\varepsilon) + \varepsilon)(1 + \sum_{k=1}^m t_k^p)^{-1/p}\right), \quad p := q/(q-1).$$

4 Relaxation

We define now the generalization for Banach spaces of the Weak Relaxed Greedy Algorithm studied in [17] (see [16] for the case of a Hilbert space).

Weak Relaxed Greedy Algorithm (WRGA).
We define $f_0^r := f_0^{r,\tau} := f$ and $G_0^r := G_0^{r,\tau} := 0$. Then for each $m \geq 1$ we inductively define:

1) $\varphi_m^r := \varphi_m^{r,\tau} \in \mathcal{D}$ is any satisfying

$$F_{f_{m-1}^r}(\varphi_m^r - G_{m-1}^r) \geq t_m \sup_{g \in \mathcal{D}} F_{f_{m-1}^r}(g - G_{m-1}^r).$$

2) Find $0 \leq \lambda_m \leq 1$ such that

$$\|f - ((1-\lambda_m)G_{m-1}^r + \lambda_m \varphi_m^r)\| = \inf_{0 \leq \lambda \leq 1} \|f - ((1-\lambda)G_{m-1}^r + \lambda \varphi_m^r)\|$$

and define $G_m^r := G_m^{r,\tau} := (1-\lambda_m)G_{m-1}^r + \lambda_m \varphi_m^r$.

3) Denote $f_m^r := f_m^{r,\tau} := f - G_m^r$.

Both, the WCGA and the WRGA use the functional $F_{f_{m-1}}$ in a search for the mth element φ_m from the dictionary to be used in approximation. The construction of the approximant in the WRGA is different from the construction in the WCGA. In the WCGA we build the approximant G_m^c in a way to maximally use the approximation power of the elements $\varphi_1, \ldots, \varphi_m$. The WRGA by its definition is designed for approximation of functions from $A_1(\mathcal{D})$. In building the approximant in the WRGA we keep the property $G_m^r \in A_1(\mathcal{D})$. We call the WRGA *relaxed* because at the mth iteration of the algorithm we use a linear combination (convex combination) of the previous approximant G^r_{m-1} and a new element φ^r_m. The relaxation parameter λ_m in the WRGA is chosen at the mth iteration depending on f. Recently, the following modification of the above idea of relaxation in greedy approximation has been studied in [1] and [25]. Let a sequence $\mathbf{r} := \{r_k\}_{k=1}^{\infty}$, $r_k \in [0,1)$, of relaxation parameters be given. Then at each iteration of our new algorithm we build the mth approximant of the form $G_m = (1-r_m)G_{m-1} + \lambda\varphi_m$. With an approximant of this form we are not limited to approximation of functions from $A_1(\mathcal{D})$ as in the WRGA. Results on the approximation properties of such an algorithm in a Hilbert space have been obtained in [1]. In [25] we studied a realization of the above idea of relaxation in the case of Banach spaces. In [25] we studied the Greedy Algorithm with Weakness parameter t and Relaxation $\mathbf{r}$ (GAWR($t,\mathbf{r}$)). We give a general definition of the algorithm in the case of a weakness sequence τ.

GAWR(τ,$\mathbf{r}$).
Let $\tau := \{t_m\}_{m=1}^{\infty}$, $t_m \in [0,1]$, be a weakness sequence. We define $f_0 := f$ and $G_0 := 0$. Then for each $m \geq 1$, we inductively define:
1) $\varphi_m \in \mathcal{D}$ is any satisfying $F_{f_{m-1}}(\varphi_m) \geq t_m \|F_{f_{m-1}}\|_{\mathcal{D}}$.
2) Find $\lambda_m \geq 0$ such that

$$\|f - ((1-r_m)G_{m-1} + \lambda_m\varphi_m)\| = \inf_{\lambda \geq 0} \|f - ((1-r_m)G_{m-1} + \lambda\varphi_m)\|$$

and define $G_m := (1-r_m)G_{m-1} + \lambda_m\varphi_m$.
3) Denote $f_m := f - G_m$.

In the case $\tau = \{t\}$, $t \in (0,1]$, we write t instead of τ in the notation. We note that in the case $r_k = 0$, $k = 1,\ldots,$ when there is no relaxation, the GAWR($\tau,\mathbf{0}$) coincides with the Weak Dual Greedy Algorithm [19, p. 66]. We will also consider here a relaxation of the X-Greedy Algorithm (see [19, p. 39]) that corresponds to $\mathbf{r} = \mathbf{0}$ in the definition that follows.

X-Greedy Algorithm with Relaxation $\mathbf{r}$ (XGAR($\mathbf{r}$)).
We define $f_0 := f$ and $G_0 := 0$. Then for each $m \geq 1$ we inductively define:
1) $\varphi_m \in \mathcal{D}$ and $\lambda_m \geq 0$ are such that (we assume existence)

$$\|f - ((1-r_m)G_{m-1} + \lambda_m\varphi_m)\| = \inf_{g \in \mathcal{D}, \lambda \geq 0} \|f - ((1-r_m)G_{m-1} + \lambda g)\|$$

and $G_m := (1-r_m)G_{m-1} + \lambda_m\varphi_m$.
2) Denote $f_m := f - G_m$.

We note that, practically, nothing is known about convergence and rate of convergence of the X-Greedy Algorithm. It has been proved in [25] that relaxation helps to prove convergence results for the XGAR($\mathbf{r}$). The following theorem from [25] gives a convergence result in any uniformly smooth Banach space.

Theorem 4.1. *Let a sequence* $\mathbf{r} := \{r_k\}_{k=1}^{\infty}$, $r_k \in [0,1)$, *satisfy the conditions*

$$\sum_{k=1}^{\infty} r_k = \infty, \quad r_k \to 0 \quad as \quad k \to \infty.$$

Then the GAWR($t, \mathbf{r}$) *and the XGAR*($\mathbf{r}$) *converge in any uniformly smooth Banach space for each* $f \in X$ *and for all dictionaries* $\mathcal{D}$.

For Banach spaces with power type modulus of smoothness we have ([25]) the following rate of convergence for algorithms with a special relaxation sequence.

Theorem 4.2. *Let* X *be a uniformly smooth Banach space with modulus of smoothness* $\rho(u) \leq \gamma u^q$, $1 < q \leq 2$. *Let* $\mathbf{r} := \{2/(k+2)\}_{k=1}^{\infty}$. *Consider the GAWR*($t, \mathbf{r}$) *and the XGAR*($\mathbf{r}$) *(for this algorithm* $t = 1$*). For a pair of functions* f, f^{ε}, *satisfying*

$$\|f - f^{\varepsilon}\| \leq \varepsilon, \quad f^{\varepsilon}/A(\varepsilon) \in A_1(\mathcal{D})$$

we have

$$\|f_m\| \leq \varepsilon + C(q,\gamma)(\|f\| + A(\varepsilon)/t)m^{-1+1/q}.$$

The following versions of a relaxed greedy algorithm have also been studied in [25].

Weak Greedy Algorithm with Free Relaxation (WGAFR).
Let $\tau := \{t_m\}_{m=1}^{\infty}$, $t_m \in [0,1]$, *be a weakness sequence. We define* $f_0 := f$ *and* $G_0 := 0$. *Then for each* $m \geq 1$ *we inductively define:*

1) $\varphi_m \in \mathcal{D}$ *is any satisfying* $F_{f_{m-1}}(\varphi_m) \geq t_m \|F_{f_{m-1}}\|_{\mathcal{D}}$.

2) Find w_m *and* $\lambda_m \geq 0$ *such that*

$$\|f - ((1-w_m)G_{m-1} + \lambda_m \varphi_m)\| = \inf_{\lambda \geq 0, w} \|f - ((1-w)G_{m-1} + \lambda \varphi_m)\|$$

and define $G_m := (1-w_m)G_{m-1} + \lambda_m \varphi_m$.

3) Denote $f_m := f - G_m$.

***X*-Greedy Algorithm with Free Relaxation (XGAFR).**
Set $f_0 := f$ *and* $G_0 := 0$. *Then for each* $m \geq 1$, *we inductively define:*

1) $\varphi_m \in \mathcal{D}$ *and* $\lambda_m \geq 0$, w_m *are such that (we assume existence)*

$$\|f - ((1-w_m)G_{m-1} + \lambda_m \varphi_m)\| = \inf_{g \in \mathcal{D}, \lambda \geq 0, w} \|f - ((1-w)G_{m-1} + \lambda g)\|$$

and $G_m := (1-w_m)G_{m-1} + \lambda_m \varphi_m$.

2) Denote $f_m := f - G_m$.

We note that in the above two algorithms with free relaxation we have more freedom in choosing an approximant than in the algorithms with fixed relaxation (GAWR($\tau, \mathbf{r}$), XGAR($\mathbf{r}$)). In the WGAFR we are optimizing over two parameters w and λ at each iteration of the algorithm. In other words we are looking for the best approximation from a 2-dimensional linear subspace at each iteration. In the two other versions of a relaxed greedy algorithm (GAWR($\tau, \mathbf{r}$), XGAR($\mathbf{r}$)) we approximate from a 1-dimensional linear subspace at each iteration of the algorithm. In the WCGA we are optimizing over all m coefficients at the mth iteration of the algorithm. Thus, we have more freedom in building an approximant in the WCGA than in the WGAFR. However, we obtained in [25] the same convergence and rate of convergence results for the WGAFR as we have for the WCGA.

Theorem 4.3. *Let X be a uniformly smooth Banach space with modulus of smoothness $\rho(u)$. Assume that a sequence $\tau := \{t_m\}_{m=1}^{\infty}$ satisfies the condition: for any $\theta > 0$ we have*

$$\sum_{m=1}^{\infty} t_m s^{-1}(\theta t_m) = \infty.$$

Then for any $f \in X$ we have for the WGAFR, $\lim_{m\to\infty} \|f_m\| = 0$.

Theorem 4.4. *Let X be a uniformly smooth Banach space. Then for any $f \in X$ we have for the XGAFR, $\lim_{m\to\infty} \|f_m\| = 0$.*

Theorem 4.5. *Let X be a uniformly smooth Banach space with modulus of smoothness $\rho(u) \le \gamma u^q$, $1 < q \le 2$. Take a number $\varepsilon \ge 0$ and two elements f, f^ε from X such that*

$$\|f - f^\varepsilon\| \le \varepsilon, \quad f^\varepsilon/A(\varepsilon) \in A_1(\mathcal{D}),$$

with some number $A(\varepsilon) > 0$. Then we have for the WGAFR and the XGAFR (for this algorithm $t_k = 1$)

$$\|f_m\| \le \max\left(2\varepsilon, C(q,\gamma)(A(\varepsilon)+\varepsilon)(1+\sum_{k=1}^{m} t_k^p)^{-1/p}\right), \quad p := q/(q-1).$$

It would be interesting to understand if the answers to the Open Problem 7 and the following Open Problem 8 are the same.

Open Problem 8. Characterize the class of Banach spaces X such that the WGAFR with the weakness sequence $\tau = \{t\}$, $t \in (0,1]$, converges for every dictionary $\mathcal{D}$ and every $f \in X$.

Open Problem 9. Characterize the class of Banach spaces X such that the XGAFR converges for every dictionary $\mathcal{D}$ and for every $f \in X$.

5 Discussion and some historical remarks

The first greedy algorithm with regard to a general dictionary was studied in a Hilbert space. This algorithm is the Pure Greedy Algorithm (PGA) (see [9], [11], [12], [5], [19]). Huber ([11]) proved convergence of the PGA in the weak topology and conjectured that the PGA converges in the strong sense (in the norm of H). Jones ([12]) proved this conjecture. It is a fundamental result in the theory of greedy approximation that guarantees convergence of the PGA for any $f \in H$ and any dictionary $\mathcal{D}$.

Generalizing a setting for a Hilbert space (see [13], [2]) the authors of [7] considered the following setting in a Banach space. We remind some definitions from [7]. An *incremental sequence* is any sequence $a_1, a_2, \dots$ of X so that $a_1 \in \mathcal{D}$ and for each $n \geq 1$ there are some $g_n \in \mathcal{D}$ and $\lambda_n \in [0, 1]$ so that

$$a_n = (1 - \lambda_n)a_{n-1} + \lambda_n g_n, \quad (a_0 = 0).$$

We say that an incremental sequence $a_1, a_2, \dots$ is ε-*greedy* (with respect to f) if $(a_0 = 0)$

$$\|f - a_n\| < \inf_{\lambda\in[0,1];g\in\mathcal{D}} \|f - ((1-\lambda)a_{n-1} + \lambda g)\| + \varepsilon_n, \quad n = 1, 2, \dots .$$

It is pointed out in [7] that the sequence $\{\lambda_n\}$ can be selected beforehand. They say that an incremental sequence $a_1, a_2, \dots$ is ε-greedy (with respect to f) with *convexity schedule* $\lambda_1, \lambda_2, \dots$ if $(a_0 = 0)$

$$\|f - a_n\| < \inf_{g\in\mathcal{D}} \|f - ((1-\lambda_n)a_{n-1} + \lambda_n g)\| + \varepsilon_n, \quad n = 1, 2, \dots .$$

This is a X-greedy type algorithm. It is proved in [7] that if X is a uniformly smooth Banach space with $\rho(u) \leq \gamma u^q$, $q \in (1, 2]$, then one can choose the convexity schedule $\lambda_n := 1/(n + 1)$ and $\varepsilon_n := \varepsilon n^{-q}$ and get the rate of convergence

$$\|f - a_n\| \leq C(q, \gamma, \varepsilon)n^{1/q-1}, \quad f \in A_1(\mathcal{D}).$$

The XGA and WDGA have been introduced in [19] and the WCGA has been introduced in [17]. We note that we made an extra assumption on the existence of a φ_m in the definition of the XGA. We eliminated this problem in the WDGA by introducing a weakness parameter $t_m \in [0, 1)$. There are modifications of greedy algorithms that are motivated by practical applications of these algorithms. First of all, in [21] we assume that our evaluations are not exact. We use a functional that is an approximant of a norming functional and we use a near best approximant instead of the best approximant. Secondly, in [20] we address the very important issue of the evaluation of $\sup_{g\in\mathcal{D}} F_f(g)$. To make this evaluation feasible we restrict our search to a finite subset $\mathcal{D}(\mathcal{N})$ of $\mathcal{D}$. In other words we evaluate $\sup_{g\in\mathcal{D}(\mathcal{N})} F_f(g)$.

We studied in [21] the following modification of the WCGA. Let three sequences $\tau = \{t_k\}_{k=1}^\infty$, $\delta = \{\delta_k\}_{k=0}^\infty$, $\eta = \{\eta_k\}_{k=1}^\infty$ of numbers from $[0, 1]$ be given.

Approximate Weak Chebyshev Greedy Algorithm (AWCGA).
Set $f_0 := f_0^{\tau,\delta,\eta} := f$. *Then for each* $m \geq 1$, *we inductively define:*

1) F_{m-1} *is a functional with properties*

$$\|F_{m-1}\| \leq 1, \qquad F_{m-1}(f_{m-1}) \geq \|f_{m-1}\|(1 - \delta_{m-1});$$

and $\varphi_m := \varphi_m^{\tau,\delta,\eta} \in \mathcal{D}$ *is any element satisfying* $F_{m-1}(\varphi_m) \geq t_m \sup_{g\in\mathcal{D}} F_{m-1}(g)$.

2) Define $\Phi_m := \operatorname{span}\{\varphi_j\}_{j=1}^m$, *and denote* $E_m(f) := \inf_{\varphi\in\Phi_m} \|f - \varphi\|$.

Let $G_m \in \Phi_m$ *be such that* $\|f - G_m\| \leq E_m(f)(1 + \eta_m)$.

3) Denote $f_m := f_m^{\tau,\delta,\eta} := f - G_m$.

The term *approximate* in this definition means that we use a functional F_{m-1} that is an approximation to the norming (peak) functional $F_{f_{m-1}}$ and also that we use an approximant $G_m \in \Phi_m$ which satisfies a weaker assumption than being a best approximant to f from Φ_m. Thus, in the *approximate* version of the WCGA, we have addressed the issue of nonexact evaluation of the norming functional and the best approximant. We did not address the issue of finding the $\sup_{g\in\mathcal{D}} F_{f^c_{m-1}}(g)$. In the paper [20] we addressed this issue. We did it in two steps. First we considered the corresponding modification of the WCGA, and then the modification of the AWCGA. These modifications are done in the style of the concept of *depth search* from [8].

We now consider a countable dictionary $\mathcal{D} = \{\pm\psi_j\}_{j=1}^{\infty}$. We denote $\mathcal{D}(N) := \{\pm\psi_j\}_{j=1}^{N}$. Let $\mathcal{N} := \{N_j\}_{j=1}^{\infty}$ be a sequence of natural numbers.

Restricted Weak Chebyshev Greedy Algorithm (RWCGA).
Set $f_0 := f_0^{c,\tau,\mathcal{N}} := f$. *Then for each* $m \geq 1$, *we inductively define:*

1) $\varphi_m := \varphi_m^{c,\tau,\mathcal{N}} \in \mathcal{D}(N_m)$ *is any element satisfying*

$$F_{f_{m-1}}(\varphi_m) \geq t_m \sup_{g\in\mathcal{D}(N_m)} F_{f_{m-1}}(g).$$

2) Define $\Phi_m := \Phi_m^{\tau,\mathcal{N}} := \operatorname{span}\{\varphi_j\}_{j=1}^{m}$, *and let* $G_m := G_m^{c,\tau,\mathcal{N}}$ *be the best approximant to* f *from* Φ_m.

3) Denote $f_m := f_m^{c,\tau,\mathcal{N}} := f - G_m$.

We formulate some results from [21], [20] in a particular case of a uniformly smooth Banach space with modulus of smoothness of power type (see [21], [20] for the general case). The following theorem was proved in [21].

Theorem 5.1. *Let* X *be a Banach space that have modulus of smoothness* $\rho(u)$ *of power type* $1 < q \leq 2$; $(\rho(u) \leq \gamma u^q)$. *Assume that* $\sum_{m=1}^{\infty} t_m^p = \infty$, $p = q/(q-1)$, *and*

$$\delta_m = o(t_m^p), \qquad \eta_m = o(t_m^p).$$

Then the AWCGA converges for any $f \in X$.

We now give a theorem from [20] on greedy algorithms with restricted search.

Theorem 5.2. *Let* X *be a Banach space that have modulus of smoothness* $\rho(u)$ *of power type* $1 < q \leq 2$; $(\rho(u) \leq \gamma u^q)$. *Assume that* $\lim_{m\to\infty} N_m = \infty$, $\sum_{m=1}^{\infty} t_m^p = \infty$, *and* $p = q/(q-1)$. *Then the RWCGA converges for any* $f \in X$.

References

[1] A. Barron, A. Cohen, W. Dahmen, and R. DeVore, *Approximation and learning by greedy algorithms*, Manuscript (2005), pp. 1–27.

[2] A. R. Barron, *Universal approximation bounds for superposition of n sigmoidal functions*, IEEE Trans. Inform. Theory 39 (1993), pp. 930–945.

[3] R. Deville, G. Godefroy, and V. Zizler, *Smoothness and Renormings in Banach Spaces*. Longman Scientific & Technical, Harlow, 1993.

[4] R. A. DeVore, *Nonlinear approximation*, Acta Numerica (1998), pp. 51–150.

[5] R. A. DeVore and V. N. Temlyakov, *Some remarks on Greedy Algorithms*, Adv. Comput. Math. 5 (1996), pp. 173–187.

[6] S. Dilworth, D. Kutzarova, and V. Temlyakov, *Convergence of some Greedy Algorithms in Banach spaces*, J. Fourier Anal. Appl. 8 (2002), pp. 489–505.

[7] M. Donahue, L. Gurvits, C. Darken, and E. Sontag, *Rate of convex approximation in non-Hilbert spaces*, Constr. Approx. 13 (1997), pp. 187–220.

[8] D. L. Donoho, *Sparse Components of Images and Optimal Atomic Decompositions*, Constr. Approx. 17 (2001), pp. 353–382.

[9] J. H. Friedman and W. Stuetzle, *Projection pursuit regression*, J. Amer. Statist. Assoc. 76 (1981), pp. 817–823.

[10] M. Ganichev and N. J. Kalton, *Convergence of the Weak Dual Greedy Algorithm in L_p-spaces*, J. Approx. Theory 124 (2003), pp. 89–95.

[11] P. J. Huber, *Projection Pursuit*, Ann. Statist. 13 (1985), pp. 435–475.

[12] L. Jones, *On a conjecture of Huber concerning the convergence of projection pursuit regression*, Ann. Statist. 15 (1987), pp. 880–882.

[13] ———, *A simple lemma on greedy approximation in Hilbert space and convergence rates for projection pursuit regression and neural network training*, Ann. Statist. 20 (1992), pp. 603–613.

[14] S. V. Konyagin and V. N. Temlyakov, *Greedy approximation with regard to bases and general minimal systems*, Serdica Math. J. 28 (2002), pp. 305–328.

[15] J. Lindenstrauss and L. Tzafriri, *Classical Banach Spaces I*. Springer-Verlag, Berlin, 1977.

[16] V. N. Temlyakov, *Weak greedy algorithms*, Adv. Comput. Math. 12 (2000), pp. 213–227.

[17] ———, *Greedy algorithms in Banach spaces*, Adv. Comput. Math. 14 (2001), pp. 277–292.

[18] ———, *Greedy expansions in Banach spaces*, IMI Preprints Series 6 (2003), pp. 1–21.

[19] ———, *Nonlinear Methods of Approximation*, Found. Comput. Math. 3 (2003), pp. 33–107.

[20] ———, *Greedy Algorithms with Restricted Depth Search*, Proc. Steklov Inst. Math. 248 (2005), pp. 255–267.

[21] ———, *Greedy-Type Approximation in Banach Spaces and Applications*, Constr. Approx. 21 (2005), pp. 257–292.

[22] ———, *Greedy algorithms with prescribed coefficients*, IMI-Preprints 7 (2006), pp. 1–18.

[23] ———, *Greedy approximations*, Proceedings of FOCM, Santander 2005, Lecture Note Ser., 331, Cambridge Univ. Press, Cambridge, 2006, pp. 371–394.

[24] ———, *Greedy approximations with regards to bases*, Proceedings of the International Congress of Mathematicians, Vol. II, Eur. Math. Soc., Zürich, 2006.

[25] ———, *Relaxation in greedy approximation*, IMI-Preprints 3 (2006), pp. 1–25.

Author information

Vladimir Temlyakov, Department of Mathematics, University of South Carolina, Columbia, SC 29208, USA.

Email: `Temlyak@math.sc.edu`

Banach Spaces and their Applications in Analysis, 209–218

Some problems in asymptotic convex geometry and random matrices motivated by numerical algorithms

Roman Vershynin

Abstract. The simplex method in Linear Programming motivates several problems of asymptotic convex geometry. We discuss some conjectures and known results in two related directions – computing the size of projections of high dimensional polytopes and estimating the norms of random matrices and their inverses

Key words. Simplex algorithm, sections of convex bodies, smoothed analysis, singular values, random matrices.

AMS classification. 52B11, 15A52, 90C05.

1 Asymptotic convex geometry and Linear Programming

Linear Programming studies the problem of maximizing a linear functional subject to linear constraints. Given an *objective vector* $z \in \mathbb{R}^d$ and *constraint vectors* $a_1, \dots, a_n \in \mathbb{R}^d$, we consider the linear program:

$$\textit{maximize } \langle z, x \rangle \ \textit{ subject to } \ \langle a_i, x \rangle \le 1, \quad i = 1, \dots, n. \tag{LP}$$

This linear program has d unknowns, represented by x, and n constraints. Every linear program can be reduced to this form by a simple interpolation argument [36]. The *feasible set* of the linear program is the polytope

$$P := \{x \in \mathbb{R}^d : \ \langle a_i, x \rangle \le 1, \quad i = 1, \dots, n\}.$$

The solution of (LP) is then a vertex of P. We can thus look at (LP) from a geometric viewpoint:

> *for a polytope P in $\mathbb{R}^d$ given by n faces, and for a vector z, find the vertex that maximizes the linear functional $\langle z, x \rangle$.*

The oldest and still the most popular method to solve this problem is the *simplex method.* It starts at some vertex of P and generates a walk on the edges of P toward the solution vertex. At each step, a *pivot rule* determines a choice of the next vertex; so there are many variants of the simplex method with different pivot rules. (We are not concerned here with how to find the initial vertex, which is a nontrivial problem in itself).

Supported by the Alfred P. Sloan Foundation and by NSF DMS grant 0401032.

1.1 The shadow-vertex pivot rule and sections of polytopes

The most widely known pivot rule maximizes the objective function $\langle z, x\rangle$ over the neighboring vertices. The resulting walk on the vertices is defined iteratively, and thus is usually hard to analyze. An alternative *shadow-vertex* pivot rule [10] defines a walk on the polytope P as a preimage of a projection of P. The resulting walk can be described in a non-iterative way, so one hopes to analyze it with the methods of asymptotic convex geometry.

Suppose we know a solution x_0 of (LP) for some other objective vector z_0. The shadow-vertex simplex method interpolates between z_0 and z by computing the solutions of (LP) for all z' in the plane $E = \text{span}(z_0, z)$.

From a geometric viewpoint, we consider the orthogonal projection $Q(P)$ of the feasible polytope P onto E. It is easily checked that the vertices x_0 and x of P will be preserved by the projection: $Q(x)$ and $Q(x_0)$ will be vertices of the polygon $Q(P)$.

The shadow-vertex simplex method thus computes the vertices of the polygon $Q(P)$ one by one, starting from $Q(x_0)$ and ending with $Q(x)$. So at the end it outputs x, which is the solution of (LP). One can express the computation of $Q(x)$ as a pivot rule, and check that each next vertex can be computed in polynomial time. The resulting walk on the polytope P is therefore the preimage of the vertices of the polygon $Q(P)$ under the projection Q.

It will be convenient to work in the dual setting. The polar of P is

$$K := P^\circ = \{x \in \mathbb{R}^d : \langle x, y\rangle \le 1 \text{ for all } y \in P\} = \text{conv}(0, a_1, \dots, a_n)$$

and the polar of the projection $Q(P)$ is the section $K \cap E$. The length of the walk in the shadow-vertex simplex method is thus bounded by the size (the number of edges) of the polygon $K \cap E$.

1.2 Complexity of the simplex method and the size of sections

The running time of the simplex method is proportional to the length of the walk on the edges of P it generates. Hirsch's conjecture states that every polytope P in $\mathbb{R}^d$ with n faces has *diameter* at most $n - d$. The diameter is the maximum of the shortest walk on the edges between any pair of the vertices. The best known bound on the diameter is $n^{\log_2 d+2}$ due to Kalai and Kleitman [15].

For every known variant of the simplex method, an example of (LP) is known for which the length of the walk on P is not polynomial in n and d.[1] For the the classical (maximizing) pivot rule, such an example was first constructed by Klee and Minty [18]: on a certain deformed cube, the walk visits each of the 2^d vertices [18].

Similar pessimistic examples are known for the the shadow-vertex simplex method: the size of the planar section $K \cap E$ that bounds the length of the walk is in general exponential in n, d. This follows for example from the seminal construction in semidefinite programming by Ben-Tal and Nemirovski [5], which yields a polytope K and

[1] Recently, a randomized polynomial time pivoting algorithm for (LP) was found by Kelner and Spielman [16]. However, their algorithm generates a walk on some other polytope related to (LP) and not on P.

a plane E such that the section $K \cap E$ is an approximation of the circle with error *exponentially small* in n, d.

Problem 1.1 (Sections of polytopes). *Let K be a polytope in $\mathbb{R}^d$ with n vertices, and E be a two-dimensional subspace of $\mathbb{R}^d$. Estimate the size (the number of edges) of the polygon $K \cap E$. Under what conditions on K and E is this number polynomial in n, d?*

This problem is somewhat opposite to the typical problems of the asymptotic convex geometry, whose ideal would be to produce the most round section (fine approximation to a cicrle). In Problem 1.1 our ideal is a section with fewest edges, thus farthest from the circle. From the viewpoint of the simplex method, "round" polytopes have high complexity, while polytopes with fewest faces have low complexity.

1.3 Smoothed analysis and randomly perturbed polytopes

Despite the known examples of exponentially long walks, on most problems that occur in practice the simplex algorithm runs in polynomial and even linear time. To explain this empirical evidence, the *average analysis* of the simplex method was developed in the eighties, where the (LP) was drawn at random from some natural distribution and the expected size of the walk was shown to be polynomial in n, d [1, 2, 3, 4, 12, 21, 28, 30, 33].

In particular, Haimovich showed ([12], see [26, Section 11.5]) that if one chooses the directions of the inequalities in (LP) uniformly at random as $\leq$ or $\geq$, then the expected length of the walk in the shadow-vertex simplex method is at most $d/2$. Note that the size does not depend on the number of inequalities n.

However, reversing inequalities is hard to justify in practice. Spielman and Teng [32] proposed to replace average analysis by a finer model, which they called *smoothed analysis*, and where the random inputs are replaced by slight *random perturbations* of arbitrary inputs. Smoothed analysis thus interpolates between the worst case analysis (arbitrary inputs) and the average analysis of Smale (random inputs).

Spielman and Teng [31] first showed that the shadow-vertex simplex method has polynomial smoothed complexity. If the polytope K is randomly perturbed, then its section $K \cap E$ will have an expected polynomial size (which in turn bounds the length of the walk in the simplex method). Their result was improved in [7] and [36], and the current best bound is as follows:

Theorem 1.2 ([36]). *Let E be a plane in $\mathbb{R}^d$ and $a_1, \dots, a_n$ be independent Gaussian vectors in $\mathbb{R}^d$ with centers of norm at most 1, and whose components have standard deviation $\sigma \leq 1/6\sqrt{d \log n}$. Then the random polytope $K = \operatorname{conv}(a_1, \dots, a_n)$ satisfies*

$$\mathbb{E}\,|\operatorname{edges}(K \cap E)| \leq C d^3 \sigma^{-4}, \tag{1.1}$$

where C is an absolute constant.

The prior weaker bound of Spielman and Teng [31] was $Cnd^3\sigma^{-6}$; the subsequent work of Deshpande and Spielman [7] improved upon the exponent of d but doubled the exponent of n.

Theorem 1.1 shows that the expected size of the section $K \cap E$ is *polylogarithmic* in n, while the previous bound were polynomial in n. Going back to the pre-dual polytope P, this indicates that random perturbations of the polytopes create short walks between any two *given* vertices.Note that for large n, this polylogarithmic bound becomes better than the bound $n - d$ in Hirsch's conjecture.

Theorem 1.1 provides a solution to Problem 1.1 for a randomly perturbed polytope and a fixed subspace. A seemingly harder problem, which is still open, is for an *arbitrary polytope and a randomly perturbed subspace.* This version would be significant for the analysis of the simplex method, because it allows one to leave the constraints intact and only perturb the objective function.

Another open problem is to estimate the diameter of randomly perturbed polytopes, rather than all polytopes as in Hirsch's conjecture.

Problem 1.3 (Spielman and Teng [31]). *Let K be a perturbed polytope as in Theorem 1.2. Estimate the expected diameter of $P = K^\circ$. Is it always polynomial in n, d and σ? Perhaps even polylogarithmic in n?*

Finally, no analog of Theorem 1.2 is known for *bounded perturbations*, i.e. for for $a_i = \bar{a}_i + \sigma\theta_i$, where $\bar{a}_i$ are arbitrary fixed vectors of norms at most 1 and θ are independent vectors chosen from $\{-1,1\}^d$ or from $[-1,1]$ uniformly at random. Such *bounded smoothed analysis* is a common model for roundoff errors when real numbers are represented as binary numbers in computers [11].

1.4 Nondegeneracy of faces and invertibility of random matrices

The approach to Theorem 1.2 developed by Spielman and Teng [31] is based on the intuition that most faces of K should be non-degenerate simplices (e.g. they have inscribed balls of polynomial radii). If the plane E intersects such a nondegenerate simplex F, the length of the interval $E \cap F$ is likely to be polynomially big (if the plane intersects a simplex, it is likely to pass through its "bulk" rather than touch the boundary only).

On the other hand, with high probability all vectors in the the perturbed polytope K have norms $O(\log n)$. (Its vertices are Gaussian perturbations of n vectors of norm at most 1). Therefore, the perimeter of the polygon $K \cap E$ can be at most $O(\log n)$. Since all edges $E \cap F$ of this polygon are polynomially big, we conclude that are at most polynomially many edges, as desired.

There are several places where this approach breaks down or is not known to succeed. One such problem is the non-degeneracy of the faces. The nondegeneracy of a simplex S is usually quantified with the smallest singular value of the matrix A that realizes the change of the basis from the standard simplex to S. For the polytope $K = \mathrm{conv}(a_1, \dots, a_n)$, each face is a simplex with vertices $(a_i)_{i \in I}$ for some d-element subset $I \subset \{1, \dots, n\}$. If the vertices a_i are Gaussian as in Theorem 1.2, the change of basis is a $d \times d$ matrix with Gaussian independent entries. Thus we need the random Gaussian matrices to be far from being singular. Quantitative theory of invertibility of random matrices is the subject of the next section.

2 Invertibility of random matrices

For a one-to-one linear operator $A : X \to Y$ between two normed spaces X and Y, two quantities are central in functional analysis: the norm $\|A\|$ and the norm of the inverse $\|A^{-1}\|$. If the operator is not onto, then the inverse norm is computed for the restriction of A onto its image; so we identify $\|A^{-1}\|$ with $\|(A|_{A(X)})^{-1}\|$. Thus

$$\|A\| = \sup_{x\in X:\ \|x\|=1} \|Ax\|, \qquad \frac{1}{\|A^{-1}\|} = \inf_{x\in X:\ \|x\|=1} \|Ax\|.$$

The operator A can be viewed as realizing an *embedding* of the space X into the space Y, and the product $\|A\|\|A^{-1}\|$ is the *distortion* of the embedding (see [13]).

The canonical example is when both X and Y are finite dimensional Euclidean spaces, say $X = \mathbb{R}^k$, $Y = \mathbb{R}^n$, where we identify the linear operator A with its $k \times n$ matrix. The *singular values* of A are the eigenvalues of $|A| = \sqrt{A^*A}$, the largest and the smallest singular values being

$$\lambda_{\max}(A) = \|A\|, \qquad \lambda_{\min}(A) = \frac{1}{\|A^{-1}\|}.$$

In the numerical linear algebra and scientific computing literature, the distortion

$$\kappa(A) = \lambda_{\max}(A)/\lambda_{\min}(A) = \|A\|\|A^{-1}\|$$

is commonly called the *condition number* of A.

We are interested estimating these quantities for random matrices A. For one reason, random matrices sometimes provide an intuition for what to expect in practice; we saw such reasoning about average analysis and smoothed analysis of the simplex method in the previous section. Random linear operators with controllable distortion (or their adjoints) also serve as handy tools in most randomized constructions in geometric functional analysis [7], geometric algorithms in theoretical computer science [34, 35], compressed sensing in information processing [6, 8], vector quantization [19] and some other fields.

2.1 Gaussian matrices

We start from the simplest case of a *Gaussian matrix*, those whose entries are i.i.d. standard normal random variables. The asymptotics of the largest and the smallest singular values is well understood in this case: for a $n \times d$ Gaussian matrix A with $n \geq d$, one has

$$\lambda_{\max}(A) \approx \sqrt{n} + \sqrt{d}, \quad \lambda_{\min}(A) \approx \sqrt{n} - \sqrt{d} \quad \text{with high probability.}$$

There is a long history of such asymptotic results. In particular, the largest and the smallest singular values converge almost surely to their corresponding values above as the dimension n grows to infinity and the aspect ratio d/n converges to a constant, see

[7]. A sharp nonasymptotic result – for every fixed n and d – follows from Gordon's inequality (see [7]):

$$\sqrt{n} - \sqrt{d} \le \mathbb{E}\lambda_{\min}(A) \le \mathbb{E}\lambda_{\max}(A) \le \sqrt{n} + \sqrt{d}.$$

Combining with the concentration of measure inequality, one deduces a deviation bound [7]: for every $t > 0$, with probability at least $1 - 2e^{-t^2/2}$ one has

$$\sqrt{n} - \sqrt{d} - t \le \lambda_{\min}(A) \le \lambda_{\max}(A) \le \sqrt{n} + \sqrt{d} + t. \tag{2.1}$$

Note that the lower bounds become meaningless for *square Gaussian matrices*, those with $n = d$. Yet this case is central in some applications: as we saw in Section 1.4, the square matrices determine the faces of the polytope K in linear programming, and nondegeneracy of such a face translates into a lower bound for $\lambda_{\min}(A)$. To guess the order of the smallest singular value, note that for an $(n-1) \times n$ matrix, the lower bound is $\sqrt{n} - \sqrt{n-1} \sim n^{-1/2}$.

Von Neumann and his associates, who used random matrices as test matrices for their algorithms, indeed speculated that for random square matrices one should have

$$\lambda_{\min}(A) \sim n^{-1/2} \quad \text{with high probability} \tag{2.2}$$

(see [37], pp. 14, 477, 555.) In a more precise form, this estimate was conjectured by Smale [29] and proved by Edelman [9] for Gaussian matrices: for every $\varepsilon \ge 0$, one has

$$\mathbb{P}\big(\lambda_{\min}(A) \le \varepsilon n^{-1/2}\big) \sim \varepsilon. \tag{2.3}$$

In particular, the smallest singular value is not concentrated: the mean and the standard deviation of $n^{1/2}\lambda_{\min}(A)$ are both of the order of a constant. This is very different from the behavior of the largest singular value, which by (2.1) is tightly concentrated around its mean.

An elegant argument by Sankar, Spielman and Teng [25] generalizes (2.3) for random Gaussian perturbations of an arbitrary matrix, i.e. for the smoothed analysis setting of Theorem 1.2:

Theorem 2.1 (Sankar, Spielman, and Teng [25]). *Let A is an $n \times n$ matrix with independent Gaussian random entries (not necessarily centered), each of variance σ^2. Then, for every $\varepsilon \ge 0$, one has*

$$\mathbb{P}\big(\lambda_{\min}(A) \le \varepsilon n^{-1/2}\big) \le C\varepsilon/\sigma,$$

where $C = 1.823$.

In applications for random polytopes such as in Section 1.4, we need *all faces* to be nondegenerate, thus all $d \times d$ submatrices of a random $n \times d$ Gaussian matrix (whose rows are the constraint vectors $a_1, \dots, a_n$ from (LP)) be nicely invertible. This motivates the following problem:

Problem 2.2. *Let A be an $n \times d$ Gaussian matrix (with i.i.d. standard normal entries, or, more generally, as in Theorem 2.1). Estimate the expected minimum of the smallest singular values of all all $d \times d$ submatrices of A.*

In particular, if $n = O(d)$, we want this minimum to be polynomially small rather than exponentially small in d.

2.2 General matrices with i.i.d. entries

Most problems we discussed become much harder once Gaussian matrices are replaced with other natural matrices with i.i.d. entries. Nevertheless, understanding of discrete matrices, whose entries can take finite set of values, is important in applications such as in numerical algorithms, which can only deal with discrete values. A survey on random discrete matrices was recently written by Vu [38].

Asymptotic theory of random matrices has developed to a point where the behavior of the largest singular value is well understood. Suppose A is an $n \times d$ matrix with i.i.d. centered entries, which have variance 1. Then the finiteness of the *fourth moment* of the entries is necessary and sufficient that

$$\lambda_{\max}(A) \to \sqrt{n} + \sqrt{d} \qquad \text{almost surely} \tag{2.4}$$

as the dimension n grows to infinity and the aspect ratio d/n converges to a constant [39]. A similar statement holds for the convergence in probability, and with a slightly weaker condition than the fourth moment [27].

Under a much stronger *subgaussian* moment assumption, which still holds discrete and gaussian random variables, a parallel non-asymptotic result is known (for all finite n and d). A random variable ξ is called *subgaussian* if its tail is dominated by that of the standard normal random variable: there exists a constant $B > 0$ such that

$$\mathbb{P}(|\xi| > t) \le 2\exp(-t^2/B^2) \qquad \text{for all } t > 0. \tag{2.5}$$

The minimal B here is called the *subgaussian moment*.[2] Gaussian random variables and all bounded random variables, in particular the symmetric ± 1 random variable, are examples of subgaussian random variables. Inequality (2.5) is often equivalently stated as a moment condition

$$(\mathbb{E}|\xi|^p)^{1/p} \le CB\sqrt{p} \qquad \text{for all } p \ge 1, \tag{2.6}$$

where C is an absolute constant.

The following non-asymptotic result follows from a more general result proved by Klartag and Mendelson ([17, Theorem 1.4]) with constant probability, which was later improved by Mendelson, Pajor, and Tomczak-Jaegermann ([22, Theorem D]) to an exponential probability:

[2] In the literature in geometric functional analysis, the subgaussian moment is often called the ψ_2-norm.

Theorem 2.3 ([17, 22]). *Let A be an $n \times d$ matrix $(n \geq d)$ with i.i.d. centered subgaussian entries with variance 1. Then, with probability at least $1 - Ce^{-m}$, one has*

$$\sqrt{n} - C\sqrt{d} \leq \lambda_{\min}(A) \leq \lambda_{\max}(A) \leq \sqrt{n} + C\sqrt{d},$$

where C depends only on the subgaussian moment of the entries.

This estimate approaches the sharp asymptotic bound (2.4) for very tall matrices (for small aspect ratios d/n). However, the lower bound becomes useless for the aspect ratios above some constant, and in particular says nothing about square matrices.

Problem 2.4. *Let A be an $n \times d$ matrix $(n \geq d)$ with i.i.d. centered subgaussian entries with variance 1. Is it true that with high probability one has $\lambda_{\min}(A) \geq c(\sqrt{n} - \sqrt{d})$, where $c > 0$ depends only on the subgaussian moment of the entries?*

In a positive direction, lower bounds valid for all aspect ratios $y := d/n < 1$ were proved by Litvak, Rudelson, Pajor and Tomczak-Jaegermann [20] with an exponential dependence on $1 - y$, and improved to a linear dependence by Rudelson [23].

Nevertheless, even the positive solution of Problem 2.4 would not say anything for *square matrices*, those with $n = d$. This problem was recently solved in the work [24], which confirmed prediction (2.2) for general matrices with independent entries. Recall that the bounded fourth moment of the entries is necessary and sufficient to controll the largest singular value as in (2.4). Then [24] proves that the fourth moment assumption (i.e. the fourth moments of the entries are uniformly bounded) is also sufficient to control the smallest singular value. For an $n \times n$ matrix A with random centered entries of variances at least 1,

Under the fourth moment assumption, prediction (2.2) *holds.*

The identical distribution of the entries is not needed in this result.

For a stronger *subgaussian* assumption on the entries, prediction (2.2) holds with exponentially high probability. This was conjectured by Spielman and Teng [31] for random ± 1 matrices:

$$\mathbb{P}\big(s_n(A) \leq \varepsilon n^{-1/2}\big) \leq \varepsilon + c^n,$$

and proved in [24] in more generality – for all matrices with subgaussian i.i.d. entries, and up to a constant factor which depends only on the subgaussian moment.

Theorem 2.5 ([24]). *Let A be an $n \times n$ matrix with i.i.d. centered subgaussian entries with variance 1. Then for every $\varepsilon \geq 0$ one has*

$$\mathbb{P}\big(\lambda_{\min}(A) \leq \varepsilon n^{-1/2}\big) \leq C\varepsilon + c^n, \tag{2.7}$$

where $C > 0$ and $c < 1$ are constants that depend (polynomially) only on the subgaussian moment of the entries.

In particular, for $\varepsilon = 0$ we deduce that any random square matrix with i.i.d. subgaussian entries with variance 1 is *singular with exponentially small probability*. For random matrices with ± 1 entries, this was proved by Kahn, Komlós and Szemerédi [14]. For more on prior work and related conjectures on the singularity probability, see [38, 24].

References

[1] I. Adler, *The expected number of pivots needed to solve parametric linear programs and the efficiency of the self-dual simplex method.* Technical Report, University of California at Berkeley, 1983

[2] I. Adler, R. M. Karp, and R. Shamir, *A simplex variant solving an $m \times d$ linear program in $O(\min(m^2, n^2))$ expected number of pivot steps*, J. Complexity 3 (1987), pp. 372–387.

[3] I. Adler and N. Megiddo, *A simplex algorithm whose average number of steps is bounded between two quadratic functions of the smaller dimension*, J. ACM 32 (1985), pp. 871–895.

[4] K.-H. Borgwardt, *The simplex method. A probabilistic analysis.* Algorithms and Combinatorics: Study and Research Texts, 1. Springer-Verlag, Berlin, 1987.

[5] A. Ben-Tal and A. Nemirovski, *On polyhedral approximations of the second-order cone*, Math. Oper. Res. 26 (2001), pp. 193–205.

[6] The Compressed Sensing website `http://www.dsp.ece.rice.edu/cs/`

[7] A. Deshpande and D. A. Spielman, *Improved smoothed analysis of the shadow vertex simplex method*, FOCS 2005 (46th Annual Symposium on Foundations of Computer Science), pp. 349–356.

[8] D. Donoho, *Compressed sensing*, IEEE Trans. Inform. Theory 52 (2006), pp. 1289–1306.

[9] A. Edelman, *Eigenvalues and condition numbers of random matrices*, SIAM J. Matrix Anal. Appl. 9 (1988), pp. 543–560.

[10] S. Gass and T. Saaty, *The computational algorithm for the parametric objective function*, Naval Res. Logist. Quart. 2 (1955), pp. 39–45.

[11] R.M.Gray and D.L.Neuhoff, *Quantization*, IEEE Trans. Inform. Theory 44 (1998), pp. 2325–2383.

[12] M. Haimovich, *The simplex algorithm is very good!: On the expected number of pivot steps and related properties of random linear programs.* Technical report, Columbia University, 1983.

[13] P. Indyk and J. Matoušek, *Low-distortion embeddings of finite metric spaces*, Handbook of discrete and computational geometry, Second edition, pp. 177–196, (J. E. Goodman and J. O'Rourke, eds). Discrete Mathematics and its Applications (Boca Raton). Chapman & Hall/CRC, Boca Raton, FL, 2004.

[14] J. Kahn, J. Komlós, and E. Szemerédi, *On the probability that a random ± 1-matrix is singular*, J. Amer. Math. Soc. 8 (1995), pp. 223–240.

[15] G. Kalai and D. J. Kleitman, *A quasi-polynomial bound for the diameter of graphs of polyhedra*, Bull. Amer. Math. Soc. 26 (1992), pp. 315–316.

[16] J. A. Kelner and D. A. Spielman, *A randomized polynomial-time simplex algorithm for Linear Programming*, submitted.

[17] B. Klartag and S. Mendelson, *Empirical Processes and Random Projections*, J. Funct. Anal. 225 (2005), pp. 229–245.

[18] V. Klee and G. J. Minty, *How good is the simplex algorithm?* In Sisha, O., editor, *Inequalities – III*, pp. 159–175, Academic Press, 1972

[19] Yu. Liubarskii and R. Vershynin, *Uncertainty principles and vector quantization*, submitted

[20] A. Litvak, A. Pajor, M. Rudelson, and N. Tomczak-Jaegermann, *Smallest singular value of random matrices and geometry of random polytopes*, Adv. Math. 195 (2005), pp. 491–523.

[21] N. Megiddo, *Improved asymptotic analysis of the average number of steps performed by the self-dual simplex algorithm*, Math. Programming 35 (1986), pp. 140–172.

[22] S. Mendelson, A. Pajor, and N. Tomczak-Jaegermann, *Reconstruction and subgaussian operators*, Geom. Funct. Anal., to appear.

[23] M. Rudelson, *Lower estimates for the singular values of random matrices*, C. R. Math. Acad. Sci. Paris, 342 (2006), pp. 247–252.

[24] M. Rudelson and R. Vershynin, *The Littlewood-Offord Problem and invertibility of random matrices*, submitted.

[25] S. Sankar, D. Spielman, and S.-H. Teng, *Smoothed analysis of the condition numbers and growth factors of matrices*, SIAM J. Matrix Anal. Appl. 28 (2006), pp. 446–476.

[26] A. Schrijver, *Theory of linear and integer programming*. Wiley-Interscience Series in Discrete Mathematics. A Wiley-Interscience Publication. John Wiley & Sons, Ltd., Chichester, 1986.

[27] J. W. Silverstein, *On the weak limit of the largest eigenvalue of a large-dimensional sample covariance matrix*, J. Multivariate Anal. 30 (1989), pp. 307–311.

[28] S. Smale, *On the average number of steps of the simplex method of linear programming*, Math. Programming 27 (1983), pp. 241–262.

[29] ———, *On the efficiency of algorithms of analysis*, Bull. Amer. Math. Soc. 13 (1985), pp. 87–121.

[30] ———, *The problem of the average speed of the simplex method*. Mathematical programming: the state of the art (Bonn, 1982), 530–539, Springer, Berlin, 1983.

[31] D. A. Spielman and S.-H. Teng, *Smoothed analysis: why the simplex algorithm usually takes polynomial time*, J. ACM 51 (2004), pp. 385–463.

[32] ———, *Smoothed analysis of algorithms*, Proceedings of the International Congress of Mathematicians, Vol. I (Beijing, 2002), pp. 597–606, Higher Ed. Press, Beijing, 2002.

[33] M. J. Todd, *Polynomial expected behavior of a pivoting algorithm for linear complementarity and linear programming problems*, Math. Programming 35 (1986), pp. 173–192.

[34] S. Vempala, *The random projection method*, Handbook of randomized computing, Vol. I, II, pp. 651–671, Comb. Optim., 9, Kluwer Acad. Publ., Dordrecht, 2001.

[35] ———, *The random projection method*. With a foreword by Christos H. Papadimitriou. DIMACS Series in Discrete Mathematics and Theoretical Computer Science, 65. American Mathematical Society, Providence, RI, 2004.

[36] R. Vershynin, *Beyond Hirsch Conjecture: walks on random polytopes and smoothed complexity of the simplex method*, FOCS 2006 (47th Annual Symposium on Foundations of Computer Science), pp. 133–142.

[37] J. von Neumann, *Collected works. Vol. V: Design of computers, theory of automata and numerical analysis*. General editor: A. H. Taub. A Pergamon Press Book The Macmillan Co., New York 1963.

[38] V. Vu, *Random discrete matrices*, submitted

[39] Y. Q. Yin, Z. D. Bai, and P. R. Krishnaiah, *On the limit of the largest eigenvalue of the large-dimensional sample covariance matrix*, Probab. Theory Related Fields 78 (1988), pp. 509–521.

Author information

Roman Vershynin, Department of Mathematics, University of California, Davis, CA 95616, USA.
Email: vershynin@math.ucdavis.edu

Banach Spaces and their Applications in Analysis, 219–227

On Kolmogorov numbers of matrix transformations

Lipi R. Acharya and Manjul Gupta

Abstract. For a Banach sequence space λ and a Banach space X, the vector-valued sequence space $\lambda(X)$ is defined by:

$$\lambda(X) = \{\overline{x} = \{x_i\} : x_i \in X, \ \forall\, i \in \mathbb{N} \ \text{ and } \{\|x_i\|_X\} \in \lambda\}.$$

We relate the Kolmogorov numbers of a matrix transformation acting between such spaces with those of its component operators. As an application, we provide a characterization of compact diagonal operators.

Key words. Kolmogorov numbers, sequence spaces, matrix transformations.

AMS classification. 47B06, 47B37, 46B45.

1 Introduction

Kolmogorov numbers which were introduced in 1936 by Kolmogorov [8] as certain diameters of sets, have been found to be useful in characterizing compactness of sets. Indeed, they entered the mathematical world as set functions defined on a class of bounded sets. As bounded linear operators defined on Banach spaces map bounded sets into bounded sets, it is possible to define a sequence of these numbers for a given operator T by restricting the size of the image of the unit ball of the domain space in relation to the unit ball and finite dimensional subspaces of the range space. This sequence is decreasing in nature and decides the measure of compactness of T. In the present work, we follow the axiomatic treatment of s-numbers (introduced by Pietsch in [12]) to study operators defined on vector valued sequence spaces. One may refer to [1, 7, 11, 13] for more detailed account of these numbers.

On the other hand, vector valued sequence spaces, which emerge as generalization of scalar valued sequence spaces, have been found useful in characterizing nuclearity of the underlying locally convex spaces [10, 11]. As every linear operator on a finite dimensional space has matrix representation and such a representation remains valid for weakly continuous linear operators defined on scalar valued sequence spaces [4], it is natural to try to represent a bounded linear operator on vector valued sequence spaces as an infinite matrix of linear operators acting between underlying spaces. An initiative in this direction was taken by Gregory [3], Gupta and Patterson [6]. However, their study was confined to the representation or characterizing compactness or nuclearity of diagonal operators between vector valued sequence spaces in terms of component operators using direct methods. In our present work, we make use of Kolmogorov numbers in order to characterize compactness of diagonal operators acting between particular types of vector valued sequence spaces. We also illustrate by examples, the importance of the underlying field for the validity of our results.

2 Preliminaries

Throughout this paper X, Y and Z represent Banach spaces defined over the same field $\mathbb{K}$ which is either the set of all real numbers $\mathbb{R}$ or the set of all complex numbers $\mathbb{C}$. The closed unit ball of X will be denoted by U_X and $\mathcal{L}(X,Y)$ denotes the collection of all bounded linear operators from X into Y. We now recall some definitions.

Definition 2.1. For $n \in \mathbb{N}$ and $S \in \mathcal{L}(X,Y)$, the nth *Kolmogorov number* of S is defined as:

$$d_n(S) := \inf\{\delta > 0 : S(U_X) \subseteq N_\delta + \delta U_Y; N_\delta \text{ is a subspace of } Y, \dim(N_\delta) < n\}.$$

An axiomatic treatment of these numbers was considered by Pietsch in [12] who termed these numbers as s-numbers. Accordingly, the following general concept will be used.

Definition 2.2. A map $s : S \to \{s_n(S)\}_{n=1}^\infty$, from $\mathcal{L}(X,Y)$ into the set w^+ of sequences of non-negative real numbers, is called an *s-number function* if it satisfies the following properties:

1. $\|S\| = s_1(S) \geq s_2(S) \geq \ldots \geq s_n(S) \geq \ldots$ for $S \in \mathcal{L}(X,Y)$;
2. $s_n(S+T) \leq \|S\| + s_n(T)$ for S and T in $\mathcal{L}(X,Y)$;
3. $s_n(RST) \leq \|R\| s_n(S) \|T\|$; for $T \in \mathcal{L}(X,Y)$, $S \in \mathcal{L}(Y,Z)$, and $R \in \mathcal{L}(Z,Y)$;
4. If $\operatorname{rank}(S) < n$, then $s_n(S) = 0$;
5. $s_n(I_X) = 1$, whenever $\dim(X) \geq n$, where I_X is the identity mapping on X.

For any given operator S in $\mathcal{L}(X,Y)$, Property 1 of Definition 2.2 indicates that the rate of decrease of $\{d_n(S)\}_{n=1}^\infty$ towards zero may be regarded as a measure of compactness of S. Indeed, it is known that S is compact if and only if $\lim_{n\to\infty} d_n(S) = 0$ (see [1]).

Let us now recall few concepts and notations from the theory of vector valued sequence spaces. Throughout, λ is a scalar valued Banach sequence space equipped with the norm $\|\cdot\|_\lambda$. Recall that the norm $\|.\|_\lambda$ of λ is said to be *monotone* if $|\alpha_i| \leq |\beta_i|$, for all $i \in \mathbb{N}$ implies $\|\{\alpha_i\}\|_\lambda \leq \|\{\beta_i\}\|_\lambda$.

For a Banach space X, we consider the vector valued sequence space $\lambda(X)$ defined by setting:

$$\lambda(X) := \Big\{\overline{x} = \{x_i\} : x_i \in X,\ \forall\, i \in \mathbb{N} \text{ and } \{\|x_i\|_X\} \in \lambda\Big\}.$$

It is well known that when equipped with the norm $\|\overline{x}\|_{\lambda(X)} = \|\{\|x_i\|_X\}\|_\lambda$, for $\overline{x} = \{x_i\} \in \lambda(X)$, the space $\lambda(X)$ is a Banach space.

We also need to consider the finite dimensional versions of the space λ. For $k \in \mathbb{N}$, let us write $\lambda_k(X) = X^k$ equipped with the norm given by $\|(x_1, x_2, \ldots, x_k)\|_{\lambda_k(X)} = \|\{z_i\}\|_{\lambda(X)}$, where $z_i = x_i$ for $1 \leq i \leq k$ and $z_i = 0$, for $i > k$. For $x \in X$, δ_i^x denotes the sequence $(0, 0, \ldots, 0, x, 0, \ldots)$, where x is placed at the ith coordinate.

In the sequel, we make use of the following maps involving $\lambda(X)$ and $\lambda_k(X)$. For each $k \in \mathbb{N}$, we consider the projection and inclusion maps $P_{k,\,\lambda(X)} : \lambda(X) \to X$,

$P^k_{\lambda(X)} : \lambda(X) \to \lambda_k(X)$, $I_{k,\,\lambda(X)} : X \to \lambda(X)$, and $I^k_{\lambda(X)} : \lambda_k(X) \to \lambda(X)$, given by:

$$\begin{aligned}
P_{k,\lambda(X)}(\overline{x}) &:= x_k, \quad \overline{x} = \{x_k\} \in \lambda(X);\\
P^k_{\lambda(X)}(\overline{x}) &:= (x_1, x_2, \ldots, x_k), \quad \overline{x} = \{x_k\} \in \lambda(X);\\
I_{k,\lambda(X)}(x) &:= \delta^x_k, \quad x \in X;\\
I^k_{\lambda(X)}(x_1, x_2, \ldots, x_k) &:= (x_1, x_2, \ldots, x_k, 0, 0, \ldots), \quad (x_1, x_2, \ldots, x_k) \in \lambda_k(X).
\end{aligned}$$

To conclude this section, we recall that the space $\lambda(X)$ is called

(i) *a GK-space* if the maps $P_{n,\,\lambda(X)}$ are continuous;

(ii) *a GAK-space* if $\lambda(X)$ is a GK-space and for each $\overline{x} = \{x_i\}$ from $\Lambda(X)$, $\overline{x}^{(n)} \to \overline{x}$ as $n \to \infty$ where $\overline{x}^{(n)} = (x_1, x_2, \ldots, x_n, 0, 0, \ldots)$.

In case $X = \mathbb{K}$, we write e^i for the element δ^1_i of λ and a GK(GAK)-space is referred to as a *K(AK)-space*. Using the fact that $\|.\|_\lambda$ is monotone and $\|e^i\|_\lambda = 1$, $\forall\, i \in \mathbb{N}$, one can easily show that the norm of the any of the maps defined above can not exceed one.

We referred to [2, 3, 5, 6, 9] for detailed study on these spaces and general vector valued sequence spaces.

3 Matrix transformations from $\lambda(X)$ into $\mu(Y)$

Let λ and μ be any two normal, Banach scalar valued sequence spaces equipped with the monotone norms $\|.\|_\lambda$ and $\|.\|_\mu$ respectively. Further we assume that μ is an AK-space and $\|e^i\|_\lambda = \|e^i\|_\mu = 1$ for all $i \in \mathbb{N}$. Note that the restrictions on the norms of λ and μ yield that these are subspaces of ℓ^∞ with the inclusion maps into ℓ^∞ being continuous.

Definition 3.1. A linear map $Z : \lambda(X) \to \mu(Y)$ is said to be a *matrix transformation* if there exists a matrix $[Z_{ij}]$ of linear maps $Z_{ij} : X \to Y$ such that for every $\overline{x} = \{x_n\}$ in $\lambda(X)$ and every $i \in \mathbb{N}$, the series $\sum_{j=1}^{\infty} Z_{ij}(x_j)$ converges in Y and

$$\sum_{j=1}^{\infty} Z_{ij}(x_j) = P_{i,\mu(Y)}(Z(\overline{x})).$$

If moreover, $Z_{ij} \equiv 0$ for $i \neq j$, then Z is called a *diagonal operator* or *diagonal transformation*.

For a matrix transformation Z and $k \in \mathbb{N}$, let us define Z^k and Z_k as the linear operators from $\lambda(X)$ to $\mu_k(Y)$ and to $\mu(Y)$ respectively, given by:

$$Z^k(\overline{x}) := \Big(\sum_{j=1}^{\infty} Z_{1j}x_j, \sum_{j=1}^{\infty} Z_{2j}x_j, \ldots, \sum_{j=1}^{\infty} Z_{kj}x_j\Big), \quad \overline{x} \in \lambda(X)$$

and

$$Z_k(\overline{x}) := \Big\{\sum_{j=1}^{\infty} Z_{1j}x_j\,, \sum_{j=1}^{\infty} Z_{2j}x_j\,, \ldots, \sum_{j=1}^{\infty} Z_{kj}x_j, 0, 0\ldots\Big\}, \quad \overline{x} \in \lambda(X).$$

Proposition 3.2. *Let $Z = [Z_{ij}]$ be a matrix transformation from $\lambda(X)$ to $\mu(Y)$. If $\{\sum_{j=1}^{\infty} \|Z_{ij}\|\}_{i=1}^{\infty} \in \mu$, then Z is a bounded linear operator from $\lambda(X)$ to $\mu(Y)$ satisfying the following inequalities:*

$$\sup_{i,\, j} \|Z_{ij}\| \leq \|Z\| \leq \Big\|\{\sum_{j=1}^{\infty} \|Z_{ij}\|\}_{i=1}^{\infty}\Big\|_{\mu}.$$

Proof. Note first that $\lambda(X) \subseteq \ell^{\infty}(X)$. Since $\sum_{j=1}^{\infty} \|Z_{ij}\| < \infty$ for each $i \in \mathbb{N}$, it is immediate that for $\overline{x} = \{x_i\} \in \lambda(X)$ the series $\sum_{j=1}^{\infty} Z_{ij}(x_j)$ is absolutely convergent in Y. For $i \in \mathbb{N}$, let $y_i = \sum_{j=1}^{\infty} Z_{ij}(x_j)$. Then $\{y_i\} \in \mu(Y)$ since $\|y_i\|_Y \leq \|\overline{x}\|_{\ell^{\infty}(X)} \sum_{j=1}^{\infty} \|Z_{ij}\|_{i=1}^{\infty}$ and μ is normal. Also we get $\|Z\| = \sup_{\|\overline{x}\|_{\lambda(X)} \leq 1} \|Z(\overline{x})\| \leq \|\{\sum_{j=1}^{\infty} \|Z_{ij}\|\}_{i=1}^{\infty}\|_{\mu}$.

On the other hand, if $x \in U_X$ and $\overline{x} = \delta_k^x$ for $k \in \mathbb{N}$ then

$$\|Z\| \geq \|Z(\delta_k^x)\|_{\mu(Y)} = \|\{\|Z_{ik}(x)\|\}_{i=1}^{\infty}\|_{\mu} \geq \|Z_{ik}(x)\|, \ \forall\, i,\ k \in \mathbb{N}.$$

Thus we get $\|Z\| \geq \sup_{i,k} \|Z_{ik}\|$. □

We now present some examples illustrating that the inequalities in Proposition 3.2 may or may not be strict.

Example 3.3. Let $Z : \ell^1 \to \ell^1$ be a matrix transformation given by $Z = [Z_{ij}]$, where for $i, j \in \mathbb{N}$, $Z_{ij} : \mathbb{C} \to \mathbb{C}$ is defined by:

$$Z_{ij}(\alpha) = \frac{\alpha}{i^2 j^2}, \quad \text{for } \alpha \in \mathbb{C}.$$

Observe that $\sup_{i,j} \|Z_{ij}\| = 1$, $\|Z\| = \sup_k \|Z(e^k)\| = \sum_{i=1}^{\infty} i^{-2}$, and

$$\|\{\sum_{j=1}^{\infty} \|Z_{ij}\|\}_{i=1}^{\infty}\|_{\ell^1} = (\sum_{i=1}^{\infty} i^{-2})^2.$$

Hence in this example, the inequalities in Proposition 3.2 are both strict.

Example 3.4. Consider $Z = [Z_{ij}] : \ell^1 \to \ell^{\infty}$, where for $i, j \in \mathbb{N}$, $Z_{ij} : \mathbb{C} \to \mathbb{C}$ is given by:

$$Z_{ij}(\alpha) = \frac{\alpha}{i + j^2}, \quad \text{for } \alpha \in \mathbb{C}.$$

Clearly we have $\sup_{i,\, j} \|Z_{ij}\| = 1/2$, $\|Z\| = 1/2$, and $\sup_i \{\sum_{j=1}^{\infty} \|Z_{ij}\|\} = 1/2 + 1/5 + 1/10 + \ldots$. Thus we get that $\sup_{i,\, j} \|Z_{ij}\| = \|Z\| < \|\{\sum_{j=1}^{\infty} \|Z_{ij}\|\}_{i=1}^{\infty}\|_{\ell^{\infty}}$.

Example 3.5. For each $i, j \in \mathbb{N}$, define $Z_{ij} : \mathbb{C} \to \mathbb{C}$ by

$$Z_{ij}(\alpha) = \frac{\alpha}{i^2 + j^2}, \quad \text{for } \alpha \in \mathbb{C}.$$

One can check that $Z = [Z_{ij}]$ defines a matrix transformation from ℓ^∞ to itself. Moreover, $\sup_{i,j} \|Z_{ij}\| = 1/2$, $\|Z\| = \sum_{j=1}^\infty 1/(1 + j^2)$, and $\sup_i(\sum_{j=1}^\infty \|Z_{ij}\|) = \sup_i(\sum_{j=1}^\infty 1/(i^2 + j^2)) = \sum_{j=1}^\infty 1/(1 + j^2)$. Hence we have

$$\sup_{i,\, j} \|Z_{ij}\| < \|Z\| = \Big\|\{\sum_{j=1}^\infty \|Z_{ij}\|\}_{i=1}^\infty\Big\|_{\ell^\infty}.$$

Example 3.6. Let $Z : \ell^\infty \longrightarrow \ell^\infty$ be the diagonal operator defined by a decreasing sequence $\{\sigma_i\}_{i=1}^\infty$ of non-negative reals i.e. $Z(\overline{x}) = \{\sigma_i x_i\}$ for $\overline{x} = \{x_i\} \in \ell^\infty$. Here Z_{ii} 's are the operators from $\mathbb{C}$ to $\mathbb{C}$ with $Z_{ii}(\alpha) = \sigma_i\ \alpha$. It is easy to check that $\|Z\| = \sigma_1$. Therefore, $\|Z\| = \sup_i \|Z_{ii}\| = \|\{\sum_{j=1}^\infty \|Z_{ij}\|\}_{i=1}^\infty\|_{\ell^\infty}$.

4 Kolmogorov numbers of matrix transformations

We assume throughout this section that $Z = [Z_{ij}]$ is a matrix transformation from $\lambda(X)$ into $\mu(Y)$ satisfying the following extra assumption:

$$\Big\{\sum_{j=1}^\infty \|Z_{ij}\|\Big\}_{i=1}^\infty \in \mu. \tag{4.1}$$

Proposition 4.1. *For $i, j, k, n \in \mathbb{N}$, we have*

(i) $d_n(Z^k) = d_n(Z_k)$;
(ii) $d_n(Z^k) \leq d_n(Z)$;
(iii) $d_n(Z_{ij}) \leq d_n(Z)$.

Proof. Items (i), (ii), and (iii) follow immediately from $Z^k = P^k_{\mu(Y)} Z_k$, $Z_k = I^k_{\mu(Y)} Z^k$, $Z^k = P^k_{\mu(Y)} Z$, and $Z_{ij} = P_{i,\, \mu(Y)} Z \cdot I_{j,\, \lambda(X)}$ together with the multiplicative property of Kolmogorov numbers. □

Corollary 4.2. *For each $n \in \mathbb{N}$, $d_n(Z) = \lim_{k\to\infty} d_n(Z_k)$.*

Proof. We use the Property 2 from Definition 2.2 to derive for every $n, k \in \mathbb{N}$, we have $0 \leq d_n(Z) - d_n(Z_k) \leq \|(0, \ldots, 0, \sum_{j=1}^\infty \|Z_{(k+1)j}(x_j)\|, \sum_{j=1}^\infty \|Z_{(k+2)j}(x_j)\|, \ldots)\|_\mu$. The conclusion follows from the fact that μ is an AK-space. □

Remark 4.3. The conclusion of Corollary 4.2 may hold without the assumption (4.1) or μ being an AK-space. Indeed, if Z is the identity mapping on ℓ^p where $1 \leq p \leq \infty$, then we have $d_n(Z) = \lim_{k\to\infty} d_n(Z_k)$. But ℓ^∞ is not an AK-space and $\{\sum_{j=1}^\infty \|Z_{ij}\|\}_{i=1}^\infty$ is not in ℓ^p for $1 \leq p < \infty$.

Corollary 4.4. *If $Z = [Z_{ij}]$ is a compact matrix transformation from $\lambda(X)$ to $\mu(Y)$, then $Z_{ij} : X \to Y$ is compact, for each $i, j \in \mathbb{N}$.*

Converse of the above result is obviously not valid even when $X = \mathbb{K}$. However, relating the Kolmogorov numbers of the sections of the matrix transformation Z with Z_{ij}'s, we have

Theorem 4.5. *For every k and n in $\mathbb{N}$,*

$$d_{kn}(Z^k) \leq \|e^{(k)}\|_\mu \max_{1\leq i\leq k} \{d_n(Z_{ii}) + \sum_{j\neq i} \|Z_{ij}\|\},$$

where $e^{(k)}$ represents the kth section of $e = \{1, 1, \ldots, 1, \ldots\}$.

Proof. Let $\epsilon > 0$. For $i, j \in \mathbb{N}$, we can find a subspace $N_\epsilon^{ij} \subseteq Y$ of dimension $m < n$ such that $Z_{ij}(U_X) \subseteq N_\epsilon^{ij} + (d_n(Z_{ij}) + \epsilon)U_Y$.

Since $Z^k(\overline{x}) = (Z_{11}(x_1), \ldots, (Z_{kk}(x_k)) + (\sum_{j\neq 1} Z_{1j}(x_j), \ldots, \sum_{j\neq k} Z_{kj}(x_j))$, it follows that

$$Z^k(\overline{x}) \in \prod_{i=1}^{k} \{N_\epsilon^{ii} + (d_n(Z_{ii}) + \epsilon)U_Y\} + \prod_{i=1}^{k} \{(\sum_{j\neq i} \|Z_{ij}\|)U_Y\}.$$

Hence

$$Z^k(\overline{x}) \in \prod_{i=1}^{k} N_\epsilon^{ii} + \|e^{(k)}\|_\mu \max_{1\leq i\leq k} (d_n(Z_{ii}) + \epsilon + \sum_{j\neq i} \|Z_{ij}\|)U_{\mu_k(Y)}.$$

Since $\dim(\prod_{i=1}^{k} N_\epsilon^{ii}) < kn$ and ϵ is arbitrary, we get the desired inequality. □

Corollary 4.6. *If $\{\alpha_i\} \in \mu$ is such that $\alpha_i = 0$ for $1 \leq i \leq k$ and $\alpha_i = \sum_{j=1}^{\infty} \|Z_{ij}\|$ for $i > k$, then*

$$d_{kn}(Z) \leq \|\{\alpha_i\}\|_\mu + \|e^{(k)}\|_\mu \max_{1\leq i\leq k} \{d_n(Z_{ii}) + \sum_{j\neq i} \|Z_{ij}\|\}.$$

A particular case of Corollary 4.6 for diagonal operator Z may now be stated:

$$d_{kn}(Z) \leq \sup_{k+1\leq i<\infty} \|Z_{ii}\| + \|e^{(k)}\|_\mu \max_{1\leq i\leq k} (d_n(Z_{ii})). \tag{4.2}$$

The above results can now be used to characterize the compactness of diagonal operators as follows:

Theorem 4.7. *The diagonal operator $Z : \lambda(X) \to \lambda(Y)$ is compact if and only if Z_{ii} is compact, for each $i \in \mathbb{N}$ and $\lim_{i\to\infty} \|Z_{ii}\| = 0$.*

Proof. If Z is compact, then from Corollary 4.4, Z_{ii} is compact for every $i \in \mathbb{N}$. Moreover, the fact that $\lim_{i\to\infty} \|Z_{ii}\| = 0$, follows on similar lines as given in [6] or [9, p. 124]. Therefore, it suffices to prove the converse. Note that for any given $\epsilon > 0$, there exists a $k_0 \in \mathbb{N}$ such that

$$\sup_{k_0+1\leq i<\infty} \|Z_{ii}\| \leq \epsilon/2.$$

By the compactness of Z_{ii}'s, there exists $n_0 \in \mathbb{N}$ depending upon k_0 such that

$$d_n(Z_{ii}) \leq \epsilon/(2\|e^{(k_0)}\|_\mu), \quad \text{for } i = 1, 2, \ldots, k_0, \text{ and } n \geq n_0.$$

Hence, from Corollary 4.6, we get $d_{k_0 n_0}(Z) \leq \epsilon$. Thus $d_j(Z) \leq \epsilon$ for every $j \geq k_0 n_0$. This proves that Z is compact. □

Example 4.8. Let $Z : \ell^p \to \ell^p$, $1 \leq p \leq \infty$ be the diagonal operator defined by a non-negative decreasing sequence $\{\sigma_i\}_{i=1}^\infty$ i.e. $Z\overline{x} = \{\sigma_i x_i\}$, for $\overline{x} = \{x_i\} \in \ell^p$. Then Z is compact if and only if $\lim_{i\to\infty} \sigma_i = 0$.

For the validity of our results, we must consider the same field for the sequence spaces and the underlying Banach space. This is supported by the next two examples.

Example 4.9. Let X be the vector space $\mathbb{C}$ over the field $\mathbb{R}$, $\ell^p(X)$ be the usual ℓ^p space over $\mathbb{R}$ and $Z : \ell^p(X) \longrightarrow \ell^p(X)$, $1 \leq p \leq \infty$, be the diagonal operator $Z\overline{x} = \{\sigma_i x_i\}$, defined from a decreasing sequence $\{\sigma_i\}$ of non-negative numbers as in the preceding example. Here Z_{ii}'s are given by $Z_{ii}(\alpha) = \sigma_i\alpha$, $\alpha \in \mathbb{C}$.

Let us now calculate $d_n(Z_{ii})$'s and $d_n(Z)$. We claim that for $n \in \mathbb{N}$,

$$d_n(Z) = \begin{cases} \sigma_{(n+1)/2} & \text{if } n \text{ is odd,} \\ \sigma_{n/2} & \text{if } n \text{ is even.} \end{cases} \tag{4.3}$$

To see this, consider first $n > 1$ be an odd number. Write $k = (n-1)/2$ and as before, let $Z_k : \ell^p(X) \to \ell^p(X)$ be the map defined by:

$$Z_k\overline{x} = (\sigma_1 x_1, \sigma_2 x_2, \ldots, \sigma_k x_k, 0, 0, \ldots), \quad \overline{x} = \{x_i\} \in \ell^p(X).$$

Then, $\operatorname{rank}(Z_k) \leq 2k < n$ and $\|Z - Z_k\| = \sigma_{k+1}$. By Property 4 of Definition 2.2, $d_n(Z_k) = 0$. Again by Property 2 of Definition 2.2, we deduce that

$$\begin{aligned} d_n(Z) &\leq \|Z - Z_k\| + d_n(Z_k) \\ &\leq \sigma_{k+1} = \sigma_{(n+1)/2}. \end{aligned}$$

For the converse inequality, let us define, for each $m \in \mathbb{N}$, a map T^m from $\ell_m^p(X)$ into itself, given by

$$T^m(\xi_1, \xi_2, \ldots, \xi_m) = (\sigma_1\xi_1, \sigma_2\xi_2, \ldots, \sigma_m\xi_m), \quad (\xi_1, \xi_2, \ldots, \xi_m) \in \ell_m^p(X).$$

Then T^m is one to one and onto. Moreover, $\|T^m\| = \sigma_1$, $\|(T^m)^{-1}\| = \sigma_m^{-1}$ (assuming $\sigma_m > 0$) and we can write $T^m = P^m_{\ell^p(X)}.Z.I^m_{\ell^p(X)}$, which implies that $d_n(T^m) \leq d_n(Z)$. Hence, with $m = (n+1)/2$, we get $d_{n+1}(I_{\ell^p_m(X)}) = 1$, since the dimension of $\ell^p_m(X)$ over $\mathbb{R}$ is $2m$. We deduce that $1 \leq \|(T^m)^{-1}\|.d_{n+1}(T^m) \leq \sigma_m^{-1} d_n(Z)$ and this shows that $d_n(Z) \geq \sigma_m$. Thus we have $d_n(Z) = \sigma_{(n+1)/2}$, when n is odd.

Assume now that n be even. Then $d_n(Z) \leq d_{n-1}(Z) = \sigma_{n/2}$. For the converse inequality, write $k = n/2$. Considering the map T^k as above we have $\|T^k\| = \sigma_1$ and $\|(T^k)^{-1}\| = \sigma_k^{-1}$. We deduce that $1 = d_n(I_{\ell^p_m(X)}) \leq (\sigma_k^{-1})d_n(T^k) \leq (\sigma_k^{-1})d_n(Z)$ and therefore $d_n(Z) \geq \sigma_{n/2}$. Thus (4.3) is verified.

Since $d_1(Z_{ii}) = d_2(Z_{ii}) = \sigma_i$ and $d_n(Z_{ii}) = 0$, $\forall\, n \geq 3$, the result $d_n(Z_{ii}) \leq d_n(Z)$, $\forall\, i,\ n \in \mathbb{N}$ is satisfied in the case of real field.

Example 4.10. In this example, we consider any decreasing sequence $\{\sigma_i\}$ of non-negative real numbers such that $\sigma_1 > \sigma_2$. Let Z be the diagonal operator from ℓ^p to itself $1 \leq p \leq \infty$, as defined in the preceding example. Here, ℓ^p is the vector space $\ell^p(\mathbb{C})$, of complex sequences over the field of complex numbers and $\mathbb{C}$ is considered as a vector space over $\mathbb{R}$. As in the previous examples, Z_{ii}'s are the maps from $\mathbb{C}$ to itself, given by $Z_{ii}(\alpha) = \sigma_i \alpha$ for $\alpha \in \mathbb{C}$. From the above example, $d_1(Z_{ii}) = d_2(Z_{ii}) = \sigma_i$ for all $i \in \mathbb{N}$. Also $d_2(Z) = \sigma_2$ by [1, p. 54]. So

$$d_2(Z_{11}) = \sigma_1 > \sigma_2 = d_2(Z).$$

Hence Proposition 4.1 (iii) is not valid in this case.

References

[1] B. Carl and I. Stephani, *Entropy, compactness and the approximation of operators*, Cambridge Univ. Press, Cambridge, 1990.

[2] N. De Grande-De Kimpe, *Generalized sequence spaces*, Bull. Soc. Math. Belg. 23 (1971), pp. 123–166.

[3] D. A. Gregory, *Vector sequence spaces*, Dissertation, Univ. Michigan, 1967.

[4] M. Gupta and P. K. Kamthan, *Sequence spaces and series*, Marcel Dekker, Inc. New York and Basel, 1981.

[5] M. Gupta and J. Patterson, *The generalized ℓ^p spaces*, Tamkang J. Math. 13 (1982), pp. 161–179.

[6] ———, *Matrix transformations on generalized sequence spaces*, J. Math. Anal. Appl. 106 (1985), pp. 54–68.

[7] C. V. Hutton, J. S. Morrell, and J. R. Retherford, *Diagonal operators, approximation numbers and Kolmogoroff diameters*, J. Approx. Theory 16 (1976), pp. 48–80.

[8] A. N. Kolmogorov, *Über die beste Annäherung von Funktionen einer gegebenen Funktionenklasse*, Ann. of Math. 37 (1936), pp. 107–110.

[9] J. Patterson, *Generalized sequence spaces and matrix transformations*, Dissertation, I. I. T. Kanpur, India, 1980.

[10] A. Pietsch, *Verallgemeinerte vollkommene Folgenräume*, Akademie-Verlag, Berlin, 1962.

[11] ———, *Nuclear locally convex spaces*, Springer-Verlag, Berlin, Heidelberg, New York, 1972.

[12] ———, *s-numbers of operators in Banach spaces*, Studia Math. 51 (1974), pp. 201–223.

[13] ———, *Operator ideals*, VEB Deutscher Verlag der Wissenschaften, Berlin, 1978.

Author information

Lipi R. Acharya, Department of Mathematics and Statistics, Indian Institute of Technology Kanpur, U.P.-208016, India.
Email: `lipi@iitk.ac.in`

Manjul Gupta, Department of Mathematics and Statistics, Indian Institute of Technology Kanpur, U.P.-208016, India.
Email: `manjul@iitk.ac.in`

Banach Spaces and their Applications in Analysis, 229–239

On boundaries for spaces of holomorphic functions on the unit ball of a Banach space

María D. Acosta and Luiza A. Moraes

Abstract. If Ω is a topological space, a subset $S \subset \Omega$ is a boundary for an algebra $\mathcal{A} \subset \mathcal{C}_b(\Omega)$ if $\|f\| = \sup_{s\in S} |f(s)|$ for every $f \in \mathcal{A}$. For a complex Banach space X, let $A_b(B_X)$ be the Banach algebra of all complex valued functions defined on the closed unit ball B_X of X which are bounded and continuous on B_X and holomorphic on the interior of B_X, endowed with the sup norm, and let $A_u(B_X)$ be the Banach subalgebra of all complex valued functions defined on B_X which are uniformly continuous on B_X and holomorphic on the interior of B_X. In this paper we survey previous results about the boundaries of these algebras for some classical Banach spaces X.

Key words. Holomorphic function, boundary, peak point, strong peak set.

AMS classification. Primary 46J15, 46B45; Secondary 46G20, 32A40.

Dedicated to Nigel Kalton on the occasion of his 60th birthday.

1 Introduction

An important theorem of Šilov [16] states that if $\mathcal{A}$ is a unital and separating subalgebra of the algebra $\mathcal{C}(K)$ (complex valued continuous functions on a compact Hausdorff space K with the sup norm), then there exists a minimal closed subset M of K having the property that $\|f\| = \max_{m\in M} |f(m)|$ for every $f \in \mathcal{A}$ [9, Theorem I.4.2]. Six years later, Bishop [5] proved that if in addition it is assumed that $\mathcal{A}$ is a Banach algebra and that K is metrizable, then there exists a minimal subset of K (not necessarily closed) satisfying the above condition. He also gave an example showing that the metrizability of K is necessary.

About twenty years after these classical results of Šilov and Bishop, Globevnik considered the problem of extending them to the setting of some subalgebras $\mathcal{A} \subset \mathcal{C}_b(\Omega)$, where Ω is a Hausdorff topological space not necessarily compact, and $\mathcal{C}_b(\Omega)$ denotes the space of continuous and bounded functions on Ω endowed with its usual sup norm.

Following Globevnik, we say that a subset $B \subset \Omega$ is a *boundary* for $\mathcal{A}$ if

$$\|f\| = \sup_{x\in B} |f(x)|, \quad \forall f \in \mathcal{A} .$$

In such a setting, a closed boundary which is contained in all the closed boundaries of $\mathcal{A}$ is called the *Šilov boundary* for $\mathcal{A}$.

First author: Supported by MEC project MTM2006–04837 and Junta de Andalucía “Proyecto de Excelencia” FQM–01438.
Second author: Partially supported by CNPq, Brazil, Research Grant 306829/2003-7.

Globevnik [11] studied the boundaries for two important algebras of holomorphic functions contained in $C_b(\Omega)$ in case Ω is the closed unit ball of the complex space c_0. Since the pioneering work of Globevnik, the problem of characterizing the boundaries for some algebras of holomorphic functions has been studied successfully by several authors who replaced c_0 by some other classical Banach spaces. The aim of this paper is to survey previous results on this topic. Some of these results can be found in the literature and some others have not yet appeared.

First of all we are going to introduce some basic notation. In this paper we will consider only complex Banach spaces unless we say explicitly the contrary. If X is a Banach space, S_X and B_X will denote, respectively, the unit sphere and the closed unit ball of X. For a Banach space X, $A_b(B_X)$ will be the Banach algebra of all complex valued functions defined on B_X which are continuous and bounded on B_X and holomorphic on the interior of B_X, endowed with the sup norm, and $A_u(B_X)$ will denote the closed subalgebra of all complex valued functions defined on B_X which are uniformly continuous on B_X and holomorphic on the interior of B_X.

Necessary and sufficient conditions for a set to be a boundary for $\mathcal{A}_u(B_X)$ have been presented in [11] for $X = c_0$ and in [4] for $X = \ell_p$ $(1 \leq p \leq \infty)$. Characterizations of the boundaries for the predual $d_*(w,1)$ of the Lorentz space $d(w,1)$ were obtained in [12] for the case $w = \{1/n\}$. The given proofs work for every weight sequence w. However, despite the necessary or sufficient conditions provided in [11], [4] and [13] for a set to be a boundary for $\mathcal{A}_b(B_X)$ when X is c_0, l_∞ and $d_*(\{1/n\},1)$, the authors were unable to give a complete description of the boundaries for these algebras.

Concerning the existence of the Šilov boundary for $\mathcal{A}_u(B_X)$, a positive answer was given by Aron, Choi, Lourenço, and Paques in [4] and by Acosta and Lourenço in [2] when X is the space ℓ_p $(1 \leq p < \infty)$ and $d(w,1)$, respectively. In both cases S_X is the Šilov boundary for $\mathcal{A}_u(B_X)$. Moreover, Acosta and Lourenço showed the existence of the Šilov boundary of $\mathcal{A}_u(B_X)$ when X is the space of the trace class operators on a Hilbert space [2]. Negative answers are more frequent.

For Banach sequence spaces, Globevnik [11], and Moraes and Romero Grados [12] showed that the Šilov boundary of $\mathcal{A}_u(B_X)$ does not exist for $X = c_0$ and $X = d_*(w,1)$ in case $w = \{1/n\}$, respectively. The proof given for $X = d_*(\{1/n\},1)$ works for every space $d_*(w,1)$. Acosta and Lourenço proved the same result for the Schreier space [2]. Aron, Choi, Lourenço and Paques [4] showed the nonexistence of Šilov boundary for $\mathcal{A}_u(B_{\ell_\infty})$. Choi, García, Kim and Maestre [6] extended this result to $\mathcal{C}(K)$ for K infinite and scattered. Acosta [1] proved the same result for every infinite-dimensional $\mathcal{C}(K)$. For spaces of operators, Acosta and Lourenço showed that there is no Šilov boundary for $\mathcal{A}_u(B_X)$ when X is the space of the compact operators on any infinite-dimensional Hilbert space [2].

In this paper we are going to present a more detailed description of the above mentioned results.

We should also mention that recently Choi and Han [7] extended to the the case of Marcinkiewicz sequence spaces results proved by Moraes and Romero Grados in [12] and [13]. Just before we finished this survey we got a copy of a recent paper from Choi, Han and Lee (see [8]), where they study the boundaries for $\mathcal{A}_u(B_X)$ and for $\mathcal{A}_b(B_X)$

in a quite general setting, but such results are not included here.

The known results suggest that the shape of the unit ball of the Banach space X involved has a lot to do with the boundaries for the considered algebras of holomorphic functions. For instance, if the unit ball of the Banach space has a "good" extremal structure like in ℓ_p $(1 \leq p < \infty)$, the Šilov boundary for these algebras of holomorphic functions is the unit sphere. But in spaces with a poor extremal structure like c_0 or $d_*(w,1)$, it is possible to construct disjoint closed boundaries.

Some classes of subsets of the closed unit ball of X play an essential role in the characterization of the boundaries for $\mathcal{A}_u(B_X)$. Next we recall some of these definitions.

If Ω is a topological space, a subset $P \subset \Omega$ is a *peak set* for a subspace $\mathcal{A}$ of $\mathcal{C}_b(\Omega)$ if there exists a function $f \in \mathcal{A}$ such that

$$f(P) = \{1\} \quad \text{and} \quad |f(x)| < 1, \ \forall x \in \Omega \backslash P\,.$$

In this case we say that *f peaks at P*.

If Ω is a metric space, then $P \subset \Omega$ is a *strong peak set* for $\mathcal{A}$ if there exists a function $f \in \mathcal{A}$ such that $f(P) = \{1\}$ and

$$\forall \varepsilon > 0 \ \exists \delta > 0 : d(x,P) > \varepsilon \quad \Longrightarrow \quad |f(x)| < 1 - \delta\,.$$

If P contains exactly one element then this element will be called a *peak point* or a *strong peak point* if P is a peak set or a strong peak set, respectively.

For a complex Banach space X, a point $x \in B_X$ is called a *complex extreme point* of B_X if

$$y \in X, \ \ \|x + \lambda y\| \leq 1, \ \ \forall \lambda \in \mathbb{C}, \ |\lambda| \leq 1 \quad \Longrightarrow \quad y = 0.$$

This is easily seen to be equivalent to requiring only $|\lambda| = 1$.

We notice that every strong peak point for a space belongs to every closed boundary. The usual way to prove the existence of the Šilov boundary for a space of holomorphic functions is to check that the unit sphere of X contains a dense set S of strong peak points.

2 Some classical results

First we will state some interesting results that hold for subalgebras of $\mathcal{C}(K)$, where K is a compact Hausdorff topological space. Here $\mathcal{C}(K)$ will be the Banach space of all complex valued continuous functions on K. The next classical result is due to Šilov.

Theorem 2.1 ([9, Theorem I.4.2]). *Let $\mathcal{A}$ be a unital and separating subalgebra of $\mathcal{C}(K)$. Then there is a minimal closed subset $M \subset K$ satisfying $\|f\| = \max_{x \in M} |f(x)|$ for every $f \in \mathcal{A}$.*

The next result, due to Bishop, proves the existence of a minimal boundary (not necessarily closed) in the compact case under some extra assumptions.

Theorem 2.2 ([5, Theorem 1]). *Let $\mathcal{A}$ be a separating Banach algebra of continuous functions on a compact metrizable Hausdorff space K. Then $\mathcal{A}$ has a minimal subset M satisfying*

$$\forall f \in \mathcal{A}\ \exists m \in M : |f(m)| = \|f\|. \tag{*}$$

In fact, M is the set of all peak points for $\mathcal{A}$.

The assumption of being metrizable is essential. Indeed, if $\mathcal{A} = \mathcal{C}(K)$ where K is equal to some uncountable cartesian product of the unit interval, then there is no minimal set M satisfying (*) (see [5, p. 632] for details). In some cases, such minimal set and the Šilov boundary exist but they do not coincide (see for example [5, p. 632]).

Bishop also proved that the minimal set M satisfying (*) is a countable intersection of open sets under the assumptions of Theorem 2.2 ([5, p. 634]).

Theorem 2.1 and Theorem 2.2 cannot be extended to the noncompact case. In fact some algebras do not have peak points. The next result due to Globevnik will provide some examples.

Proposition 2.3 ([10, Theorem 1]). *If X is a complex Banach space such that B_X has no complex extreme points, then $\mathcal{A}_b(B_X)$ (and so $\mathcal{A}_u(B_X)$) is a function algebra without peak points.*

The above result can be applied to the complex spaces c_0 and to the canonical predual $d_*(w,1)$ of the Lorentz sequence space, for instance (see Section 4 for the definition of this space).

3 Boundaries for $\mathcal{A}_u(B_{c_0})$

Let us denote by S_n the set of all n-tuples of complex numbers with modulus one. From the fact that the Šilov boundary of the polydisc algebra $\mathcal{A}_b(B_{\ell_\infty^n})$ is the torus S_n, Globevnik deduced that $\cup_n S_n$ is a boundary for $\mathcal{A}_u(c_0)$. The same argument works for $\cup_k S_{n_k}$, where $\{n_k\}$ is some strictly increasing sequence of nonnegative integers. The next result implies that this is essentially the only kind of boundaries for the algebra $\mathcal{A}_u(B_{c_0})$.

Theorem 3.1 ([11, Theorem 1.1 and Corollary 1.6]). *A set $S \subset B_{c_0}$ is a boundary of $\mathcal{A}_u(B_{c_0})$ if and only if for every finite dimensional coordinate projection P, the closure of $P(S)$ contains the extreme points of the unit ball of the subspace $P(c_0)$.*

From the above characterization result it follows:

Corollary 3.2 ([11, Theorem 1.8]). *Every closed boundary for $\mathcal{A}_u(B_{c_0})$ contains a proper closed subset which is a boundary for $\mathcal{A}_u(B_{c_0})$.*

In view of Proposition 2.3, B_{c_0} lacks peak points (and consequently strong peak points) for $\mathcal{A}_u(B_{c_0})$. However, B_{c_0} does have strong peak sets for $\mathcal{A}_u(B_{c_0})$. In fact, Globevnik obtained the following characterization of boundaries for this algebra:

Theorem 3.3 ([11, Theorem 1.9]). *A set* $S \subset B_{c_0}$ *is a boundary for* $\mathcal{A}_u(B_{c_0})$ *if and only if* dist $(S, P) = 0$ *for every strong peak set* P *for* $\mathcal{A}_u(B_{c_0})$.

In the proof of the above result the following notion of face is used. If $x \in S_{c_0}$ has finite support and it is an extreme point of the unit ball of the subspace $[e_i : i \in \text{supp } x]$, the face of B_{c_0} through x is the set

$$\{x + y : y \in B_{c_0} \text{ and supp } x \cap \text{supp } y = \emptyset\}.$$

In fact, every peak set for $\mathcal{A}_u(B_{c_0})$ contains a face.

The following example of boundary for $\mathcal{A}_u(B_{c_0})$ is also due to Globevnik:

Example 3.4 ([11, Example 2.1]). There is a countable set $\{x_n : n \in \mathbb{N}\}$ in S_{c_0} satisfying

$$\|x_i - x_j\| \geq 1, \quad \forall i \neq j$$

and such that $\{x_n : n \in \mathbb{N}\}$ is a boundary for $\mathcal{A}_u(B_{c_0})$.

The sequence $\{x_n\}$ is constructed in the following way. Let $\{u_{mn} : n \in \mathbb{N}\}$ be a sequence which is dense in the set of extreme points of the unit ball of the subspace $[e_1, \ldots, e_m]$. Define

$$v_{mn} = u_{mn} + e_{\Phi(m,n+n_m)},$$

where $\Phi : \mathbb{N}^2 \longrightarrow \mathbb{N}$ is injective and n_m is such that $\Phi(m, n) > m$ for $n \geq n_m$.

By using the density of $\{u_{mn}\}$ in the set of extreme points of the unit ball of $[e_1, \ldots, e_m]$ and Theorem 3.1, $\{v_{mn} : m, n \in \mathbb{N}\}$ is a boundary of $\mathcal{A}_u(B_{c_0})$.

As Φ is injective, the definition of the sequence $\{v_{mn}\}$ implies that $\|v_{mn} - v_{st}\| \geq 1$ for $(m, n) \neq (s, t)$. Thus $\{v_{mn} : m, n \in \mathbb{N}\}$ satisfies the above requirements.

In Section 2 of [11], there are also some other examples of boundaries for $\mathcal{A}_u(B_{c_0})$ and $\mathcal{A}_b(B_{c_0})$. In fact, an example is given of a boundary for $\mathcal{A}_u(B_{c_0})$, which is a null set of some non-zero function $f \in \mathcal{A}_b(B_{c_0})$. To the knowledge of the authors, no characterization of the boundaries of the algebra $\mathcal{A}_b(B_{c_0})$ is known.

4 Boundaries for $A_u(B_{d_*(w,1)})$

Let $w \in c_0 \backslash \ell_1$ be a decreasing sequence of positive real numbers such that $w_1 = 1$. For each bounded complex sequence z we may define

$$\phi_n(z) = \sup_{|J|=n} \left\{ \left(\sum_{j=1}^{n} w_j \right)^{-1} \sum_{j \in J} |z_j| \right\}$$

where $J \subset \mathbb{N}$ and $|J|$ denotes the cardinal of the set J. We denote by $d_*(w, 1)$ the complex Banach space of the complex sequences z such that $\lim_{n \to \infty} \phi_n(z) = 0$, endowed with the norm given by

$$\|z\| = \max_{n \in \mathbb{N}} \phi_n(z) \quad (z \in d_*(w, 1)).$$

It is easy to check that $d_*(w,1) \subset c_0$ as a set and $\{e_i\}$ is a normalized Schauder basis in $d_*(w,1)$ (where $e_i = (\delta_{ij})_{j=1}^{\infty}$ $\forall i \in \mathbb{N}$). The space $d_*(w,1)$ is a predual of the Lorentz space $d(w,1)$ (we will recall the definition of $d(w,1)$ later).

From now on, the set of all non empty finite subsets of $\mathbb{N}$ will be denoted by $\mathbb{N}_f$. For each $n \in \mathbb{N}$, we will write $W_n = \sum_{j=1}^{n} w_j$.

The following definition of torus plays an important role in the study of boundaries for $\mathcal{A}_u(B_{d_*(w,1)})$.

Definition 4.1. Let $F \in \mathbb{N}_f$. The torus in $d_*(w,1)$ associated to F is the set $\mathcal{T}_F$ of all $z \in d_*(w,1)$ such that

(t1) $F = \operatorname{supp} z$;
(t2) $\sum_{j\in F} |z_j| = W_{|F|}$;
(t3) $\sum_{j\in J} |z_j| \le W_{|J|}$ for every $J \in \mathbb{N}_f$ satisfying $J \subset F$.

It is clear that $\mathcal{T}_F \subset S_{d_*(w,1)}$. Let $P_F : d_*(w,1) \to d_*(w,1)$ be the norm one projection defined by

$$P_F(z) = \sum_{j\in F} z_j\, e_j \quad (z \in d_*(w,1)).$$

We call P_F *the* F*-projection.* A projection P is a finite dimensional coordinate projection if P is the F-projection for some $F \in \mathbb{N}_f$.

We are ready now to state the following important characterization of the boundaries for $\mathcal{A}_u(B_{d_*(w,1)})$.

Theorem 4.2. *Let $S \subset B_{d_*(w,1)}$ and $(P_{F_n})_{n=1}^{\infty}$ be a sequence of finite dimensional coordinate projections. Suppose that every finite subset of $\mathbb{N}$ is contained in some F_n. Then S is a boundary for $\mathcal{A}_u(B_{d_*(w,1)})$ if and only if for each $n \in \mathbb{N}$ the closure of $P_{F_n}(S)$ contains the torus associated to F_n.*

This theorem was proved in [12, Theorem 3.4] for the case $w = \{1/n\}$ and the proof can be easily adapted to the general case.

To show the reverse implication in the above theorem, it is enough to prove that $\bigcup_{F\in\mathbb{N}_f} \mathcal{T}_F$ is a boundary for $\mathcal{A}_u(B_{d_*(w,1)})$ and this follows by using the density in $S_{d_*(w,1)}$ of the finitely supported vectors of the unit sphere S_X, the uniform continuity on $B_{d_*(w,1)}$ of the elements of $\mathcal{A}_u(B_{d_*(w,1)})$ and [11, Lemma 1.4].

In the other direction the proof uses the following fact that will play an important role in all this section.

Proposition 4.3. *Given $F \in \mathbb{N}_f$ and $x \in \mathcal{T}_F$, there exists a polynomial f on X such that $f(x) = 1$ and $|f(z)| < 1$ for all $z \in B_{d_*(w,1)}$ satisfying $P_F(z) \neq P_F(x)$.*

We sketch the proof of this fact. If $|F| = 1$ write $F = \{j\}$ and define

$$\phi(z) = 1 + e^{i\theta_j} z_j \quad (z \in B_{d_*(w,1)}),$$

where $\theta_j = -\arg x_j$. Otherwise, define $\phi : B_{d_*(w,1)} \longrightarrow \mathbb{C}$ by

$$\phi(z) := \sum_{j\in F}\Big[\Big(1+\frac{1}{b-|x_j|}e^{i\theta_j}z_j\Big)\Big(1+\frac{1}{|x_j|}\sum_{k\in F\setminus\{j\}}e^{i\theta_k}z_k\Big)\Big]$$

where $\theta_j = -\arg x_j$ for all $j \in F$ and $b = W_{|F|}$.

Now, by using the same idea of [12, Theorem 4.3] it can be proved that $\phi(x) > 0$ and the function $f = \phi/\phi(x)$ is a complex valued non-homogeneous polynomial on X such that $f(x) = 1$ and $|f(z)| < 1$ for all $z \in B_{d_*(w,1)}$ satisfying $P_F(z) \neq P_F(x)$.

A consequence of the above proposition is the fact that $\mathcal{T}_F$ is a strong peak set for the algebra of polynomials on $d_*(w,1)$ whenever $F \in \mathbb{N}_f$. Consequently, for every boundary S for $\mathcal{A}_u(B_X)$ we have that the closure of $P_{F_n}(S)$ contains the torus associated to F_n for each $n \in \mathbb{N}$.

From Theorem 4.2 it can be deduced that every closed boundary for $\mathcal{A}_u(B_{d_*(w,1)})$ contains a proper closed subset which is a boundary for $\mathcal{A}_u(B_{d_*(w,1)})$. More explicitly, by using Theorem 4.2 it can be shown that if $x \in S$ and $0 < r < 1$ then the set $S \setminus (z + r\mathring{B}_{d_*(w,1)})$ is a boundary for $\mathcal{A}_u(B_{d_*(w,1)})$ whenever S is a boundary for $\mathcal{A}_u(B_{d_*(w,1)})$. As a consequence, there is no minimal boundary for $\mathcal{A}_u(B_{d_*(w,1)})$. The proof of this fact can be found in [3, Corollary 3.2].

The next theorem, proved in [3], provides a characterization of the boundaries for $\mathcal{A}_u(B_{d_*(w,1)})$ in terms of the distance to the strong peak sets for this algebra.

Theorem 4.4. *A subset $S \subset B_{d_*(w,1)}$ is a boundary for $A_u(B_{d_*(w,1)})$ if and only if* dist $(S,P) = 0$ *for every strong peak set P for $A_u(B_{d_*(w,1)})$.*

The proof of the necessity follows easily from the definition of strong peak set. The proof of the sufficiency uses the following notion of face through a normalized element of $d_*(w,1)$.

Definition 4.5. Given $J \in \mathbb{N}_f$ and $x \in \mathcal{T}_J$, the face of $B_{d_*(w,1)}$ through x is the set

$$\mathcal{F}_{x,J} = \big\{x + y \in B_{d_*(w,1)} : J \cap \operatorname{supp} y = \emptyset\big\}.$$

The next proposition and Theorem 4.7 show that the above definition provides a family of strong peak sets of $\mathcal{A}_u(B_{d_*(w,1)})$ and that the boundaries for this algebra can be characterized in terms of the distance to the elements of this family.

Proposition 4.6. *For every $F \in \mathbb{N}_f$ and $x \in \mathcal{T}_F$, the face of $B_{d_*(w,1)}$ through x is a strong peak set of $\mathcal{A}_u(B_{d_*(w,1)})$.*

Sketch of the proof. By Proposition 4.3, there exists $f \in \mathcal{A}_u(B_{d_*(w,1)})$ such that $f(x) = 1$ and $|f(z)| < 1$ for all $z \in B_{d_*(w,1)}$ satisfying $P_F(z) \neq P_F(x)$. In order to show that the face $\mathcal{F}_{x,J}$ is a strong peak set of $\mathcal{A}_u(B_{d_*(w,1)})$, it is enough to prove that this function f satisfies the following condition:

$$u_n \in B_{d_*(w,1)}, \forall n,\ \{|f(u_n)|\} \to 1 \implies \{d(u_n, \mathcal{F}_{x,F})\} \to 0.$$

For details, see [3, Proposition 3.9]. □

The proof of [3, Theorem 3.3] contains the proof of the fact that every strong peak set contains some $\mathcal{F}_{x,F}$. It is not known if the class of the strong peak sets coincide with the class of the faces $\mathcal{F}_{x,F}$ for $F \in \mathbb{N}_f$ and $x \in \mathcal{T}_F$. But the next theorem shows that even if these two classes of sets are different it is possible to characterize the boundaries for $\mathcal{A}_u(B_{d_*(w,1)})$ in terms of the distance to all faces $\mathcal{F}_{x,F}$ ($F \in \mathbb{N}_f$ and $x \in \mathcal{T}_F$).

Theorem 4.7. *A subset* $S \subset B_{d_*(w,1)}$ *is a boundary for* $\mathcal{A}_u(B_{d_*(w,1)})$ *if and only if* $\operatorname{dist}(S, \mathcal{F}_{x,F}) = 0$ *for every* $F \in \mathbb{N}_f$ *and* $x \in \mathcal{T}_F$.

Proof. See the proof of [3, Theorem 3.3]. □

5 Boundaries for $\mathcal{A}_u(B_{\ell_p})$ and $\mathcal{A}_u(B_{d(w,1)})$

The case of ℓ_p ($1 \le p < \infty$) was first considered by Aron, Choi, Lourenço, and Paques [4].

Theorem 5.1 ([4, Corollary 11]). *For* $1 \le p < \infty$, *the Šilov boundary for* $\mathcal{A}_u(B_{\ell_p})$ *is* S_{ℓ_p}.

Every point $x \in S_{\ell_p}$ ($1 < p < \infty$) is strongly exposed, i.e., there is $x^* \in S^*_{\ell_p}$ satisfying $\{x_n\}$ converges to x whenever $\{x_n\}$ is a sequence in B_{ℓ_p} and $\{x^*(x_n)\} \to 1$. If $\mathcal{A} \subset \mathcal{C}_b(B_X)$ is a subalgebra containing the constant functions and X^*, it is clear that every strongly exposed point of the unit ball of the Banach space X is a strong peak point for this algebra. The proof of Theorem 5.1 easily follows from the above fact for $1 < p < \infty$.

The proof of the result for the case $p = 1$ requires more work. Aron, Choi, Lourenço, and Paques proved that all the finitely supported sequences in S_{ℓ_1} are strong peak points for $\mathcal{A}_u(B_{\ell_1})$. In [2, Theorem 2.6] it is proved that every point in the unit sphere of ℓ_1 is in fact a strong peak point for such algebra. As a consequence, it follows that any closed boundary for $\mathcal{A}_u(B_{\ell_1})$ is S_{ℓ_1}.

The above result remains true for the Lorentz sequence spaces. Given a decreasing sequence w of positive real numbers satisfying $w \in c_0 \setminus \ell_1$ and $w_1 = 1$, the *complex Lorentz sequence space* $d(w, 1)$ is the set of complex sequences x satisfying

$$\sup\Big\{\sum_{n=1}^{\infty} |x(\sigma(n))| w_n,\ \sigma : \mathbb{N} \longrightarrow \mathbb{N} \text{ is injective}\Big\} < \infty.$$

This space is a Banach space when endowed with the norm

$$\|x\| = \sup\Big\{\sum_{n=1}^{\infty} w_n |x(\sigma(n))|,\ \sigma : \mathbb{N} \longrightarrow \mathbb{N} \text{ is injective}\Big\}, \quad (x \in d(w,1)).$$

Theorem 5.2 ([2, Theorem 2.4]). *The set of strong peak points for the space of the polynomials of degree less or equal than two on $d(w,1)$ is the unit sphere. Hence $S_{d(w,1)}$ is the Šilov boundary for $\mathcal{A}_u(B_{d(w,1)})$ (and so for $\mathcal{A}_b(B_{d(w,1)})$).*

The case of ℓ_∞ is more difficult than the case ℓ_p $(1 \leq p < \infty)$. In fact, there is no characterization of the closed boundaries for $\mathcal{A}_u(B_{\ell_\infty})$. First we will mention some results due to Aron, Choi, Lourenço, and Paques.

Theorem 5.3 ([4, Theorem 2]). *The set $\mathbb{T} = \{x \in S_{\ell_\infty} : |x_n| = 1, \forall n\}$ is a closed boundary for $\mathcal{A}_b(B_{\ell_\infty})$, and hence for $\mathcal{A}_u(B_{\ell_\infty})$.*

The fact that ℓ_∞ has the Dunford–Pettis property, and consequently the polynomial Dunford–Pettis property, i.e. scalar valued polynomials on ℓ_∞ preserve weakly convergent sequences, is the key result to show:

Theorem 5.4 ([4, Theorem 4 and Theorem 6]). *Given any sequence $\{x_n\}$ in $\mathbb{T}$, the set $\mathbb{T} \setminus \cup_n(x_n + 2B_{\ell_\infty})$ is a boundary for $\mathcal{A}_u(B_{\ell_\infty})$ and, for every $0 < \gamma < 2$, the set $B_{\ell_\infty} \setminus \cup_n(x_n + \gamma B_{\ell_\infty})$ is a closed boundary for $\mathcal{A}_b(B_{\ell_\infty})$.*

Theorem 5.5 ([4, Proposition 5]). *(a) If F is a boundary for $\mathcal{A}_u(B_{\ell_\infty})$, then F is w^*-dense in $\mathbb{T}$.*
(b) If F is weakly sequentially dense in $\mathbb{T}$, then F is a boundary for $\mathcal{A}_u(B_{\ell_\infty})$.

The same authors proved that there is a closed boundary for $\mathcal{A}_b(B_{\ell_\infty})$, (so also for $\mathcal{A}_u(B_{\ell_\infty})$) disjoint from $\mathbb{T}$.

6 Results for $\mathcal{C}(K)$ and some other Banach spaces

Choi, Kim, García and Maestre [6] proved that for every scattered compact topological space K, the set of extreme points of the unit ball of $\mathcal{C}(K)$ is a boundary for $\mathcal{A}_u(B_{\mathcal{C}(K)})$. This result was generalized by Acosta [1, Theorem 4.2] by using similar ideas to the arguments used in [4] for the space ℓ_∞. For every compact Hausdorff topological space K, the set of extreme points in the unit ball of $\mathcal{C}(K)$ is a boundary for $\mathcal{A}_b(B_{\mathcal{C}(K)})$.

By using that $\mathcal{C}(K)$ has the Dunford–Pettis property, the following result is also proved.

Theorem 6.1 ([1, Proposition 4.1]). *Let $S \subset B_{\mathcal{C}(K)}$ be balanced and weakly sequentially dense in a boundary for $\mathcal{A}_b(B_{\mathcal{C}(K)})$. Then S is also a boundary for $\mathcal{A}_b(B_{\mathcal{C}(K)})$.*

The same result is true for $\mathcal{A}_u(B_{\mathcal{C}(K)})$.

We should also mention a few results for some other classical Banach spaces. For the Schreier space a result similar to the c_0 case is obtained. A subset $E = \{n_1 < n_2 < \cdots < n_k\}$ of the natural numbers $\mathbb{N}$ is said to be *admissible* if $k \leq n_1$. The *Schreier space* $\mathcal{S}$ is the completion of the space c_{00} of all finitely supported scalar sequences with respect to the norm $\|x\| = \sup \sum_{j \in E} |x_j|$, where the supremum is taken over all admissible sets of natural numbers E.

Theorem 6.2 ([2, Theorem 3.1]). *Let $\mathcal{S}$ be the Schreier space and B be a boundary for $\mathcal{A}_b(B_{\mathcal{S}})$. If $x_0 \in B$ and $0 < r < 1$, then $B \setminus (x_0 + rB_{\mathcal{S}})$ is a boundary for $\mathcal{A}_b(B_{\mathcal{S}})$. As a consequence, there is no Šilov boundary for $\mathcal{A}_b(B_{\mathcal{S}})$.*

Before stating the next result we recall the definition of the trace class operators on a Hilbert space. Let H be a complex Hilbert space. An operator $T : H \longrightarrow H$ is called a *trace class operator* if there are orthonormal sequences (e_n) and (f_n) in H, and a sequence (λ_n) in ℓ_1 such that $T(x) = \sum_{n=1}^{\infty} \lambda_n < x, e_n > f_n$ for every $x \in H$. In such a case, the norm of T is given by $\|T\| = \sum_{n=1}^{\infty} |\lambda_n|$ and it is independent of the sequences (e_n) and (f_n). We denote by $C_1(H)$ the Banach space of all trace class operators on H.

For $X = C_1(H)$ the following result is known.

Theorem 6.3 ([2, Theorem 4.1]). *If H is a complex Hilbert space, there is Šilov boundary for $\mathcal{A}_u(B_{C_1(H)})$ and $\mathcal{A}_b(B_{C_1(H)})$ and both coincide.*

The boundaries for $\mathcal{A}_b(B_X)$, where the Banach space X is the space of all compact operators on ℓ_p, for $1 \leq p < \infty$, have been considered too and the following result has been obtained.

Theorem 6.4 ([2, Theorem 5.1]). *If $1 \leq p \leq q < \infty$, then there is no Šilov boundary for $\mathcal{A}_b(B_{K(\ell_p,\ell_q)})$. In fact, if B is a boundary for $\mathcal{A}_b(B_{K(\ell_p,\ell_q)})$, $0 < r < 1$ and $x_0 \in B$, the set $B \setminus (x_0 + rB_{K(\ell_p,\ell_q)})$ is also a boundary for $\mathcal{A}_b(B_{K(\ell_p,\ell_q)})$.*

The same assertions also hold for $\mathcal{A}_u(B_{K(\ell_p,\ell_q)})$. It is also true that there are closed boundaries A and B for $\mathcal{A}_b(B_{K(\ell_p,\ell_q)})$ such that $\text{dist}\,(A, B) \geq 1$.

7 Open questions

Problem 7.1. *Is it true that $\mathcal{A}_u(B_X)$ has a Šilov boundary whenever X is a reflexive Banach space?*

Problem 7.2. *When is it true that every complex extreme point of B_X is a peak point of $\mathcal{A}_b(B_X)$?*

Problem 7.3. *Characterize the boundaries for $\mathcal{A}_u(B_{\mathcal{C}(K)})$.*

Problem 7.4. *Characterize the boundaries for $\mathcal{A}_b(B_X)$ in case that X is a classical space.*

References

[1] M. D. Acosta, *Boundaries for spaces of holomorphic functions on $\mathcal{C}(K)$*, Publ. Res. Inst. Math. Sci. 42 (2006), pp. 27–44.

[2] M. D. Acosta and M. L. Lourenço, *Šilov boundary for holomorphic functions on some classical Banach spaces*, Studia Math. 179 (2007), pp. 27–39.

[3] M. D. Acosta, L. A. Moraes, and L. Romero Grados, *On boundaries on the predual of the Lorentz sequence space*, J. Math. Anal. Appl., to appear.

[4] R. M. Aron, Y. S. Choi, M. L. Lourenço, and O. W. Paques, *Boundaries for algebras of analytic functions on infinite dimensional Banach spaces*. Banach Spaces (B.L. Lin and W.B. Johnson, eds.), Contemporary Mathematics 144, pp. 15–22. American Mathematical Society, 1993.

[5] E. Bishop, *A minimal boundary for function algebras,* Pacific J. Math. 9 (1959), pp. 629–642.

[6] Y. S. Choi, D. García, S. G. Kim, and M. Maestre, *Norm or numerical radius attaining polynomials on $C(K)$*, J. Math. Anal. Appl. 295 (2004), pp. 80–96.

[7] Y. S. Choi and K. H. Han, *Boundaries for algebras of holomorphic functions on Marcinkiewicz sequence spaces*, J. Math. Anal. Appl. 323 (2006), pp. 1116–1133.

[8] Y. S. Choi, K. H. Han, and H. J. Lee, *Boundaries for algebras of holomorphic functions on Banach spaces*, Illinois J. Math., to appear.

[9] T. W. Gamelin, *Uniform Algebras*, Chelsea, New York, 1984.

[10] J. Globevnik, *On interpolation by analytic maps in infinite dimensions*, Math. Proc. Cambridge Philos. Soc. 83 (1978), pp. 243–252.

[11] J. Globevnik, *Boundaries for polydisc algebras in infinite dimensions*, Math. Proc. Cambridge Philos. Soc. 85 (1979), pp. 291–303.

[12] L. A. Moraes and L. Romero Grados, *Boundaries for algebras of holomorphic functions*, J. Math. Anal. Appl. 281 (2003), pp. 575–586.

[13] ———, *Boundaries for an algebra of bounded holomorphic functions*, J. Korean Math. Soc. 41 (2004), pp. 231–242.

[14] W. Rudin, *Function Theory in Polydiscs*, Benjamin, New York, 1969.

[15] ———, *Real and Complex Analysis*, McGraw-Hill, 1974.

[16] G. E. Šhilov, *On the decomposition of a commutative normed ring in a direct sum of ideals*, Mat. Sbornik 32 (1953), pp. 353–364.

Author information

María D. Acosta, Departamento de Análisis Matemático, Universidad de Granada, 18071 Granada, Spain.
Email: dacosta@ugr.es

Luiza A. Moraes, Instituto de Matemática Universidade Federal do Rio de Janeiro CP 68530 - CEP 21945-970 Rio de Janeiro, RJ, Brasil.
Email: luiza@im.ufrj.br

Banach Spaces and their Applications in Analysis, 241–250

Some equivalent norms on the Hilbert space

George Androulakis and Frank Sanacory

Abstract. We present a family of some new Tsirelson-type norms for the separable Hilbert space. Our results extend some results of Bellenot, Bernués, Deliyanni and Argyros and provide candidates for distorted norms on the Hilbert space.

Key words. Hilbert space norms, distortion problem.

AMS classification. 46B03.

In this note we present a family of some new Tsirelson-type norms for the separable Hilbert space ℓ_2. The motivation for presenting these norms is the following question of Odell and Schlumprecht [5] which is also mentioned by Gowers [4]:

Question 1. Is it possible, for $\lambda > 0$ to explicitly define an equivalent norm $|\cdot|$ on ℓ_2 such that every infinite dimensional subspace Y of ℓ_2 contains two vectors y_1 and y_2 with $\|y_1\|_2 = \|y_2\|_2 = 1$ (where $\|\cdot\|_2$ denotes the usual norm of ℓ_2) and $|y_1|/|y_2| > \lambda$?

An implicitly defined norm with the above property exists by the solution of the famous distortion problem by Odell and Schlumprecht [5, 6]. The family of norms that we present gives candidates for the solution of Question 1. Some of the norms of our family were first presented by Bellenot [2] and recently Pelczar [7] proved that these norms do not answer Question 1. Another purpose of the present note is to define equivalent norms on the Hilbert space more general than previously known norms as well as improve the known equivalence constants (for previous results see [1, 2, 3]).

In order to define the new norms on ℓ_2 we first introduce some notation. For $x = (x(i)) \in \ell_2$ and $E \subseteq \mathbb{N}$ we denote by Ex the natural projection of x on E, i.e. $Ex = ((Ex)(i))$ where $(Ex)(i) = x(i)$ for all $i \in E$ and $(Ex)(i) = 0$ otherwise. Let c_{00} be the vector space of scalar sequences with finite support. Set

$$\mathcal{A} = \left\{ y = (y(1), y(2), \ldots, y(N)) \in (0,\infty)^N : N \in \mathbb{N},\ N \geq 2,\ \sum_{i=1}^{N} |y(i)|^2 = 1 \right\}.$$

We define a sequence of equivalent norms $\|\cdot\|_y$, for $y \in \mathcal{A}$, on ℓ_2 as follows. Fix $y = (y(i))_{i=1}^N \in \mathcal{A}$. Then $\|\cdot\|_y$ is the unique norm which satisfies

$$\|x\|_y = \|x\|_\infty \vee \sup \sum_{i=1}^{N} y(i) \|E_i x\|_y \text{ for every } x \in c_{00}, \tag{1}$$

where the supremum is taken with respect to any sequence of sets $E_1 < E_2 < \cdots < E_N$. Notice that in the definition of $\|\cdot\|_y$ we allow the sets E_i to be empty and we

Second author: The present paper is part of the Ph.D thesis of the second author which is prepared at the University of South Carolina under the supervision of the first author.

adopt the convention that "$A < \emptyset$" and "$\emptyset < B$" are valid for any $A, B \subseteq \mathbb{N}$ (similarly, we can define $E < F$ for $E, F \subseteq \mathbb{R}$). It is standard to show that for every $y \in \mathcal{A}$ there exists a unique norm $\|\cdot\|_y$ which satisfies (1). We provide two equivalent definitions of $\|\cdot\|_y$. Firstly, for every fixed $y \in \mathcal{A}$ we define a sequence of norms $\|\cdot\|_y^{(n)}$ for $n \in \mathbb{N} \cup \{0\}$ which increases to the norm $\|\cdot\|_y$, as follows. Let $\|\cdot\|_y^{(0)} = \|\cdot\|_\infty$ (the ℓ_∞ norm). If for $n \in \mathbb{N} \cup \{0\}$ the $\|\cdot\|_y^{(n)}$ has been defined, then we define

$$\|x\|_y^{(n+1)} = \|x\|_y^{(n)} \vee \sup \sum_{i=1}^{N} y(i) \|E_i x\|_y^{(n)} \text{ for } x \in c_{00}, \tag{2}$$

where the supremum is taken with respect to any sequence of sets $E_1 < E_2 < \cdots < E_N$. Then $\|\cdot\|_y$ is the (pointwise) limit of $\|\cdot\|_y^{(n)}$ for $n \in \mathbb{N} \cup \{0\}$, and it satisfies (1).

Secondly, we define a sequence of subsets K_n, $n = 0, 1, 2, \ldots$ of c_{00} as follows (where (e_j) denotes the standard unit vector basis of ℓ_2). Let

$$K_0 = \{\lambda e_i : i \in \mathbb{N}, |\lambda| = 1\}.$$

If K_n has been defined for some $n \in \mathbb{N} \cup \{0\}$ then

$$K_{n+1} = K_n \cup \{\sum_{i=1}^{N} y(i) x_i : x_1 < x_2 < \cdots < x_N \text{ and } x_i \in K_n \text{ for all } i \in \{1, \ldots, N\}\}.$$

Let $K = \cup_{n=1}^{\infty} K_n$. Then, for every $x = (x(i)) \in c_{00}$ we have that

$$\|x\|_y = \sup_{z=(z(i)) \in K} \sum_{i=1}^{\infty} x(i) z(i).$$

By Cauchy–Schwartz inequality it is easy to show by induction on $n \in \mathbb{N} \cup \{0\}$ that $\|x\|_y^{(n)} \leq \|x\|_2$ for every $x \in c_{00}$, or $\|z\|_2 \leq 1$ for all $z \in K_n$. Thus we have that $\|x\|_y \leq \|x\|_2$ for every $x \in c_{00}$. We will also show in Theorem 5 that $\|\cdot\|_y$ is in fact equivalent to $\|\cdot\|_2$.

For the proof of our main result we will need the notions of the *N-tree*, the *tree decomposition of an interval* and the *tree decomposition of a function.*

Definition 2. Let N be an integer larger than 1. An *N-tree* is a subset of $\{\emptyset\} \cup \cup_{k=1}^{\infty} \{1, 2, \ldots, N\}^k$ endowed with an order $\prec$, satisfying the following:

(i) $\emptyset \in \mathcal{T}$ and $\emptyset \prec t$ for all $t \in \mathcal{T} \setminus \{\emptyset\}$.

(ii) If $1 \leq k \leq \ell$, $n_i \in \{1, \ldots, N\}$ for $1 \leq i \leq \ell$ and $(n_1, \ldots, n_\ell) \in \mathcal{T}$ then $(n_1, \ldots, n_k) \in \mathcal{T}$.

(iii) If $k, \ell \in \mathbb{N}$ and $n_i, m_i \in \{1, \ldots, N\}$ then $(n_1, \ldots, n_k) \prec (m_1, \ldots, m_\ell)$ if and only if $k < \ell$ and $m_i = n_i$ for $1 \leq i \leq k$.

A tree may be finite (respectively infinite) if it has finitely (respectively infinitely) many nodes. A node $t \in \mathcal{T}$ will be called *maximal* if there is no $s \in \mathcal{T}$ with $t \prec s$. If

$t \in \mathcal{T} \setminus \{\emptyset\}$ we denote by t^- the unique immediate predecessor of $\mathcal{T}$. We denote by $\max(\mathcal{T})$ the set of maximal nodes of $\mathcal{T}$. If $t \in \mathcal{T} \setminus \max(\mathcal{T})$ then t^+ will denote the set of the immediate successors of t, namely $t^+ = \{s \in \mathcal{T} : t \prec s$ and there is no $t' \in \mathcal{T}$ with $t \prec t' \prec s\}$.

Definition 3. Let $\mathcal{T}$ be an N-tree for some $N \in \mathbb{N}$, $N > 1$. By a *tree decomposition* or $\mathcal{T}$*- decomposition* of an interval $E \subset \mathbb{R}$ we mean a family of intervals $(E_t)_{t \in \mathcal{T}}$ indexed by the tree $\mathcal{T}$, satisfying the following:

(i) $E_\emptyset = E$.
(ii) Let $1 \leq n < m \leq N$. If $(n), (m) \in \mathcal{T}$ then $E_{(n)} < E_{(m)}$. If $k \in \mathbb{N}$, $n_1, \ldots, n_k \in \{1, \ldots, N\}$ and $(n_1, \ldots, n_k, n), (n_1, \ldots, n_k, m) \in \mathcal{T}$ then $E_{(t,n)} < E_{(t,m)}$, where for $t = (n_1, \ldots, n_k) \in \mathcal{T}$ and $n \in \{1, \ldots, N\}$ we write (t, n) to denote $(n_1, \ldots, n_k, n)$.
(iii) $E_t \supseteq \cup\{E_{(t,i)} : (t,i) \in \mathcal{T},\ i \in \{1, \ldots, N\}\}$.

Definition 4. Let $\mathcal{T}$ be an N-tree for some $N \in \mathbb{N}$, $N > 1$. By a *tree decomposition* or $\mathcal{T}$*-decomposition* of a function $g : [0, \infty) \to \mathbb{R}$ we mean a family of functions $(g_t)_{t \in \mathcal{T}}$ indexed by the tree $\mathcal{T}$, satisfying the following two properties.

(i) $g_\emptyset = g$.
(ii) If $t \in \mathcal{T} \setminus \max(\mathcal{T})$ and $t^+ = \{(t,i) : i \in F\}$ for some $F \subseteq \{1, \ldots, N\}$ then $g_t = \sum_{i \in F} y(i) g_{(t,i)}$ with $\operatorname{supp}(g_{t,i}) < \operatorname{supp}(g_{t,j})$ for every $i, j \in F$ with $i < j$.

If g has a $\mathcal{T}$-decomposition $(g_t)_{t \in \mathcal{T}}$ and $I \subseteq \max(\mathcal{T})$ then there is a unique tree $\widetilde{\mathcal{T}} \subseteq \mathcal{T}$ such that $\max(\widetilde{\mathcal{T}}) = I$. Let $S := \cup_{t \in I} \operatorname{supp}(g_t)$ and $f := g|_S$ be the restriction of g on S. Then f has a $\widetilde{\mathcal{T}}$-tree decomposition $(f_t)_{t \in \widetilde{\mathcal{T}}}$ which is naturally inherited by the $\mathcal{T}$-decomposition of g, if we set $f_t := g_t$ for all $t \in I$ and $f_t := g_t|_S$ for all $t \in \widetilde{\mathcal{T}} \setminus \max(\widetilde{\mathcal{T}})$.

Theorem 5. *For every $y \in \mathcal{A}$, $\|\cdot\|_y$ is equivalent to the usual norm of ℓ_2. In particular, if $y = (y(i))_{i=1}^N \in \mathcal{A}$, $a := \min_{1 \leq i \leq N} y(i)$, $b := \max_{1 \leq i \leq N} y(i)$ and $M \in \mathbb{N}$ is such that $b^M \leq a$, then*

$$\|\cdot\|_y \leq \|\cdot\|_2 \leq (2MN + 2)\|\cdot\|_y.$$

Proof. Fix $y = (y(i))_{i=1}^N \in \mathcal{A}$. We consider the isometric embedding $\phi : (\ell_2, \|\cdot\|_2) \to (L_2[0,\infty), \|\cdot\|_2)$ defined by

$$\phi(x) = \sum_{i=1}^{\infty} x(i) \chi_{[i-1,i)}$$

where $x = (x(i)) \in \ell_2$ and χ_E denotes the characteristic function of a set E. Here $\|\cdot\|_2$ denotes the usual norm on both ℓ_2 and $L_2[0,\infty)$. Let $\mathcal{B}$ denote the set of functions in $L_2[0,\infty)$ with bounded support, where for $f \in L_2[0,\infty)$ the support of f, $\operatorname{supp}(f)$, is defined to be the essential support of f. Obviously we have that $\phi(c_{00}) \subset \mathcal{B}$. We divide the proof into three steps. In Step 1 we define a norm $|\cdot|_y$ on $\mathcal{B}$ such that

$\phi|_{c_{00}} : (c_{00}, \|\cdot\|_y) \to (\mathcal{B}, |\cdot|_y)$ is an isometric embedding. In Step 2 we define a norm $|||\cdot|||_y$ on $L_2[0,\infty)$ and we prove that it is equal to the usual norm of $L_2[0,\infty)$. In Step 3 we prove that the norms $|\cdot|_y$ and $|||\cdot|||_y$ are equivalent. Of course, these three steps finish the proof.

Step 1: Define a norm $|\cdot|_y$ on $L_2[0,\infty)$ in terms of its dual ball, as follows. Let

$$L_0 = \{f \in L_2[0,\infty) : \|f\|_2 \leq 1 \text{ and } \operatorname{supp}(f) \subseteq [n-1, n) \text{ for some } n \in \mathbb{N}\}.$$

If $n \in \mathbb{N} \cup \{0\}$ and L_n has been defined, then

$$\begin{aligned} L_{n+1} &= L_n \cup \{\textstyle\sum_{i=1}^N y(i) f_i : f_i \in L_n \text{ for all } i \text{ and there exists } (n_i) \subseteq \mathbb{N} \text{ such that} \\ &\quad \operatorname{supp}(f_1) < n_1 \leq \operatorname{supp}(f_2) < n_2 \leq \operatorname{supp}(f_3) < \ldots < n_{N-1} \leq \operatorname{supp}(f_N)\}. \end{aligned}$$

Let $L = \cup_{n=0}^{\infty} L_n$ and for $f \in L_2[0,\infty)$ define

$$|f|_y = \sup_{g \in L} \int_0^\infty fg.$$

Notice that $L \subset \mathcal{B}$ and for every $g \in L$ we have that $\|g\|_2 \leq 1$. Thus $|\cdot|_y$ is well defined and $|\cdot|_y \leq \|\cdot\|_2$. We now prove that $\phi|_{c_{00}} : (c_{00}, \|\cdot\|_y) \to (\mathcal{B}, |\cdot|_y)$ is an isometric embedding. For this purpose we need the following lemma.

Lemma 6. *(i) For $x = (x(i))$ and $z = (z(i))$ in c_{00}, we have*

$$\sum_{i=1}^{\infty} x(i) z(i) = \int_0^\infty \phi(x)\phi(z).$$

(ii) For every $n \in \mathbb{N} \cup \{0\}$, $\phi(K_n) \subset L_n$.

(iii) For every $n, m, \ell \in \mathbb{N} \cup \{0\}$ with $m < \ell$, and $x \in c_{00}$, we have

$$\sup_{z \in K_n} \int_m^\ell \phi(x)\phi(z) = \sup_{g \in L_n} \int_m^\ell \phi(x) g.$$

Since ϕ is an isometry, item (i) is immediate. In order to prove (ii) we proceed by induction on n. For $n = 0$ and $z \in K_n$ we have that $z = \lambda e_k$ where $|\lambda| = 1$ and $k \in \mathbb{N}$. Thus $\phi(z) = \lambda \chi_{[k-1,k)} \in L_0$. Assume that (ii) is valid for some $n \in \mathbb{N} \cup \{0\}$ and let $z \in K_{n+1}$. There exist $z_1 = (z_1(i)) < z_2 = (z_2(i)) < \cdots < z_N = (z_N(i))$ in K_n such that $z = \sum_{i=1}^N y(i) z_i = \sum_{i=1}^N y(i) \sum_{j=1}^\infty z_i(j) e_j$. Then $\phi(z) = \sum_{i=1}^N y(i) \sum_{j=1}^\infty z_i(j) \chi_{[j-1,j)}$. Notice that for all $i \in \{1, \ldots, N\}$, $\sum_{j=1}^\infty z_i(j) \chi_{[j-1,j)} = \phi(z_i) \in K_n$ and

$$\begin{aligned} &\sum_{j=1}^\infty z_1(j)\chi_{[j-1,j)} < \max(\operatorname{supp}(z_1)) \leq \sum_{j=1}^\infty z_2(j)\chi_{[j-1,j)} < \max(\operatorname{supp}(z_2)) \leq \cdots \\ &\qquad \leq \sum_{j=1}^\infty z_N(j)\chi_{[j-1,j)}. \end{aligned}$$

Thus $\phi(z) \in L_{n+1}$ which finishes the inductive proof of (ii).

Notice that by (ii) we obtain "$\leq$" in (iii). In order to prove "$\geq$" in (iii) we use induction on n. Let $n = 0$, $g \in L_n$ and $m < \ell$ in $\mathbb{N}$. There exists $k \in \mathbb{N}$ such that $\operatorname{supp}(g) \subseteq [k-1, k)$ and $\|g\|_2 \leq 1$. Thus $\int_m^\ell \phi(x) g = x(k) \int_{[m,\ell)\cap[k-1,k)} g$. The last expression is equal to zero if $[m,\ell) \cap [k-1,k) = \emptyset$. If $[m,\ell) \cap [k-1,k) \neq \emptyset$ then $[k-1,k) \subseteq [m,\ell)$ and therefore

$$|\int_m^\ell \phi(x) g| \leq |x(k)| \left(\int_{[k-1,k)} |g|^2 \right)^{1/2} \leq |x(k)| = \int_{[k-1,k)} x(k)\phi(z) = \int_m^\ell \phi(x)\phi(z)$$

where $z = (z(i)) \in K_0$ is defined by $z(i) = 0$ for all $i \neq k$, $z(k) = |x(k)|/x(k)$ if $x(k) \neq 0$, $z(k) = 0$ if $x(k) = 0$.

Assume that (iii) is valid for some $n \in \mathbb{N} \cup \{0\}$. Let $g \in L_{n+1}$ and $m < \ell$ in $\mathbb{N}$. There exist $g_1, g_2, \ldots, g_N \in L_n$ and $(n_i)_{i=1}^{N-1} \subset \mathbb{N}$ such that $\operatorname{supp}(g_1) < n_1 \leq \operatorname{supp}(g_2) < n_2 \leq \cdots < n_{N-1} \leq \operatorname{supp}(g_N)$ and $g = \sum_{i=1}^N y(i) g_i$. Thus if we set $n_0 = 0$ and we adopt the convention that $[s,t) = \emptyset$ whenever $s \geq t$, then

$$\begin{aligned}
\int_{[m,\ell)} \phi(x) g &= \sum_{i=1}^N y(i) \int_{[m,\ell)} \phi(x) g_i \\
&= \sum_{i=1}^N y(i) \int_{[m \vee n_{i-1}, \ell \wedge n_i)} \phi(x) g_i \\
&\leq \sum_{i=1}^N y(i) \sup_{z_i \in K_n} \int_{[m \vee n_{i-1}, \ell \wedge n_i)} \phi(x)\phi(z_i),
\end{aligned}$$

where the last inequality follows by the inductive hypothesis. Thus we continue our estimates as follows:

$$\begin{aligned}
\int_{[m,\ell)} \phi(x) g &= \sum_{i=1}^N y(i) \sup \Big\{ \int_{[m,\ell)} \phi(x)\phi(z_i) : z_i \in K_n, \operatorname{supp}(z_i) \in [n_{i-1}+1, n_i] \Big\} \\
&\leq \sup \Big\{ \int_{[m,\ell)} \phi(x)\phi\big(\sum_{i=1}^N y(i) z_i\big) : z_i \in K_n \text{ for all } i, \text{ and} \\
&\quad \operatorname{supp}(z_1) \leq \tilde{n}_1 < \operatorname{supp}(z_2) \leq \tilde{n}_2 < \cdots \leq \operatorname{supp}(z_N) \text{ for some } (\tilde{n}_i) \subseteq \mathbb{N} \Big\} \\
&= \sup_{z \in K_{n+1}} \int_{[m,\ell)} \phi(x)\phi(z)
\end{aligned}$$

which finishes the inductive proof of (iii).

Now using Lemma 6, we obtain that for every $x = (x(i)) \in c_{00}$,

$$\begin{aligned}\|x\|_y &= \sup_{z=(z(i))\in K} \sum_{i=1}^{\infty} x(i)z(i) \\ &= \sup_{z=(z(i))\in K} \int_0^{\infty} \phi(x)\phi(z) \text{ (by Lemma 6 (i))} \\ &= \sup_{g\in L} \int_0^{\infty} \phi(x)g \text{ (by Lemma 6 (iii))} \\ &= |\phi(x)|_y.\end{aligned}$$

Thus $\phi|_{c_{00}} : (c_{00}, \|\cdot\|_y) \to (\mathcal{B}, |\cdot|_y)$ is an isometric embedding.
Step 2: We define a norm $|||\cdot|||_y$ on $L_2[0,\infty)$ in terms of its dual ball as follows. Let

$$L_0' = \{f \in L_2[0,\infty) : \|f\|_2 \leq 1, \text{supp}(f) \subseteq [\alpha, \alpha+1) \text{ for some } \alpha \geq 0\}.$$

If $n \in \mathbb{N} \cup \{0\}$ and L_n' has been defined then

$$L_{n+1}' = L_n' \cup \{\sum_{i=1}^{\infty} y(i) f_i : f_i \in L_n' \text{ for all } i, \text{supp}(f_1) < \text{supp}(f_2) < \cdots < \text{supp}(f_N)\}.$$

Let $L' = \cup_{n=0}^{\infty} L_n'$ and for a function $f \in L_2[0,\infty)$ define

$$|||f|||_y = \sup_{g\in L'} \int_0^{\infty} fg.$$

Notice that L' is a subset of the unit ball of $(L_2[0,\infty), \|\cdot\|_2)$ thus $|||\cdot|||_y \leq \|\cdot\|_2$. Also notice that for every $g \in L'$ there exists a finite tree $\mathcal{T}$ and functions $(g_t)_{t\in\mathcal{T}}$ with

$$g_\emptyset = g, \tag{3}$$

$$\begin{aligned}&g_t = \textstyle\sum_{i=1}^{N} y(i) g_{t_i} \text{ with supp}(g_{t_1}) < \text{supp}(g_{t_2}) < \cdots < \text{supp}(g_{t_N}) \\ &\text{for every } t \in \mathcal{T} \setminus \max(\mathcal{T}), \text{ where } t^+ = \{t_1, \ldots, t_N\},\end{aligned} \tag{4}$$

and

$$\begin{aligned}&\|g_t\|_2 \leq 1 \text{ and there exists } \alpha_t \geq 0 \text{ such that supp}(g_t) \subseteq [\alpha_t, \alpha_t + 1) \\ &\text{for every } t \in \max(\mathcal{T}).\end{aligned} \tag{5}$$

We now show that L' is a dense subset of the unit ball of $(L_2[0,\infty), \|\cdot\|_2)$. This would imply that $|||\cdot|||_y$ is equal to $\|\cdot\|_2$.

Let $f \in \mathcal{B}$ with $\|f\|_2 \leq 1$. We define a (possibly infinite) tree $\mathcal{T}$, a $\mathcal{T}$-decomposition $(E_t)_{t\in\mathcal{T}}$ of $[\inf(\text{supp}(f)), \sup(\text{supp}(f)))$, and for every $t \in \mathcal{T}$ we define $f_t \in L_2[0,\infty)$ with $\text{supp}(f_t) \subseteq E_t$ as follows. Let $f_\emptyset = f$ and $E_\emptyset = [\inf(\text{supp}(f)), \sup(\text{supp}(f)))$. Assume that f_t has been defined for some $t \in \mathcal{T}$, and $\|f_t\|_2 = \|f\|_2$. Then t is a maximal node of $\mathcal{T}$ if there exists $\alpha \geq 0$ such that $\mathrm{supp}\,(f_t) \subseteq [\alpha, \alpha+1)$. Otherwise, let $\alpha_0 = \inf(E_t)$ and define $\alpha_0 < \alpha_1 < \alpha_2 < \cdots < \alpha_N = \sup(E_t)$ (where $N = \#\mathrm{supp}\,(y)$) such that $\|f_t \chi_{[\alpha_{i-1},\alpha_i)}\|_2 = y(i)\|f_t\|_2$ for all $i = 1, \ldots, N$. This task is feasible since $\sum_{i=1}^n y(i)^2 = 1$ and the function $\alpha \mapsto \|f_t \chi_{[\alpha_{i-1},\alpha)}\|_2$ is continuous. Then define $E_{t_i} = [\alpha_{i-1}, \alpha_i)$ and $f_{t_i} = \|f_t\|_2 f_t \chi_{[\alpha_{i-1},\alpha_i)} / \|f_t \chi_{[\alpha_{i-1},\alpha_i)}\|_2$ for $i = 1, \ldots, N$ and $t^+ = \{t_1, \ldots, t_N\}$. Notice that $\|f_{t_i}\|_2 = \|f_t\|_2 = \|f\|_2$ for all i, and

$$\sum_{i=1}^N y(i) f_{t_i} = \sum_{i=1}^N y(i)\|f_t\|_2 \frac{f_t \chi_{[\alpha_{i-1},\alpha_i)}}{\|f_t \chi_{[\alpha_{i-1},\alpha_i)}\|_2} = \sum_{i=1}^N f_t \chi_{[\alpha_{i-1},\alpha_i)} = f_t.$$

It is easy to see by induction on $n = 0, 1, 2, \ldots$ that for every $t \in \mathcal{T}$ which has n predecessors, $\|f\chi_{E_t}\|_2 = \|f\|_2 y(j_1) y(j_2) \cdots y(j_n)$ for some $j_1, j_2, \ldots, j_n \in \{1, \ldots, N\}$ (which depend on t). Thus

$$\|f\chi_{E_t}\| \to 0 \text{ as } n \text{ (the number of predecessors of } t\text{) tends to infinity.} \tag{6}$$

Therefore for any $n \in \mathbb{N}$ the number of t's in $\mathcal{T}$ with n predecessors and the length of E_t greater than 1 does not exceed $\sup(E_\emptyset) - \inf(E_\emptyset)$. Thus by (6) we obtain that

$$g := f - \sum_{\substack{(\text{length of } E_t) > 1 \\ t \text{ has } n \text{ predecessors}}} f\chi_{E_t}$$

is a good approximant of f with n sufficiently big. Moreover, working as above with g we obtain a *finite* tree $\mathcal{T}$ and functions $(g_t)_{t\in\mathcal{T}}$ which satisfy (3), (4) and (5). Thus $g \in L'$ which implies that L' is dense in $\{f \in \mathcal{B} : \|f\|_2 \leq 1\}$ and also in the unit ball of $(L_2[0,\infty), \|\cdot\|_2)$.

Step 3: We now prove that $|||\cdot|||_y$ is equivalent to $|\cdot|_y$. Obviously we have that $|\cdot|_y \leq |||\cdot|||_y$. Let $a := \min_{1\leq i\leq N} y(i)$ and $b := \max_{1\leq i\leq N} y(i)$. Let $M \in \mathbb{N}$ such that $b^M \leq a$. We will show that every function $g \in L'$, the norming set of $|||\cdot|||_y$, can be written as a sum of at most $(2MN+2)$ many functions from L which will finish the proof.

Fix $g \in L'$ and consider an N-tree $\mathcal{T}$ and functions $(g_t)_{t\in\mathcal{T}} \subset L_2[0,\infty)$ which satisfy (3), (4) and (5). In particular, $(g_t)_{t\in\mathcal{T}}$ is a $\mathcal{T}$-decomposition of g. By (5), for every $t \in \max(\mathcal{T})$ there exists $\alpha_t \geq 0$ such that $\mathrm{supp}\,(g_t) \subseteq [\alpha_t, \alpha_t+1)$. Notice that if $\alpha_t \in \mathbb{N}\cup\{0\}$ for every $t \in \max(\mathcal{T})$ and moreover for every $s, t \in \max(\mathcal{T})$ with $s \neq t$ we have that $\alpha_t \neq \alpha_s$ then $g \in L$.

For every $i = 1, 2, \ldots, N$ let $\mathcal{T}_i \subseteq \mathcal{T}$ to be the smallest tree which contains all maximal nodes of $\mathcal{T}$ of the form $(n_1, \ldots, n_k, i)$. Thus all maximal nodes of $\mathcal{T}_i$ have the form $(n_1, \ldots, n_k, i)$ for some $k \in \mathbb{N}$ and $n_j \in \{1, \ldots, N\}$ for $1 \leq j \leq k$. For every $t \in \mathcal{T}$ let E_t to be the smallest interval (closed from the left, open from the right) containing the support of g_t. For $t \in \mathcal{T}\setminus\max(\mathcal{T})$ we have that the length of E_t is larger than 1. For every $i \in \{1, \ldots, N\}$ let $A_i, B \subseteq \max(\mathcal{T})$ with $A_1\cup\cdots\cup A_N \cup B = \max(\mathcal{T})$ be defined by $A_i = \{t \in \max(\mathcal{T}_i) : E_t \cap \mathbb{N} = \emptyset\}$ and $B = \max(\mathcal{T}) \setminus \cup_{i=1}^N A_i$. For every $t \in B$, let $n(t) \in \mathbb{N}\cap E_t$(such $n(t)$ is unique since the length of E_t is at most equal to 1). We have that $\text{supp}(g) \subset \Omega$ where

$$\Omega = \bigcup_{i=1}^{N} \Big\{ \bigcup \{E_t : t \in A_i\} \Big\} \cup \bigcup \{E_t \cap [0, n(t)) : t \in B\} \cup \bigcup \{E_t \cap [n(t), \infty) : t \in B\}.$$

For $i \in \{1, \dots, N\}$ let F_i be the restriction of g on $\cup\{E_t : t \in A_i\}$, G be the restriction of g on $\cup\{E_t \cap [0, n(t)) : t \in B\}$ and H be the restriction of g on $\cup\{E_t \cap [n(t), \infty) : t \in B\}$. Notice any two intervals of the form $E_t \cap [0, n(t))$, where $t \in B$, are separated with an integer. The same is true for any two intervals of the form $E_t \cap [n(t), \infty)$, where $t \in B$. Thus G and H have a $\mathcal{T}$-tree decomposition (naturally inherited from the $\mathcal{T}$-tree decomposition of g), such that if $(G_t)_{t \in \mathcal{T}}$ and $(H_t)_{t \in \mathcal{T}}$ are the tree decompositions of G and H respectively, then the following is satisfied.

For any $s, t \in \max(\mathcal{T})$ with $s \neq t$ and $G_t, G_s \neq 0$, we have that the supports of G_s and G_t are separated by an integer. Similarly, for any $s, t \in \max(\mathcal{T})$ with $s \neq t$ and $H_t, H_s \neq 0$, we have that the supports of H_s and H_t are separated by an integer. Thus G, H belong to L, the norming set for $|\cdot|_y$.

Unfortunately we cannot say the same about F_i. Indeed for $i \in \{1, \dots, N\}$, let $(F_{i,t})_{t \in \mathcal{T}_i}$ be the $\mathcal{T}_i$-tree decomposition of F_i which is naturally inherited by the $\mathcal{T}$-tree decomposition of g. In general for $s, t \in \max(\mathcal{T}_i)$ with $s \neq t$, the supports of $F_{i,s}$ and $F_{i,t}$ may not be separated by an integer. For instance consider the following scenario. Consider two maximal nodes s, t of $\mathcal{T}_i$ with $s^- \prec t^-$. Since $\text{supp}(\tilde{F}_{i,t^-}) \subset \text{supp}(\tilde{F}_{i,s^-})$, and $\text{supp}(\tilde{F}_{i,t^-})$ is not contained in any integer interval we have that $\text{supp}(\tilde{F}_{i,s^-})$ is not contained in any integer interval either. Obviously the inclusions $\text{supp}(\tilde{F}_{i,s}) \subseteq \text{supp}(\tilde{F}_{i,s^-})$ and $\text{supp}(\tilde{F}_{i,t}) \subseteq \text{supp}(\tilde{F}_{i,t^-})$ do not imply that the sets $\text{supp}(\tilde{F}_{i,s})$ and $\text{supp}(\tilde{F}_{i,t})$ are separated by an integer.

The special nodes s, t that we just considered have to be treated in a special way, and that is what we do in what follows.

A sequence $(t_1, t_2, \dots, t_k) \subset \mathcal{T}_i$, $(k \geq 1)$, is called *i-special* if the following are satisfied:

(a) $t_j \in \max(\mathcal{T}_i)$ for $1 \leq j \leq k$.
(b) $t_1^- \prec t_2^- \prec \cdots \prec t_k^-$ (trivially satisfied if $k = 1$).
(c) The sets $\text{supp}(F_{i,t_j})$ for $j \in \{1, \dots, k\}$ are all contained in the same integer interval $[m, m+1)$ for some $m \in \mathbb{N} \cup \{0\}$. In this case we will say that the sequence $(t_1, t_2, \dots, t_k)$ is contained in the integer interval $[m, m+1)$.
(d) The sequence $(t_1, t_2, \dots, t_k)$ is maximal with the properties (a), (b) and (c) (i.e. it is not properly contained in any sequence which satisfies (a), (b) and (c)).

Properties (b), (c), (d) and the fact that $t_j \in A_i$ for all $j \in \{1, 2, \dots, k\}$ ensure that any two i-special sequences intersect in at most the initial node (i.e. if $(t_1, t_2, \dots, t_k)$ and $(s_1, s_2, \dots, s_\ell)$ are two i-special sequences in the same integer interval then they are either disjoint or $s_1 = t_1$). Also (c), (d), the tree structure and our assumption that for all $t \in \mathcal{T} \setminus \max \mathcal{T}$ we have $|E_t| \geq 1$, imply that each interval will contain at most two i-special sequences.

For each $i \in \{1, \dots, N\}$ we will split $\mathcal{T}_i$ into two subtrees $\mathcal{T}_i^L$ the nodes with a leftmost i-special sequences (within each integer interval) and $\mathcal{T}_i^R$ the nodes with a

rightmost i-special sequences (within each integer interval). Thus for any node t in $\mathcal{T}_i^L$ we have that either $E_t = \emptyset$ or $E_t \subseteq [m, m+1)$ for some $m \in \mathbb{N} \cup \{0\}$ and for every other node $s \in \mathcal{T}_i^L$ $E_s \cap [m, m+1) = \emptyset$, and similarly for $\mathcal{T}_i^R$. In other words if $(t_1, t_2, \ldots, t_k) \subset \mathcal{T}_i$ is an i-special sequence in some integer interval $[m, m+1)$ with corresponding supports $E_{t_1}, E_{t_2}, \ldots, E_{t_k}$ then the following holds:
If there exists an i-special sequence with supports in $[m, \min\{E_{t_i} : 1 \leq i \leq k\})$ then $(t_1, t_2, \ldots, t_k) \subset \mathcal{T}_i^R$; otherwise $(t_1, t_2, \ldots, t_k) \subset \mathcal{T}_i^L$.
So $\mathcal{T}_i^L \cap \mathcal{T}_i^R = \emptyset$ and $\mathcal{T}_i^L \cup \mathcal{T}_i^R = \mathcal{T}_i$. Also define

$$L_i = F_i|_{\cup_{t \in \mathcal{T}_i^L} E_t} \text{ and } R_i = F_i|_{\cup_{t \in \mathcal{T}_i^R} E_t}.$$

Thus $F_i = L_i + R_i$.

For $j \in \{1, \ldots, M-1\}$ let $\mathcal{T}_{i,j}^L$ be the smallest subtree of $\mathcal{T}_i^L$ whose maximal nodes are the elements t_j of any i-special sequence $(t_1, \ldots, t_k)$ with $k \geq j$. By the definition of $\mathcal{T}_{i,j}^L$ for $1 \leq j \leq M-1$ we have that if $(t_1, \ldots, t_k)$ is an i-special sequence and $1 \leq j \leq k$ then $t_\ell \notin \mathcal{T}_{i,j}^L$ for $\ell \in \{1, \ldots, k\} \setminus \{j\}$. For every $j \in \{1, \ldots, M-1\}$ let $L_{i,j}$ be the restriction of L_i on $\cup\{E_t : t \in A_i \cap \mathcal{T}_{i,j}\}$. Hence $L_{i,j}$ has a $\mathcal{T}_{i,j}^L$-tree decomposition $(L_{i,j,t})_{t \in \mathcal{T}_{i,j}^L}$ which is naturally inherited by the $\mathcal{T}_i^L$-tree decomposition of L_i. Moreover, by remarks above, we have that no two maximal nodes of $\mathcal{T}_{i,j}^L$ belong in the same i-special sequence. Hence for every two maximal nodes s, t of $\mathcal{T}_{i,j}^L$ we have that the supports of $L_{i,j,s}$ and $L_{i,j,t}$ are separated by an integer. Thus $L_{i,j}$ belongs to the norming set L of $|\cdot|_y$ for every $j \in \{1, \ldots, M-1\}$.

Repeat similarly for $\mathcal{T}_{i,j}^R$ and R_i to get $\mathcal{T}_{i,j}^R$ and $R_{i,j}$ with a $\mathcal{T}_{i,j}^R$-tree decomposition such that each $R_{i,j}$ belongs to the norming set L of $|\cdot|_y$ for every $j \in \{1, \ldots, M-1\}$.

Notice that $\cup_{j=1}^{M-1} \operatorname{supp}(L_{i,j})$ may be a strict subset of $\operatorname{supp}(L_i)$ and thus L_i may not be equal to $\sum_{j=1}^{M-1} L_{i,j}$. This will be the case if there are i-special sequences in $\mathcal{T}_i^L$ that have length $k \geq M$. Now we take care of such long i-special sequences. Fix an i-special sequence $(t_1, \ldots, t_k)$ with $k \geq M$. Let $[m, m+1)$ be the unique integer interval with $\operatorname{supp}(L_{i,t_j}) \subseteq [m, m+1)$ for all $j \in \{1, \ldots, k\}$. Let $L_{i,t_1^-} \restriction_{[m,m+1)}$ and $L_{i,t_M^-} \restriction_{[m,m+1)}$ be the restrictions of L_{i,t_1^-} and L_{i,t_M^-} respectively on $[m, m+1)$. Thus we can write

$$L_{i,t_1^-}|_{[m,m+1)} = \left(\prod y(n_\alpha)^{m_\alpha}\right) L_{i,t_M^-}|_{[m,m+1)}$$

where $n_\alpha \in \{1, \ldots, N\}, m_\alpha \in \mathbb{N}$ and the product $\prod y(n_\alpha)^{m_\alpha}$ has at least M-terms counting multiplicities. Let $\mathcal{T}_{i,M}^L$ be the subtree of $\mathcal{T}_{i,1}^L$ whose maximal nodes are the nodes t_1 for any i-special sequence $(t_1, \ldots, t_k)$ with $k \geq M$. Let

$$\begin{aligned} S_i := \cup\{E_t : t \in \mathcal{T}_i^L &\text{ and there exists an } i\text{-special sequence } (t_1, \ldots, t_k) \text{ with } k \geq M \\ &\text{and } M \leq j \leq k \text{ such that } t = t_j\}. \end{aligned}$$

Thus $L_{i,M} := L_i|_{S_i}$ has a $\mathcal{T}_{i,M}^L$-tree decomposition $(L_{i,M,t})_{t \in \mathcal{T}_{i,M}^L}$, where for any i-special sequence $(t_1, \ldots, t_k)$ with $k \geq M$ we set

$$L_{i,M,t_1} := \frac{(\prod y(n_\alpha)^{m_\alpha})}{y(i)} L_{i,t_M^-}|_{[m,m+1)},$$

(where $[m, m+1)$ is the unique integer interval which contains the sets supp (L_{i,t_j}) for $j \in \{1, \ldots, k\}$) and $L_{i,M,t_1^-} := L_{i,t_1^-}|_{[m,m+1)}$. Recall that in the beginning of Step 3 we defined a, b and M by: $a := \min_{1 \leq \ell \leq N} y(\ell)$, $b := \max_{1 \leq \ell \leq N} y(\ell)$ and $M \in \mathbb{N}$ such that $b^M \leq a$. Since the product $\prod y(n_\alpha)^{m_\alpha}$ has at least M-terms counting multiplicities, $\prod y(n_\alpha)^{m_\alpha} \leq b^M \leq a \leq y(i)$. Hence $\|L_{i,t_1^-}|_{[m,m+1)}\|_2 \leq y(i)\|L_{i,t_M^-}|_{[m,m+1)}\|_2$ and $\|L_{i,t_M^-}|_{[m,m+1)}\|_2 \leq \|L_{i,t_M^-}\|_2 \leq 1$ by (5). Hence $L_{i,M}$ belongs to L, the norming set of $|\cdot|_y$. Similarly we can create $R_{i,M}$.

We have decomposed the arbitrary element g of the norming set L' of $|||\cdot|||_y$ into the sum of the functions G, H, $L_{i,j}$, $R_{i,j}$ (where $i \in \{1, \ldots, N\}$ and $j \in \{1, \ldots, M\}$), a total of $(2MN+2)$ many functions of the norming set L of $|\cdot|_y$. This implies that $|||\cdot|||_y \leq (2MN+2)|\cdot|_y$ and finishes the proof. □

References

[1] S. A. Argyros and I. Deliyanni, *Banach spaces of the type of Tsirelson*, Banach Archive arXiv:math.FA (1992).

[2] S. F. Bellenot, *Tsirelson superspaces and l_p*, J. Funct. Anal. 69 (1986), pp. 207–228.

[3] J. Bernués and I. Deliyanni, *Families of finite subsets of $\mathbb{N}$ of low complexity and Tsirelson type spaces*, Math. Nachr. 222 (2001), pp. 15–29.

[4] W. T. Gowers, *Ramsey methods in Banach spaces*, Handbook of the geometry of Banach spaces, Vol. 2, North-Holland, Amsterdam, 2003, pp. 1071–1097.

[5] E. Odell and Th. Schlumprecht, *The distortion problem*, Acta Math. 173 (1994), pp. 259–281. MR MR1301394 (96a:46031)

[6] ———, *Distortion and stabilized structure in Banach spaces; new geometric phenomena for Banach and Hilbert spaces*. Proceedings of the International Congress of Mathematicians, Vol. 1, 2 (Zürich, 1994), pp. 955–965. Birkhäuser, Basel, 1995.

[7] A. M. Pelczar, *Stabilization of Tsirelson-type norms on ℓ_p-spaces*, Proc. Amer. Math. Soc. 135 (2007), pp. 1365–1375.

Author information

George Androulakis, Department of Mathematics, University of South Carolina, Columbia, SC 29208, USA.
Email: `giorgis@math.sc.edu`

Frank Sanacory, Department of Mathematics, University of South Carolina, Columbia, SC 29208, USA.
Email: `sanacory@math.sc.edu`

Banach Spaces and their Applications in Analysis, 251–264

Ball proximinality in Banach spaces

Pradipta Bandyopadhyay, Bor-Luh Lin, and T. S. S. R. K. Rao

Abstract. In this paper we study the notion of ball proximinality. We exhibit several classes of Banach spaces that are ball proximinal. We show that the unit ball of an M-ideal is proximinal in the unit ball of the whole space. We also define a stronger notion of ball proximinality and show that in a locally uniformly convex space, the unit ball is strongly ball proximinal. We show that strong proximinality does not imply ball proximinality. We also consider ball proximinality for spaces of vector-valued continuous and Bochner integrable functions

Key words. Ball proximinality, proximinality.

AMS classification. Primary 41A65, 41A50; Secondary 46B20.

Dedicated to Professor Nigel Kalton on his 60th birthday!

1 Introduction

Let X be a Banach space and let $K \subseteq X$ be a closed set. For $x \in X$ we denote the distance function of K at x by $d(x, K) = \inf\{\|x - k\| : k \in K\}$. The metric projection of x onto K is $P_K(x) = \{k \in K : \|x - k\| = d(x, K)\}$. The set K is called *proximinal* (respectively, *Chebyshev*) if for every $x \in X \setminus K$, $P_K(x)$ is nonempty (respectively, singleton).

Proximinality for a subspace $Y \subseteq X$ is a well studied concept in Banach space theory, see the monograph [10]. Simple examples of proximinal subspaces include finite dimensional subspaces, reflexive subspaces, weak*-closed subspaces of dual spaces and hyperplanes defined by norm-attaining functionals. For first three sets of examples, by compactness, one has also that the closed unit ball B_Y is proximinal in X.

Saidi [9] recently proved that there are Banach spaces X with a closed proximinal hyperplane Y for which B_Y is not proximinal. Motivated by this we consider the following notion of ball proximinality, which was called 'locally proximinal' in [8].

Definition 1.1. A closed subspace $Y \subseteq X$ is said to be *ball proximinal* if B_Y is proximinal in X.

It is easy to see (Proposition 2.4) that any ball proximinal subspace is proximinal. In this paper we undertake a detailed study of ball proximinality. We show that several stronger notions of proximinality studied in the literature turn out to be ball proximinal. We also consider two other notions of proximinality closely related to ball proximinality. We show that if $Y \subseteq X$ is a M-ideal then B_Y is proximinal in B_X.

Research partially supported by a DST-NSF project grant DST/INT/US(NSF-RPO-0141)/2003 and NSF/OISE-03-52523.

Motivated by the notion of strong proximinality of subspaces studied in [3], we introduce the notion of strong ball proximinality and show that it implies strong proximinality. We give an example to show that strong proximinality does not imply ball proximinality. We show that in a strictly convex dual space with the Namioka–Phelps property, any weak*-closed subspace is strongly ball proximinal. We also study the continuity property of the metric projection in proximinal and ball proximinal situation.

An interesting question related to the notion of strong ball proximinality is, for which Banach spaces X is B_X a strongly proximinal set? We show that this is the case in particular for locally uniformly convex spaces.

Saidi's example [9, Theorem 1] is a proximinal hyperplane that fails to be ball proximinal. Thus it is interesting to investigate situations where proximinal subspaces of finite co-dimension or more generally factor reflexive (*i.e.*, the quotient space X/Y is reflexive) proximinal subspaces are ball proximinal. In this direction we show that for the c_0-direct sum of any family of reflexive Banach spaces, any factor reflexive proximinal subspace is ball proximinal.

As for the stability of ball proximinality, we note that it is preserved by c_0-direct sums and also by vector-valued continuous functions when the metric projection has a continuous selection. We do not know if ball proximinality is preserved by ℓ^1 sums. We give some examples from the theory of L-embedded spaces [4] where it does get preserved in the space of Bochner integrable functions.

It may be noted that several standard techniques from the theory of best approximations by closed subspaces, like translating by a vector from the subspace Y or the mini-max theorem, $d(x,Y) = \|x|_{Y^\perp}\|$ are no longer available in the case of ball proximinality. However we do have a mini-max result of a restricted nature (Lemma 2.1).

Our investigations lead to several natural questions about proximinality that are scattered throughout the paper.

2 General results

We work with *real* scalars. For a Banach space X, B_X and S_X will denote the closed unit ball and the unit sphere of X, respectively. We will denote by $B[x,r]$ (respectively, $B(x,r)$) the closed (respectively, open) ball of radius $r > 0$ centered at $x \in X$. By a subspace, we mean a norm closed linear subspace.

Let $Y \subseteq X$ be a subspace and $x_0 \in X \setminus B_Y$. We begin with a minimax type expression for the distance $d(x_0, B_Y)$ and use it to obtain a necessary and sufficient condition for $z_0 \in P_{B_Y}(x_0)$.

Lemma 2.1. *Let $Y \subseteq X$ be a subspace and $x_0 \in X \setminus B_Y$.*

(a) $d(x_0, B_Y) = \sup\{f(x_0) - \|f|_Y\| : f \in B_{X^*}\}$.

(b) $z_0 \in P_{B_Y}(x_0)$ if and only if there exists $f_0 \in S_{X^*}$ such that f_0 attains its norm on X at $x_0 - z_0$ and $f_0|_Y$ attains its norm at z_0.

Proof. *(a)* For $f \in B_{X^*}$ and $y \in B_Y$, $f(x_0) - f(y) \leq \|x_0 - y\|$. Taking the infimum over $y \in B_Y$, we get that for any $f \in B_{X^*}$, $f(x_0) - \|f|_Y\| \leq d(x_0, B_Y)$. Hence

$$\sup\{f(x_0) - \|f|_Y\| : f \in B_{X^*}\} \leq d(x_0, B_Y).$$

On the other hand, by definition, $B(x_0, d(x_0, B_Y))$ is disjoint from B_Y, therefore, there exists $f_0 \in S_{X^*}$ such that $\inf f_0(B(x_0, d(x_0, B_Y))) \geq \sup f_0(B_Y)$. Therefore,

$$f_0(x_0) - d(x_0, B_Y) \geq \|f_0|_Y\|.$$

Thus, equality holds.

(b) If $z_0 \in P_{B_Y}(x_0)$ and $f_0 \in S_{X^*}$ is as above, it is clear that f_0 attains its norm on X at $x_0 - z_0$ and $f_0|_Y$ attains its norm at z_0.

Conversely, if there exists $f_0 \in S_{X^*}$ such that f_0 attains its norm on X at $x_0 - z_0$ and $f_0|_Y$ attains its norm at z_0, then by *(a)*,

$$d(x_0, B_Y) \geq f_0(x_0) - \|f_0|_Y\| = f_0(x_0) - f_0(z_0) = \|x_0 - z_0\| \geq d(x_0, B_Y).$$

Hence, $z_0 \in P_{B_Y}(x_0)$. □

Remark 2.2. From the above proof it follows that:

(a) If $z_0 \in P_{B_Y}(x_0) \setminus P_Y(x_0)$, then $\|z_0\| = 1$. Indeed, if $\|z_0\| < 1$, then $f_0|_Y \equiv 0$ and hence, $d(x_0, B_Y) = d(x_0, Y)$ and $z_0 \in P_Y(x_0)$.

(b) If X is strictly convex and $x_0 \in X \setminus Y$, then $\|f_0|_Y\| < \|f_0\|$. Otherwise, f_0 attains its norm both at $x_0 - z_0$ and z_0, forcing $x_0 = \alpha z_0$ for some α.

We now show that ball proximinality implies proximinality. We need the following lemma, which is also of independent interest.

Lemma 2.3. *Let Y be a subspace of a Banach space X.*

(a) For all $x \in X$ and $\lambda > 0$, $\lambda P_{B_Y}(\lambda^{-1}x) = P_{\lambda B_Y}(x)$.

(b) For all $x \in X$ and $\lambda \geq \|x\| + d(x, Y)$, $P_Y(x) \subseteq P_{\lambda B_Y}(x)$.

(c) For all $x \in X$ and $\lambda > \|x\| + d(x, Y)$, $P_Y(x) = P_{\lambda B_Y}(x)$.

(d) If P_{B_Y} is continuous then P_Y is continuous.

Proof. *(a)* Let $x \in X$, $\lambda > 0$, and $y_0 \in P_{B_Y}$. For any $y \in B_Y$, $\|\lambda^{-1}x - y_0\| \leq \|\lambda^{-1}x - y\|$ is equivalent to $\|x - \lambda y_0\| \leq \|x - \lambda y\|$. Hence $\lambda P_{B_Y}(\lambda^{-1}x) = P_{\lambda B_Y}(x)$.

(b) Let $\lambda \geq \|x\| + d(x, Y)$. For $y \in P_Y(x)$, $\|y\| \leq \|x\| + \|x - y\| = \|x\| + d(x, Y) \leq \lambda$. Hence $P_Y(x) \subseteq P_{\lambda B_Y}(x)$.

(c) Let $x \in X$ and $d = d(x, Y)$. Let $\lambda > \|x\| + d$ and $y_0 \in P_{\lambda B_Y}(x)$. It is easy to see that $d(x, Y) = \inf\{\|x - y\| : y \in Y \cap B[x, d + \delta]\}$ for any $\delta > 0$. In particular, we may take $\delta = \lambda - (\|x\| + d)$.

For any $y \in Y$, if $y \in Y \cap B[x, d + \delta]$, then $y \in \lambda B_Y$. Hence $\|x - y\| \geq \|x - y_0\|$. It follows that $y_0 \in P_Y(x)$. Thus, $P_{\lambda B_Y}(x) \subseteq P_Y(x)$. Equality follows from *(b)*.

(d) Follows since P_{B_Y} is continuous if and only if $P_{\lambda B_Y}$ is continuous. □

Proposition 2.4. *Let Y be a subspace of a Banach space X. Consider the following statements:*

(a) Y is ball proximinal in X.

(b) B_Y is proximinal in B_X.

(c) Y is proximinal in X.

(d) B_Y is proximinal in $2^{-1}B_X \cup Y$.

Then (a) $\Rightarrow$ (b) $\Rightarrow$ (c) $\Leftrightarrow$ (d).

Proof. Clearly, *(a)* $\Rightarrow$ *(b)*.

(b) $\Rightarrow$ *(c)*. Let $x \in X$ and $\lambda > \|x\| + d(x, Y)$. By *(b)*, λB_Y is proximinal in λB_X. As $x \in \lambda B_X$ and $P_{\lambda B_Y}(x) \neq \emptyset$, by Lemma 2.3*(c)*, $P_Y(x) = P_{\lambda B_Y}(x) \neq \emptyset$.

(c) $\Rightarrow$ *(d)*. As noted above, if $\|x\| \leq 1/2$ and $y \in P_Y(x)$, then $\|y\| \leq d(x, Y) + \|x\| \leq 2\|x\| \leq 1$. Hence *(d)* follows.

(d) $\Rightarrow$ *(c)*. Let $x \in X \setminus Y$. We may assume that $\|x\| > d(x, Y)$. By *(d)*, $P_{B_Y}(x/(2\|x\|)) \neq \emptyset$, and hence, $P_{2\|x\|B_Y}(x) \neq \emptyset$. Since $2\|x\| > \|x\| + d(x, Y)$, by Lemma 2.3*(c)*, $P_Y(x) = P_{2\|x\|B_Y}(x) \neq \emptyset$. □

Example 2.5. We note that in Saidi's example [9], Y is not ball proximinal in X but B_Y is proximinal in B_X and therefore, *(b)* $\not\Rightarrow$ *(a)*.

Indeed, it is easy to see from the definition of the set K in [9] that if $y + \lambda z_0 \in K$, then $y \in B_Y$ and $|\lambda| \leq 1$. If $x = y + \lambda z_0$ and $\|x\| \leq 1$, then by [9, Lemma 1], $y \in P_Y(x)$ and $\|y\| \leq 1$.

Remark 2.6. We conjecture that the other reverse implication, that is, *(c)* $\Rightarrow$ *(b)*, is also in general false.

The following lemma extends [9, Proposition 1] and shows that several natural summands are ball proximinal. We recall that when $\|x\| \geq 1$, $d(x, B_X) = \|x\| - 1 = \left\|x - \|x\|^{-1}x\right\|$.

Lemma 2.7. *Let $P : X \to X$ be a continuous linear projection with range Y.*

(a) If $\|I - P\| = 1$, then Y is proximinal in X.

(b) If P is bicontractive, i.e., *$\|P\| = \|I - P\| = 1$, then B_Y is proximinal in B_X.*

(c) If there exists a monotone map $\varphi : \mathbb{R}^+ \times \mathbb{R}^+ \to \mathbb{R}^+$ with $\|x\| = \varphi(\|P(x)\|, \|x - P(x)\|)$ for all $x \in X$, then Y is ball proximinal.

Proof. *(a)* Let $x \in X$. For any $y \in Y$, $\|x - y\| \geq \|(I - P)(x - y)\| = \|x - P(x)\|$. Hence $P(x) \in P_Y(x)$.

(b) Let $x \in B_X$. By *(a)*, $P(x) \in P_Y(x)$ and since $\|P\| = 1$, $P(x) \in B_Y$. Thus $P(x) \in P_{B_Y}(x)$.

(c) Let $x \notin B_Y$. If $P(x) \in B_Y$, we are done. Otherwise, we will show that $P(x)/\|P(x)\| \in P_{B_Y}(x)$. Let $y \in B_Y$, by the monotonicity of φ, we have

$$\begin{aligned}\|x-y\| &= \varphi(\|P(x)-y\|, \|x-P(x)\|)\\ &\geq \varphi(\|P(x)-\frac{P(x)}{\|P(x)\|}\|, \|x-P(x)\|) = \|x-\frac{P(x)}{\|P(x)\|}\|,\end{aligned}$$

as desired. □

Remark 2.8. It is clear from the above lemma that if $Y \subseteq X$ is a complemented subspace (in particular, if Y is of finite dimension or co-dimension) then X can be renormed so that Y is ball proximinal in the new norm. We do not know in general how to renorm X (even with additional conditions) so that Y is ball proximinal in the new norm. We however have the following.

Proposition 2.9. *Let Y be a subspace of a Banach space X. Then there is a renorming of X that does not affect the norm on Y and B_Y is proximinal in B_X.*

Proof. Given a subspace $Y \subseteq X$ and $\varepsilon > 0$, define $\|x\|_\varepsilon = \max\{\|x\|, (1+\varepsilon)d(x,Y)\}$. It is easy to see that Y is proximinal in X in $\|\cdot\|_\varepsilon$ and the norm on Y remains unaffected. Thus, we may assume that Y is proximinal in X.

Define a new norm on X as $\|x\|_1 = \|x\| + d(x,Y)$, where $\|\cdot\|$ is the original norm on X. Then, clearly, the norm on Y remains unaffected. And for any $x \in X\backslash Y$ and $y \in Y$, $\|x-y\|_1 = \|x-y\| + d(x,Y)$. Thus, $d_1(x,Y) =: \inf\{\|x-y\|_1 : y \in Y\} = 2d(x,Y)$. Since Y is proximinal in the $\|\cdot\|$-norm, there is $z \in Y$ such that $\|x-z\| = d(x,Y)$. It follows that $\|x-z\|_1 = \|x-z\| + d(x,Y) = 2d(x,Y) = d_1(x,Y)$. Moreover, $\|z\|_1 = \|z\| \leq \|x\| + d(x,Y) = \|x\|_1$. Hence the result. □

Summands as in Lemma 2.7 can be used to get more examples of ball proximinal subspaces as the following proposition illustrates.

Proposition 2.10. *Let $Y \subseteq X$ be the kernel of a norm one projection P. Let $Z \subseteq P(X)$ be a subspace. If $Y + Z$ is ball proximinal in X then Z is ball proximinal in $P(X)$.*

Proof. Let $x \in P(X)$. Then $d(x, B_Z) \geq d(x, B_{(Y+Z)}) = \|x - y_0 - z_0\|$ for some $y_0 + z_0 \in B_{(Y+Z)}$. Now $\|x - y_0 - z_0\| \geq \|P(x - y_0 - z_0)\| = \|x - z_0\|$. Also $\|z_0\| = \|P(y_0 + z_0)\| \leq 1$. Therefore Z is ball proximinal in $P(X)$. □

Remark 2.11. Let $Y \subseteq X$ be a subspace and $\pi : X \to X/Y$ be the quotient map. Suppose $Z \subseteq X/Y$ is a subspace such that $\pi^{-1}(Z)$ is proximinal. Then it is well known that Z is proximinal [6, Lemma 2.11]. The above is a ball proximinal analogue of this simple proximinality result, under additional hypothesis. We do not know whether ball proximinality of $\pi^{-1}(Z)$ implies the ball proximinality of Z, even if Y is reflexive.

It is well known that if Y is an M-ideal in X, then Y is proximinal [4, Proposition II.1.1]. In fact, more is true.

Proposition 2.12. *If Y is an M-ideal in X, then B_Y is proximinal in B_X.*

Proof. We adapt the proof of [4, Proposition II.1.1]. Let $x \in B_X \setminus B_Y$ and $d = d(x, B_Y)$.

Claim: Let $\varepsilon > 0$. Let $y \in B_Y$ be such that $\|y - x\| \leq d + \varepsilon$. Then, there is a $z \in B_Y$ such that

$$\|y - z\| \leq 3\varepsilon/4 \quad \text{and} \quad \|z - x\| \leq d + 3\varepsilon/4.$$

To prove the claim, consider the closed balls $B[x, d + \varepsilon/2]$, $B[y, \varepsilon/2]$ and $B_X = B[0, 1]$. Since the distance between the centers of the first two balls is less than the sum of the radii, they intersect. Moreover, since $x, y \in B_X$, the line segment $[x, y] \subseteq B_X$. And since there must be a point of intersection on $[x, y]$, all the three balls intersect. And clearly, all the three balls also intersect Y. Thus, by the 3-ball property [4, Theorem I.2.2], there is

$$u \in Y \cap B[x, d + 5\varepsilon/8] \cap B[y, 5\varepsilon/8] \cap B[0, 1 + \varepsilon/8].$$

If $u \in B_Y$, $z = u$ works. If $1 < \|u\| \leq 1 + \varepsilon/8$, put $z = u/\|u\|$. Then, $\|u - z\| = \|u\| - 1 \leq \varepsilon/8$. It follows that $\|y - z\| \leq \|y - u\| + \|u - z\| \leq 5\varepsilon/8 + \varepsilon/8 = 3\varepsilon/4$ and similarly, $\|x - z\| \leq d + 3\varepsilon/4$, proving the claim.

Repeatedly applying the claim with $y = y_n$ and $\varepsilon = (3/4)^{n-1}$ and putting $y_{n+1} = z$, we get a sequence $\{y_n\} \subseteq B_Y$ such that for all $n \geq 1$,

$$\|y_{n+1} - y_n\| \leq \left(\frac{3}{4}\right)^n \quad \text{and} \quad \|y_n - x\| \leq d + \left(\frac{3}{4}\right)^{n-1}.$$

Now, the first inequality implies $\{y_n\}$ is a Cauchy sequence with limit $y_0 \in B_Y$ (say), while the second implies that $\|y_0 - x\| \leq d$, so that $y_0 \in P_{B_Y}(x)$. □

Question 2.13. *If Y is an M-ideal in X, then is Y ball proximinal in X?*

3 Strong proximinality vis-a-vis ball proximinality

The notion of strongly proximinal subspaces was defined in [3]. It was extended to subsets in [1] as follows:

Definition 3.1. For a closed set $K \subseteq X$, $x \in X$ and $\delta > 0$, let

$$P_K(x, \delta) = \{z \in K : \|z - x\| < d(x, K) + \delta\}.$$

A closed set $K \subseteq X$ is *strongly proximinal* at $x \in X \setminus K$ if K is proximinal at x and for any $\varepsilon > 0$, there exists $\delta > 0$ such that $P_K(x, \delta) \subseteq P_K(x) + B(0, \varepsilon)$. If this holds for every $x \in X \setminus K$, then K is said to be strongly proximinal in X.

Proposition 3.2. *Let Y be a subspace of a Banach space X.*

(a) For all $x \in X$, $\delta > 0$ and $\lambda > 0$, $P_{\lambda B_Y}(x, \delta) = \lambda P_{B_Y}(\lambda^{-1}x, \lambda^{-1}\delta)$.

(b) For all $x \in X$ and $\lambda > 0$, λB_Y is strongly proximinal at $x \in X$ if and only if B_Y is strongly proximinal at x/λ.

(c) If B_Y is strongly proximinal in X, then so is Y.

Proof. *(a)* Follows from arguments similar to Lemma 2.3.

(b) Easily follows from *(a)*.

(c) Let $x \in X$ and $\lambda > \|x\| + d(x, Y)$. If B_Y is strongly proximinal at x/λ, then λB_Y is strongly proximinal at $x \in X$. Arguing as in Lemma 2.3, for $0 < \delta < \lambda - (\|x\| + d(x, Y))$, $P_Y(x, \delta) \subseteq \lambda B_Y$ and hence, $P_Y(x, \delta) = P_{\lambda B_Y}(x, \delta)$. Moreover, $P_{\lambda B_Y}(x) = P_Y(x)$. Hence Y is strongly proximinal at x. □

Example 3.3. The example in [9] is strongly proximinal, and therefore, strong proximinality does not imply ball proximinality.

Proof. Let Y be a non-reflexive Banach space with a non-norm attaining functional $y^* \in S_{Y^*}$. Let $A = 2B_Y \cap \{y \in Y : y^*(y) = 1\}$. Then, $d(A, B_Y) = 0$, but $A \cap B_Y = \emptyset$. Let $a \in A$ and $R = A - a$. Choose $\gamma \in \mathbb{R}^+$ such that $\gamma^{-1}(\pm R) \subseteq B_Y$. Let $X = Y \oplus \mathbb{R}$. Let $z_0 = 1$. In X, consider the set $K = \overline{co}\{\gamma^{-1}R - z_0, B_Y, -\gamma^{-1}R + z_0\}$ and let the norm on X be defined by the Minkowski functional of K. In this norm, [9, Lemma 1] proves that Y is proximinal in X with $P_Y(y + \lambda z_0) = y + \gamma^{-1}|\lambda|R$ and $d(y + \lambda z_0, Y) = |\lambda|$. Moreover, Y is not ball proximinal.

Clearly, $P_Y(y + \lambda z_0) = y + \lambda P_Y(z_0)$ and $P_Y(y + \lambda z_0, \delta) = y + \lambda P_Y(z_0, \delta/|\lambda|)$. Thus, to show that Y is strongly proximinal in X, it suffices to show that Y is strongly proximinal at z_0. Recall that $P_Y(z_0) = \gamma^{-1}R$ and $d(z_0, Y) = 1$. Let $\varepsilon > 0$ and $\delta > 0$ such that $\delta < \varepsilon$. We will show that $P_Y(z_0, \delta) \subseteq P_Y(z_0) + B(0, \varepsilon)$, or, equivalently, for any $y_0 \in Y$ with $\|y_0 - z_0\| < 1 + \delta$, $d(y_0, \frac{1}{\gamma}R) < \varepsilon$.

So, let $y_0 \in Y$ and $\rho = \|y_0 - z_0\| < 1 + \delta$. We may assume that $\rho > 1$. Then $(y_0 - z_0)/\rho \in K$. Hence, as in the proof of [9, Lemma 1], there exists $\alpha_n, \beta_n, \nu_n \in [0, 1]$ with $\alpha_n + \beta_n + \nu_n = 1$ and $u_n \in \gamma^{-1}R$, $u_n^- \in -\gamma^{-1}R$ and $y_n \in B_Y$ such that

$$\frac{y_0 - z_0}{\rho} = \lim_n [\alpha_n u_n + \beta_n u_n^- + \nu_n y_n + (\beta_n - \alpha_n) z_0].$$

Passing to subsequences, if necessary, we may assume that $\alpha_n \to \alpha$, $\beta_n \to \beta$ and $\nu_n \to \nu$, where $\alpha, \beta, \nu \in [0, 1]$ and $\alpha + \beta + \nu = 1$. It follows that

$$\alpha - \beta = \frac{1}{\rho} \tag{3.1}$$

and

$$y_0 = \lim_n \rho[\alpha u_n + \beta u_n^- + \nu y_n]. \tag{3.2}$$

By (3.1), $1 \geq \alpha = 1/\rho + \beta \geq 1/\rho$. Now since $u_n, u_n^-, y_n \in B_Y$, we get from (3.2) that,

$$\begin{aligned} d(y_0, \frac{1}{\gamma}R) &\leq \inf_n \|y_0 - u_n\| \leq \liminf_n \|\rho[\alpha u_n + \beta u_n^- + \nu y_n] - u_n\| \\ &\leq \liminf_n \|(\rho\alpha - 1)u_n + \rho\beta u_n^- + \rho\nu y_n\| \\ &\leq (\rho\alpha - 1) + \rho\beta + \rho\nu = \rho - 1 < \delta < \varepsilon. \end{aligned}$$

This completes the proof. □

We next exhibit some examples of subspaces for which B_Y is strongly proximinal. We recall some definitions from [1].

Definition 3.4. A sequence $\{z_n\} \subseteq K$ is called a *minimizing sequence* for $x \in X \setminus K$, if $\|x - z_n\| \to d(x, K)$.

Let $K \subseteq X$ be a closed subset and $x_0 \in X \setminus K$. We say that K is *strongly Chebyshev* (respectively, *approximatively compact*) at x_0 if every minimizing sequence $\{z_n\} \subseteq K$ for x_0 is convergent (respectively, has a convergent subsequence).

If K is strongly Chebyshev (or, approximatively compact) at every $x \in X \setminus K$, we say K is strongly Chebyshev (respectively, approximatively compact) in X.

The following result is observed in [1, Theorems 2.2, 2.3, 2.5].

Theorem 3.5. *Let $K \subseteq X$ be a closed subset and $x_0 \in X \setminus K$. Then*

(a) K is strongly Chebyshev at x_0 if and only if K is strongly proximinal at x_0 and $P_K(x_0)$ is a singleton. In this case, the metric projection P_K is continuous at x_0.

(b) K is approximatively compact at x_0 if and only if K is strongly proximinal at x_0 and $P_K(x_0)$ is compact.

Theorem 3.6. *Suppose X^* is a strictly convex space with the Namioka–Phelps property, i.e., weak* and norm topologies coincide on S_{X^*}) and $Y \subseteq X^*$ is a w^*-closed subspace. Then B_Y is strongly proximinal in X^* and P_{B_Y} is continuous on X^*.*

Proof. By Theorem 3.5*(a)*, it suffices to show that B_Y is strongly Chebyshev in X^*.

Let $f \in X^* \setminus Y$. Let $\{f_n\} \subseteq B_Y$ be a minimizing sequence for f. If $f_0 \in B_Y$ is a w^*-cluster point of $\{f_n\}$, then by the w^*-lower semicontinuity of the norm,

$$d(f, B_Y) \leq \|f - f_0\| \leq \liminf \|f - f_n\| = d(f, B_Y). \tag{3.3}$$

Thus $f_0 \in P_{B_Y}(f)$. By strict convexity of X^*, $P_{B_Y}(f) = \{f_0\}$. That is, f_0 is the unique w^*-cluster point of $\{f_n\}$. Thus, $f_n \to f_0$ in the w^*-topology. Moreover, by (3.3), $\|f - f_n\| \to \|f - f_0\|$. Consequently, by the Namioka–Phelps property, $f_n \to f_0$ in norm. □

It also follows from Theorem 3.5 that any finite dimensional subspace is strongly ball proximinal. Similarly, we get the following corollary. We recall that a Banach space X is said to have the *Property G* if every point of S_X is a denting point. Equivalently these are strictly convex spaces such that weak and norm topologies agree on S_X. Notice in particular that a LUR space has Property G.

Corollary 3.7. *If X has Property G and $Y \subseteq X$ is a reflexive subspace, then B_Y is strongly proximinal in X and P_{B_Y} is continuous on X.*

Since for a Banach space X, B_X is always proximinal, it is interesting to consider situations when it is strongly proximinal. The following result is immediate from [1, Theorem 2.11]. We include a self-contained proof. We note that Banach spaces satisfying the hypothesis below lie strictly between LUR Banach spaces and spaces with Property G [1, Theorem 3.4].

Proposition 3.8. *Let X be a Banach space such that every norm attaining functional in X^* is a strongly exposing functional. Then B_X is strongly Chebyshev, and hence, strongly proximinal.*

Proof. Let $x_0 \in X \setminus B_X$. Recall that $d(x_0, B_X) = \|x_0\| - 1 = \|x_0 - x_0/\|x_0\|\|$. As in the proof of Lemma 2.1, the open ball $B(x_0, \|x_0\| - 1)$ is disjoint from B_X. Therefore, there exists $f_0 \in S_{X^*}$ such that $f_0(x_0) - (\|x_0\| - 1) \geq 1$, so that $f_0(x_0) = \|x_0\|$. Thus, f_0 is norm attaining, and by hypothesis, it strongly exposes $x_0/\|x_0\|$.

Let $\{z_n\} \subseteq B_X$ be a minimizing sequence. Then

$$\|x_0\| - 1 \leq \|x_0\| - \|z_n\| \leq \|x_0\| - f_0(z_n) \leq f_0(x_0 - z_n) \leq \|x_0 - z_n\| \to \|x_0\| - 1.$$

It follows that $f_0(z_n) \to 1$, and therefore, $z_n \to x_0/\|x_0\|$. □

Question 3.9. *Is the converse true? That is, if B_X is strongly Chebyshev, is every norm attaining functional in X^* strongly exposing?*

Question 3.10. *If X is LUR, is every proximinal subspace of X ball proximinal?*

Recall that Saidi's example [9] is a proximinal subspace $Y \subseteq X$ with a $x \in X$ such that $d(x, Y) = d(x, B_Y)$, the distance is attained at a point of Y, but not at B_Y. Thus it is interesting to consider local conditions on x to ensure that this distance is attained at B_Y as well.

Proposition 3.11. *Let $x \in X$. Suppose $d(x, B_Y) = d(x, Y)$. Assume that Y is strongly proximinal at x and $P_Y(x)$ is a compact set. Then B_Y is strongly proximinal at x.*

Proof. We observe that, by Theorem 3.5, the hypothesis implies Y is approximatively compact at x.

Note that if $d(x, B_Y) = d(x, Y)$ and $\{y_n\}$ is a minimizing sequence for x in B_Y, then $\{y_n\}$ is a minimizing sequence for x in Y as well. Now, if Y is approximatively compact, $\{y_n\} \subseteq B_Y$ has a convergent subsequence. Hence, B_Y is approximatively compact, and therefore, strongly proximinal at x. □

In fact, in Saidi's example [9], the given $x \in X$ satisfies the stronger property that $d(P_Y(x), B_Y) = 0$, but $P_Y(x) \cap B_Y = \emptyset$. Observe that if $d(P_Y(x), B_Y) = 0$ and $P_Y(x)$ is weakly compact, then $P_Y(x) \cap B_Y \neq \emptyset$, so that B_Y is proximinal at x. In particular, such examples are not possible in strictly convex spaces. See [4, Proposition IV.1.12] for more examples of Banach spaces for which $P_Y(x)$ is weakly compact.

Remark 3.12. It is easy to see that a Banach space is reflexive if and only if every hyperplane (and hence every subspace) is ball proximinal. On the other hand, Saidi's example [9] shows that a Banach space X is reflexive if and only if X is ball proximinal in every superspace where it isometrically embeds as a hyperplane. In case of hyperplanes, we do not know necessary and sufficient conditions for ball proximinality in terms of the determining functionals.

4 Stability results

We now show the ball proximinality of factor reflexive proximinal subspaces in c_0-direct sums. The following proposition is well-known and is included here to indicate an alternate proof of a result of Garkavi [10, Chapter III] that we will be using below.

Proposition 4.1. *Let $Z \subseteq Y \subseteq X$ be subspaces such that Y/Z is proximinal in X/Z. If Z is proximinal in X then Y is proximinal in X.*

Proof. Let $x \in X$. Suppose $d(\pi(x), Y/Z) = \|\pi(x) - \pi(y_0)\|$ for some $y_0 \in Y$. Now $\|\pi(x) - \pi(y_0)\| = \|\pi(x - y_0)\| = d(x - y_0, Z) = \|x - y_0 - z_0\|$ for some $z_0 \in Z$ as Z is proximinal in X. For $y \in Y$, $\|x - y\| \geq \|\pi(x - y)\| \geq \|x - (y_0 + z_0)\|$. As $y_0 + z_0 \in Y$, Y is proximinal in X. □

Corollary 4.2 (Garkavi). *Let $Z \subseteq Y \subseteq X$ be factor reflexive subspaces of X. If Z is proximinal in X then Y is proximinal in X.*

Proof. By our assumption of factor reflexivity, Y/Z is a proximinal subspace of the reflexive space X/Z. □

Question 4.3. *We do not know if a ball proximinal versions of the above proposition or corollary is true?*

Remark 4.4. There exists a Banach space X with subspaces $Y \subseteq Z \subseteq X$, Y is a hyperplane in Z and Z is a hyperplane in X, Z is ball proximinal in X but Y is not. To see this one starts with a Banach space Z and a hyperplane Y that is not ball proximinal in Z as in [9, Lemma 1]. Let $X = Z \oplus_\infty \mathbb{R}$. As noted in Lemma 2.7, Z is ball proximinal in X. Clearly Y is not ball proximinal in X.

We next exhibit examples where proximinal factor reflexive subspaces are ball proximinal.

Theorem 4.5. *Let $\{X_\alpha\}_{\alpha\in I}$ be a family of reflexive spaces and $X = (\oplus_{\alpha\in I} X_\alpha)_{c_0}$ be the c_0-direct sum. Let $Y' \subseteq Y \subseteq X$ be any proximinal, factor reflexive subspaces. Then Y is ball proximinal in X and Y' is ball proximinal in Y. Also the intersection of finitely many ball proximinal factor reflexive subspaces of X is again ball proximinal.*

Proof. Let $f \in Y^\perp$. Since $Y \subseteq \ker(f) \subseteq X$, Y is proximinal and factor reflexive, by Corollary 4.2, we get that $\ker(f)$ is a proximinal subspace and hence f attains its norm.

Since $X^* = (\oplus_{\alpha\in I} X_\alpha^*)_{\ell^1}$, it is easy to see that f has only finitely many non-zero coordinates. Since $Y^\perp$ is a Banach space, by a simple application of the Baire Category theorem we see that there is a finite set $A \subseteq I$ such that $f_\alpha = 0$ for all $\alpha \notin A$ and $f \in Y^\perp$.

Thus it is easy to see that $B_Y = B_{Y\cap(\oplus_{\alpha\in A} X_\alpha)_\infty} \oplus_\infty B_{(\oplus_{\alpha\notin A} X_\alpha)_{c_0}}$.

Consider now the corresponding splitting of $X = (\oplus_{\alpha\in A} X_\alpha)_\infty \oplus_\infty (\oplus_{\alpha\notin A} X_\alpha)_{c_0} = X_1 \oplus_\infty X_2$ (say). As A is a finite set, X_1 is reflexive and hence, $B_{Y\cap X_1}$ is proximinal in X_1. Since $B_{Y\cap X_2} = B_{X_2}$, it is proximinal in X_2. Therefore B_Y is proximinal in X.

We note that we have in particular proved that any factor reflexive proximinal subspace of X is again a c_0-sum of reflexive spaces. Thus if $Y' \subseteq Y$ is a proximinal factor reflexive subspace, then Y' is ball proximinal in Y.

Also it is clear from the above arguments that if $Y \subseteq X$ is a subspace such that $Y^\perp \subseteq (\oplus_{\alpha\in A} X_\alpha^*)_{\ell^1}$ for a finite set A, then Y is ball proximinal. Note that any set of the form $(\oplus_{\alpha\in A} X_\alpha^*)_{\ell^1}$ is a w^*-closed subset of X^*. Thus if $Y_1, \ldots, Y_n$ are ball proximinal factor reflexive subspaces of X then, since $(\bigcap_{1\le i\le n} Y_i)^\perp \subseteq (\oplus_{\alpha\in A} X_\alpha^*)_{\ell^1}$ for some finite set A, we get that $\bigcap_{1\le i\le n} Y_i$ is ball proximinal. □

The following lemma is a reformulation of [4, Lemma IV.1.11].

Lemma 4.6. *Let P be a contractive projection in X^*. Suppose $Y \subseteq P(X^*)$ is a subspace such that $P(\overline{B_Y}^*) \subseteq Y$ (w^*-closure). Then Y is ball proximinal in $P(X^*)$.*

Proof. Let $f \in P(X^*)$. Since $\overline{B_Y}^*$ is w^*-compact, there is $g_0 \in \overline{B_Y}^*$ such that $d(f, \overline{B_Y}^*) = \|f - g_0\|$. Now $d(f, B_Y) \ge d(f, \overline{B_Y}^*) = \|f - g_0\| \ge \|P(f - g_0)\| = \|f - P(g_0)\|$. As $P(g_0) \in B_Y$, we have the required conclusion. □

We recall from [4] that X is said to be L-embedded if under the canonical embedding of X in X^{**}, there exists a linear onto projection $P : X^{**} \to X$ such that $\|\Lambda\| = \|P(\Lambda)\| + \|\Lambda - P(\Lambda)\|$ for all $\Lambda \in X^{**}$. These spaces are natural analogues of reflexive spaces. See [4, Chapter IV] for several examples and geometric properties of these spaces.

Corollary 4.7. *$Y \subseteq X$, such that both Y and X are L-embedded. Then Y is ball proximinal in X.*

Proof. It is known that if $Y \subseteq X$ and both are L-embedded then the L-projection P in X^{**} with range X has the property that $P(\overline{Y}^*) \subseteq Y$ [4, Proposition IV.1.10]. Hence B_Y is proximinal in X. □

The importance of these notions lies in the fact that being L-embedded is preserved by ℓ^1-sums and in several cases the space of Bochner integrable functions with values in a L-embedded space is again a L-embedded space.

Corollary 4.8. *Let μ, ν be positive measures. Let $Y \subseteq L^1(\mu)$ be a reflexive subspace. $L^1(\nu, Y)$ is ball proximinal in $L^1(\nu \otimes \mu)$.*

Proof. It is well known that $L^1(\nu, Y)$ is a L-embedded subspace of $L^1(\nu, L^1(\mu)) = L^1(\nu \otimes \mu)$ ([4, p. 200]). Thus the conclusion follows from Corollary 4.7. □

We next consider the stability of ball proximinality for spaces of vector-valued function spaces. We follow the notations and terminology of [2].

It is easy to see that if Y_α is a family of ball proximinal subspaces in X_α, then $(\oplus_\alpha Y_\alpha)_{c_0}$ is also ball proximinal in $(\oplus_\alpha X_\alpha)_{c_0}$.

We first consider the space $C(K, X)$ of X-valued continuous functions on a compact set K.

Proposition 4.9. *Suppose $Y \subseteq X$ is a ball proximinal subspace and $P : X \to B_Y$ is single-valued and continuous or has a continuous selection. For any compact Hausdorff space K, $B_{C(K,Y)}$ is proximinal in $C(K, X)$.*

Proof. Let $f \in C(K, X)$, put $f_0 = P \circ f$. Then $f_0 \in B_{C(K,Y)}$ and for any $g \in B_{C(K,Y)}$ and $k \in K$, $\|f(k) - f_0(k)\| \leq \|f(k) - g(k)\|$. Thus $d(f, B_{C(K,Y)}) = \|f - f_0\|$.

Similar arguments work in the case P has a continuous selection. □

The following theorem extends, for $p = \infty$, the known result [5] that for a separable proximinal subspace $Y \subseteq X$, and $1 \leq p \leq \infty$, $L^p(\mu, Y)$ is proximinal in $L^p(\mu, X)$. To keep the measure theory simple we assume that the domain is $[0, 1]$ with the Lebesgue measure μ. Standard measure theoretic arguments can be used to extend it to positive measures on completions of countably generated σ-fields.

Theorem 4.10. *Let $Y \subseteq X$ be a separable ball proximinal subspace. Then for the Lebesgue measure μ on $[0, 1]$, $L^\infty(\mu, Y)$ is ball proximinal in $L^\infty(\mu, X)$.*

Proof. Let $f \in L^\infty(\mu, X)$. We assume without loss of generality that the functions are defined everywhere. Let $G = \{(t, y) \in [0, 1] \times B_Y : d(f(t), B_Y) \geq \|f(t) - y\|\}$. Since B_Y is proximinal, the projection of G to the first coordinate is $[0, 1]$.

Let $\{y_n\}_{n \geq 1}$ be a dense sequence in B_Y. Then $G = \cap_n \{(t, y) \in [0, 1] \times B_Y : \|f(t) - y_n\| \geq \|f(t) - y\|\}$. Since f is measurable, G is a measurable set.

Thus as a consequence of the von Neumann selection theorem [7, Theorem 7.2] we get a measurable function $g_0 : [0, 1] \to B_Y$ such that $(t, g_0(t)) \in G$ for a.e t. In particular, $g_0(t) \in P_{B_Y}(f(t))$ a.e.

Clearly, $g_0 \in B_{L^\infty(\mu,Y)}$. For any $g \in B_{L^\infty(\mu,Y)}$,

$$\|f - g\|_\infty = \sup \|f(t) - g(t)\| \geq \sup \|f(t) - g_0(t)\| = \|f - g_0\|_\infty.$$

Therefore, $g_0 \in P_{B_{L^\infty(\mu,Y)}}(f)$. □

We do not know if ball proximinality is preserved by ℓ^1-sums. However, arguments similar to the above can be used to prove a weaker proximinality result in the case of Bochner integrable functions. We do not know for a separable ball proximinal $Y \subseteq X$, if $L^1(\mu, Y)$ is ball proximinal in $L^1(\mu, X)$.

Corollary 4.11. *Let $Y \subseteq X$ be a separable ball proximinal subspace. Then for the Lebesgue measure μ on $[0,1]$, $L^1(\mu, B_Y)$ is a proximinal set in $L^1(\mu, X)$.*

Proof. Let $f \in L^1(\mu, X)$. Arguing as above, there is a measurable function $g_0 : [0,1] \to B_Y$ such that $g_0(t) \in P_{B_Y}(f(t))$ a.e. Since Y is separable, $g_0 \in L^1(\mu, B_Y)$.

For any $g \in L^1(\mu, B_Y)$, $\|f - g\|_1 = \int_0^1 \|f(t) - g(t)\| dt \geq \int_0^1 \|f(t) - g_0(t)\| dt = \|f - g_0\|_1$. □

Acknowledgments. Most of this work was done during the visits of the first and third authors to the University of Iowa in the summers of 2005 and 2006. We would like to thank the second author and the UI for the warm hospitality.

References

[1] P. Bandyopadhyay, B. L. Lin, Y. Li, and D. Narayana, *Proximinality in Banach Spaces*, preprint (2006).

[2] J. Diestel and J. J. Uhl, Jr., *Vector measures*, Mathematical Surveys, No. 15, American Mathematical Society, Providence, R.I., 1977.

[3] G. Godefroy and V. Indumathi, *Strong proximinality and polyhedral spaces*, Rev. Mat. Complut. 14 (2001), pp. 105–125.

[4] P. Harmand, D. Werner, and W. Werner, *M-ideals in Banach spaces and Banach algebras*, Lecture Notes in Math. 1547, Springer, Berlin, 1993.

[5] J. Mendoza, *Proximinality in $L_p(\mu, X)$*, J. Approx. Theory 93 (1998), pp. 331–343.

[6] D. Narayana and T. S. S. R. K. Rao, *Transitivity of proximinality and norm attaining functionals*, Colloq. Math. 104 (2006), pp. 1–19.

[7] T. Parthasarathy, *Selection theorems and their applications*, Lecture Notes in Mathematics 263, Springer-Verlag, Berlin New York, 1972.

[8] F. B. Saidi, *On the smoothness of the metric projection and its application to proximinality in $L_p(S, X)$*, J. Approx. Theory 83 (1995), pp. 205–219.

[9] ———, *On the proximinality of the unit ball of proximinal subspaces in Banach spaces: a counterexample*, Proc. Amer. Math. Soc. 133 (2005), pp. 2697–2703.

[10] I. Singer, *Best approximation in normed linear spaces by elements of linear subspaces*, Springer-Verlag, New York Berlin, 1970.

Author information

Pradipta Bandyopadhyay, Stat–Math Division, Indian Statistical Institute, 202, B. T. Road, Kolkata 700108, India.
Email: pradipta@isical.ac.in

Bor-Luh Lin, Department of Mathematics, University of Iowa, Iowa City, IA 52242, USA.
Email: bllin@math.uiowa.edu

T. S. S. R. K. Rao, Stat–Math Unit, Indian Statistical Institute, R. V. College P.O., Bangalore 560059, India.
Email: tss@isibang.ac.in

Banach Spaces and their Applications in Analysis, 265–283

Discretization versus transference for bilinear operators

Earl Berkson, Oscar Blasco, María J. Carro and Thomas A. Gillespie

Abstract. A very general transference method for bilinear operators is presented and used to show that discretization techniques can also be obtained from transference methods. It is applied to show the boundedness of the discrete version of the bilinear fractional operator and the bisublinear Hardy–Littlewood maximal operator. Also a method for bilinear vector-valued transference is presented.

Key words. Bilinear Hilbert transform, maximal operator, transference, discretization.

AMS classification. Primary 42A45; Secondary 37A50, 42A50, 42B20, 42B25.

1 Introduction

In 1977, a very general and abstract method of transference was introduced by Coifman and Weiss in [8]. Their procedure showed that if a "convolution type" operator defined on a group is bounded on $L^p(G)$ and the group G is represented in the space of bounded linear operators $\mathcal{B}(L^p(\mu))$ for some measure μ then a transferred operator, defined by means of the representation, is also bounded on the corresponding $L^p(\mu)$ spaces. Their method relies on the following result:

Theorem 1.1. *Let G be an amenable group with left Haar measure m, $1 \leq p < \infty$, and $u \to R_u$ be strongly continuous uniformly bounded representations of G in $\mathcal{B}(L^p(\mu))$. If $K \in L^1(G)$ is compactly supported and the operator*

$$T(g)(v) = \int_G g(u^{-1}v)K(u)dm(u),$$

has norm $N_p(K)$ in $\mathcal{B}(L^p(G))$ then the transferred operator

$$\tilde{T}(f)(x) = \int_G (R_u f)(x)K(u)dm(u),$$

defined for f in some dense subset of $L^p(\mu)$, extends to bounded operator on $L^p(\mu)$ with norm bounded by $CN_p(K)$.

Since then, this method has been developed and extended by many other people (see for example [3] or [4]) and has shown to be an extremely useful tool to prove boundedness of many operators defined in the setting of measure spaces assuming that we know the boundedness of appropriately related convolution operators in the context of amenable groups.

Second author: Supported in part by Proyecto MTM2005-08350-C03-03.
Third author: Supported in part by Project MTM2007-60500 and 2005SGR00556.

In 1996, Grafakos and Weiss (see [13]) proved a first result concerning a transference method for multilinear operators. They considered a multilinear operator T defined on an amenable group G by

$$T(g_1,\dots,g_k)(v)=\int_{G^k}K(u_1,\dots,u_k)g_1(u_1^{-1}v)\dots g_k(u_k^{-1}v)dm(u_1)\dots dm(u_k),$$

with g_j in some dense subset of $L^{p_j}(G)$ and where K is a kernel on G^k which may not be integrable. They were able to transfer the boundedness of $T: L^{p_1}(G)\times\cdots\times L^{p_k}(G)\to L^{p_0}(G)$ whenever $1/p_0=1/p_1+\cdots+1/p_k$ to the boundedness of operator $\tilde{T}: L^{p_1}(\mu)\times\cdots\times L^{p_k}(\mu)\to L^{p_0}(\mu)$ where (M,μ) is a measure space and

$$\tilde{T}(f_1,\dots,f_k)(x)=\int_{G^k}K(u_1,\dots,u_k)(R^1_{u_1}f_1)(x)\dots(R^k_{u_k}f_k)(x)dm(u_1)\dots dm(u_k),$$

where f_j is in some dense subset of $L^{p_j}(\mu)$, and $R^j: G\to\mathcal{B}(L^{p_j}(\mu))$ $(j=0,1,\dots,k)$ are representations which are connected through $R^0_vR^j_u=R^j_{uv}$ for all $u,v\in G$ and $1\le j\le k$, and satisfy certain boundedness conditions.

Recently, pursuing the transference to other groups and measure spaces of the results obtained for the bilinear Hilbert transform and other bilinear multipliers some methods have been developed. In particular, the reader is referred to [5, 7, 10, 11] for some different approaches, using DeLeeuw type methods, which also allow to transfer the boundedness of bilinear multipliers from one group to another one.

A technique extending Coiffman–Weiss transference method was introduced in [6] for the bilinear situation. Namely, if G is a locally compact abelian group with Haar measure m, $K\in L^1(G)$ is a kernel with compact support, $0<p_1,p_2,p_3<\infty$ with $1/p_3=1/p_1+1/p_2$ and the bilinear map

$$B_K(\phi,\psi)(v)=\int_G\phi(u^{-1}v)\psi(uv)K(u)du,$$

is bounded from $L^{p_1}(G)\times L^{p_2}(G)\to L^{p_3}(G)$ then one can define the transference bilinear map $T_K: L^{p_1}(\mu)\times L^{p_2}(\mu)\to L^{p_3}(\mu)$ by

$$T_K(f,g)(x)=\int_G(R^1_{u^{-1}}f)(x)(R^2_ug)(x)K(u)du,$$

where $R^j: G\to\mathcal{B}(L^{p_j}(\mu))$ are strongly continuous and uniformly bounded representations for $j=1,2$ and (M,μ) is a σ-finite measure space.

Observe that if $p_3\ge 1$, then $T_K(f,g)(x)$ is well defined at almost every x, but this is not the case if $p_3<1$ since it is not true, in general, that if $f\in L^{p_1}(\mu)$, $g\in L^{p_2}(\mu)$ and $K\in L^1(G)$, then $T_K(f,g)\in L^{p_3}(\mu)$ or even that it is well defined. Hence, in this case, we may have to assume something else in the mappings R^j such as, for example, that our operators R^j act also continuously in $L^2(\mu)$. Moreover, whenever $p_3<1$, we have to consider $f\in L^{p_1}(\mu)\cap L^2(\mu)$ and $g\in L^{p_2}(\mu)\cap L^2(\mu)$ in order to have that the transferred operator is well defined.

The following result was shown and applied to obtain some new results acting on other groups or measure spaces.

Theorem 1.2 ([6]). *Under the above conditions, if for $j = 1, 2$ and every $v \in G$, there exist $A_j > 0$ such that*

$$\|R_v^j f\|_{L^{p_j}} \leq A_j \|f\|_{L^{p_j}} \tag{1.1}$$

and there exists a strongly continuous mapping $R^3 : G \to \mathcal{B}(L^{p_3}(\mu))$ satisfying that, for every $u, v \in G$ and every $f \in L^{p_1}(\mu)$ and $g \in L^{p_2}(\mu)$,

$$R_v^3(R_{u^{-1}}^1 f R_u^2 g) = R_{vu^{-1}}^1 f R_{vu}^2 g, \tag{1.2}$$

and such that, for every $v \in G$, there exists $B > 0$ satisfying

$$\|f\|_{L^{p_3}(\mu)} \leq B \|R_v^3 f\|_{L^{p_3}(\mu)}, \tag{1.3}$$

then, the bilinear operator $T_K : L^{p_1}(\mu) \times L^{p_2}(\mu) \to L^{p_3}(\mu)$ is bounded and it has norm bounded by $N_{p_1,p_2}(K) A_1 A_2 B$ where $N_{p_1,p_2}(K)$ stands for the norm of the bilinear map B_K in the corresponding spaces.

One of the basic aims of the transference methods is to provide machinery for translating estimates in harmonic analysis into discretized counterparts for ergodic operator theory. For the bilinear setting the procedure can, in principle, take the form of direct discretization of bilinear operators initially defined for the real line, and then application of abstract results such as Theorem 1.2 to transfer individual discrete bilinear operators, along with their bounds, to the ergodic theory setting. In [6] direct discretization techniques were initiated for the bilinear Hilbert transform, and this approach was advanced in [1], where general discretization and transference of bilinear maximal estimates were developed. In particular, the discretization techniques in [1] were used to obtain the following counterpart for the integers of the bilinear Hilbert transform for the real line [19].

Theorem 1.3 ([1, 6]). *Let $1 < p_1, p_2 < \infty$ and $1/p_1 + 1/p_2 = 1/p_3 < 3/2$. Then for $a \equiv \{a_j\}_{j=-\infty}^{\infty} \in \ell^{p_1}(\mathbb{Z})$, $b \equiv \{b_j\}_{j=-\infty}^{\infty} \in \ell^{p_2}(\mathbb{Z})$, the series*

$$(H_{\mathbb{Z}}(a,b))(k) \equiv \sum_{j=-\infty, j\neq 0}^{\infty} \frac{a_{k+j} b_{k-j}}{j} \tag{1.4}$$

converges absolutely for each $k \in \mathbb{Z}$, and the bilinear operator $H_{\mathbb{Z}}$ defined on $\ell^{p_1}(\mathbb{Z}) \times \ell^{p_2}(\mathbb{Z})$ satisfies

$$\|H_{\mathbb{Z}}(a,b)\|_{\ell^{p_3}(\mathbb{Z})} \leq \mathfrak{A}_{p_1,p_2} \|a\|_{\ell^{p_1}(\mathbb{Z})} \|b\|_{\ell^{p_2}(\mathbb{Z})},$$

for all $a \in \ell^{p_1}(\mathbb{Z})$, $b \in \ell^{p_2}(\mathbb{Z})$, where $\mathfrak{A}_{p_1,p_2}$ is a constant depending only on p_1 and p_2.

For $a \equiv \{a_j\}_{j=-\infty}^{\infty} \in \ell^{p_1}(\mathbb{Z})$, $b \equiv \{b_j\}_{j=-\infty}^{\infty} \in \ell^{p_2}(\mathbb{Z})$, and $N \in \mathbb{N}$, let

$$H_{N,\mathbb{Z}}(a,b)(k) = \sum_{0<|j|\leq N} \frac{a_{k+j} b_{k-j}}{j}, \tag{1.5}$$

for all $k \in \mathbb{Z}$. Then $\{H_{N,\mathbb{Z}}(a,b)\}_{N=1}^{\infty}$ converges to $H_{\mathbb{Z}}(a,b)$ in the metric topology of $\ell^{p_3}(\mathbb{Z})$.

In [1] the discretization of the bisublinear maximal operators of [16] furnishes the following extension of Theorem 1.3.

Theorem 1.4 ([1]). *Let* $1 < p_1, p_2 < \infty$ *and* $1/p_1 + 1/p_2 = 1/p_3 < 3/2$. *For* $a \equiv \{a_j\}_{j=-\infty}^{\infty} \in \ell^{p_1}(\mathbb{Z})$, $b \equiv \{b_j\}_{j=-\infty}^{\infty} \in \ell^{p_2}(\mathbb{Z})$, *let*

$$\mathcal{H}_{\mathbb{Z}}(a,b)(k) = \sup_{N\in\mathbb{N}} \Big| \sum_{0<|j|\leq N} \frac{a_{k+j}b_{k-j}}{j} \Big|, \quad \textit{for all } k \in \mathbb{Z}. \tag{1.6}$$

$$\mathcal{M}_{\mathbb{Z}}(a,b)(k) = \sup_{N\geq 1} \frac{1}{2N+1} \sum_{n=-N}^{N} |a_{m-n}|.|b_{m+n}|, \quad \textit{for all } k \in \mathbb{Z}. \tag{1.7}$$

Then there are constants $\mathfrak{B}_{p_1,p_2}$ *and* $\mathfrak{C}_{p_1,p_2}$, *depending only on* p_1 *and* p_2, *such that*

$$\|\mathcal{H}_{\mathbb{Z}}(a,b)\|_{\ell^{p_3}(\mathbb{Z})} \leq \mathfrak{B}_{p_1,p_2} \|a\|_{\ell^{p_1}(\mathbb{Z})} \|b\|_{\ell^{p_2}(\mathbb{Z})} \tag{1.8}$$

and

$$\|\mathcal{M}_{\mathbb{Z}}(a,b)\|_{\ell^{p_3}(\mathbb{Z})} \leq \mathfrak{C}_{p_1,p_2} \|a\|_{\ell^{p_1}(\mathbb{Z})} \|b\|_{\ell^{p_2}(\mathbb{Z})} \tag{1.9}$$

for all $a \in \ell^{p_1}(\mathbb{Z})$, $b \in \ell^{p_2}(\mathbb{Z})$.

Remark 1.5. The boundedness result for the discrete bisublinear Hardy–Littlewood maximal operator in (1.9) is included in [9, Proposition 14.1], which is an article devoted to the treatment of generalized multisublinear Hardy–Littlewood maximal operators, and their transference by measure-preserving transformations to discrete dynamical systems.

The main goal of this paper is to show that the boundedness of some of the discrete versions previously mentioned can be also seen as particular cases of the general method of transference from $G = \mathbb{R}$ to operators acting on $\ell^p(\mathbb{Z})$ when replacing the use of representations by general measurable functions. To motivate this approach, let us point out that if $G = \mathbb{R}$ and we assume that the bilinear map

$$B_K(\phi,\psi)(x) = \int_{\mathbb{R}} \phi(x-y)\psi(x+y)K(y)dy,$$

given by an integrable kernel K is bounded from $L^{p_1}(\mathbb{R}) \times L^{p_2}(\mathbb{R}) \to L^{p_3}(\mathbb{R})$ for some values of $0 < p_1, p_2, p_3 < \infty$ then we can consider the map $R : \mathbb{R} \to \mathcal{B}(\ell^p(\mathbb{Z}))$ given by $R_u(x_n) = x_{n+[u+1/2]}$ for which the corresponding transferred bilinear map T_K becomes

$$T_K((\alpha_n),(\beta_n))(m) = \sum_n \alpha_{m-n}\beta_{m+n}K_n$$

for finite sequences (α_n) and (β_n) where $K_n = \int_{[n-1/2,n+1/2)} K(u)du$. However neither $u \to R_u$ is a representation of $\mathbb{R}$ in the space of operators on $\ell^p(\mathbb{Z})$, nor the map $u \to R_u((\alpha_n))$ is continuous from $\mathbb{R}$ into $\ell^p(\mathbb{Z})$ for a given $(\alpha_n) \in \ell^p(\mathbb{Z})$. Nevertheless it is still measurable in the strong operator topology.

In this paper we shall present a generalization of Theorem 1.2 where the assumptions are relaxed to obtain the discretization actually as special cases of the general transference principle. A careful look to the proof in Theorem 1.2 allows to see that there are three aspects of the result that can be exploited better and are relevant for applications. First of all the continuity in the strong operator topology of the map $u \to R_u$ is not really needed and actually the fact that R_u are representations is not important once condition (1.2) is assumed. Second of all the conditions (1.1) and (1.3) can be weakened up to new conditions which can be seen as $u \to R_u$ belonging to certain vector-valued spaces. Although only (1.1) and (1.3) will be needed for applications in this paper we believe that the new and more general conditions can be used in further applications. Finally one observes that the setting where it has been used, that is transferring bounded bilinear operators acting from $L^{p_1}(G) \times L^{p_2}(G) \to L^{q_3}(G)$ where $q_3 = p_3$ and $1/p_3 = 1/p_1 + 1/p_2$ to the case of operators $\mathcal{B}(L^p(\mu))$ can be easily extended, under the assumption $q_3 \geq p_3$, to the case of operators $\mathcal{B}(\ell^p(\mathbb{Z}))$ or, under the assumption $q_3 \leq p_3$, to the case of operators $\mathcal{B}(L^p(\mu))$ whenever $\mu(M) < \infty$.

One of the new applications of our result is the discrete version of boundedness of the bilinear fractional integration obtained by Kenig and Stein ([15]) (see also the work of Grafakos and Kalton [12]).

The paper is organized as follows. We first establish and present the general transference method for bilinear maps and obtain some corollaries in the case of positive kernels. Later we present similar approach to transfer also maximal bisublinear operators and recover the results about the discrete version of the bisublinear maximal Hardy–Littlewood operator. Finally we present another general transference method in the setting of general bilinear maps acting on Banach spaces, whose application to L^p-spaces allows to recover the result in [6].

Throughout the paper G stands for a locally compact abelian topological group, m denotes the Haar measure, we use either $m(A)$ or $|A|$ for the measure of a given set A, $\int_A f(u)dm(u)$ or $\int_A f(u)du$ for the integral of a given function f, $1 < p_1, p_2 < \infty$, $1/p_3 = 1/p_1 + 1/p_2$, and C stands for a constant that may vary from line to line.

2 A general bilinear transference method

Let us start by giving the following definition, which is related to the amenability condition in the classical theory and that will be convenient for our general framework.

Definition 2.1. A collection $\mathcal{V}$ of measurable sets in G is said to be *complete* if the following condition holds: for every $\varepsilon > 0$ and every compact set C (that we shall always assume to be symmetric and contain the unit e), there exist $V_0 \in \mathcal{V}$ and $V_1 \in \mathcal{V}$ such that $V_0 C \subset V_1$ and $1 \leq |V_1|/|V_0| \leq 1 + \varepsilon$.

Examples. *1. If G is a compact group, then every $\mathcal{V}$ containing G is complete.*

2. Let $G = (\mathbf{R^n}, +)$ and let $\mathcal{V}_M = \{(-R, R)^n, R > M\}$. Then $\mathcal{V}_M$ is complete, for every $M > 0$.

3. Let $G = (\mathbf{R}^+, .)$ and let $\mathcal{V} = \{(1/R, R), R > 1\}$.

4. Let $G = (\mathbb{Z}, +)$ and let $\mathcal{V} = \{[-N, N] \cap \mathbb{Z}, N \geq 1\}$.
5. If the group is amenable the collection of neighborhoods of zero is a complete class.

For our theorems we shall see that we can restrict ourselves to complete families of measurable sets $\mathcal{V}$.

Definition 2.2. Let $0 < p < \infty$, X be a quasi-Banach space, and $\mathcal{V}$ be a complete collection of measurable sets in G. We denote by $A^p(X) = A^p(G, \mathcal{V}; X)$ the space of (strongly) measurable functions $F : G \to X$ such that

$$\sup_{V \in \mathcal{V}} \Big(\frac{1}{|V|}\int_V ||F(u)||_X^p du\Big)^{1/p} = ||F||_{A^p(X)} < \infty.$$

If $X = \mathcal{B}(Y)$ for a quasi-Banach space Y we denote by $A^p_s(Y) = A^p_s(G, \mathcal{V}; Y)$ the space of functions $F : G \to \mathcal{B}(Y)$ such that $u \to F(u)(y)$ measurable for all $y \in Y$, and satisfy

$$\sup_{V \in \mathcal{V}} \sup_{||y||=1} \Big(\frac{1}{|V|}\int_V ||F(u)(y)||_Y^p du\Big)^{1/p} = ||F||_{A^p_s(Y)} < \infty.$$

Of course these classes satisfy $A^{p_2}(X) \subset A^{p_1}(X)$ (respectively $A^{p_2}_s(Y) \subset A^{p_1}_s(Y)$) for all $0 < p_1 \leq p_2 < \infty$. Also it is clear that for all $0 < p < \infty$ and any $\mathcal{V}$ one has

$$L^\infty(G, \mathcal{B}(Y)) \subset A^p(G, \mathcal{V}, \mathcal{B}(Y)) \subset A^p_s(G, \mathcal{V}, Y).$$

Examples. *1. Let $G = \mathbb{R}$, $X = \mathbb{C}$ and denote by B^p the space of functions such that*

$$\lim_{R \to \infty} \frac{1}{2R}\int_{-R}^{R} |F(u)|^p du < \infty.$$

Hence for $A^p(\mathbb{R}, \mathcal{V}, \mathbb{C}) \subset B^p$, for $\mathcal{V} = \{(-R, R); R > 1\}$. Particular examples of functions in B^p are almost periodic functions $F(x) = \hat{\mu}(x)$ for a finite Borel measure on $\mathbb{R}$ with finite support, say $F(x) = \sum \alpha_t e^{itx}$. If $F \in A^p(\mathbb{R}, \mathbb{C})$ then $\alpha = \sum \alpha_t \chi_t$ belongs to $L^p(\hat{\mathbf{D}})$.

This follows from the fact that $||F||_{B^p} = ||\alpha||_{L^p(\hat{\mathbf{D}})}$(see [20]) where $\mathbf{D}$ stands for the group $\mathbb{R}$ with the discrete topology, $\hat{\mathbf{D}}$ stands for the dual group of $\mathbf{D}$, which coincides with the Bohr compactification of $\mathbb{R}$ (see [21], 1.8) and $\alpha = \sum \alpha_t \chi_t$ where χ_t stands for the corresponding character in $\mathbf{D}$.

2. Let $\mathcal{V} = \{\mathbb{Z} \cap (-N, N); N > 1\}$. A sequence (x_n) in X belongs to $A^p(\mathbb{Z}, \mathcal{V}, X)$ if

$$\sup_{N \in \mathbb{N}} \Big(\frac{1}{2N}\sum_{-N}^{N} ||x_n||_X^p\Big)^{1/p} < \infty.$$

A sequence of operators (T_n) in $\mathcal{B}(X)$ belongs to $A^p_s(\mathbb{Z}, \mathcal{V}, \mathcal{B}(X))$ if

$$\sup_{N \in \mathbb{N}} \sup_{||y||=1} \Big(\frac{1}{2N}\sum_{-N}^{N} ||T_n(y)||_Y^p\Big)^{1/p} < \infty.$$

Definition 2.3. Let (M,μ) be a σ-finite measure space and $0 < p_1, p_2, q_3 < \infty$. Let $K \in L^1(G)$ compactly supported and denote

$$B_K(\phi,\psi)(v) = \int_G \phi(u^{-1}v)\psi(uv)K(u)du,$$

for $\phi \in L^{p_1}(G) \cap L^\infty(G)$ and $\psi \in L^{p_2}(G) \cap L^\infty(G)$.

We assume that B_K extends to a bounded bilinear operator from $L^{p_1}(G) \times L^{p_2}(G)$ to $L^{q_3}(G)$ with norm $N(K)$.

Let $R^i : G \to \mathcal{B}(L^{p_i}(\mu))$ be functions which are measurable in the strong operator topology of $\mathcal{B}(L^{p_i}(\mu))$ for $i = 1,2$, i.e. $u \to R^i_u f$ is measurable for any $f \in L^{p_i}(\mu)$. Assume also that, for all measurable sets A with $\mu(A) < \infty$, one has that $R^i_u \chi_A \in L^2(\mu)$ for $u \in G$ and $u \to \|R^i_u \chi_A\|_2$ is bounded over compact sets.

Now define the *transference operator* $T_K : L^{p_1}(\mu) \times L^{p_2}(\mu) \to L^{q_3}(\mu)$ by setting for simple functions f and g,

$$T_K(f,g)(x) = \int_G (R^1_{u^{-1}} f)(x)(R^2_u g)(x)K(u)du.$$

Note that $u \to (R^1_{u^{-1}} f)(R^2_u g)K(u)$ belongs to $L^1(G, L^1(\mu))$ if f and g are simple functions, and then $T_K(f,g) \in L^1(\mu)$ in this case.

Let us now state the main result of the paper.

Theorem 2.4. *Let (M,μ) be a σ-finite space, $0 < p_1, p_2, q_3 < \infty$, $1/p_3 = 1/p_1 + 1/p_2$, $K \in L^1(G)$ compactly supported and B_K and T_K are defined as above where*

1. *$q_3 = p_3$ for the general case,*
2. *$q_3 \geq p_3$ in the case $(\mathbb{Z},\nu)$ for the counting measure ν,*
3. *$q_3 \leq p_3$ in the case $\mu(M) < \infty$.*

Let us denote $X_1 = \mathcal{B}(L^{p_1}(\mu))$, $X_2 = \mathcal{B}(L^{p_2}(\mu))$ and $X_3 = \mathcal{B}(L^{p_3}(\mu))$ and assume that:

(i) *there exist bounded functions ϕ_i with $\operatorname{supp}(\phi_i) = G_i$ such that $\sum_{i-1}^n \phi_i(u) = 1$ for any $u \in G$ and there exists a complete family $\mathcal{V}$ in G and $\gamma > 0$ for which $|V| \leq \gamma |V \cap G_i|$ for all i and for all $V \in \mathcal{V}$;*

(ii) *there exist functions (measurable in the strong operator topology) $R : G \to X_3$, $S^i : G \to X_1$ and $T^i : G \to X_2$ satisfying that, for every $f \in L^{p_1}(\mu)$ and $g \in L^{p_2}(\mu)$,*

$$R_v(R^1_{u^{-1}} f R^2_u g) = S^i_{vu^{-1}} f T^i_{vu} g, \qquad u, v \in G_i, \tag{2.1}$$

where R_u are invertible operators for all $u \in G$ and $R^{-1} \in A^p(G,\mathcal{V},X_3)$ for some $0 < p \leq 1$, where $R^{-1}(u) = R_u^{-1}$, $S^i \in A^{p_1}_s(G,\mathcal{V},L^{p_1})$ and $T^i \in A^{p_2}_s(G,\mathcal{V},L^{p_2})$ for $i = 1,2,\ldots,n$.

Then, the bilinear operator T_K can be extended to a bounded operator

$$T_K : L^{p_1}(\mu) \times L^{p_2}(\mu) \to L^{q_3}(\mu)$$

with norm bounded by $C(n,\gamma)A_1A_2A_3 \sup_{1\le i\le n} N(K\phi_i)$ *where* $A_1 = \|R^{-1}\|_{A^p(X_3)}$, $A_2 = \sup_{1\le i\le n}\|S^i\|_{A_s^{p_1}(L^{p_1})}$, *and* $A_3 = \sup_{1\le i\le n}\|T^i\|_{A_s^{p_2}(L^{p_2})}$.

Proof. Let f, g be simple functions and let $V \in \mathcal{V}$ and denote $k_i = K\phi_i$, $V_i = V \cap G_i$ and $C_i = C \cap G_i$, with $C = \mathrm{supp}K$. Now, for fixed $(v_1,\dots,v_n) \in V_1 \times \cdots \times V_n$, one has

$$\begin{aligned} T_K(f,g) &= \sum_{i=1}^n \int_{G_i} R^1_{u^{-1}} f R^2_u g k_i(u)du \\ &= \sum_{i=1}^n R_{v_i}^{-1}\Big(\int_{G_i} R_{v_i}(R^1_{u^{-1}} f R^2_u g) k_i(u)du\Big) \\ &= \sum_{i=1}^n R_{v_i}^{-1}\Big(\int_{G_i} S^i_{v_i u^{-1}} f T^i_{v_i u} g k_i(u)du\Big) \\ &= \sum_{i=1}^n R_{v_i}^{-1}\Big(\int_{G} S^i_{v_i u^{-1}} f \chi_{V_i C_i^{-1}}(v_i u^{-1}) T^i_{v_i u} g \chi_{V_i C_i}(v_i u) k_i(u)du\Big). \end{aligned}$$

Hence, for every $(v_1,\dots,v_n) \in V_1 \times \cdots \times V_n$, we have that

$$T_K(f,g) = \sum_{i=1}^n R_{v_i}^{-1} B_{k_i}((S^i_u f)\chi_{V_iC_i}, (T^i_u g)\chi_{V_iC_i})(v_i).$$

Therefore, if $q_3 \ge 1$,

$$\|T_K(f,g)\|_{q_3} \le \sum_{i=1}^n \|R_{v_i}^{-1}\|_{X_3} \|B_{k_i}((S^i_u f)\chi_{V_iC_i^{-1}}, (T^i_u g)\chi_{V_iC_i})(v_i)\|_{L^{q_3}(\mu)}$$

and for $0 < q_3 < 1$,

$$\|T_K(f,g)\|_{q_3}^{q_3} \le \sum_{i=1}^n \|R_{v_i}^{-1}\|_{X_3}^{q_3} \|B_{k_i}((S^i_u f\chi_{V_iC_i^{-1}}), (T^i_u g)\chi_{V_iC_i})(v_i)\|_{L^{q_3}(\mu)}^{q_3}.$$

In particular, for any $0 < \alpha \le \min\{1, q_3\}$,

$$\|T_K(f,g)\|_{q_3}^{\alpha} \le \sum_{i=1}^n \|R_{v_i}^{-1}\|_{X_3}^{\alpha} \|B_{k_i}((S^i_u f\chi_{V_iC_i^{-1}}), (T^i_u g)\chi_{V_iC_i})(v_i)\|_{L^{q_3}(\mu)}^{\alpha}.$$

Let $q = 1 + p/q_3$ and $\alpha = pq_3/(p+q_3)$. Clearly, $0 < \alpha \le \min\{1, q_3\}$, $q > 1$, $1/q + \alpha/q_3 = 1$, and $q\alpha = p$. Now integrate over $V_1 \times \cdots \times V_n$ and denote $\beta = \prod_{j=1}^n |V_j|$ and $\beta_i = \prod_{j\ne i} |V_j|$. Hence, we can write

$$\|T_K(f,g)\|_{q_3}^{\alpha} \le \frac{1}{\beta}\sum_{i=1}^{n}\beta_i \int_{V_i} \|R_v^{-1}\|_{X_3}^{\alpha}\|B_{k_i}((S_u^i f\chi_{V_iC_i^{-1}}),(T_u^i g)\chi_{V_iC_i})(v)\|_{L^{q_3}(\mu)}^{\alpha}dv$$

$$= n\sum_{i=1}^{n}\frac{\beta_i|V_i|}{n\beta}\frac{1}{|V_i|}\int_{V_i}\|R_v^{-1}\|_{X_3}^{\alpha}\|B_{k_i}((S_u^i f\chi_{V_iC_i^{-1}}),(T_u^i g)\chi_{V_iC_i})(v)\|_{L^{q_3}(\mu)}^{\alpha}dv$$

$$\le n\sup_{1\le i\le n}\frac{1}{|V_i|}\Big(\int_{V_i}\|R_v^{-1}\|_{X_3}^{\alpha q}dv\Big)^{1/q}$$

$$\cdot\Big(\int_{V_i}\|B_{k_i}((S_u^i f\chi_{V_iC_i^{-1}}),(T_u^i g)\chi_{V_iC_i})(v)\|_{L^{q_3}(\mu)}^{q_3}dv\Big)^{\alpha/q_3}$$

$$\le n\|R^{-1}\|_{A^p(X_3)}^{\alpha}\sup_{1\le i\le n}\Big(\frac{1}{|V_i|}\int_G\|B_{k_i}((S_u^i f\chi_{V_iC_i^{-1}}),(T_u^i g)\chi_{V_iC_i})(v)\|_{L^{q_3}(\mu)}^{q_3}dv\Big)^{\alpha/q_3}$$

$$= n\|R^{-1}\|_{A^p(X_3)}^{\alpha}$$

$$\cdot\sup_{1\le i\le n}\Big(\int_M\frac{1}{|V_i|}\int_G|B_{k_i}((S_u^i f\chi_{V_iC_i^{-1}})(x),(T_u^i g)(x)\chi_{V_iC_i})(v)|^{q_3}dvd\mu(x)\Big)^{\alpha/q_3}$$

$$\le n\|R^{-1}\|_{A^p(X_3)}^{\alpha}\sup_{1\le i\le n}N^{\alpha}(k_i)\Big(\int_M\Big(\frac{1}{|V_i|}\int_{V_iC_i^{-1}}|S_u^i f(x)|^{p_1}du\Big)^{q_3/p_1}$$

$$\cdot\Big(\frac{1}{|V_i|}\int_{V_iC_i}|T_u^i g(x)|^{p_2}du\Big)^{q_3/p_2}d\mu(x)\Big)^{\alpha/q_3}.$$

For each $1\le i\le n$, we denote

$$I_i=\int_M\Big(\frac{1}{|V_i|}\int_{V_iC_i^{-1}}|S_u^i f(x)|^{p_1}du\Big)^{q_3/p_1}\Big(\frac{1}{|V_i|}\int_{V_iC_i}|T_u^i g(x)|^{p_2}du\Big)^{q_3/p_2}d\mu(x).$$

Hence

$$\|T_k(f,g)\|_{q_3}\le n^{1/\alpha}\|R^{-1}\|_{A^p(X_3)}\sup_{1\le i\le n}N(k_i)I_i^{1/q_3}. \tag{2.2}$$

Now consider three cases.

Case 1: $q_3=p_3$. Using that $1/p_3=1/p_1+1/p_2$ and Hölder inequality one gets

$$I_i \le \Big(\frac{1}{|V_i|}\int_{V_iC_i^{-1}}\|S_u^i f\|_{p_1}^{p_1}du\Big)^{p_3/p_1}\Big(\frac{1}{|V_i|}\int_{V_iC_i}\|T_u^i g\|_{p_2}^{p_2}du\Big)^{p_3/p_2}.$$

Case 2: $q_3\ge p_3$ *and* $(M,\mu)=(\mathbb{Z},\nu)$. Write $q_3=\delta p_3$ for some $\delta\ge 1$. Hence

$$I_i\le\Big(\int_M\Big(\frac{1}{|V_i|}\int_{V_iC_i^{-1}}|S_u^i f(x)|^{p_1}du\Big)^{p_3/p_1}\Big(\frac{1}{|V_i|}\int_{V_iC_i}|T_u^i g(x)|^{p_2}du\Big)^{p_3/p_2}d\mu(x)\Big)^{\delta}.$$

This shows that

$$I_i^{1/\delta}\le\Big(\frac{1}{|V_i|}\int_{V_iC_i^{-1}}\|S_u^i f\|_{p_1}^{p_1}du\Big)^{p_3/p_1}\Big(\frac{1}{|V_i|}\int_{V_iC_i}\|T_u^i g\|_{p_2}^{p_2}du\Big)^{p_3/p_2}.$$

Case 3: $q_3 \le p_3$ and $\mu(M) < \infty$. Write $q_3 = \rho p_3$ for some $\rho \le 1$. Hence

$$I_i^{1/\rho} \le \mu(M)^{1/\rho-1}\Big(\frac{1}{|V_i|}\int_{V_iC_i^{-1}} \|S_u^i f\|_{p_1}^{p_1} du\Big)^{p_3/p_1}\Big(\frac{1}{|V_i|}\int_{V_iC_i} \|T_u^i g\|_{p_2}^{p_2} du\Big)^{p_3/p_2}.$$

In all three cases

$$\begin{aligned} I_i^{1/q_3} &\le A\Big(\frac{\gamma}{|V|}\int_{VC} \|S_u^i(\frac{f}{\|f\|_{p_1}})\|_{p_1}^{p_1} du\Big)^{1/p_1} \\ &\cdot\Big(\frac{\gamma}{|V|}\int_{VC} \|T_u^i(\frac{g}{\|g\|_{p_2}})\|_{p_2}^{p_2} du\Big)^{1/p_2} \|f\|_{p_1}\|g\|_{p_2}. \end{aligned}$$

Finally for every $\varepsilon > 0$, let $V_0, V_1 \in \mathcal{V}$ such that $V_0 C \subset V_1$ and $|V_1|/|V_0| \le 1+\varepsilon$. Therefore, applying the previous estimates for V_0, one gets

$$\begin{aligned} \|T_K(f,g)\|_{q_3} &\le C(n,\gamma)\sup_{1\le i\le n} N(k_i)(1+\varepsilon)^{1/p_3}\|R^{-1}\|_{A^p(X_3)} \\ &\cdot(\sup_{1\le i\le n}\|S^i\|_{A_s^{p_1}(L^{p_1})}\|T^i\|_{A_s^{p_2}(L^{p_2})})\|f\|_{p_1}\|g\|_{p_2}. \end{aligned}$$

Taking limits as ε goes to zero completes the proof. □

Let us formulate now a corollary from which one can actually get most applications in this paper.

Corollary 2.5. *Let $1 < p_1, p_2 < \infty$ and (M,μ) be either a finite measure space or $\mathbb{Z}$ with the counting measure. Let K be positive, integrable and with compact support defining a bounded bilinear map $B_K : L^{p_1}(G) \times L^{p_2}(G) \to L^{q_3}(G)$ where $q_3 \le p_3$ (in case $\mu(M) < \infty$) or $p_3 \le q_3$ for $M = \mathbb{Z}$. Let us consider a bounded and measurable in the strong operator topology function $R : G \to \mathcal{B}(L^{p_i}(\mu))$ for $i = 1,2,3$ and where R_u are invertible operators for all $u \in G$ such that $R_u^{-1} = R_{u^{-1}}$.*

(1) Assume that there exist $G = \cup_{i=1}^n G_i$ for some pairwise disjoint measurable sets and a complete family $\mathcal{V}$ in G and $\gamma > 0$ for which $m(V) \le \gamma m(V \cap G_i)$ for all i and for all $V \in \mathcal{V}$.

(2) Assume that there exist bounded and measurable in the strong operator topology functions $S^i : G \to \mathcal{B}(L^{p_1}(\mu))$ and $T^i : G \to \mathcal{B}(L^{p_2}(\mu))$ satisfying that for every $f \in L^{p_1}(M)$ and $g \in L^{p_2}(M)$,

$$R_v(R_{u^{-1}} f R_u g) = S^i_{vu^{-1}} f T^i_{vu} g, \qquad u, v \in G_i. \tag{2.3}$$

Then, the bilinear operator $T_K(f,g) = \int_G R_u f R_{u^{-1}} g K(u) dm(u)$ can be extended to a bounded operator $T_K : L^{p_1}(\mu) \times L^{p_2}(\mu) \to L^{q_3}(\mu)$ with

$$\|T_K\| \le C(n,\gamma) A_1 A_2 A_3 \sup_{1\le i\le n} N(K)$$

where $A_1 = \sup_{u\in G, 1\le i\le n}\|S_u^i\|_{\mathcal{B}(L^{p_1}(\mu))}$, $A_2 = \sup_{u\in G, 1\le i\le n}\|T_u^i\|_{\mathcal{B}(L^{p_2}(\mu))}$, and $A_3 = \sup_{u\in G}\|R_u\|_{\mathcal{B}(L^{q_3}(\mu))}$.

In particular one has the following application:

Corollary 2.6. *Let* $q_3 \geq p_3$*, let* K *be positive, integrable and with compact support defined in* $\mathbb{R}$ *such that*

$$B_K(\phi,\psi)(v) = \int_{\mathbb{R}} \phi(v-u)\psi(v+u)K(u)dm(u),$$

is bounded from $L^{p_1}(\mathbb{R}) \times L^{p_2}(\mathbb{R}) \to L^{q_3}(\mathbb{R})$.

Define $K_n = \int_{A_n} K(u)du$ *where* $A_0 = [-1/4, 1/4]$, $A_n = (n-3/4, n+1/4]$ *for* $n \in \mathbb{N}$ *and* $A_{-n} = -A_n$*. Then the "discrete bilinear" transform associated to* K

$$T_{\mathbb{Z},K}(a,b)(m) = \sum_{n\in\mathbb{Z}} a_{m+n}b_{m-n}K_n$$

is bounded from $\ell_{p_1}(\mathbb{Z}) \times \ell_{p_2}(\mathbb{Z})$ *to* $\ell_{q_3}(\mathbb{Z})$ *and* $\|T_{\mathbb{Z},K}\| \leq CN(K)$.

Proof. We shall apply Corollary 2.5 for $G = \mathbb{R}$. Denote $I_k = [k-1/4, k+1/4]$ for $k \in \mathbb{Z}$, $J_k = (k-3/4, k-1/4)$ and $J'_k = (-k+1/4, -k+3/4)$ for $k \in \mathbb{N}$. Define

$$G_1 = \cup_{k\in\mathbb{Z}} I_k,\ G_2 = \cup_{k\in\mathbb{N}} J_k \text{ and } G_3 = \cup_{k\in\mathbb{N}} J'_k.$$

Consider $\mathcal{V} = \{(-N,N) : N \in \mathbb{N}\}$. It is clear that $m((-N,N)\cap G_2) = m((-N,N)\cap G_3) = m((-N,N)\cap G_1)/2 = m((-N,N))/4$. This gives $\gamma = 2$.

Let us define $R : \mathbb{R} \to \mathcal{B}(\ell_{p_i}(\mathbb{Z}))$ given by

$$R_u = S^{(u)}\chi_{G_1}(u) + S^{(u+\frac{1}{2})}\chi_{G_2}(u) + S^{(u-\frac{1}{2})}\chi_{G_3}(u)$$

where S stands for the Shift operator $S((x_n)) = (x_{n+1})$ and (u) stands for the closest integer to u respectively.

Observe that, for $k \in \mathbb{Z}$, and $u \in I_k$ then $R_u = S^k$. Also, for $k \in \mathbb{N}$, if $u \in J_k$ then $R_u = S^k$ and if $u \in J'_k$ then $R_u = S^{-k}$. Also $R_u(ab) = R_u(a)R_u(b)$ for any sequences a and b. If $u, v \in G_1$ then $-u \in G_1$ and one has $(v+u) = (v)+(u)$ and $(v-u) = (v)+(-u)$. Hence $R_v(R_u a R_{-u} b) = S^{(u+v)}aS^{(v-u)}b$ for $u, v \in G_1$. This allows to take $S^1_u = T^1_u = S^{(u)}$.

If $u, v \in G_2$ then $-u \in G_3$ and we have that $-u-1/2, u+1/2, v+1/2 \in G_1$. Therefore $(v+u+1) = (v+1/2)+(u+1/2)$ and $(v-u) = (v+1/2)+(-u-1/2)$. Hence $R_v(R_u a R_{-u} b) = S^{(u+v+1)}aS^{(v-u)}b$ for $u, v \in G_2$. This allows to take $S^2_u = S^{(u+1)}$ and $T^2_u = S^{(u)}$.

If $u, v \in G_3$ then $-u \in G_2$ and we have that $-u+1/2, u-1/2, v-1/2 \in G_1$. Therefore $(v+u-1) = (v-1/2)+(u-1/2)$ and $(v-u) = (v-1/2)+(-u+1/2)$. Hence $R_v(R_u a R_{-u} b) = S^{(u+v-1)}aS^{(v-u)}b$ for $u, v \in G_3$. This allows to take $S^3_u = S^{(u-1)}$ and $T^3_u = S^{(u)}$.

Since all operators appearing are norm 1 on $\ell_p(\mathbb{Z})$ for any value of p and for any $u \in \mathbb{R}$, one gets, using Corollary 2.5, that T_K is bounded from $\ell_{p_1}(\mathbb{Z}) \times \ell_{p_2}(\mathbb{Z})$ to $\ell_{q_3}(\mathbb{Z})$ and $\|T_K\| \leq 2N(K)$.

Let us finally compute T_K in this case

$$
\begin{aligned}
T_K(a,b) &= \int_{\mathbb{R}} R_{-u}aR_u bK(u)du \\
&= \sum_{k\in\mathbb{Z}} \int_{I_k} R_{-u}aR_u bK(u)du + \sum_{k\in\mathbb{N}} \int_{J_k\cup J_k'} R_{-u}aR_u bK(u)du \\
&= ab\int_{I_0} K(u)du + \sum_{k\in\mathbb{N}} S^{-k}aS^k b\int_{I_k\cup J_k} K(u)du + \sum_{k\in\mathbb{N}} S^k aS^{-k}b\int_{I_{-k}\cup J_k'} K(u)du \\
&= ab\int_{I_0} K(u)du + \sum_{k\in\mathbb{N}} S^{-k}aS^k b\int_{(k-\frac{3}{4},k+\frac{1}{4}]} K(u)du \\
&\quad + \sum_{k\in\mathbb{N}} S^k aS^{-k}b\int_{[-k-\frac{1}{4},-k+\frac{3}{4})} K(u)du \\
&= \sum_{k\in\mathbb{Z}} S^k aS^{-k}b\int_{A_k} K(u)du
\end{aligned}
$$

and therefore $T_K(a,b)(m) = \sum_{n\in\mathbb{Z}} a_{m+n}b_{m-n}K_n$. □

Now one can obtain the following application.

Theorem 2.7. *Let* $1 < p_1, p_2 < \infty$, $1/p_3 = 1/p_1 + 1/p_2$, $0 < \alpha < \min\{1, 1/p_3\}$, *and* $1/q_3 = 1/p_3 - \alpha$. *For* $a \equiv \{a_j\}_{j=-\infty}^{\infty} \in \ell^{p_1}(\mathbb{Z})$, $b \equiv \{b_j\}_{j=-\infty}^{\infty} \in \ell^{p_2}(\mathbb{Z})$, *let*

$$\mathcal{I}_\alpha(a,b)(m) = \sum_{n\in\mathbb{N}} \frac{a_{m+n}b_{m-n}}{n^{1+\alpha}}.$$

Then there is a constant $\mathfrak{D}_{p_1,p_2}$, *depending only on* p_1 *and* p_2, *such that*

$$\|\mathcal{I}_\alpha(a,b)\|_{\ell^{q_3}(\mathbb{Z})} \le \mathfrak{D}_{p_1,p_2}\|a\|_{\ell^{p_1}(\mathbb{Z})}\|b\|_{\ell^{p_2}(\mathbb{Z})}.$$

Proof. Assume first that $p_3 \ge 1/(\alpha+1)$, that is $q_3 \ge 1$. This case follows from the vector-valued inequality

$$
\begin{aligned}
\|\mathcal{I}_\alpha(a,b)\|_{\ell^{q_3}(\mathbb{Z})} &\le \sum_{n\in\mathbb{N}} \frac{\|a_{\cdot+n}b_{\cdot-n}\|_{\ell^{q_3}(\mathbb{Z})}}{n^{1+\alpha}} \le \sum_{n\in\mathbb{N}} \frac{\|a_{\cdot+n}b_{\cdot-n}\|_{\ell^{p_3}(\mathbb{Z})}}{n^{1+\alpha}} \\
&\le \|a\|_{\ell^{p_1}(\mathbb{Z})}\|b\|_{\ell^{p_2}(\mathbb{Z})}\sum_{n\in\mathbb{N}} \frac{1}{n^{1+\alpha}} \le C\|a\|_{\ell^{p_1}(\mathbb{Z})}\|b_{\cdot-n}\|_{\ell^{p_2}(\mathbb{Z})}.
\end{aligned}
$$

In the case $p_3 < 1/(\alpha+1)$ we use transference. It was shown by Kenig and Stein that for $0 < \alpha < \min\{1, 1/p_3\}$ the operator $I_\alpha : L^{p_1}(\mathbb{R}) \times L^{p_2}(\mathbb{R}) \to L^{q_3}(\mathbb{R})$ defined by

$$I_\alpha(f,g)(x) = \int_{\mathbb{R}} \frac{f(x-t)g(g+t)}{|t|^{\alpha+1}}dt$$

is bounded for $1/q_3 = 1/p_3 - \alpha$. Apply Corollary 2.6 to the kernels $K^N(t) = 1/|t|^{1+\alpha}\chi_{(-N,N)}(t)$ for $G = \mathbb{R}$ and notice that $K_n^N \approx 1/|n|^{1+\alpha}$ for $0 < |n| \le N$. □

3 Transference for maximal operators

In this section we do not give complete proofs since the arguments are quite similar to the previous ones. For a complete treatment of maximal bisublinear discretization and transference without the special assumptions used below, see [1].

Theorem 3.1. *Let us assume the hypotheses of Theorem 2.4 for the case $q_3 = p_3$ and that R_u^{-1} are positive operators. Let $\{K_j\}_j$ be a family of kernels in $L^1(G)$ with compact supports $\{C_j\}_j$ and assume that, for $i = 1, \dots, n$, the corresponding bisublinear maximal operator*

$$B_K^*(\phi, \psi)(v) = \sup_{j\in\mathbb{N}, 1\le i\le n} \Big| \int_G \phi(vu^{-1})\psi(vu)K_j(u)\phi_i(u)dm(u) \Big|, \tag{3.1}$$

is bounded from $L^{p_1}(G)\times L^{p_2}(G)$ to $L^{p_3}(G)$ with norm less than or equal to $N(\{K_j\}_j)$. Then we have that the maximal operator

$$T^*(f,g) = \sup_j |T_{K_j}(f,g)| = \sup_j \Big| \int_G R_u^1 f R_{u^{-1}}^2 g K_j(u)dm(u)\Big|$$

is bounded from $L^{p_1}(\mu)\times L^{p_2}(\mu)$ into $L^{p_3}(\mu)$ with $\|T^*\| \le C(n,p)A_1A_2A_3N(\{K_j\}_j)$ where A_i for $i = 1, 2, 3$ are the same constants appearing in Theorem 2.4.

Proof. Denote $k_i^j(u) = K_j(u)\phi_i(u)$. As shown in Theorem 2.4, for $(v_1, \dots, v_n) \in V_1 \times \cdots \times V_n$, and $j \in \mathbb{N}$ we have that

$$T_{K_j}(f,g) = \sum_{i=1}^n R_{v_i}^{-1} B_{k_i^j}((S_u^i f)\chi_{V_iC_i}, (T_u^i g)\chi_{V_iC_i})(v_i).$$

Hence, using the positivity of $R_{v_i}^{-1}$ one has

$$\sup_{1\le j\le N} |T_{K_j}(f,g)| \le \sum_{i=1}^n R_{v_i}^{-1} \sup_{1\le j\le N} |B_{k_i^j}((S_u^i f)\chi_{V_iC_i}, (T_u^i g)\chi_{V_iC_i})(v_i))|.$$

Therefore $T^*(f,g) \le \sum_{i=1}^n R_{v_i}^{-1}(B_K^*((S_u^i f)\chi_{V_iC_i}, (T_u^i g)\chi_{V_iC_i})(v_i))$. Now repeat the argument from Theorem 2.4. □

Similarly it is not difficult to show the following maximal version of Corollary 2.6.

Theorem 3.2. *Let $\bar{K} = \{K_j\}_j$ be a family of positive and integrable functions defined in $\mathbb{R}$ such that*

$$B_{\bar{K}}^*(\phi,\psi)(v) = \sup_j \Big| \int_{\mathbb{R}} \phi(v-u)\psi(v+u)K_j(u)du \Big|$$

is bounded from $L^{p_1}(\mathbb{R}) \times L^{p_2}(\mathbb{R})$ into $L^{p_3}(\mathbb{R})$, with norm $N(B^*)$.

Define $K_n^j = \int_{n-1/4}^{n+1/4} K_j(u)du$. *Then the maximal "discrete bisublinear" transform associated to* $\bar{K}$

$$T_{\bar{K}}^*(a,b)(m) = \sup_j \Big| \sum_{n\in\mathbb{Z}} a_{m-n} b_{m+n} K_n^j \Big|$$

is bounded from $\ell_{p_1}(\mathbb{Z}) \times \ell_{p_2}(\mathbb{Z})$ *into* $\ell_{p_3}(\mathbb{Z})$ *and* $\|T_{\bar{K}}^*\| \leq CN(B^*)$.

Then one can transfer the bisublinear Hardy–Littlewood maximal operator in $\mathbb{R}$. It was shown by Lacey (see [16]) that

$$M(f,g)(x) = \sup_{R>0} \frac{1}{2R} \int_R^R |f(x+t)||g(x-t)|dt$$

maps $L^{p_1}(\mathbb{R}) \times L^{p_2}(\mathbb{R})$ into $L^{p_3}(\mathbb{R})$ for $p_1, p_2 > 1$ and $1/p_1 + 1/p_2 = 1/p_3 < 3/2$. The reader should be aware that the case $p_3 > 1$ is elementary, and only the case $p_3 \leq 1$ is relevant.

We can now give the following alternative proof of (1.8) whose statement we repeat as the next corollary.

Corollary 3.3. *Let* $p_1, p_2 > 1$ *and* $1/p_1 + 1/p_2 = 1/p_3 < 3/2$. *Then*

$$M(a,b)(m) = \sup_{N\geq 1} \frac{1}{2N} \sum_{n=-N}^{N} |a_{m-n}||b_{m+n}|$$

is bounded from $\ell_{p_1}(\mathbb{Z}) \times \ell_{p_2}(\mathbb{Z})$ *into* $\ell_{p_3}(\mathbb{Z})$.

Proof. Let us consider $K_j = \frac{1}{2j}\chi_{(-j-1/4, j+1/4)}$. Clearly

$$\begin{aligned} B_K^*(\phi,\psi)(v) &= \sup_{j\in\mathbb{N}, 1\leq i\leq n} \Big| \int_{G_i} \phi(v-u)\psi(v+u)K_j(u)du \Big| \\ &\leq \sup_{j\in\mathbb{N}} \Big| \frac{1}{2j} \int_{-j-\frac{1}{4}}^{j+\frac{1}{4}} \phi(v-u)\psi(v+u)K_j(u)du \Big| \\ &\leq 2M(\phi,\psi)(v). \end{aligned}$$

Notice that $K_n^j = \int_{A_n} K_j(u)du = \frac{1}{2j}\chi_{\{|n|\leq j\}}(n)m(A_n)$ where $A_0 = [-1/4, 1/4]$, $A_n = (n-3/4, n+1/4]$ and $A_{-n} = [-n-1/4, -n+3/4)$ for $n \in \mathbb{N}$. Therefore

$$T_{\mathbb{Z},K_j}(a,b) = \frac{1}{2j} \sum_{|k|\leq j} S^k a S^{-k} b - \frac{1}{4j} ab.$$

Then for $a, b \geq 0$, the "maximal discrete bilinear" transform can be estimated as follows:

$$\begin{aligned} M(a,b)(m) &= \sup_{j\geq 1} \frac{1}{2j} \sum_{k=-j}^{j} |a_{m-k}||b_{m+k}| \\ &\leq \sup_j |T_{\mathbb{Z},K_j}(a,b)(m)| + a(m)b(m). \end{aligned}$$

And the result follows from Theorem 3.2. □

In turn, Corollary 3.3 can be transferred so as to yield the bisublinear ergodic maximal operator, which we formulate here as special case of [1, Theorem 4.3].

Theorem 3.4. *Let (Ω, Σ, μ) be a σ-finite measure space, $\tau : \Omega \to \Omega$ be an invertible measure-preserving transformation and define $T(f) = f \circ \tau$. Then the "bilinear ergodic maximal transform"*

$$M_\tau(f,g)(x) = \sup_{N\in\mathbb{N}} \frac{1}{2N} \sum_{|n|\leq N} |T^n f(x)||T^{-n} g(x)|$$

is bounded from $L^{p_1}(\Omega) \times L^{p_2}(\Omega)$ into $L^{p_3}(\Omega)$ whenever $p_1, p_2 > 1$ and $1/p_1 + 1/p_2 = 1/p_3 < 3/2$.

In particular, let A be a matrix with $|\det(A)| = 1$ and consider $Tf(x) = f(Ax)$ for $x \in \mathbb{R}^n$ one obtains the following:

Corollary 3.5 ([1]). *The maximal transform*

$$M_A(f,g)(x) = \sup_{N\in\mathbb{N}} \frac{1}{2N} \sum_{|n|\leq N} |f(A^n x)||g(A^{-n} x)|$$

is bounded from $L^{p_1}(\mathbb{R}^n) \times L^{p_2}(\mathbb{R}^n)$ into $L^{p_3}(\mathbb{R}^n)$ whenever $p_1, p_2 > 1$ and $1/p_1 + 1/p_2 = 1/p_3 < 3/2$.

4 Bilinear vector-valued transference

Throughout this section X, Y, and Z are arbitrary Banach spaces, and β is a bounded bilinear mapping of $X \times Y$ into Z, G is an arbitrary locally compact abelian group with given Haar measure m (sometimes abbreviated by du), and K is an arbitrary m-integrable complex-valued function on G. When (Ω, μ) is a measure space and $1 \leq p < \infty$, we shall denote by $L^p_X(\mu)$ the usual Lebesgue space of X-valued μ-measurable functions ψ such that

$$\|\psi\|^p_{L^p_X(\mu)} \equiv \int_\Omega \|\psi\|^p_X \, d\mu < \infty.$$

If μ is the Haar measure m of G (respectively, if X is the field of complex numbers $\mathbb{C}$), then $L^p_X(\mu)$ will be denoted by $L^p_X(G)$ (respectively, by $L^p(\mu)$).

$R^{(1)}_{(\cdot)}$, $R^{(2)}_{(\cdot)}$, and $R^{(3)}_{(\cdot)}$ will designate given functions defined on G which take values in $\mathcal{B}(X)$, $\mathcal{B}(Y)$, and $\mathcal{B}(Z)$ respectively and satisfy the following hypotheses:

(a) For $j = 1, 2, 3$, $R^{(j)}_{(\cdot)}$ is a strongly continuous function on G.

(b) For $j = 1, 2$,

$$A_j := \sup_{u\in G} \|R^{(j)}_u\| < \infty. \tag{4.1}$$

(c) There is a positive real constant A_3 such that

$$\|z\| \leq A_3 \left\| R_u^{(3)} z \right\|, \quad \text{for all } z \in Z,\, u \in G. \tag{4.2}$$

(d) For all $u \in G$, $v \in G$, $x \in X$, and $y \in Y$,

$$R_v^{(3)} \left(\beta(R_{u^{-1}}^{(1)} x, R_u^{(2)} y) \right) = \beta\left(R_{vu^{-1}}^{(1)} x, R_{vu}^{(2)} y \right). \tag{4.3}$$

Under the foregoing assumptions and notation, we now use Z-valued Bochner integration to define the bilinear mapping $H_K : X \times Y \to Z$ by putting

$$H_K(x,y) = \int_G \beta\left(R_{u^{-1}}^{(1)} x, R_u^{(2)} y\right) K(u)\, du, \text{ for all } x \in X,\, y \in Y. \tag{4.4}$$

Notice that H_K is a bounded bilinear mapping, with

$$\|H_K\| \leq \|\beta\|\, A_1 A_2\, \|K\|_{L^1(G)}\,. \tag{4.5}$$

Since L^1-norms of integration kernels tend to have higher orders of magnitude than corresponding integration operators defined by them, it is desirable to replace the factor $\|K\|_{L^1(G)}$ in the majorant of (4.5) with a quantity which has a milder size in principle. This will be accomplished in our main transference result below (Theorem 4.3), where vector-valued transference methods effectively replace $\|K\|_{L^1(G)}$ in (4.5) with the norm of the bilinear mapping $B_{K,\beta}$ defined as follows.

Definition 4.1. Suppose that $1 < p_1, p_2 < \infty$, $1/p_3 = 1/p_1 + 1/p_2$ and $1 \leq p_3 < \infty$. This notation will be tacitly in effect henceforth. We use Z-valued Bochner integration to define the bilinear mapping $B_{K,\beta} : L_X^{p_1}(G) \times L_Y^{p_2}(G) \to L_Z^{p_3}(G)$ by writing for all $f \in L_X^{p_1}(G)$, $g \in L_Y^{p_2}(G)$,

$$B_{K,\beta}(f,g)(v) = \int_G \beta\left(f(u^{-1}v), g(uv)\right) K(u)\, du. \tag{4.6}$$

Remark 4.2. It is straightforward to see that the integral on the right hand side of (4.6) exists for m-almost all $v \in G$ and defines a Z-valued m-measurable function of v satisfying the crude estimate

$$\|B_{K,\beta}(f,g)\|_{L_Z^{p_3}(G)} \leq \|K\|_{L^1(G)} \|\beta\| \|f\|_{L_X^{p_1}(G)} \|g\|_{L_Y^{p_2}(G)}. \tag{4.7}$$

In the special case where X, Y, and Z coincide with $\mathbb{C}$, and $\beta(x,y) = xy$, we shall denote the bounded bilinear mapping $B_{K,\beta} : L^{p_1}(G) \times L^{p_2}(G) \to L^{p_3}(G)$ by s_K. When K has compact support, s_K coincides with the bilinear operator B_K defined in Section 1.

We are now ready to take up the result of this section, which is stated as follows (compare with [2, Theorem 3.2]).

Theorem 4.3. *Let p_1, p_2, and p_3 be as in Definition 4.1. Then in terms of the above hypotheses and notation, we have for all $x \in X$ and $y \in Y$,*

$$\|H_K(x, y)\| \leq A_1 A_2 A_3 \|B_{K,\beta}\| \|x\| \|y\| . \tag{4.8}$$

Proof. In view of (4.5) and (4.7) together with standard approximations in $L^1(G)$, it suffices to establish (4.8) in the special case where K is compactly supported (which we now assume). Let C be a compact subset of G such that K vanishes outside of C. Temporarily fix vectors $x \in X$, $y \in Y$. By (4.2) and (4.3), we see that

$$\|H_K(x, y)\|^{p_3} \leq A_3^{p_3} \Big\| \int_G \beta(R^{(1)}_{vu^{-1}} x, R^{(2)}_{vu} y) K(u)\, du \Big\|^{p_3}, \text{ for all } v \in G. \tag{4.9}$$

Let $\varepsilon > 0$ be arbitrary. Use that G is a l.c.a. group to get an open neighborhood V of the identity in G such that V has compact closure, and

$$\frac{|m\left(V(C \cup C^{-1})\right)|}{m(V)} < 1 + \varepsilon. \tag{4.10}$$

Denote by χ the characteristic function, defined on G, of $V(C \cup C^{-1})$. Integrating (4.9) over V with respect to dv, we see that

$$\|H_K(x, y)\|^{p_3} \leq \frac{A_3^{p_3}}{m(V)} \int_V \Big\| \int_G \beta(R^{(1)}_{vu^{-1}} x, R^{(2)}_{vu} y) \chi(vu^{-1}) \chi(vu) K(u)\, du \Big\|^{p_3} dv. \tag{4.11}$$

Next, let us define $f \in L^{p_1}_X(G)$ and $g \in L^{p_2}_Y(G)$ by writing, for all $u \in G$,

$$f(u) = \chi(u) R^{(1)}_u x \quad \text{and } g(u) = \chi(u) R^{(2)}_u y. \tag{4.12}$$

We can accordingly rewrite (4.11) in the following form:

$$\|H_K(x, y)\|^{p_3} \leq \frac{A_3^{p_3}}{m(V)} \int_V \|B_{K,\beta}(f, g)(v)\|^{p_3}\, dv.$$

Consequently,

$$\begin{aligned} \|H_K(x, y)\| &\leq \frac{A_3}{[m(V)]^{1/p_3}} \|B_{K,\beta}(f, g)\|_{L^{p_3}_Z(G)} \\ &\leq \frac{A_3}{[m(V)]^{1/p_3}} \|B_{K,\beta}\| \|f\|_{L^{p_1}_X(G)} \|g\|_{L^{p_2}_Y(G)} . \end{aligned} \tag{4.13}$$

By (4.12),

$$\|f\|_{L^{p_1}_X(G)} \leq A_1 \|x\| [m(V(C \cup C^{-1}))]^{1/p_1}$$

$$\|g\|_{L^{p_2}_Y(G)} \leq A_2 \|y\| [m(V(C \cup C^{-1}))]^{1/p_2}.$$

Applying these estimates to (4.13), we see directly that

$$\|H_K(x,y)\| \le \frac{A_1A_2A_3}{[m(V)]^{1/p_3}} \|B_{K,\beta}\| \left[m\left(V\left(C \cup C^{-1}\right)\right)\right]^{1/p_3} \|x\| \, \|y\| \,,$$

and hence by (4.10), $\|H_K(x,y)\| \le (1+\varepsilon)^{1/p_3} A_1A_2A_3 \|B_{K,\beta}\| \|x\| \|y\|$. We now let $\varepsilon \to 0$ to obtain (4.8), and thereby complete the proof. □

We now specialize our last result to $L^p(\mu)$-spaces. Actually, we show that the estimate in the general transference result for bilinear maps (Theorem 4.3) can be refined when we specialize the general Banach spaces X, Y, and Z to be, respectively, $L^{p_1}(\mu)$, $L^{p_2}(\mu)$, and $L^{p_3}(\mu)$. This refinement is accomplished by the following lemma which can be demonstrated by suitably adapting the reasoning of [2, Lemma 4.2].

Lemma 4.4. *Let p_1, p_2, p_3 be as in Definition 4.1, and (Ω, μ) be an arbitrary measure space. Specialize the preceding hypotheses and notation surrounding the arbitrary function $K \in L^1(G)$ to the case where $X = L^{p_1}(\mu)$, $Y = L^{p_2}(\mu)$, and $Z = L^{p_3}(\mu)$, and the bounded bilinear form $\beta : X \times Y \to Z$ be defined as the pointwise product on Ω: $\beta(f,g) = fg$ (in particular, it is automatic that $\|\beta\| \le 1$ here). Then, in terms of the bilinear mapping s_K defined in Remark 4.2 above, we have*

$$\|B_{K,\beta}\| \le \|s_K\| \,. \tag{4.14}$$

Remark 4.5. When the hypotheses of Theorem 4.3 are specialized in accordance with the statement of Lemma 4.4, the theorem and lemma combine to furnish the following transference estimate in the resultant measure-theoretic context ([6, Theorem 2.1]):

$$\|H_K\| \le A_1A_2A_3 \|s_K\| \,.$$

This estimate has the pleasant feature that $\|s_{K,p_1,p_2}\|$ is independent of the abstract measure μ (in contrast to $\|B_{K,\beta}\|$).

References

[1] E. Berkson, O. Blasco, M. Carro, and T. A. Gillespie, *Discretization and transference of bisublinear maximal operators*, J. Fourier Anal. Appl. 12 (2006), pp. 447–481.

[2] E. Berkson and T. A. Gillespie, *Spectral decompositions and vector-valued transference*, Analysis at Urbana, Vol. II (Urbana, IL, 1986–1987), pp. 22–51, London Math. Soc. Lecture Note Ser. 138, Cambridge Univ. Press, Cambridge, 1989.

[3] ———, *Mean-boundedness and Littlewood-Paley for separation preserving operators*, Trans. Amer. Math. Soc. 349 (1997), pp. 1169–1189.

[4] E. Berkson, M. Paluszyński, and G. Weiss, *Transference couples and their applications to convolution operators and maximal operators*, in "Interaction between Functional Analysis, Harmonic Analysis, and Probability", Proceedings of Conference at U. of Missouri-Columbia (May 29–June 3, 1994), Lecture Notes in Pure and Applied Math., vol. 175, Marcel Dekker, Inc., New York, 1996, pp. 69–84.

[5] O. Blasco, *Bilinear multipliers and transference*, Int. J. Math. Math. Sci. 2005 (2005), pp. 545–554.

[6] O. Blasco, M. Carro, and T. A. Gillespie, *Bilinear Hilbert transform on measure spaces*, J. Fourier Anal. Appl. 11 (2005), pp. 459–470.

[7] O. Blasco and F. Villarroya, *Transference of bilinear multipliers on Lorentz spaces*, Illinois J. Math. 47, pp. 1327–1343.

[8] R. Coifman and W. Weiss, *Transference Methods in Analysis*, Regional Conf. Series in Math., No. 31, Amer. Math. Soc., Providence, [1977].

[9] C. Demeter, T. Tao, and C. Thiele, *Maximal multilinear operators* (Submitted).

[10] D. Fan and S. Sato, *Transference of certain multilinear multipliers operators*, J. Aust. Math. Soc. 70 (2001), pp. 37–55.

[11] L. Grafakos and P. Honzík, *Maximal transference and summability of multilinear Fourier series*, J. Aust. Math. Soc. 80 (2006), pp. 65–80.

[12] L. Grafakos and N. Kalton, *Some remarks on multilinear maps and interpolation*, Math. Annalen 319 (2001), pp. 151–180.

[13] L. Grafakos and G. Weiss, *Transference of multilinear operators*, Illinois J. Math. 40 (1996), pp. 344–351.

[14] E. Hewitt and K. Ross, *Abstract Harmonic Analysis I*, Grundlehren der Math. Wissenschaften (Band 115), Academic Press, New York, 1963.

[15] C. E. Kenig and E. M. Stein, *Multilinear estimates and fractional integration*, Math. Res. Lett. 6 (1999), pp. 1–15.

[16] M. Lacey, *The bilinear maximal function maps into L^p for $\frac{2}{3} < p \leq 1$*, Ann. Math. (2) 151 (2000), pp. 35–57.

[17] M. Lacey and C. Thiele, *L^p estimates on the bilinear Hilbert transform for* $2 < p < \infty$, Ann. Math. (2) 146 (1997), pp. 693–724.

[18] ———, *Weak bounds for the bilinear Hilbert transform on L^p*, Documenta Mathematica, extra volume ICM 1–1000, 1997.

[19] ———, *On Calderón's conjecture*, Ann. Math. (2) 149 (1999), pp. 475–496.

[20] L. Loomis, *Abstract Harmonic Analysis*, Van Nostrand, New York 1953.

[21] W. Rudin, *Fourier Analysis on Groups*, Interscience, New York 1962.

Author information

Earl Berkson, Department of Mathematics, University of Illinois, 1409 West Green Street, Urbana, IL 61801, USA.
Email: `berkson@math.uiuc.edu`

Oscar Blasco, Departmento de Análisis Matemático, Universidad de Valencia, Burjassot (Valencia) 46100, Spain.
Email: `oblasco@uv.es`

María J. Carro, Departament de Matematica Aplicada i Analisis, Universitat de Barcelona, Barcelona 08071, Spain.
Email: `carro@ub.es`

Thomas A. Gillespie, Department of Mathematics and Statistics, University of Edinburgh, Edinburgh EH9 3JZ, Scotland, UK.
Email: `alastair@maths.ed.ac.uk`

Banach Spaces and their Applications in Analysis, 285–302

Implicit function theorems for nonregular mappings in Banach spaces. Exit from singularity

Olga A. Brezhneva and Alexey A. Tret'yakov

Abstract. We consider the equation $F(x, y) = 0$, where $F : X \times Y \to Z$ is a smooth mapping, and X, Y and Z are Banach spaces. In the case when $F(x^*, y^*) = 0$ and the mapping F is regular at (x^*, y^*), i.e., when $F'_y(x^*, y^*)$, the derivative of F with respect to y, is invertible, the classical implicit function theorem guarantees the existence of a smooth mapping ϕ defined on a neighborhood of x^* such that $F(x, \phi(x)) = 0$ and $\phi(x^*) = y^*$. We are interested in the case when the mapping F is nonregular and the classical implicit function theorem is not applicable. We present generalizations of the implicit function theorem for this case. The results are illustrated by some examples, including differential equations.

Key words. Implicit function theorem, nonlinear equation, nonlinear boundary-value problem, nonregular mapping.

AMS classification. 47J07, 58C15, 34B15.

1 Introduction

We consider the equation

$$F(x, y) = 0, \quad F : X \times Y \to Z, \tag{1.1}$$

and the problem of existence of a mapping $\varphi(x) : U \to Y$ such that

$$F(x, \varphi(x)) = 0, \tag{1.2}$$

for all $x \in U$, where U is some neighborhood of the point x^* and $F(x^*, y^*) = 0$.

There are numerous books and papers devoted to the implicit function theorem, amongst which are [7, 12]. However, most of the work covers the regular case, when $F'_y(x^*, y^*)$, the derivative of F with respect to y, is onto, and, hence, the classical implicit function theorem guarantees the existence of a mapping $\varphi(x)$ defined on a neighborhood of x^* such that $F(x, \varphi(x)) = 0$ and $\varphi(x^*) = y^*$.

We are interested in the case when the operator $F'_y(x^*, y^*)$ is not surjective and the classical implicit function theorem cannot be applied. To our knowledge, the first generalization of the implicit function theorem for nonregular mappings was formulated in [15]. Generalizations of the implicit function theorem for 2-regular mappings were obtained in [3, 4, 9, 10, 11]. This paper presents generalized higher order implicit function theorems applicable to nonregular mappings and can be considered as an extension of the results presented in [5, 10, 11, 15]. Some of the theorems introduced in the paper are applicable to new classes of the nonregular problems. All results obtained in the paper are based on the constructions of the p-regularity theory whose basic concepts and main results are described in [10, 11, 17]. We illustrate the results by examples, including differential equations.

In the regular case, existence of the mapping $\varphi(x)$ is based on the existence of the inverse operator $\{F'_y(x^*, y^*)\}^{-1}$. Since in the nonregular case the operator $F'_y(x^*, y^*)$ is not onto, we instead use another operator that is onto. Such an operator is called the *p-factor-operator* and its construction is based on the derivatives of F up to some order as well as on some element h. The presence of h, which we call the *exit from singularity*, yields a family of the p-factor-operators "replacing" the nonregular operator $F'_y(x^*, y^*)$. In the paper, we also introduce generalized operators, whose definitions are based on a sequence of vectors $\{h_1, \dots, h_q\}$ rather than on a unique vector h. As a result, we obtain a family of generalized implicit function theorems, which are applicable to different nonregular problems.

The organization of the paper is as follows. In the next section, we recall main definitions and concepts of the p-regularity theory. In Section 3, we formulate and prove the implicit function theorems for nonregular mappings. We start with consideration of Theorem 3.2, Theorem 3.5, Theorem 3.8, and Theorem 3.10, which are based on the construction of the p-factor-operator. Then we define generalized operators, which allow us to introduce Theorem 3.13 and Theorem 3.17. We illustrate the theorems by some examples, including ordinary differential equations.

Notation. We use $\mathcal{H}(S_1, S_2)$ to denote the *Hausdorff distance* between two sets S_1 and S_2:

$$\mathcal{H}(S_1, S_2) = \max\Big\{\sup_{x\in S_1} \operatorname{dist}(x, S_2),\ \sup_{y\in S_2} \operatorname{dist}(y, S_1)\Big\}.$$

We denote the space of all continuous linear operators from X to Y by $\mathcal{L}(X, Y)$.

Let p be a natural number and let $B : X \times X \times \cdots \times X$ (with p copies of X) $\to Y$ be a continuous symmetric p-multilinear mapping. The p-form associated to B is the map $B[\cdot]^p : X \to Y$ defined by

$$B[x]^p = B(x, x, \dots, x),$$

for $x \in X$. Alternatively, we may simply view $B[\cdot]^p$ as a homogeneous polynomial $Q : X \to Y$ of degree p, i.e., $Q(\alpha x) = \alpha^p Q(x)$. The space of continuous homogeneous polynomials $Q : X \to Y$ of degree p will be denoted by $\mathcal{Q}^p(X, Y)$.

For a differentiable mapping $F : X \times Y \to Z$, its Fréchet derivative with respect to y at a point $(x, y) \in X \times Y$ will be denoted $F'_y(x, y) : Y \to Z$. If F is of class C^p, we let $F^{(p)}_{y\dots y}(x, y)$ be the pth derivative of F with respect to y at the point (x, y) (a symmetric multilinear map of p copies of Y to Z) and the associated p-form, also called the pth-order mapping, is

$$F^{(p)}_{y\dots y}(x, y)[h]^p = F^{(p)}_{y\dots y}(x, y)(h, h, \dots, h).$$

We also use the notation

$$\operatorname{Ker}^p F^{(p)}(x, y) = \{h \in X \times Y \mid F^{(p)}(x, y)\,[h]^p = 0\}$$

and refer to it as the p-kernel of the pth-order mapping.

2 The p-factor operator

Consider a nonlinear mapping $F : X \times Y \to Z$, where X, Y and Z are Banach spaces. Assume that for some point $(x^*, y^*) \in X \times Y$,

$$\operatorname{Im} F'_y(x^*, y^*) \neq Z.$$

For the purpose of describing nonlinear problems, the concept of p-regularity was introduced in [13, 14, 15] using the notion of a p-factor operator. The p-factor operator is constructed under the assumption that the space Z is decomposed into the direct sum

$$Z = Z_1 \oplus \cdots \oplus Z_p,$$

where $Z_1 = \overline{\operatorname{Im}} F'_y(x^*, y^*)$, the closure of the image of the first partial derivative of F with respect to y evaluated at (x^*, y^*), and the remaining spaces are defined as follows. Let W_2 be a closed complementary subspace to Z_1 (we are assuming that such a closed complement exists) and let $P_{W_2} : Z \to W_2$ be the projection operator onto W_2 along Z_1. Let Z_2 be the closed linear span of the image of the quadratic map $P_{W_2} F''_{yy}(x^*, y^*)[\cdot]^2$. More generally, define inductively,

$$Z_i = \overline{\operatorname{span}}(\operatorname{Im} P_{W_i} F^{(i)}_{y\ldots y}(x^*, y^*)[\cdot]^i) \subseteq W_i, \quad i = 2, \ldots, p-1,$$

where W_i is a choice of closed complementary subspace for $(Z_1 \oplus \cdots \oplus Z_{i-1})$ with respect to Z, $i = 2, \ldots, p$, and $P_{W_i} : Z \to W_i$ is the projection operator onto W_i along $(Z_1 \oplus \cdots \oplus Z_{i-1})$ with respect to Z, $i = 2, \ldots, p$. Finally, let $Z_p = W_p$.

Define the following mappings (see [15])

$$f_i(x, y) : X \times Y \to Z_i, \quad f_i(x, y) = P_{Z_i} F(x, y), \quad i = 1, \ldots, p, \tag{2.1}$$

where $P_{Z_i} : Z \to Z_i$ is the projection operator onto Z_i along $(Z_1 \oplus \cdots \oplus Z_{i-1} \oplus Z_{i+1} \oplus \cdots \oplus Z_p)$ with respect to Z. Then the mapping F can be represented as

$$F(x, y) = f_1(x, y) + \cdots + f_p(x, y).$$

Throughout the paper we will not make any distinction between $Z_1 \times \cdots \times Z_p$ and $Z_1 \oplus \cdots \oplus Z_p$. Also any function F into $Z_1 \times \cdots \times Z_p$ may be represented as $f_1(x, y) + \cdots + f_p(x, y)$ or $(f_1(x, y), \ldots, f_p(x, y))$.

The class of operators below plays a central role in p-regularity theory.

Definition 2.1. The operator $\Psi_p(h) \in \mathcal{L}(Y, Z_1 \times \cdots \times Z_p)$, for $h \in Y$, defined by

$$\Psi_p(h) = \left(f'_{1y}(x^*, y^*), f''_{2\,yy}(x^*, y^*)[h], \ldots, \frac{1}{(p-1)!} f^{(p)}_{p\;y\ldots y}(x^*, y^*)[h]^{p-1} \right), \tag{2.2}$$

is called the *p-factor operator*.

We also introduce the corresponding inverse multivalued operator Ψ_p^{-1}:

$$\{\Psi_p(h)\}^{-1}(z) =$$
$$\left\{ \xi \in Y : (f'_{1y}(x^*, y^*)[\xi], \ldots, \frac{1}{(p-1)!} f^{(p)}_{p\;y\ldots y}(x^*, y^*)[h^{p-1}, \xi]) = z \right\},$$

where $z = (z_1, \ldots, z_p)$ with $z_i \in Z_i$ for $i = 1, \ldots, p$.

Definition 2.2. The mapping $F(x,y)$ is called *p-regular* at the point (x^*,y^*) with respect to the vector h if $\operatorname{Im}\Psi_p(h) = Z_1 \times \cdots \times Z_p$.

Definition 2.3. The mapping $F(x,y)$ is called *uniformly p-regular* over the set M if

$$\sup_{h\in M} \|\{\Psi_p(\bar h)\}^{-1}\| < \infty, \quad \bar h = \frac{h}{\|h\|}, \quad h \neq 0,$$

where $\|\{\Psi_p(\bar h)\}^{-1}\| = \sup_{\|z\|=1} \inf\{\|y\| : \Psi_p(h)[y] = z\}$.

3 Implicit function theorems for nonregular mappings

Consider the equation

$$F(x,y) = 0,$$

where $F : X \times Y \to Z$ is a sufficiently smooth mapping, and X, Y and Z are Banach spaces. Assume that for some (x^*,y^*), $F(x^*,y^*) = 0$, and that the mapping F is nonregular, that is,

$$\operatorname{Im} F'_y(x^*,y^*) \neq Z.$$

In this section we present generalizations of the implicit function theorem for the mapping $F(x,y)$.

Introduce the mapping $\Phi_p : Y \to Z_1 \times \cdots \times Z_p$ defined by

$$\Phi_p[y]^p = \Big(f'_{1\,y}(x^*,y^*)[y], \frac{1}{2} f''_{2\,yy}(x^*,y^*)[y]^2, \dots, \frac{1}{p!} f^{(p)}_{p\,y\dots y}(x^*,y^*)[y]^p\Big), \quad y \in Y.$$

Under the assumption that $Z = Z_1 \oplus \cdots \oplus Z_p$, we also introduce the corresponding inverse multivalued operator Φ_p^{-1}:

$$\Phi_p^{-1}(z) = \Big\{\eta \in Y : (f'_{1y}(x^*,y^*)[\eta], \dots, \frac{1}{p!} f^{(p)}_{p\,y\dots y}(x^*,y^*)[\eta]^p) = (z_1, z_2, \dots, z_p)\Big\},$$

where $z_i \in Z_i$, $i = 1, \dots, p$.

The following theorem is used to prove the generalized implicit function theorems for nonregular mappings.

Theorem 3.1 (Multivalued contraction mapping theorem, [8]). *For a Banach space* Ω *and* $w_0 \in \Omega$, *let* $\Lambda : B_{r_1}(w_0) \to 2^\Omega$ *be a multivalued mapping defined on some ball* $B_{r_1}(w_0) \subset \Omega$. *Assume that* $\Lambda(w) \neq \emptyset$ *for any* $w \in B_{r_1}(w_0)$. *Assume also that there exists a number* $\alpha \in (0,1)$ *such that*

1. $\mathcal{H}(\Lambda(w_1), \Lambda(w_2)) \le \alpha \|w_1 - w_2\|$, *for all* $w_1, w_2 \in B_{r_1}(w_0)$;

2. $\operatorname{dist}(w_0, \Lambda(w_0)) < (1-\alpha) r_1$.

Then for any r_2 *such that* $\operatorname{dist}(w_0,\Lambda(w_0)) < r_2 < (1-\alpha)r_1$, *there exists* $\bar w \in B_{r_3}(w_0)$ *with* $r_3 = r_2/(1-\alpha)$ *such that*

$$\bar w \in \Lambda(\bar w). \tag{3.1}$$

Moreover, among the points $\bar{w}$ *satisfying* (3.1), *there exists a point such that*

$$\|\bar{w} - w_0\| \le \frac{2}{1-\alpha} \operatorname{dist}(w_0, \Lambda(w_0)).$$

In a slightly modified form, the next theorem was introduced in [5] (without a proof).

Theorem 3.2 (The p-order implicit function theorem). *Let X, Y, and Z be Banach spaces, $N(x^*)$ and $N(y^*)$ be sufficiently small neighborhoods of $x^* \in X$ and $y^* \in Y$ respectively, $F \in C^{p+1}(X \times Y)$, and $F(x^*, y^*) = 0$. Assume that the mappings $f_i(x, y)$, $i = 1, \dots, p$, introduced by* (2.1), *satisfy the following conditions:*

1. Singularity condition:

$$f^{(r)}_{i\,\underbrace{x\dots x}_{q}\,\underbrace{y\dots y}_{r-q}}(x^*, y^*) = 0, \quad r = 1, \dots, i-1, \; q = 0, \dots, r-1, \; i = 1, \dots, p,$$

$$f^{(i)}_{i\,\underbrace{x\dots x}_{q}\,\underbrace{y\dots y}_{i-q}}(x^*, y^*) = 0, \quad q = 1, \dots, i-1, \; i = 1, \dots, p;$$

2. P-factor-approximation: *for all* $y_1, y_2 \in (N(y^*) - y^*)$,

$$\Big\| f_i(x, y^* + y_1) - f_i(x, y^* + y_2) - \frac{1}{i!} f^{(i)}_{i\,y\dots y}(x^*, y^*)[y_1]^i + \frac{1}{i!} f^{(i)}_{i\,y\dots y}(x^*, y^*)[y_2]^i \Big\|$$
$$\le \varepsilon\big(\|y_1\|^{i-1} + \|y_2\|^{i-1}\big)\|y_1 - y_2\|, \qquad i = 1, \dots, p,$$

where $\varepsilon > 0$ is sufficiently small;

3. Banach condition: *there exist $c_1 \in (0, \infty)$ and a nonempty set $\Gamma(x^*) \subset N(x^*)$ in X such that for any sufficiently small γ, $(\Gamma(x^*) \cap N_\gamma(x^*)) \setminus \{x^*\} \neq \emptyset$ and for $x \in \Gamma(x^*)$, there exists $\bar{h}(x)$ such that*

$$\Phi_p[\bar{h}(x)]^p = -\frac{F(x, y^*)}{\|F(x, y^*)\|}, \quad \|\bar{h}(x)\| \le c_1; \tag{3.2}$$

4. Uniform p-regularity condition of the mapping F over the set $\Phi_p^{-1}(-F(x, y^*))$.

Then there exists a constant $k > 0$, $\delta > 0$, and a mapping $\varphi : \Gamma(x^*) \cap N_\delta(x^*) \to N(y^*)$ such that the following holds for $x \in \Gamma(x^*) \cap N_\delta(x^*)$ and some $\theta(x)$:

$$\varphi(x) = y^* + h(x) + \theta(x); \tag{3.3}$$

$$\varphi(x^*) = y^*, \quad h(x^*) = 0, \quad \theta(x^*) = 0, \quad h(x) = \bar{h}(x)\|F(x, y^*)\|^{1/p};$$
$$F(x, \varphi(x)) = 0, \quad \|\theta(x)\| = o(\|h(x)\|);$$
$$\|\varphi(x) - y^*\|_Y \le k \sum_{r=1}^{p} \|f_r(x, y^*)\|_{Z_r}^{1/r}.$$

Sketch of the proof. Set $\Lambda(y) = y - \{\Psi_p(h)\}^{-1}F(x, y^*+h+y)$, $y \in (N(y^*)-y^*)$ with $h = h(x)$, $x \in \Gamma(x^*)$. One can verify that all conditions of Theorem 3.1 are satisfied by $\Lambda(y)$ for some $B_{r_1}(0) \subset (N(y^*) - y^*)$ and $y_0 = 0$. By Theorem 3.1, for $x \in \hat{\Gamma}$, there exists $\theta(x)$ such that $\theta(x) \in \Lambda(\theta(x))$ and $\|\theta(x)\| \leq 2\|\Lambda(0)\|/(1-\alpha) \leq o(\|h(x)\|)$, which implies

$$F(x, y^* + h + \theta(x)) = 0.$$

Let $\varphi(x) = y^* + h(x) + \theta(x)$. Then $F(x, \varphi(x)) = 0$ and other statements follow from assumptions of the theorem. □

Remark 3.3. One of the differences between the regular case and the nonregular case is in the form of the mapping $\varphi(x)$. In the nonregular case, the parameter h, which we call *the "exit from singularity"* is used in the definition of $\varphi(x)$:

$$\varphi(x) = y^* + h(x) + \theta(x).$$

Example 3.4 ([5]). Consider the boundary-value problem:

$$y''(t) + y(t) + y^2(t) = a \sin t, \qquad y(0) = y(\pi) = 0, \tag{3.4}$$

where a is a constant. We are concerned with the existence of a nontrivial solution $y(t)$ of BVP (3.4). Introduce the mapping $F : X \times Y \to Z$:

$$F(x, y) = y'' + y + y^2 - x, \tag{3.5}$$

where $X = \{x \in C[0, \pi] \mid x(0) = x(\pi) = 0\}$, $Y = \{y \in C^2[0, \pi] \mid y(0) = y(\pi) = 0\}$, and $Z = C[0, \pi]$. Then we can rewrite BVP (3.4) as

$$F(x, y) = 0. \tag{3.6}$$

Our assumptions imply that $(0, 0)$ is a solution of (3.6). Without loss of generality, we may restrict our attention to some neighborhood $U \times V \subset X \times Y$ of the point $(0, 0)$. Then, for a sufficiently small $a \geq 0$, the problem of existence of a solution of BVP (3.4) is equivalent to the problem of existence of an implicit function $\varphi(x) : U \to Y$ such that $y = \varphi(x)$ and $F(x, \varphi(x)) = 0$. Since the operator $F'_y(0, 0)$ is not surjective, the classical implicit function theorem cannot be applied. However, all conditions of Theorem 3.2 are satisfied for the mapping $F(x, y)$ defined by (3.5) with $x^*(t) = 0$, $y^*(t) = 0$, and $p = 2$. Namely, for the mapping F, the operator Φ_2 is given by

$$\Phi_2[h]^2 = h'' + h + \frac{2}{\pi} \sin t \int_0^{\pi} \sin \tau h^2(\tau) d\tau,$$

and the element $\bar{h} = \bar{h}(t)$ is defined as a solution of the equation:

$$\frac{2}{\pi} \sin t \int_0^{\pi} \sin \tau h^2(\tau) d\tau = a \sin t,$$

which has a solution

$$\bar{h}(t) = \sqrt{\frac{3\pi a}{8}} \sin t, \quad t \in [0, \pi],$$

for $a > 0$. Moreover, the 2-factor-operator $\Psi_2(h)$ for the mapping F can be defined as

$$\Psi_2(h)\xi = \xi'' + \xi + \frac{4}{\pi}\sin t \int_0^{\pi} \sin\tau h(\tau)\xi(\tau)d\tau.$$

With $x = a\sin t$, the set $\Phi_2^{-1}(-F(x,y^*)) = \{\bar{h}\}$ and F is uniformly p-regular over $\Phi_2^{-1}(-F(x,y^*))$ since the operator $\Psi_2(\bar{h})$ is surjective [11].

By Theorem 3.2, for a sufficiently small $a > 0$, there exists a solution $y(t)$, $t \in [0,\pi]$, of BVP (3.4) such that

$$\|y(t)\| \le k\|a\sin t\|^{1/2} \le ka^{1/2}, \quad k > 0.$$

To motivate the next result, consider the problem from Example 3.4 with modified boundary conditions:

$$y''(t) + y(t) + y^2(t) = a\sin t, \qquad y(0) = y(2\pi) = 0. \tag{3.7}$$

Theorem 3.2 is not applicable to (3.7) even for $a = 0$. This leads to the consideration of the following theorem (a proof of a slightly modified form can be found in [10] and [16]).

Theorem 3.5 (The p-order implicit function theorem for nontrivial kernel).

Let X, Y, and Z be Banach spaces, $F \in C^{p+1}(X \times Y)$, and $F(x^*,y^*) = 0$, Let $f_i(x,y)$, $i = 1,\dots,p$, be defined by (2.1) and the operator $\Psi_p(h)$ be given by (2.2). Assume that there exists an element $\bar{h} \in \bigcap_{r=1}^{p} \operatorname{Ker}^r f_{r\,y}^{(r)}(x^*,y^*)$, $\|\bar{h}\| = 1$, such that $\operatorname{Im}\Psi_p(\bar{h}) = Z$, that is the mapping F is p-regular with respect to the vector $\bar{h}$.

Then for a sufficiently small $\varepsilon > 0$, $\nu > 0$, and $\delta = \varepsilon\nu^p$, there exists a mapping $\varphi(x) : N_\delta(x^*) \to N_\varepsilon(y^*)$, and constants $k > 0$ and $c_1 > 0$ such that the following holds:

a) $\varphi(x^*) = y^*$;

b) $F(x,\varphi(x)) = 0$ for all $x \in N_\delta(x^*)$;

c) $\varphi(x) = y^* + h(x) + \bar{y}(x)$, where $h(x)$ (the exit from singularity) is defined by

$$h(x) = \gamma(x)\bar{h}$$

and

$$c_1\|x - x^*\|^{1/p} \le \|\gamma(x)\| \le \nu.$$

Moreover, $\bar{y}(x)$ satisfies

$$\|\bar{y}(x)\|_Y \le k\sum_{r=1}^{p} \frac{\|f_r(x, y^* + h(x)\|_{Z_r}}{\|\gamma(x)\|^{(r-1)}}, \quad x \in N_\delta(x^*),\ \gamma(x) \ne 0. \tag{3.8}$$

Remark 3.6. Estimate (3.8) can be replaced by the following:

$$\|\bar{y}(x)\|_Y \le K\sum_{r=1}^{p} \|f_r(x, y^* + h(x))\|_{Z_r}^{1/r},\ x \in N_\delta(x^*),\ x \ne x^*,\ K > 0.$$

Example 3.7. We consider Problem (3.7) with $a = 0$. We show how Theorem 3.5 can be applied to Problem (3.7). Introduce the mapping $F : X \times Y \to Z$,

$$F(x, y) = y''(t) + y(t) + y^2(t),$$

where $X = \{x \in C[0, 2\pi] \mid x(0) = x(2\pi) = 0\}$, $Y = \{y \in C^2[0, 2\pi] \mid y(0) = y(2\pi) = 0\}$, and $Z = C[0, 2\pi]$. Consider the point $(x^*, y^*) = (0, 0)$. Then

$$\mathrm{Ker}^1 f'_{1\,y}(0,0) \cap \mathrm{Ker}^2 f''_{2\,y}(0,0) = C \sin t, \quad C \in \mathbb{R}.$$

The 2-factor-operator $\Psi_2(\bar{h})$ is defined by

$$\Psi_2(\bar{h})\xi = \xi'' + \xi + \frac{\sin t}{\pi} \int_0^{2\pi} \sin \tau \bar{h}(\tau)\xi(\tau) d\tau, \quad \bar{h} = \sin t,$$

and is surjective [11]. Then, by Theorem 3.5, there exists $\nu > 0$ and a nontrivial solution $y(t)$, $t \in [0, 2\pi]$, of Problem (3.7) such that

$$y(t) = \nu \sin t + w(\nu, t), \quad \|w(\nu, t)\| = o(\nu), \quad t \in [0, 2\pi]. \qquad \square$$

There are more nonregular problems where the previously discussed theorems can not be applied. For example,

$$F(x, y) = xy^{p-1} + g(x, y) = 0, \tag{3.9}$$

where $g(0, 0) = 0$, $g'(0, 0) = 0$, …, $g^{(p)}(0, 0) = 0$. Another example is the initial-value problem

$$\varepsilon y^{(k)} + g(y, \varepsilon) = 0, \quad y(0) = y'(0) = \cdots = y^{(k-1)} = 0, \tag{3.10}$$

where $g(0, 0) = 0$, $g'(0, 0) = 0$, …, $g^{(k)}(0, 0) = 0$.

Theorem 3.8 and Theorem 3.10 below can be used to solve problems such as (3.9) and (3.10). To state these theorems, a modified form of the p-factor-operator is introduced. In a slightly modified form, Theorem 3.8 was given in [5].

Theorem 3.8. *Let $F(x, y) \in C^{p+1}(X \times Y)$, $F : X \times Y \to Z$, X, Y and Z be Banach spaces. Let mappings $f_i(x, y)$, $i = 1, \ldots, p$, be defined by* (2.1). *Assume that $F(x^*, y^*) = 0$ and that F is p-regular with respect to y along an element $h \in \bigcap_{k=1}^{p} \mathrm{Ker}^k f_k^{(k)}(x^*, y^*)$, $h = (\bar{x}, 0)$, $\bar{x} \neq 0$, that is*

$$\left\{f_1'(x^*, y^*) + f_2''(x^*, y^*)[h] + \cdots + \frac{1}{(p-1)!} f_p^{(p)}(x^*, y^*)[h]^{p-1}\right\} \cdot (\{0\}_X \times Y) = Z. \tag{3.11}$$

Then there exist a constant $C > 0$, a sufficiently small $\varepsilon > 0$ and a mapping

$$\varphi(x) = y^* + th + \omega(t, h)$$

such that the following holds for $t \in [0,\varepsilon)$ *and* $x = x^* + t\bar{x}$*:*

$$F(x,\varphi(x)) = 0, \quad \|\omega(t,h)\| = o(t),$$

and

$$\|\varphi(x) - y^*\|_Y \le C \sum_{r=1}^{p} \|f_r(x,y^*)\|_{Z_r}^{1/r}. \tag{3.12}$$

Proof. We proceed as in the proof of Theorem 3.2 but with the mapping Λ defined as

$$\Lambda(\xi) = \xi - \left\{f_1'(x^*,y^*) + \cdots + \frac{1}{(p-1)!} f_p^{(p)}(x^*,y^*)[h]^{p-1}\right\}_Y^{-1} \cdot F(x, y^* + \xi).$$

□

Remark 3.9. In Theorem 3.8, the exit from singularity is given by $h = (\bar{x}, 0)$ and $(x,\varphi(x)) = (x^*,y^*) + th + \omega(t,h)$ with $\|\omega(t,h)\| = o(t)$.

The following theorem is a modification of Theorem 3.8.

Theorem 3.10. *Let* $F(x,y) \in C^{p+1}(X \times Y)$*,* $F: X \times Y \to Z$*, where* X*,* Y *and* Z *are Banach spaces. Let mappings* $f_i(x,y)$*,* $i = 1,\ldots,p$*, be defined by* (2.1)*. Assume that* $F(x^*,y^*) = 0$ *and* F *is* p*-regular with respect to* y *along* $h \in \bigcap_{k=1}^{p} \operatorname{Ker}^k f_k^{(k)}(x^*,y^*)$*,* $h = (h_x, h_y)$*,* $\|h_x\|_X \neq 0$*,* $\|h_y\|_Y \neq 0$*, that is*

$$\left\{f_1'(x^*,y^*) + f_2''(x^*,y^*)[h] + \cdots + \frac{1}{(p-1)!} f_p^{(p)}(x^*,y^*)[h]^{p-1}\right\} \cdot (\{0\}_X \times Y) = Z. \tag{3.13}$$

Then for $x = x^* + th_x$*,* $t \in [0,\varepsilon)$*,* $\varepsilon > 0$*, there exists* $\varphi(x) = y^* + th_y + \omega(t,h)$ *such that*

$$F(x,\varphi(x)) = 0, \quad \|\omega(t,h)\| = o(t),$$

and

$$\|\varphi(x) - y^*\|_Y \le C \sum_{r=1}^{p} \|f_r(x,y^* + th_y)\|_{Z_r}^{1/r}, \tag{3.14}$$

where $C > 0$ *is an independent constant.*

Proof. We proceed as in the proof of Theorem 3.2 but with the mapping Λ defined as

$$\Lambda(\xi) = \xi - \left\{f_1'(x^*,y^*) + \cdots + \frac{1}{(p-1)!} f_p^{(p)}(x^*,y^*)[h]^{p-1}\right\}_Y^{-1} \cdot F(x, y^* + th_y + \xi).$$

□

Remark 3.11. Estimates (3.12) and (3.14) can be replaced by the following:

$$\|\varphi(x) - y^*\| \le C\|x - x^*\|,$$

where $C > 0$ is an independent constant.

Example 3.12 ([5]). Consider the problem:

$$\varepsilon y'(t) + y^3(t) + \varepsilon^3 = 0, \qquad y(0) = 0. \tag{3.15}$$

Introduce the mapping $F(\varepsilon, y) = \varepsilon y' + y^3 + \varepsilon^3$. (In this example, we replaced x by ε.) Then all conditions of Theorem 3.8 are satisfied for the mapping $F(\varepsilon, y)$ with $p = 2$, $y^* = 0$, $\varepsilon^* = 0$, and $h = (h_\varepsilon, 0)^T$, $h_\varepsilon \in \mathbb{R}$. Hence, by Theorem 3.8 and Remark 3.11, there exist $C > 0$ and $\delta > 0$ such that for a sufficiently small ε and $t \in (-\delta, \delta)$ there exists a solution $y(\varepsilon, t)$ of (3.15) such that the following holds:

$$\|y(\varepsilon, t)\| \leq C\varepsilon, \qquad y(\varepsilon, 0) = 0.$$

The p-factor-operator defined by (2.2) and its modifications introduced in (3.11) and (3.13) are based on the derivatives of F up to some order as well as on some element h. However, there are nonregular nonlinear problems, for which the corresponding p-factor operator based on a unique vector h is not onto, and, hence, Theorem 3.2, Theorem 3.5, Theorem 3.8, and Theorem 3.10 are not applicable. For such problems, we introduce generalized operators, whose definitions are based on sequences of vectors $\{h_1, \ldots, h_q\}$. These *generalized operators* allow us to introduce Theorem 3.13 and Theorem 3.17, which guarantee existence of implicit functions where F is neither regular nor p-regular.

Theorem 3.13. *Let $F \in C^{p+1}(X \times Y)$, $F : X \times Y \to Z$, where X, Y and Z are Banach spaces. Assume that $F^{(k)}(x^*, y^*) = 0$, $k = 0, 1, \ldots, p-1$. Assume that there exist elements $h_x \in X$ and $h_y \in Y$ such that for $\bar{h}_x = (h_x, 0)^T$, $\bar{h}_y = (0, h_y)^T$, the following holds:*

$$F^{(p)}(x^*, y^*)[\bar{h}_x]^p = 0, \quad F^{(p)}(x^*, y^*)[\bar{h}_y]^p = 0. \tag{3.16}$$

Introduce the subspaces $Z_1, \ldots, Z_p$ as

$$Z_k = \overline{\operatorname{Im}}(P_k F^{(p)}(x^*, y^*)[\bar{h}_y]^{p-k}[\bar{h}_x]^{k-1})) \cdot (0_X \times Y), \quad k = 1, \ldots, p,$$

where P_k is a projector onto a complement of $Z_1 \oplus \cdots \oplus Z_{k-1}$ in Z. For $k = 1, \ldots, p$, let $f_k(x, y) = P_k F(x, y)$ and

$$V \cdot (0_X \times Y) = Z \tag{3.17}$$

for $V = f_1^{(p)}(x^*, y^*)[\bar{h}_y]^{p-1} + f_2^{(p)}(x^*, y^*)[\bar{h}_y]^{p-2}[\bar{h}_x] + \cdots + f_p^{(p)}(x^*, y^*)[\bar{h}_x]^{p-1}$. Assume also that there exist $\varepsilon > 0$, a sufficiently small $\delta > 0$ and $C > 0$ such that for $t \in [0, \delta)$,

$$\|f_k(x^* + t^{1+\varepsilon} h_x, y^* + Cth_y)\| = o(t^{p-1+\varepsilon(k-1)}), \quad k = 1, 2, \ldots, p. \tag{3.18}$$

Then for $x = x^* + t^{1+\varepsilon} h_x$, $t \in [0, \delta)$, there exists $\varphi(x)$ such that

$$F(x, \varphi(x)) = 0, \ \varphi(x) = y^* + Cth_y + \omega(x, h_x, h_y)$$

and

$$\|\omega(x,h_x,h_y)\| \le m\sum_{k=1}^{p}\|f_k(x,y^*+Cth_y)\|^{1/k} \quad \text{or} \quad \|\omega(x,h_x,h_y)\| \le o(\|x-x^*\|),$$

where $m>0$ is an independent constant.

Proof. The proof is similar to the proof of Theorem 4.1 in [6]. □

Remark 3.14. Condition (3.17) can be considered as a generalization of the condition of p-regularity of the mapping F in the degenerate case $F^{(k)}(x^*,y^*)=0$, $k=0,1,\dots,p-1$.

Example 3.15. Consider the mapping $F:\mathbb{R}\times\mathbb{R}\to\mathbb{R}$ defined by:

$$F(x,y)=xy^2-x^5+y^4 \tag{3.19}$$

and let $(x^*,y^*)=(0,0)$. We have $F'(0,0)=(0,0)^T$, $F''(0,0)=0_{2\times 2}$, and

$$F'''(0,0)[h]^2=\begin{pmatrix} 2h_y^2 \\ 4h_xh_y \end{pmatrix}, \quad h=\begin{pmatrix} h_x \\ h_y \end{pmatrix}. \tag{3.20}$$

Moreover, $\operatorname{Ker}^3 F'''(0,0)=\{\alpha h \mid h=(1,0)^T \text{ or } h=(0,1)^T\}$. For $p=3$, we define

$$f_3'''(0,0)[h]^2=F'''(0,0)[h]^2=\begin{pmatrix} 2h_y^2 \\ 4h_xh_y \end{pmatrix}.$$

Observe that neither (3.11) nor (3.13) is satisfied with $h\in\operatorname{Ker}^3 F'''(0,0)$. Hence, Theorem 3.8 and Theorem 3.10 are not applicable to Problem (3.19). We show how Theorem 3.13 can be applied to this problem.

In this example, $p=3$ and $\operatorname{Ker}^3 F'''(0,0)=\{\alpha h \mid h=(1,0)^T \text{ or } h=(0,1)^T\}$. Then by (3.20) we get with $\bar h_x=(-1,0)^T$ and $\bar h_y=(0,1)$,

$$F'''(0,0)[\bar h_x]^2=(0,0), \quad F'''(0,0)[\bar h_y]^2=(2,0), \quad F'''(0,0)[\bar h_y,\bar h_x]=(0,-2).$$

Then $F'''(0,0)[\bar h_x]^3=0$, $F'''(0,0)[\bar h_y]^3=0$, and, hence, (3.16) holds. Moreover, $f_1(x,y)=0$, $f_3(x,y)=0$, $f_2(x,y)=F(x,y)$, and (3.17) reduces to

$$f_2'''(0,0)[\bar h_y,\bar h_x](0,y)^T=(0,-2)(0,y)^T=\mathbb{R}, \quad y\in\mathbb{R}.$$

Hence, (3.17) holds. In the notation of Theorem 3.13, $h_x=-1$ and $h_y=1$. Now, we will verify that (3.18) is satisfied with $C=1$, $t\in(0,\delta)$ (δ is sufficiently small), and $(x^*+t^{1+\varepsilon}h_x, y^*+Cth_y)=(-t^{1+\varepsilon},t)$. Namely,

$$f_2(-t^{1+\varepsilon},t)=F(-t^{1+\varepsilon},t)=-t^{1+\varepsilon}t^2+t^4+t^{5(1+\varepsilon)}=-t^{3+\varepsilon}+t^4+t^{5(1+\varepsilon)}$$

and for $\varepsilon = 1$ we get

$$\|f_2(-t^{1+\varepsilon}, t)\| = t^{10} = o(t^3).$$

Hence, since $f_1 = 0$ and $f_3 = 0$, (3.18) holds. Thus the assumptions of Theorem 3.13 are satisfied.

Then Theorem 3.13 implies that for $x = -t^2$ there exists $\varphi(x) = y^* + t + \omega(t) = 0 + t + \omega(t)$, such that $\|\omega(t)\| = o(t^2)$ and $F(x, \varphi(x)) = 0$.

Example 3.16. Consider the equation, which describes the *Padera* or *foot surface*:

$$F(x, y_1, y_2) = xy_1y_2 - (x^2 + y_1^2 + y_2^2)^3 = 0. \tag{3.21}$$

In this example, $F : \mathbb{R} \times \mathbb{R}^2 \to \mathbb{R}$. We consider a sufficiently small neighborhood of the point $(x^*, y_1^*, y_2^*) = (0, 0, 0)$. Define $\bar{h}_x = (1, 0, 0)^T$ and $\bar{h}_y = (0, 1, 0)^T$. Then $F'''(0, 0, 0)[\bar{h}_x]^2 = (0, 0, 0)$, $F'''(0, 0, 0)[\bar{h}_y]^2 = (0, 0, 0)$, and $F'''(0, 0, 0)[\bar{h}_y, \bar{h}_x] = (0, 0, 1)$. Hence, (3.16) and (3.17) hold with $p = 3$. In the notation of Theorem 3.13, $h_x = 1$ and $h_y = (1, 0)^T$. One can verify that (3.18) holds with $\varepsilon = 1/4$, $C = 1$, and $t \in (0, \delta)$ (δ is sufficiently small), that is with

$$x = x^* + t^{1+\varepsilon} h_x = t^{5/4}$$

and

$$y^* + Cth_y = (t, 0)^T.$$

Then Theorem 3.13 implies that for $x = t^{5/4}$, $t \in [0, \delta)$, there exists $\varphi(x) = (t + \omega_1(x), \omega_2(x))$ such that $\|\omega(x)\| = o(t^{5/4})$ and $F(x, \varphi(x)) = 0$. □

In Theorem 3.13, we use two vectors $h_1 = \bar{h}_x$ and $h_2 = \bar{h}_y$. In the rest of the paper, we consider the case of $p = 2$ with $F'(x^*, y^*) = 0$ and introduce a set of vectors $\{h_1, \ldots, h_q\} \in X \times Y$. We assume again that the space Z is decomposed into the direct sum $Z = Z_1 \oplus \cdots \oplus Z_q$, but with a different definition of the subspaces Z_i. Namely, we define Z_i a closed subspace such that

$$Z_1 = \overline{\mathrm{Im}}\,(F''(x^*, y^*)[h_1]) \cdot (0_X \times Y),$$

$$Z_k = \overline{\mathrm{Im}}\bar{P}_k(F''(x^*, y^*)[h_k]) \cdot (0_X \times Y), \quad k = 2, \ldots, q,$$

where $\bar{P}_k$ is the projection operator onto $Z \backslash (Z_1 \oplus \cdots \oplus Z_{k-1})$ along $(Z_1 \oplus \cdots \oplus Z_{k-1})$ with respect to Z, $k = 2, \ldots, q$.

Define the following mappings

$$f_k(x, y) : X \times Y \to Z_k, \quad f_k(x, y) = P_k F(x, y), \quad k = 1, \ldots, q, \tag{3.22}$$

where $P_k : Z \to Z_k$ is the projection operator onto Z_k along $(Z_1 \oplus \cdots \oplus Z_{k-1} \oplus Z_{k+1} \oplus \cdots \oplus Z_q)$ with respect to Z, $k = 1, \ldots, q$. Then the mapping F can be represented as

$$F(x, y) = f_1(x, y) + \cdots + f_q(x, y).$$

Theorem 3.17. *Let $F \in C^3(X \times Y)$, $F : X \times Y \to Z$, where X, Y, and Z are Banach spaces. Assume that $F(x^*, y^*) = 0$ and $F'(x^*, y^*) = 0$. Assume that there exist a set of vectors $\{h_1, \dots, h_q\} \in X \times Y$, $h_k = (h_{k_X}, h_{k_Y})$, $h_k \neq 0$, $h_{k_X} \in X$, $h_{k_Y} \in Y$, $k = 1, \dots, q$, such that $Z = Z_1 \oplus \cdots \oplus Z_q$ and for the mappings $f_k(x, y)$ defined by (3.22) the following holds:*

$$f_k''(x^*, y^*)[h_{k+r}, h_{k+p}] = 0, \quad 1 \le k \le q, \quad 0 \le r \le (q-k), \quad 0 \le p \le (q-k-r), \tag{3.23}$$

and

$$(f_1''(x^*, y^*)[h_1] + \cdots + f_q''(x^*, y^*)[h_q]) \cdot (0_X \times Y) = Z. \tag{3.24}$$

Then there exists a sufficiently small $\delta > 0$ and $C > 0$ such that for $t \in [0, \delta)$, $\varepsilon = 1/(2q)$,

$$\bar{x}(t) = th_{1_X} + \cdots + t^{1+(q-1)\varepsilon} h_{q_X}, \quad \bar{y}(t) = th_{1_Y} + \cdots + t^{1+(q-1)\varepsilon} h_{q_Y}, \tag{3.25}$$

and $x = x^* + \bar{x}(t)$ there exists $\varphi(x) = y^* + \bar{y}(t) + \omega(t)$ such that

$$F(x, \varphi(x)) = 0, \quad \|\omega(t)\| = o(t^{1+(q-1)\varepsilon})$$

and

$$\|\omega(t)\| \le C \sum_{k=1}^{q} \|f_k(x^* + \bar{x}(t), y^* + \bar{y}(t))\|^{\frac{1}{2+(k-1)\varepsilon}}.$$

Remark 3.18. Condition (3.24) can be considered as a generalization of the condition of p-regularity of the mapping F in the degenerate case of $F(x^*, y^*) = 0$ and $F'(x^*, y^*) = 0$.

To prove Theorem 3.17, we need to introduce some additional definitions and notation. For a surjective operator $A \in \mathcal{L}(Y, Z)$, we define the right inverse operator $A^{-1} : Z \to 2^Y$ as

$$A^{-1}z = \{y \in Y \mid Ay = z\}.$$

Then, by definition,

$$\|A^{-1}\| = \sup_{\|z\|=1} \inf\{\|y\| \mid Ay = z\}.$$

The reader should observe that A^{-1} is the standard inverse of the isomorphism induced by A between the spaces $Y/\mathrm{Ker}A$ and Z. Therefore, it is clear that $\|A^{-1}\| < \infty$. Moreover, there is a constant $C > 0$ such that

$$C\|z\| \le \|A^{-1}z\| \le \|A^{-1}\| \cdot \|z\|, \text{ for all } z \in Z.$$

Lemma 3.19. *Assume that there exist operators $A_k \in \mathcal{L}(Y, Z)$, $k = 1, \dots, q$, such that $\mathrm{Im}A_k = Z_k$ where Z_k is a closed subspace of Z and $Z = Z_1 \oplus \cdots \oplus Z_q$. Assume that the operator $A = A_1 + \cdots + A_q$ is onto. Then the following holds:*

$$\|A^{-1}z\| \le C(\|A_1^{-1}z_1\| + \cdots + \|A_q^{-1}z_q\|), \quad z = (z_1, \dots, z_q). \tag{3.26}$$

Proof. For each $k = 1, \dots, q$, fix $c_k > 0$ such that

$$c_k\|z_k\| \le \|A_k^{-1} z_k\| \le \|A_k^{-1}\| \cdot \|z_k\|, \quad \forall z_k \in Z_k. \tag{3.27}$$

Assume to the contrary that there exists a sequence $\{z^n\}_{n=0}^{\infty}$, with $\|z^n\| = 1$ and such that

$$\|A^{-1}z^n\| = C_n(\|A_1^{-1}z_1^n\| + \cdots + \|A_q^{-1}z_q^n\|), \quad z^n = (z_1^n, \dots, z_q^n),$$

where $C_n \to \infty$ as $n \to \infty$. Then, for any z^n, there exists z_k^n such that $\|z_k^n\| \ge 1/q$ and by (3.27),

$$\begin{aligned}\|A^{-1}\| &\ge C_n(\|A_1^{-1}z_1^n\| + \cdots + \|A_q^{-1}z_q^n\|) \ge C_n(c_1\|z_1^n\| + \cdots + c_q\|z_q^n\|) \\ &\ge \frac{C_n}{q} \min_{i=1,\dots,q} c_i > C_n C,\end{aligned}$$

where $C > 0$ and $C_n \to \infty$ as $n \to \infty$. We get a contradiction and therefore, the desired inequality holds. □

Proof of Theorem 3.17. Let $\varepsilon = 1/(2q)$, $t > 0$ be sufficiently small, and $\{h_1, \dots, h_q\}$ be the set of vectors from $X \times Y$ with $h_k = (h_{k_X}, h_{k_Y})$ satisfying (3.23). Introduce the operators

$$A_k = f_k''(x^*, y^*)[h_k], \quad k = 1, \dots, q,$$

and

$$A = (tA_1, \dots, t^{1+(q-1)\varepsilon}A_q).$$

Moreover, in accordance with (3.25), define $\bar{x}(t)$ and $\bar{y}(t)$ as

$$\bar{x}(t) = th_{1_X} + \cdots + t^{1+(q-1)\varepsilon}h_{q_X}, \quad \bar{y}(t) = th_{1_Y} + \cdots + t^{1+(q-1)\varepsilon}h_{q_Y}.$$

Consider the mapping (in general, it is multivalued):

$$\Lambda(y) = y - A^{-1}\Big(f_1(x^* + \bar{x}(t), y^* + \bar{y}(t) + y), \dots, f_q(x^* + \bar{x}(t), y^* + \bar{y}(t) + y)\Big).$$

We will verify that all conditions of Theorem 3.1 are satisfied for $\Lambda(y)$ with some $B_{r_1}(0) \subset (N(y^*) - y^*)$ and $y_0 = 0$.

We start from verifying condition 2) of Theorem 3.1, which is done by estimating $\|\Lambda(0)\|$. Using the definition of $\Lambda(0)$, there exists $\bar{C} \ge 0$ such that

$$\|\Lambda(0)\| \le \bar{C}\|A^{-1}F(x^* + \bar{x}(t), y^* + \bar{y}(t))\|.$$

Then by Lemma 3.19, there exists $C \ge 0$ such that

$$\|\Lambda(0)\| \le C\Big(\frac{\|f_1(x^* + \bar{x}(t), y^* + \bar{y}(t))\|}{t} + \cdots + \frac{\|f_q(x^* + \bar{x}(t), y^* + \bar{y}(t))\|}{t^{1+(q-1)\varepsilon}}\Big). \tag{3.28}$$

Using Taylor expansion, assumptions $F(x^*, y^*) = 0$ and $F'(x^*, y^*) = 0$, and definitions of $\bar{x}(t)$ and $\bar{y}(t)$, we get for $k = 1, \dots, q$,

$$\|f_k(x^* + \bar{x}(t), y^* + \bar{y}(t))\| = \|f_k''(x^*, y^*)[th_1 + \cdots + t^{1+(q-1)\varepsilon}h_q]^2 + \omega_k(t)\|, \tag{3.29}$$

where $\|\omega_k(t)\| \le Ct^3$.

The definition of mappings f_i implies

$$f_k''(x^*, y^*)[h_i, h_j] = 0 \quad \text{if} \quad i < k \quad \text{or} \quad j < k. \tag{3.30}$$

By (3.30) and (3.23), we get that

$$\frac{\|f_k(x^* + \bar{x}(t), y^* + \bar{y}(t))\|}{t^{1+(k-1)\varepsilon}} = o(t^{1+(q-1)\varepsilon}), \quad k = 1, \ldots, q.$$

Then for $\varepsilon = 1/(2q)$ we obtain from the last relation and (3.28) that

$$\|\Lambda(0)\| = o(t^{1+(q-1)\varepsilon}) = o(t^{1+(q-1)/(2q)}).$$

For a sufficiently small t we get with some $\alpha \in (0, 1)$ and $r_1 = o(t^{1+(q-1)\varepsilon})$ that

$$\|\Lambda(0)\| < (1 - \alpha)r_1,$$

which proves condition 2) of Theorem 3.1.

Now we will show that condition 1) of Theorem 3.1 holds for all $y_1, y_2 \in B_{r_1}(0)$, that is

$$\mathcal{H}(\Lambda(y_1), \Lambda(y_2)) \le \alpha\|y_1 - y_2\|, \quad 0 \le \alpha < 1.$$

By the definition of $\Lambda(y)$, we have with $\tilde{x}(t) = x^* + \bar{x}(t)$ and $\tilde{y}(t) = y^* + \bar{y}(t)$,

$$\begin{aligned}\mathcal{H}(\Lambda(y_1),\Lambda(y_2)) &= \inf\{\|z_1 - z_2\| \mid z_i \in \Lambda(y_i), \quad i = 1, 2\} \\ &= \inf\{\|z_1 - z_2\| \mid Az_i = Ay_i - F(\tilde{x}(t), \tilde{y}(t) + y_i), \quad i = 1, 2\} \\ &= \inf\{\|z\| \mid Az = A(y_1 - y_2) - F(\tilde{x}(t), \tilde{y}(t) + y_1) + F(\tilde{x}(t), \tilde{y}(t) + y_2)\} \\ &\le \|A^{-1}\left(A(y_1 - y_2) - F(\tilde{x}(t), \tilde{y}(t) + y_1) + F(\tilde{x}(t), \tilde{y}(t) + y_2)\right)\|.\end{aligned}$$

By Lemma 3.19, we deduce that

$$\begin{aligned}&\mathcal{H}(\Lambda(y_1), \Lambda(y_2)) \\ &\quad \le C\Big(\frac{\|f_1''(x^*, y^*)[th_1](y_1 - y_2) - f_1(\tilde{x}(t), \tilde{y}(t) + y_1) + f_1(\tilde{x}(t), \tilde{y}(t) + y_2)\|}{t} + \cdots \\ &\quad + \frac{\|f_q''(x^*, y^*)[t^{1+(q-1)\varepsilon}h_q](y_1 - y_2) - f_q(\tilde{x}(t), \tilde{y}(t) + y_1) + f_q(\tilde{x}(t), \tilde{y}(t) + y_2)\|}{t^{1+(q-1)\varepsilon}}\Big).\end{aligned}$$

Using Mean Value Theorem and Taylor expansion, for $k = 1, \ldots, q$, we get that there exists $C_k \ge 0$ such that

$$\begin{aligned}&\frac{\|f_k''(x^*, y^*)[t^{1+(k-1)\varepsilon}h_k](y_1 - y_2) - f_k(\tilde{x}(t), \tilde{y}(t) + y_1) + f_k(\tilde{x}(t), \tilde{y}(t) + y_2)\|}{t^{1+(k-1)\varepsilon}} \\ &\qquad \le C_k \sup_{\xi \in B_{r_1}(0)} \|\xi\| \frac{\|y_1 - y_2\|}{t^{1+(k-1)\varepsilon}}.\end{aligned}$$

Hence, with $r_1 = o(t^{1+(q-1)\varepsilon})$, we get $\mathcal{H}(\Lambda(y_1), \Lambda(y_2)) \le \alpha \|y_1 - y_2\|$, which proves condition 1) of Theorem 3.1.

By Theorem 3.1 there exists $\omega(t)$ such that $\omega(t) \in \Lambda(\omega(t))$, which is equivalent to

$$0 \in \{A\}^{-1} \left(F(x^* + \bar{x}(t), y^* + \bar{y}(t) + w(t)) \right).$$

Hence,

$$F(x^* + \bar{x}(t), y^* + \bar{y}(t) + w(t)) = 0$$

with

$$\|w(t)\| \le C_1 \|\Lambda(0)\| = o(t^{1+(q-1)\varepsilon}) = o(t^{1+(q-1)/(2q)}).$$

One can show that (3.28) and $\|w(t)\| \le C_1 \|\Lambda(0)\|$ also yield the following estimate:

$$\|\omega(t)\| \le C \sum_{k=1}^{q} \|f_k(x^* + \bar{x}(t), y^* + \bar{y}(t))\|^{\frac{1}{2+(k-1)\varepsilon}}.$$

Taking $\varphi(x) = y^* + \bar{y}(t) + w(t)$ finishes the proof. □

Example 3.20. Consider the mapping $F : \mathbb{R} \times \mathbb{R}^3 \to \mathbb{R}^2$,

$$F(x, y) = \begin{pmatrix} xy_1 - y_2^2 + x^3 \\ y_2 y_3 + x y_3^2 \end{pmatrix}.$$

Let $x^* = 0$, $y^* = (y_1^*, y_2^*, y_3^*)^T = (0, 0, 0)^T$, $h_1 = (1, 0, 0, 0)^T$ and $h_2 = (0, 0, 1, 0)^T$. Then $F(x^*, y^*) = (0, 0)^T$ and $F'(x^*, y^*) = 0_{2\times 2}$. We define $f_k(x, y)$ according to (3.22). We have $Z = Z_1 \oplus Z_2$,

$$P_1 F''(0,0)h_1 = f_1''(0,0)h_1 = \begin{pmatrix} 0 & 1 & 0 & 0 \\ 0 & 0 & 0 & 0 \end{pmatrix}, \quad P_1 = \begin{pmatrix} 1 & 0 \\ 0 & 0 \end{pmatrix},$$

$$P_2 = \begin{pmatrix} 0 & 0 \\ 0 & 1 \end{pmatrix}, \quad P_2 F''(0,0)h_2 = f_2''(0,0)h_2 = \begin{pmatrix} 0 & 0 & 0 & 0 \\ 0 & 0 & 0 & 1 \end{pmatrix}.$$

Condition (3.23) is satisfied since

$$f_1''(0,0)[h_1]^2 = f_1''(0,0)[h_1, h_2] = f_2''(0,0)[h_2]^2 = (0,0)^T.$$

Assumption (3.24) reduces to

$$(f_1''(0,0)h_1 + f_2''(0,0)h_2) \cdot (0 \times \mathbb{R}^3) = \begin{pmatrix} 0 & 1 & 0 & 0 \\ 0 & 0 & 0 & 1 \end{pmatrix} \begin{pmatrix} 0 \\ y_1 \\ y_2 \\ y_3 \end{pmatrix} = \begin{pmatrix} z_1 \\ z_2 \end{pmatrix}.$$

Hence, (3.24) holds.

In the notation of Theorem 3.17, $h_{1_X} = (h_1)_X = 1$, $h_{2_X} = (h_2)_X = 0$, $h_{1_Y} = (h_1)_Y = (0,0,0)^T$ and $h_{2_Y} = (h_2)_Y = (0,1,0)^T$. Then, by Theorem 3.17, there exists a sufficiently small $\delta > 0$ and $C > 0$ such that for $t \in [0, \delta)$, $\varepsilon = 1/4$, and

$$x = x^* + th_{1_X} + t^{1+1/4}h_{2_X} = t,$$

there exist $\omega(t) = (\omega_1(t), \omega_2(t), \omega_3(t))^T$ and

$$\varphi(x) = y^* + th_{1_Y} + t^{1+\varepsilon}h_{2_Y} + \omega(t) = t^{5/4}(0,1,0)^T + (\omega_1(t), \omega_2(t), \omega_3(t))^T$$

such that $\|\omega(t)\| = o(t^{5/4})$ and $F(x, \varphi(x)) = 0$. □

Acknowledgments. The authors thank the anonymous reviewer for valuable suggestions that helped us improve the presentation of this paper, and N. Randrianantoanina for his great help preparing the final version of the paper.

References

[1] R. Abraham, J. E. Marsden, and T. S. Ratiu, *Manifolds, Tensor Analysis and Applications*, Vol. 75, Applied Mathematical Sciences. Springer-Verlag, New York, 1988.

[2] V. M. Alekseev, V. M. Tikhomirov, and S. V. Fomin, *Optimal Control*, Consultants Bureau, New York and London, 1987.

[3] A. V. Arutyunov, *The implicit function theorem and abnormal points* (in Russian), Dokl. Akad. Nauk 368 (1999), pp. 586–589. English translation: Doklady Math. 60 (1999), pp. 231–234.

[4] E. R. Avakov, *Theorems on estimates in a neighborhood of a singular point of a mapping* (in Russian), Mat. Zametki 47 (1990), pp. 3–13. English translation: Math. Notes 47 (1990), pp. 425–432.

[5] O. A. Brezhneva, Y. G. Evtushenko, and A. A. Tret'yakov, *New numerical methods and some applied aspects of the p-regularity theory* (in Russian), Zh. Vychisl. Mat. Mat. Fiz. 46 (2006), pp. 1987-2000; English translation: Comput. Math. Math. Phys. 46 (2006), pp. 1896–1909.

[6] O. A. Brezhneva and A. A. Tret'yakov, *Optimality conditions for degenerate extremum problems with equality constraints*, SIAM J. Control Optim. 42 (2003), pp. 729–745.

[7] R. S. Hamilton, *The inverse function theorem of Nash and Moser*, Bull. Amer. Math. Soc. 7 (1982), pp. 65–222.

[8] A. D. Ioffe and V. M. Tikhomirov, *Theory of extremal problems*, North-Holland, Amsterdam, 1979.

[9] A. F. Izmailov, *Theorems on the representation of families of nonlinear mappings and implicit function theorems* (in Russian), Mat. Zametki 67 (2000), pp. 57–68. English translation: Math. Notes 67 (2000), pp. 45–54.

[10] A. F. Izmailov and A. A. Tret'yakov, *Factor-analysis of nonlinear mappings* (in Russian), Nauka, Moscow, 1994.

[11] ———, *2–regular solutions of nonregular problems* (in Russian), Fizmatlit, Moscow, 1999.

[12] S. G. Krantz and H. R. Parks, *The implicit function theorem: history, theory, and applications*, Birkhäuser, Boston, Basel, Berlin, 2002.

[13] A. A. Tret'yakov, *Necessary conditions for optimality of p-th order* (in Russian), Control and Optimization, Moscow, MSU, 1983, pp. 28–35.

[14] ———, *Necessary and sufficient conditions for pth order optimality* (in Russian), Zh. Vychisl. Mat. i Mat. Fiz. 24 (1984), pp. 203–209. English translation: USSR Comput. Math. and Math. Phys. 24 (1984), pp. 123–127.

[15] ———, *The implicit function theorem in degenerate problems* (in Russian), Uspekhi Mat. Nauk 42 (1987), pp. 215–216. English translation: Russ. Math. Surv. 42 (1987), pp. 179–180.

[16] ———, *Structures of nonlinear degenerate mappings and their application to the construction of numerical methods*, doctoral dissertation, Moscow, 1987.

[17] A. A. Tret'yakov and J. E. Marsden, *Factor-analysis of nonlinear mappings: p-regularity theory*, Commun. Pure Appl. Anal. 2 (2003), pp. 425–445.

Author information

Olga A. Brezhneva, Department of Mathematics and Statistics, 123 Bachelor Hall, Miami University, Oxford, OH 45056, USA.
Email: brezhnoa@muohio.edu

Alexey A. Tret'yakov, Dorodnicyn Computing Center of the Russian Academy of Sciences, Vavilova 40, 119991 Moscow GSP-1, Russia; System Research Institute, Polish Academy of Sciences, Newelska 6, 01-447 Warsaw, Poland; and University of Podlasie in Siedlce, 3 Maja 54, 08-110 Siedlce, Poland.
Email: tret@ap.siedlce.pl

Banach Spaces and their Applications in Analysis, 303–306

A general Baker superstability criterion for the D'Alembert functional equation

Stefan Czerwik and Maciej Przybyła

Abstract. We prove a very general Baker superstability criterion for the D'Alembert functional equation in the spirit of Székelyhidi utilizing the so called A-conjugate linear spaces with translation property.

Key words. D'Alembert functional equation, general Baker superstability criterion.

AMS classification. 39B82, 39B62.

1 Introduction

The following problem of stability of functional equations has been formulated by Ulam [10]. Let X be a group and Y a group with a metric d. Let $\varepsilon > 0$ be a given number. Does there exist $\delta > 0$ such that if $f\colon X \to Y$ satisfies

$$d[f(x+y), f(x) + f(y)] < \delta \quad \text{for all } x, y \in X,$$

then there exists a homomorphism $A\colon X \to Y$ with

$$d[f(x), A(x)] < \varepsilon \quad \text{for all } x \in X?$$

In other words, it means that if a mapping f is almost a homomorphism, then there exists a homomorphism A close to f with small error. In the case of affirmative answer, we say that the equation of homomorphism

$$A(x+y) = A(x) + A(y), \quad x, y \in X,$$

is stable.

One can ask a similar question for other important functional equations. The first partial solution of this problem was given by Hyers [6]. During the last decade, the stability problems for different types of functional equations have been investigated by many authors (see [2], [3], [4], [7], [8]). Gruber ([5]) observed that Ulam's problem is of particular interest in probability theory. Also the stability properties of functional equations may have applications to some other problems (for example see the paper of Zhou [11]).

The stability problem for the D'Alembert functional equation (called also the cosine equation)

$$f(x+y) + f(x-y) = 2f(x)f(y) \tag{1.1}$$

has been solved by Baker [1]. For this equation we have a different situation and therefore a different kind of stability, called superstability (i.e. if f satisfies some inequality, then f satisfies the equation).

Baker proved the following famous result.

Theorem 1.1. *Let $\delta > 0$ and G be an abelian group and $f\colon G \to \mathbb{C}$ (where $\mathbb{C}$ denotes the set of complex numbers) be a function satisfying the inequality*

$$|f(x+y) + f(x-y) - 2f(x)f(y)| \leq \delta \quad \text{for all } x, y \in G. \tag{1.2}$$

Then either f is bounded or satisfies the D'Alembert's functional equation (1.1).

We have a similar situation for the multiplicative Cauchy equation

$$f(xy) = f(x)f(y). \tag{1.3}$$

For this equation the following very general result concerning the superstability was proved by Székelyhidi [9].

Theorem 1.2. *Let G be a semigroup and V be a right invariant vector space of complex-valued mappings on G. If $f, m\colon G \to \mathbb{C}$ are mappings such that the mapping*

$$f(xy) - f(x)m(y)$$

in variable x belongs to V for each $y \in G$, then $f \in V$ or m is multiplicative.

Until now, we don't have such general and interesting result for the D'Alembert functional equation. In this paper we will fill up this big gap in the theory of functional equations.

2 Main result

Let G be a group and $\mathbb{C}$ the set of complex numbers. We denote for $f, g\colon G \to \mathbb{C}$

$$\begin{aligned}
g_a(x) &:= g(x+a), \quad x, a \in G, \\
A(f)(x,y) &:= f(x+y) + f(x-y) - 2f(x)f(y), \quad x, y \in G, \\
U^1 &:= \{g \mid g\colon G \to \mathbb{C}\}, \\
U^2 &:= \{H \mid H\colon G \times G \to \mathbb{C}\}.
\end{aligned}$$

We have the following very useful lemma whose proof consists of straightforward calculations.

Lemma 2.1. *Let G be an abelian group and F be a field. Let $f\colon G \to F$ be a function. Then we have for all $x, u, v \in G$*

$$\begin{aligned}
2f(x)A(f)(u,v) &= A(f)(x+u,v) - A(f)(x,u+v) \\
&\quad - A(f)(x,u-v) + A(f)(x-u,v) \\
&\quad + 2f(v)A(f)(x,u).
\end{aligned} \tag{2.1}$$

Now we can prove

Proposition 2.2. *Let $f\colon G \to \mathbb{C}$ be a function. Let U^1 be a linear space over $\mathbb{C}$. If, moreover, for every $(u,v) \in G \times G$ the function*

$$f(\cdot)A(f)(u,v) \in U^1, \tag{2.2}$$

then

$$f \in U^1 \quad \text{or} \quad A(f) = 0. \tag{2.3}$$

Proof. Assume that $A(f) \neq 0$. Then there exists $(u,v) \in G \times G$ such that $A(f)(u,v) \neq 0$. Now if

$$g(x) := f(x)A(f)(u,v), \quad x \in G,$$

then

$$f(x) = \frac{g(x)}{A(f)(u,v)}, \quad x \in G$$

and hence clearly $f \in U^1$. □

Remark 2.3. If U^1 has the property: if $g \in U^1$ then $g_a \in U^1$, then we say that U^1 has the "translation property" (i.e. U^1 is the right invariant vector space).

The main result of the paper reads as follows.

Theorem 2.4. *Let $f\colon G \to \mathbb{C}$ be a function. Let U^1 be a linear space over $\mathbb{C}$ with the "translation property". If for every $u \in G$, $A(f)(\cdot,u) \in U^1$, then*

$$f \in U^1 \quad \text{or} \quad A(f) = 0.$$

Proof. By Lemma 2.1, since U^1 is a linear space with the translation property, we have that for fixed $(u,v) \in G \times G$, the function $g\colon G \to \mathbb{C}$ defined by

$$g(x) := 2f(x)A(f)(u,v)$$

satisfies the condition: $g \in U^1$. Therefore, in view of Proposition 2.2, we get our assertion and the proof is complete. □

Remark 2.5. Linear spaces U^1, U^2 satisfying the additional conditions

(i) $A(f) \in U^2$,
(ii) for every $u \in G$, $A(f)(\cdot,u) \in U^1$,

are called A-conjugate spaces.

Remark 2.6. We get the famous Baker result on superstability of the D'Alembert functional equation for

$$\begin{aligned} U^1 &= B(G,\mathbb{C}),\\ U^2 &= B(G \times G,\mathbb{C}), \end{aligned}$$

where U^1 and U^2 denote the A-conjugate linear spaces of bounded functions on G and $G \times G$ respectively with values on complex numbers.

References

[1] J. A. Baker, *The stability of the cosine equation*, Proc. Amer. Math. Soc. 80 (1980), pp. 411–416.

[2] S. Czerwik, *Functional Equations and Inequalities in Several Variables*, World Scientific, New Jersey, London, 2002.

[3] S. Czerwik and K. Dłutek, *Superstability of the equation of quadratic functionals in L^p spaces*, Aequationes Math. 63 (2002), pp. 210–219.

[4] ______, *Stability of the quadratic functional equation in Lipschitz spaces*, J. Math. Anal. Appl. 293 (2004), pp. 79–88.

[5] P. M. Gruber, *Stability of isometries*, Trans. Amer. Math. Soc. 245 (1978), pp. 263–277.

[6] D. H. Hyers, *On the stability of the linear functional equation*, Proc. Nat. Acad. Sci. 27 (1941), pp. 222–224.

[7] D. H. Hyers, G. Isac, and Th. M. Rassias, *Stability of Functional Equations in Several Variables*, Birkhäuser Verlag, 1998.

[8] Th. M. Rassias, *On the stability of the linear mapping in Banach spaces*, Proc. Amer. Math. Soc. 72 (1978), pp. 297–300.

[9] L. Székelyhidi, *On a theorem of Baker, Lawrence and Zorzitto*, Proc. Amer. Math. Soc. 84 (1992), pp. 95–96.

[10] S. M. Ulam, *A collection of the mathematical problems* Interscience Publ., New York, 1960.

[11] Ding-Xuan Zhou, *On a conjecture of Z. Ditzian*, J. Approx. Theory 69 (1992), pp. 167–172.

Author information

Stefan Czerwik, Institute of Mathematics, Silesian University of Technology, Kaszubska 23, 44-100 Gliwice, Poland.
Email: stefan.czerwik@polsl.pl

Maciej Przybyła, Institute of Mathematics, Silesian University of Technology, Kaszubska 23, 44-100 Gliwice, Poland.
Email: przybyla@polsl.pl

Banach Spaces and their Applications in Analysis, 307–330

Metric derived numbers and continuous metric differentiability via homeomorphisms

Jakub Duda and Olga Maleva

Abstract. We define the notions of unilateral metric derivatives and "metric derived numbers" in analogy with Dini derivatives (also referred to as "derived numbers") and establish their basic properties. We also prove that the set of points where a path with values in a metric space with continuous metric derivative is not "metrically differentiable" (in a certain strong sense) is σ-symmetrically porous and provide an example of a path for which this set is uncountable. In the second part of this paper, we study the continuous metric differentiability via a homeomorphic change of variable.

Key words. Metric derivatives, derived numbers, Dini derivatives, porosity, differentiation via homeomorphisms.

AMS classification. Primary 26A24; Secondary 14H50.

This paper is dedicated to Nigel Kalton on the occasion of his sixtieth birthday.

1 Introduction

The main aim of this paper is to study analogues of the usual notion of differentiability which work for mappings with values in metric spaces. Let (X, ρ) be a metric space and $f : [a, b] \to X$ be any mapping. As every metric space isometrically embeds in some Banach space (see e.g. [3, Lemma 1.1]), we can suppose that the distance in X is in fact generated by a complete norm $\|\cdot\|$. Define

$$md_{\pm}(f, x) = \lim_{t \to 0+} \frac{\|f(x \pm t) - f(x)\|}{t}$$

to be the *unilateral right* (*resp. left*) *metric derivative* of the mapping f at x. If $md_+(f, x)$ and $md_-(f, x)$ exist, and are equal, then we call $md(f, x) := md_+(f, x)$ the *metric derivative* of f at the point x.

We say that f *is metrically differentiable at* x provided $md(f, x)$ exists and

$$\|f(y) - f(z)\| - md(f, x)|y - z| = o(|y - x| + |z - x|), \text{ when } (y, z) \to (x, x). \quad (1.1)$$

Note that in this terminology, the existence of the "metric derivative" $md(f, x)$ of f at x does not necessarily imply that f is metrically differentiable at x! The basic example of such mapping would be $f(t) = |t| : \mathbb{R} \to \mathbb{R}$ and $x = 0$.

Metric derivatives were introduced by Kirchheim in [13] (see also [1, 5, 15]), and were studied by several authors (see e.g. [2, 6, 7, 8]). In [2], the authors work with a slightly weaker version of metric differentiability.

First author: Supported in part by ISF.

We start section 3 by noting that the set of points where $md_{\pm}(f,x)$ exist, but $md_{+}(f,x) \neq md_{-}(f,x)$, is countable; see Theorem 3.1. This is analogous to a similar theorem for unilateral derivatives of real-valued functions.

There is a well established theory of derived numbers (or Dini derivatives) of real-valued functions $f : \mathbb{R} \to \mathbb{R}$ (see e.g. [4]). In section 3, we generalize theorems about relationships among the Dini derivatives to the context of metric derived numbers $mD^{\pm}, mD_{\pm}$.

In Theorem 3.2, we prove that the set of "angular" points of each $f : \mathbb{R} \to X$, i.e. points $x \in \mathbb{R}$ where either $mD_{-}(f,x) > mD^{+}(f,x)$ or $mD_{+}(f,x) > mD^{-}(f,x)$, is countable. Theorem 3.3 (resp. Theorem 3.4 if f is pointwise-Lipschitz) shows that the sets of points $x \in \mathbb{R}$ where $mD^{+}(f,x) \neq mD^{-}(f,x)$ (resp. $mD_{+}(f,x) \neq mD_{-}(f,x)$) is σ-porous. Theorem 3.5 (see also Corollary 3.6) is a metric analogue of the so-called Denjoy–Young–Saks theorem about Dini derivatives (see e.g. [4, Theorem 4.4]).

In section 4, we show that if $md(f,\cdot)$ is a continuous function, then the set of points x, where f is not metrically differentiable, is σ-symmetrically porous (Theorem 4.7). In Theorem 4.9, we show that this set is not necessarily countable. This means that the properties of metric derivatives are different from the properties of standard ones; in the latter case, the set considered in section 4, would necessarily be countable (if say $md(f,\cdot) \equiv 1$ for a real-valued f then the standard unilateral derivatives of f are equal to ± 1 at all points).

In section 5, we discuss sufficient conditions for a mapping to be metrically differentiable at a point. This is closely related to the notion of bilateral metric regularity.

In a recent paper [8], L. Zajíček together with the first author characterized those mappings $f : [a,b] \to X$ that allow a metrically differentiable (resp. boundedly metrically differentiable) parametrization. In section 6, we study the situation when f allows a continuously metrically differentiable parametrization (by this we mean that for a suitable homeomorphism h, the composition $f \circ h$ is metrically differentiable and its metric derivative is continuous), or just a parametrization with continuous metric derivative; see Theorems 6.2 and 6.1 for more details.

2 Preliminaries

By λ we denote the 1-dimensional Lebesgue measure on $\mathbb{R}$, and by $\mathcal{H}^1$ the 1-dimensional Hausdorff measure. In the following, X is always a real Banach space.

The following is a version of the Sard's theorem. For a proof see e.g. [8, Lemma 2.2].

Lemma 2.1. *Let $f : [a,b] \to X$ be arbitrary. Then $\mathcal{H}^1(f(\{x \in [a,b] : md(f,x) = 0\})) = 0$.*

By $B(x,r)$, we denote the open ball in X with center $x \in X$ and radius $r > 0$. Let $M \subset \mathbb{R}$, $x \in M$, and $R > 0$. Then we define $\gamma(x,R,M)$ to be the supremum of all $r > 0$ for which there exists $z \in \mathbb{R}$ such that $B(z,r) \subset B(x,R) \setminus M$. Also, we define $S\gamma(x,R,M)$ to be the supremum of all $r > 0$ for which there exists $z \in \mathbb{R}$ such that

$B(z,r) \cup B(2x - z, r) \subset B(x, R) \setminus M$. Further, we define the *upper porosity* of M at x as

$$\overline{p}(M, x) := 2 \limsup_{R \to 0+} \frac{\gamma(x, R, M)}{R},$$

and the *symmetric upper porosity* of M at x as

$$S\overline{p}(M, x) := 2 \limsup_{R \to 0+} \frac{S\gamma(x, R, M)}{R}.$$

We say that M is *porous*[1] (resp. *symmetrically porous*) provided $\overline{p}(M, x) > 0$ for all $x \in M$ (resp. $S\overline{p}(M, x) > 0$ for all $x \in M$). We say that $N \subset \mathbb{R}$ is *σ-porous* (resp. *σ-symmetrically porous*) provided it is a countable union of porous (resp. symmetrically porous) sets. For more information about porous sets, see a recent survey [16].

Let $f : [a, b] \to X$. Then we say that f *has finite variation* or that f *is BV*, provided $\bigvee_a^b f < \infty$. (Recall that $\bigvee_a^b f = \sup_D \sum_{i=0}^{n(D)-1} \|f(x_i) - f(x_{i+1})\|$, where the supremum is taken over all partitions $D = \{a = x_0 < x_1 < \cdots < x_n = b\}$, of $[a, b]$ and $n(D) = \#D - 1$.) We define $\bigvee_v^u f := -\bigvee_u^v f$ for $a \le u < v \le b$. We will denote $v_f(x) := \bigvee_a^x f$ for $x \in [a, b]$.

We say that $f : \mathbb{R} \to X$ is *pointwise-Lipschitz* if $\limsup_{y \to x} \|f(x) - f(y)\|/|x - y| < \infty$ for every $x \in \mathbb{R}$.

A considerable part of the present article is devoted to metric analogues of derived numbers (Dini derivatives). Now, we give a definition of metric derived numbers. Let $f : \mathbb{R} \to X$. Define

$$mD^{\pm}(f, x) = \limsup_{t \to 0+} \frac{\|f(x \pm t) - f(x)\|}{t},$$

and

$$mD_{\pm}(f, x) = \liminf_{t \to 0+} \frac{\|f(x \pm t) - f(x)\|}{t},$$

to be the *unilateral upper (resp. lower) metric derived numbers* (we also allow the value $+\infty$).

Note that if all four metric derived numbers of a mapping $f : \mathbb{R} \to X$ agree at a point x, then $md(f, x)$ exists, but still f is not necessarily metrically differentiable at x.

3 Unilateral metric derivatives

It is well known that the set where the standard unilateral derivatives of a real function of a real variable exist but are not equal is countable (see e.g. [12, Theorem 7.2]). The following theorem shows that it is also true for unilateral metric derivatives.

Theorem 3.1. *Let $f : \mathbb{R} \to X$. Then the set of points $x \in \mathbb{R}$ where $md_+(f, x)$, $md_-(f, x)$ exist but $md_+(f, x) \neq md_-(f, x)$, is countable.*

[1] In the terminology of [16], this corresponds to M being "an upper-porous set".

Proof. The proof is similar to the proof of [12, Theorem 7.2] and thus we omit it. □

It is well known that for a real function of a real variable the set of angular points (i.e. points where $D_- f > D^+ f$ or $D_+ f > D^- f$; $D^\pm f$, $D_\pm f$ are the standard derived numbers) is countable; see e.g. [12, Theorem 7.2]. The following theorem shows what happens for metric derived numbers.

Theorem 3.2. *Let $f : \mathbb{R} \to X$. Then the set of points $x \in \mathbb{R}$ where either $mD_-(f,x) > mD^+(f,x)$ or $mD_+(f,x) > mD^-(f,x)$ is countable.*

Proof. By symmetry, it is enough to prove that the set $E = \{x \in \mathbb{R} : mD_-(f,x) > mD^+(f,x)\}$ is countable. Let $h < k$ be two positive rational numbers. For a positive integer n let E_{hkn} be the set of points $x \in E$ for which $\|f(\xi) - f(x)\|/|\xi - x| < h$ and $\|f(\xi') - f(x)\|/|\xi' - x| > k$ whenever $0 < \xi - x < 1/n$ and $0 < x - \xi' < 1/n$. Then $E_{hkn} \cap (x - 1/n, x + 1/n) = \{x\}$. Suppose that is not true, and there is a point $x_1 \in E_{hkn} \cap (x - 1/n, x + 1/n)$ such that $x_1 \neq x$. Then assuming $x > x_1$, say, we get $\|f(x_1) - f(x)\|/|x_1 - x| < h$ and $\|f(x) - f(x_1)\|/|x - x_1| > k$, a contradiction. Thus all points of E_{hkn} are isolated, and E_{hkn} is countable. Because $E \subset \bigcup_{h,k,n} E_{hkn}$, we obtain the conclusion of the theorem. □

We have the following two theorems concerning the points where unilateral lower and upper metric derivatives differ. In the proofs, we use similar ideas as in [9, Theorem 1].

Theorem 3.3. *Let X be a Banach space, and $f : \mathbb{R} \to X$ be arbitrary. Then the set*

$$\{x \in \mathbb{R} : mD^+(f,x) \neq mD^-(f,x)\}$$

is σ-porous.

Proof. We will only prove that the set

$$A = A_f = \{x \in \mathbb{R} : mD^-(f,x) < mD^+(f,x)\},$$

is σ-porous (and notice that $\{x \in \mathbb{R} : mD^-(f,x) > mD^+(f,x)\}$ is σ-porous as it is equal to $A_{f(-\cdot)}$). To that end, it is enough to establish that

$$A_{rs} = \{x \in A : mD^-(f,x) < r < s < mD^+(f,x)\},$$

is σ-porous for all $r < s$ pairs of positive rational numbers. Define

$$A_{rsn} = \left\{x \in A_{rs} : \frac{\|f(x) - f(y)\|}{|x - y|} < r \text{ for } y \in (x - 1/n, x)\right\}.$$

We easily see that $A_{rs} = \bigcup_n A_{rsn}$. We will prove that A_{rsn} is $(\delta - 1)/\delta$-porous, where $\delta = \min(2, (s + r)/2r)$. Let $x \in A_{rsn}$. Then there exist $x_k \to x+$ such that

$\|f(x)-f(x_k)\|/|x-x_k| > s$. Choose k large enough such that $|x-x_k| < 1/n$. Define $w_k = x + \delta(x_k - x)$, and let $y \in [x_k, w_k] \cap A_{rsn}$. Then

$$\begin{aligned}\|f(x)-f(y)\| &\geq \|f(x)-f(x_k)\| - \|f(x_k)-f(y)\| \\ &\geq s\,|x-x_k| - r\,|x_k-y| \geq s\,|x-x_k| - r\,|x_k-w_k| \\ &= s\,|x-x_k| - r\,(\delta-1)\,|x_k-x| \\ &= |x-x_k|\,(s-r\,(\delta-1)) = |w_k-x|\,\frac{(s-r\,(\delta-1))}{\delta} \\ &\geq r\,|x-y|,\end{aligned}$$

by the choice of δ (we used that $w_k - x = \delta\,(x_k - x)$, and $w_k - x_k = (\delta-1)\,(x_k - x)$). Thus $y \notin A_{rsn}$, and $[x_k, w_k] \cap A_{rsn} = \emptyset$. Finally, note that $(w_k - x_k)/(w_k - x) = (\delta-1)/\delta > 0$. □

Theorem 3.4. *Let X be a Banach space, and $f : \mathbb{R} \to X$ be pointwise-Lipschitz. Then the set*

$$\{x \in \mathbb{R} : mD_+(f,x) \neq mD_-(f,x)\}$$

is σ-porous.

Proof. We will only prove that the set

$$B = B_f = \{x \in \mathbb{R} : mD_-(f,x) < mD_+(f,x)\},$$

is σ-porous, and notice that $\{x \in \mathbb{R} : mD_-(f,x) > mD_+(f,x)\}$ is σ-porous as it is equal to $B_{f(-\cdot)}$. We will prove that B_f is σ-porous for f that is pointwise-Lipschitz. To that end, it is enough to establish that $B_{rs} = \{x \in B : mD_-(f,x) < r < s < mD_+(f,x)\}$, is σ-porous for all $r < s$ pairs of positive rational numbers. For $n \in \mathbb{N}$, define

$$\begin{aligned}B_{rsn} = \Big\{x \in B_{rs} : &\frac{\|f(x)-f(y)\|}{|x-y|} > s \text{ for } y \in (x, x+1/n), \\ &\text{and } \frac{\|f(x)-f(z)\|}{|x-z|} < n \text{ whenever } 0 < |z-x| < 1/n\Big\}.\end{aligned}$$

Since f is pointwise-Lipschitz, we easily see that $B_{rs} = \bigcup_n B_{rsn}$. We will prove that B_{rsn} is $(\delta-1)/\delta$-porous, where $\delta = \min(((s-r)/n)+1, 2)$. Let $x \in B_{rsn}$. Then there exist $x_k \to x-$ such that $\|f(x)-f(x_k)\|/|x-x_k| < r$. Choose k large enough such that $|x-x_k| < 1/n$. Define $w_k = x - \delta\,(x - x_k)$, and let $y \in [w_k, x_k] \cap B_{rsn}$. Then

$$\begin{aligned}\|f(x)-f(y)\| &\leq \|f(x)-f(x_k)\| + \|f(x_k)-f(y)\| \\ &\leq r\,|x-x_k| + n\,|x_k-y| \leq r\,|x-x_k| + n\,|x_k-w_k| \\ &= r\,|x-x_k| + n\,(\delta-1)\,|x_k-x| = |x-x_k|\,(r+n\,(\delta-1)) \\ &\leq s\,|x-y|,\end{aligned}$$

by the choice of δ (we used that $x - w_k = \delta\,(x - x_k)$, $x_k - w_k = (\delta - 1)\,(x - x_k)$, and $|x_k - w_k| < 1/n$). Thus $y \notin B_{rsn}$, and $[x_k, w_k] \cap B_{rsn} = \emptyset$. Finally, note that $(x_k - w_k)/(x - w_k) = (\delta - 1)/\delta > 0$. □

The following theorem asserts that outside of a set of measure 0, the fact that $mD^+(f,x) < \infty$ already implies that $md(f,x)$ exists.

Theorem 3.5. *Let $f : \mathbb{R} \to X$ be arbitrary. Then there exists a set N with Lebesgue measure zero such that if $x \in \mathbb{R} \setminus N$ and $mD^+(f,x) < \infty$, then $md(f,x)$ exists, and $md(f,x) = mD^+(f,x)$.*

Proof. Let N_1 be the set of points $x \in \mathbb{R}$ where $mD^-(f,x) \neq mD^+(f,x)$. Then, by Theorem 3.3 N_1 is σ-porous. Therefore, by the Lebesgue density theorem, its Lebesgue measure $\lambda(N_1)$ is zero. Let

$$A_n = \{x \in \mathbb{R} : \|f(x+h) - f(x)\| \leq nh \text{ for } 0 < h < 1/n\}.$$

Let A be the set of points x such that $mD^+(f,x) < \infty$. Then $A = \bigcup_n A_n$. Let $A_{n,j}$ be subsets of A_n, such that $A_n = \bigcup A_{n,j}$, and $\operatorname{diam}(A_{n,j}) < 1/n$. Then $f|_{A_{n,j}}$ is n-Lipschitz, and thus, by Kirszbraun theorem, see [14], it can be extended to an n-Lipschitz function $f_{n,j}$ defined on the whole real line. By [7, Theorem 2.7], we obtain that $f_{n,j}$ is metrically differentiable at all $x \in D_{n,j}$, where $\lambda(\mathbb{R} \setminus D_{n,j}) = 0$. Let $E_{n,j} \subset D_{n,j} \cap A_{n,j}$ be the set of points of density of $D_{n,j} \cap A_{n,j}$. By the Lebesgue density theorem we have that $\lambda(D_{n,j} \cap A_{n,j} \setminus E_{n,j}) = 0$. We shall prove that $md(f,x)$ exists and is equal to $mD^+(f,x)$ at all points $x \in E_{n,j}$ for all $n, j \in \mathbb{N}$. This will conclude the proof, as the set $N = \bigcup_{n,j}(A_{n,j} \setminus E_{n,j})$ has Lebesgue measure 0.

To finish the proof, let $x \in E_{n,j}$. Fix $\varepsilon > 0$. Find $\delta > 0$ such that $\lambda(E_{n,j} \cap (x, x + t)) \geq (1 - (\varepsilon/(4n)))t$ for $0 < t < \delta$, and $|\|f_{n,j}(x+t) - f_{n,j}(x)\|/|t| - md(f_{n,j}, x)| \leq \varepsilon$, whenever $0 < |t| < \delta$. Thus for each $0 < h < \delta$ there exists $y \in E_{n,j} \cap (x, x+h)$ such that $|y - (x+h)| \leq (\varepsilon h)/(2n)$. Now,

$$\begin{aligned}\|f(x+h) - f(x)\| &\leq \|f(y) - f(x)\| + \|f(x+h) - f(y)\|\\ &\leq (md(f_{n,j}, x) + \varepsilon)(y - x) + \varepsilon h\\ &\leq (md(f_{n,j}, x) + \varepsilon)h,\end{aligned}$$

since x and y belong to $E_{n,j} \subset A_n$ and $y > x$. On the other hand,

$$\begin{aligned}\|f(x+h) - f(x)\| &\geq \|f(y) - f(x)\| - \|f(x+h) - f(y)\|\\ &\geq (md(f_{n,j}, x) - \varepsilon)(y - x) - \varepsilon h\\ &\geq ((md(f_{n,j}, x) - \varepsilon)(1 - \varepsilon h \cdot (2n)^{-1} - \varepsilon)h.\end{aligned}$$

Thus $md_+(f,x) = md(f_{n,j}, x) = mD^+(f,x)$.

A similar argument shows that $md_-(f,x) = md(f_{n,j}, x) = mD^+(f,x)$ for $x \in E_{n,j}$, and thus $md(f,x)$ exists for all $x \in A \setminus N$. □

Theorem 3.5 has the following corollary.

Corollary 3.6. *Let $f : \mathbb{R} \to X$ be arbitrary. Then there exists a set $N \subset \mathbb{R}$ with $\lambda(N) = 0$, such that if $x \in \mathbb{R} \setminus N$, and $\min(mD^-(f,x), mD^+(f,x)) < \infty$, then $md(f,x)$ exists.*

Corollary 3.6 together with [7, Theorem 2.6] imply the following:

Corollary 3.7. *Let $f : \mathbb{R} \to X$ be arbitrary. Then there exists a set $M \subset \mathbb{R}$ with $\lambda(M) = 0$, such that if $x \in \mathbb{R} \setminus M$, and $\min(mD^-(f,x), mD^+(f,x)) < \infty$, then f is metrically differentiable at x.*

4 Points of metric non-differentiability

We will use following lemma proved in [8, Lemma 2.4].

Lemma 4.1. *Let $f : [c,d] \to X$, $x \in [c,d]$. Then the following hold.*
(i) If $md(f,x) = 0$, then f is metrically differentiable at x.
(ii) If $h : [a,b] \to [c,d]$ is differentiable at $w \in [a,b]$, $h(w) = x$, and f is metrically differentiable at x, then $f \circ h$ is metrically differentiable at w, and $md(f \circ h, w) = md(f,x) \cdot |h'(w)|$.

Lemma 4.2. *Let X be a Banach space, and let $f : [a,b] \to X$. If $md(f,\cdot)$ is continuous at $x \in [a,b]$, then there exists $\delta > 0$ such that*

$$\bigvee_s^t f = \int_s^t md(f,y)\, dy \quad \textit{for all } s < t,\ s,t \in [x-\delta, x+\delta] \cap [a,b].$$

Proof. Let $\delta > 0$ be chosen such that for all $s \in [x-\delta, x+\delta] \cap [a,b]$ we have that $md(f,s)$ exists and $|md(f,x) - md(f,s)| \le 1$. It follows from [10, §2.2.7] that $f|_{[x-\delta,x+\delta]\cap[a,b]}$ is Lipschitz. We obtain that

$$\int_s^t md(f,y)\, dy = \int_{f([s,t])} N(f|_{[s,t]}, y)\, d\mathcal{H}^1(y) = \bigvee_s^t f,$$

for all $s < t$, $s,t \in [x-\delta, x+\delta] \cap [a,b]$ (here, $N(f|_{[s,t]}, y)$ is the multiplicity with which the function $f|_{[s,t]}$ assumes a value y). The first equality follows from [13, Theorem 7], the second equality follows from [10, Theorem 2.10.13]. □

Let $f : [a,b] \to X$, $I = [a,b]$. We say that $x \in I$ is *metrically regular point of the function* f, provided

$$\lim_{\substack{t\to 0\\ x+t\in I}} \frac{\|f(x+t) - f(x)\|}{|\bigvee_x^{x+t} f|} = 1.$$

Lemma 4.3. *Let X be a Banach space, $g : [a,b] \to X$, $x \in [a,b]$, $md(g,x) > 0$, and $md(g,\cdot)$ is continuous at x. Then x is metrically regular point of the function g.*

Proof. Let $\varepsilon > 0$. Find $\delta_0 > 0$ such that $(1-\varepsilon)\, md(g,x)|t| \leq \|g(x+t) - g(x)\|$ and $md(g, x+t) < (1+\varepsilon)\cdot md(g,x)$, whenever $|t| < \delta_0$ and $x+t \in [a,b]$. Using Lemma 4.2, we can find $0 < \delta < \delta_0$ such that for all $|t| < \delta$ we have

$$\left(\frac{1-\varepsilon}{1+\varepsilon}\right)\left|\bigvee_x^{x+t} g\right| = \left(\frac{1-\varepsilon}{1+\varepsilon}\right)\left|\int_x^{x+t} md(g,s)\,ds\right|$$
$$\leq (1-\varepsilon)\cdot md(g,x)\,|t| \leq \|g(x+t)-g(x)\| \leq \left|\bigvee_x^{x+t} g\right|.$$

If $t \neq 0$, by dividing by $|\bigvee_x^{x+t} g|$ (which is strictly positive), we obtain $(1-\varepsilon)/(1+\varepsilon) \leq \|g(x+t) - g(x)\|/|\bigvee_x^{x+t} g| \leq 1$, and thus x is metrically regular point of f. □

The following lemma shows that the condition (1.1) is satisfied "unilaterally" at a point x provided $md(f,\cdot)$ is continuous at x.

Lemma 4.4. *Let X be a Banach space, and let $f : [a,b] \to X$. If $md(f,\cdot)$ is continuous at $x \in [a,b]$, then*

$$\|f(y) - f(z)\| - md(f,x)|y-z| = o(|x-z| + |x-y|), \tag{4.1}$$

whenever $(y,z) \to (x,x)$, and $\operatorname{sign}(z-x) = \operatorname{sign}(y-x)$.

Proof. If $md(f,x) = 0$, then the conclusion follows from Lemma 4.1, so we can assume that $md(f,x) > 0$. Lemma 4.3 implies that x is metrically regular point of f. Now we will prove that f satisfies (4.1) at x. Let $0 < \varepsilon < 1$. Using Lemma 4.2, find $\delta > 0$ such that for all t with $x+t \in [a,b]\cap[x-\delta, x+\delta]$ we have that $(1-\varepsilon)\,|\bigvee_x^{x+t} f| \leq \|f(x+t) - f(x)\|$,

$$(1-\varepsilon)\, md(f,x) \leq md(f, x+t) \leq (1+\varepsilon)\, md(f,x),$$

and $\bigvee_y^z f = \int_y^z md(f,s)\,ds$ for all $y,z \in [a,b]\cap[x-\delta, x+\delta]$. Let $y,z \in [a,b]\cap[x-\delta, x+\delta]$ with $\operatorname{sign}(z-x) = \operatorname{sign}(y-x)$. Without any loss of generality, we can assume that $z > x$, and $|z-x| \geq |y-z|$. We obtain that

$$\begin{aligned}
\|f(y) - f(z)\| &\geq \|f(z) - f(x)\| - \|f(y) - f(x)\| \\
&\geq (1-\varepsilon)\bigvee_x^z f - \|f(y) - f(x)\| \\
&\geq (1-\varepsilon)\int_x^z md(f,t)\,dt - \bigvee_x^y f \\
&\geq (1-\varepsilon)\int_x^z md(f,t)\,dt - \left|\int_x^y md(f,t)\,dt\right| \\
&\geq (1-\varepsilon)^2 md(f,x)(z-x) - (1+\varepsilon)\, md(f,x)|y-x| \\
&= md(f,x)|z-y| - \varepsilon\cdot \underbrace{((2-\varepsilon)(z-x) + |y-x|)\cdot md(f,x)}_{\eta(\varepsilon,y,z)}.
\end{aligned}$$

It is easy to see that $\eta(\varepsilon, y, z)/(|z-x|+|y-x|)$ is bounded from above by $2 \cdot md(f,x)$ for all $\varepsilon \in (0,1)$. For the other inequality, note that

$$\|f(y) - f(z)\| \le \bigvee_y^z f = \int_y^z md(f,s)\, ds \le (1+\varepsilon)\, md(f,x)|z-y|, \tag{4.2}$$

and the conclusion easily follows. □

We now show that if the metric derivative of f exists at each point and is continuous, then the mapping is metrically differentiable on a large set of points. We prove this in several steps.

Proposition 4.5. *Let X be a Banach space, $f : [a,b] \to X$ be such that $md(f,x) = 1$ for each $x \in [a,b]$. Then the set of points $x \in [a,b]$ such that f is not metrically differentiable at x, is σ-symmetrically porous.*

Proof. Let A be the set of points $x \in (a,b)$ such that f is not metrically differentiable at x. By Lemma 4.4, we see that the condition (1.1) is satisfied unilaterally at each $x \in [a,b]$.

Suppose that $x \in A$. We claim that there exist $\delta_j = \delta_j(x) \to 0+$ such that

$$\liminf_{j\to\infty} \frac{\|f(x+\delta_j) - f(x-\delta_j)\|}{2\delta_j} < 1. \tag{4.3}$$

To see this, note that because $x \in A$, there exist sequences $(y_j)_j$, $(z_j)_j$ such that $y_j < x < z_j$ (because (1.1) is satisfied unilaterally at x), $\lim_j y_j = \lim_j z_j = x$, and $\liminf_{j\to\infty} \|f(z_j) - f(y_j)\|/(z_j - y_j) < 1 - \varepsilon$, for some $\varepsilon > 0$. Without any loss of generality, we can assume that $z_j - x \le x - y_j$. Let $\tilde{y}_j = 2x - y_j$, and note that $z_j \le \tilde{y}_j$. If $\tilde{y}_j = z_j$, take $\delta_j = z_j - x$, otherwise note that for $j \in \mathbb{N}$ large enough we have

$$\begin{aligned}\|f(y_j) - f(\tilde{y}_j)\| &\le \|f(y_j) - f(z_j)\| + \|f(z_j) - f(\tilde{y}_j)\| \\ &\le (1-\varepsilon)(z_j - y_j) + (\tilde{y}_j - z_j).\end{aligned}$$

Now, as $\tilde{y}_j - z_j \le z_j - y_j$, we obtain $(\tilde{y}_j - z_j) - (\varepsilon/2)(z_j - y_j) \le (1-(\varepsilon/2))(\tilde{y}_j - z_j)$, and thus $\|f(y_j) - f(\tilde{y}_j)\| \le (1-(\varepsilon/2))(\tilde{y}_j - y_j)$. Now define $\delta_j = \tilde{y}_j - x = x - y_j$, and (4.3) follows.

Let A_{nm} be the set of all $x \in A$ such that

- there exists a sequence $(\delta_j)_j$, such that $\delta_j \to 0+$, and $\|f(x-\delta_j) - f(x+\delta_j)\| \le (1-(1/m))\, 2\,\delta_j$,
- for each $t \in [0,1]$ with $0 < |x-t| < 1/n$ we have $(1-(1/(2m)))|x-t| < \|f(t) - f(x)\|$.

By the above argument, it is easy to see that $A = \bigcup_{n,m} A_{nm}$.

Fix $n, m \in \mathbb{N}$. Let $x \in A_{nm}$. There exists $j_0 \in \mathbb{N}$ such that for all $j \ge j_0$ we have $0 < \delta_j < (4n)^{-1}$. Let $z_j := x + \delta_j$, and $y_j := x - \delta_j$. Fix $j \ge j_0$ and suppose that

$w \in [z_j, z_j + 2\delta_j]$. Then $|w - y_j| < 1/n$, and we have that

$$\|f(y_j) - f(w)\| \leq \|f(y_j) - f(z_j)\| + \|f(z_j) - f(w)\| \\ \leq \left(1 - \frac{1}{m}\right) 2\delta_j + |w - z_j|.$$

By the choice of w we have $-(2\delta_j)/(2m) + (w - z_j) \leq (1 - (1/(2m)))(w - z_j)$, and thus $\|f(y_j) - f(w)\| \leq (1 - (1/(2m)))(w - y_j)$. This implies that $w \notin A_{nm}$. We obtained that $[z_j, z_j + 2\delta_j] \cap A_{nm} = \emptyset$. Similarly, $[y_j - 2\delta_j, y_j] \cap A_{nm} = \emptyset$, and the symmetric porosity of A_{nm} follows. □

We will need the following auxiliary lemma.

Lemma 4.6. *Let $B \subset [a, b]$ be symmetrically porous and $h : [a, b] \to \mathbb{R}$ be continuously differentiable and bilipschitz. Then $h(B)$ is symmetrically porous.*

Proof. Let $L > 0$ be such that $L^{-1}|x-y| \leq |h(x)-h(y)| \leq L|x-y|$ for all $x, y \in [a, b]$. Let $x \in B$. Let $\delta_n, \alpha_n > 0$ be such that $B(x-\delta_n, \alpha_n) \cup B(x+\delta_n, \alpha_n) \subset \mathbb{R} \setminus B$, $\alpha_n \to 0$, and $c\delta_n \leq \alpha_n$. First, we will show that

$$B(h(x \pm \delta_n), \alpha_n/(2L)) \cap h(B) = \emptyset. \tag{4.4}$$

Let $z \in B(h(x \pm \delta_n), \alpha_n/(2L))$. If $y \in [a, b]$ is such that $h(y) = z$, then

$$|x \pm \delta_n - y| \leq L|h(x \pm \delta_n) - h(y)| \leq \alpha_n/2,$$

and thus $z \notin h(B)$ (since h is one-to-one), and (4.4) holds.

Note that since $h'(x) \neq 0$,

$$\left|1 - \frac{h(x + \delta_n) - h(x)}{h(x) - h(x - \delta_n)}\right| = \left|1 - \frac{h'(x)\delta_n + o(\delta_n)}{h'(x)\delta_n + o(\delta_n)}\right| \to 0,$$

as $n \to \infty$, and thus

$$\begin{aligned} |h(x + \delta_n) - h(x) - (h(x) - h(x-\delta_n))| &\leq c/(4L^2)|h(x - \delta_n) - h(x)| \\ &\leq c\delta_n/(4L) \leq \alpha_n/(4L) \end{aligned} \tag{4.5}$$

for n large enough. Now, we will show that $2h(x) - z \in B(h(x + \delta_n), \alpha_n/(2L))$ whenever $z \in B(h(x - \delta_n), \alpha_n/(4L))$, and n is large enough. Together with (4.4), this easily implies that $S\overline{p}(h(B), h(x)) > 0$.

Assume that $z \in B(h(x - \delta_n), \alpha_n/(4L))$, and that (4.5) holds. Then

$$\begin{aligned} |2h(x) - z - h(x + \delta_n)| &\leq |h(x + \delta_n) - h(x) + (h(x - \delta_n) - h(x))| \\ &\quad + |h(x - \delta_n) - z| \\ &\leq \alpha_n/(4L) + \alpha_n/(4L) = \alpha_n/(2L), \end{aligned}$$

and thus $2h(x) - z \in B(h(x + \delta_n), \alpha_n/(2L))$, and the conclusion follows. □

We have the following:

Theorem 4.7. *Let $f : [a, b] \to X$ be such that $md(f, \cdot)$ is continuous on $[a, b]$. Then the set of points, where f is not metrically differentiable, is σ-symmetrically porous.*

Proof. Let $A \subset [a, b]$ be the set where f is not metrically differentiable. Lemma 4.1 implies that if $x \in A$, then $md(f, x) > 0$. Let $A = \bigcup_n A_n$, where $A_n = \{x \in A : md(f, x) > 1/n\}$. It is enough to show that each A_n is σ-symmetrically porous. Because $md(f, \cdot)$ is continuous, we have that each A_n is open. Let (c, d) be an open component of A_n, let $g = f|_{[c,d]}$, $G = g \circ v_g^{-1}$ (see section 2 for the definition of v_g). Using Lemma 4.2, it is easy to see that $md(G, x) = 1$ for all $x \in v_g((c, d))$. Then Proposition 4.5 implies that G is metrically differentiable outside a σ-symmetrically porous set B. Because v_g is continuously differentiable and bilipschitz, by Lemmas 4.1 and 4.6, we obtain that $g = G \circ v_g$ is metrically differentiable outside a σ-symmetrically porous set $v_g^{-1}(B)$. □

Remark 4.8. It is easy to see that if f is a real-valued function and $md(f, \cdot)$ is continuous on $[a, b]$, then the set of points where f is not metrically differentiable is at most countable. However, in Theorem 4.9 below, we show that already in a 2-dimensional situation such a set may be uncountable. Thus, Theorem 4.9 shows that Theorem 4.7 cannot be strengthened to make the exceptional set countable.

Theorem 4.9. *For any norm $\|\cdot\|$ in the 2-dimensional plane, there exists a curve $\gamma : [0, \ell] \to (\mathbb{R}^2, \|\cdot\|)$ with $md(\gamma, x) = 1$ for all $x \in [0, \ell]$, but such that the set of points where γ is not metrically differentiable is uncountable.*

We will give a detailed proof of this theorem for $\|\cdot\|$ being the Euclidean norm. In Remark 4.14, we explain how this case reflects the most general situation. Note however, that if one uses a "polygonal" norm (for example, the ℓ_1-norm), then much simpler constructions are possible. We explain this in Remark 4.15.

Before we start the proof of Theorem 4.9, let us establish the following property of logarithmic spirals, which will be used in the proof of Lemma 4.11.

Lemma 4.10. *Assume $S_{a,b}$ is a planar curve defined in polar coordinates (r, ϕ) by the equation $r = ae^{b\phi}$ with $a > 0$, $b \neq 0$ (logarithmic spiral). Then the length of the arc of $S_{a,b}$ between the origin and the point with modulus r_0 and argument ϕ_0 is equal to $(\sqrt{b^2+1}/|b|)r_0$.*

In other words, if $S_{a,b} : [0, +\infty) \to \mathbb{C}$ is the arc-length parametrization of this logarithmic spiral such that $S_{a,b}(0) = 0$, then

$$\frac{|S_{a,b}(t)|}{t} = \frac{|b|}{\sqrt{b^2+1}} \tag{4.6}$$

for all $t > 0$.

Proof. A routine computation of the length of the logarithmic spiral with the given equation in polar coordinates proves the lemma. □

Lemma 4.11. *For any angle $\alpha \in (0, \pi/2)$ and a constant $q \in (0,1)$ there is a piecewise smooth planar curve such that its arc-length parametrization $g = g_{q,\alpha} : \mathbb{R} \to \mathbb{C}$ has the following properties:*

(a) $g([0,1])$ is a horizontal interval and there exists $L_{q,\alpha} > 0$ so that $g([1+L_{q,\alpha}, +\infty))$ and $g((-\infty, -L_{q,\alpha}])$ are horizontal rays;

(b) there exists $t_{q,\alpha} > 1/2$ such that the arguments of $z_\pm = g(1/2 \pm t_{q,\alpha}) - g(1/2)$ are equal to $(-\alpha)$ and $(\pi + \alpha)$ resp.;

(c) $|g(t) - g(s)|/|t-s| > q$ for all $s \in [0,1]$ and $t \neq s$.

Proof. Let $B > 0$ be large enough so as to ensure that

$$\frac{B}{\sqrt{B^2+1}} > q \quad \text{and} \quad -B\sin\alpha + \cos\alpha < 0. \tag{4.7}$$

In (4.13), we will impose another condition on B which also bounds B from below. Fix $b > B$ and denote $k = b/\sqrt{b^2+1}$.

We first construct a piecewise smooth planar curve $f = f_{q,\alpha} : [0, +\infty) \to \mathbb{C}$ such that

$$g(t) = \begin{cases} f(t), & \text{if } t \geq 0, \\ 1 - \overline{f(1-t)}, & \text{if } t < 0, \end{cases} \tag{4.8}$$

has the desired properties.

For $t \in [0,1]$ we set $f(t) = t + 0i$. Now let $S_{1,-b} : [0, +\infty) \to \mathbb{C}$ be the arc-length parametrization of the logarithmic spiral from Lemma 4.10. Identity (4.6) implies that the point $S_{1,-b}(k^{-1})$ has modulus 1, therefore, it coincides with $f(1)$.

For $t \in [1, 1 + k^{-1}(e^{b\alpha} - 1)]$ we put $f(t) = S_{1,-b}(t + k^{-1} - 1)$. Then for every $s \geq 0$ one has:

$$\frac{1+s}{|f(1+s) - f(0)|} < \frac{k^{-1}+s}{|f(1+s) - f(0)|} = \frac{k^{-1}+s}{|S_{1,-b}(k^{-1}+s)|} = k^{-1}. \tag{4.9}$$

Let $s_0 = k^{-1}(e^{b\alpha} - 1)$. Then the point $f(1+s_0) = S_{1,-b}(k^{-1}e^{b\alpha})$ has modulus $e^{b\alpha}$ and argument $-\alpha$.

Now let $S_{e^{2b\alpha},b} : [0, +\infty) \to \mathbb{C}$ be another logarithmic spiral parametrized by the arc-length. For $t \in [1 + s_0, 1 + s_0 + s_1]$ (where s_1 is defined below), let $f(t) = S_{e^{2b\alpha},b}(t + k^{-1} - 1)$. Again, note that $S_{1,-b}(t + k^{-1} - 1)$ and $S_{e^{2b\alpha},b}(t + k^{-1} - 1)$ are equal at $t = 1 + s_0$, since by (4.6) the lengths of the arcs of both logarithmic spirals between the origin and the point with modulus $e^{b\alpha}$ and argument $-\alpha$ are equal to $k^{-1}e^{b\alpha} = k^{-1} + s_0$. Furthermore, for every $s \geq 0$ one has:

$$\frac{1+s_0+s}{|f(1+s_0+s) - f(0)|} < \frac{k^{-1}+s_0+s}{|S_{e^{2b\alpha},b}(k^{-1}+s_0+s)|} = k^{-1}. \tag{4.10}$$

Let us find the slope of the tangent to the logarithmic spiral $S_{e^{2b\alpha},b}$ at the point with modulus $e^{b\alpha}$ and argument $-\alpha$. If we denote by $z(\phi) = e^{2b\alpha}e^{b\phi}e^{i\phi}$ the polar parametrization of $S_{e^{2b\alpha},b}$, then $\operatorname{Im} dz/d\phi(-\alpha)$ is equal to $e^{b\alpha}(-b\sin\alpha + \cos\alpha) < 0$.

Therefore, the y-coordinate of $f(t)$ continues to decrease as ϕ increases from $-\alpha$ to some $-\beta \in (-\alpha, 0)$ such that $-b\sin\beta + \cos\beta = 0$ (i.e., $\tan\beta = 1/b$). Let s_1 be such that $f(1+s_0+s_1) = S_{e^{2b\alpha},b}(e^{2b\alpha}e^{-b\beta})$ is the point with modulus $e^{2b\alpha-b\beta}$ and argument $-\beta$, i.e., $s_1 = k^{-1}e^{b\alpha}(e^{b(\alpha-\beta)} - 1)$.

For $t \geq 1 + s_0 + s_1$, define $f(t)$ as $f(1 + s_0 + s_1) + (t - 1 - s_0 - s_1)$. Then one easily checks that since $\cos\beta = k$, the law of cosines for the triangle with vertices in $f(0)$, $f(1 + s_0 + s_1)$ and $f(1 + s_0 + s_1 + s)$ guarantees that the inequality

$$\frac{(1 + s_0 + s_1 + s)^2}{|f(1 + s_0 + s_1 + s) - f(0)|^2} \leq k^{-2} \tag{4.11}$$

holds for all $s \geq 0$.

Inequalities (4.9), (4.10), (4.11) imply that

$$\frac{t}{|f(t) - f(0)|} \leq k^{-1} \tag{4.12}$$

for all $t > 0$.

Note that if we now define g as in (4.8), then property *(a)* in the lemma holds for $L_{q,\alpha} = s_0 + s_1$.

The argument of $f(1 + s_0) - f(0)$ is equal to $(-\alpha)$. Then the arguments of $g(1/2 \pm (1/2 + s_0)) - g(1/2)$ are equal to $(-\alpha')$ and $(\pi + \alpha')$ respectively, where $\alpha' > \alpha$. Since the argument of $g(1/2 + t) - g(1/2)$ is continuous in t, there is a value $t_{q,\alpha}$ between $1/2$ and $1/2 + s_0$ such that property *(b)* in the present lemma holds for $t_{q,\alpha}$.

We have already proved, see (4.12), that property *(c)* in the present lemma holds for $s = 0$ and all $t > 0$ (as $g(t) = f(t)$ for $t \geq 0$). If $s \in [0, 1]$ and $1 \leq t \leq 1 + s_0 + s_1$, then $g(s) = s$ and

$$\begin{aligned} t - s &= (t - 1) + (1 - s) = (k^{-1}|g(t)| - k^{-1}) + (1 - s) \\ &= k^{-1}(|g(t)| - s) - (k^{-1} - 1)(1 - s) < k^{-1}|g(t) - g(s)|. \end{aligned}$$

If $t \geq 1 + s_0 + s_1$, then $|g(t) - g(0)|/t \geq k$, and therefore,

$$\frac{|g(t) - g(s)|}{t - s} > \frac{|g(t)| - s}{t - s} \geq \frac{k - x}{1 - x},$$

where $x = s/t \leq 1/(1 + s_0 + s_1)$. Then $(k - x)/(1 - x) \geq k - (1 - k)/(s_0 + s_1) \geq k - (k(1 - k))/(e^{b\alpha} - 1)$. Note that the latter expression is an increasing function of b (as k is a function of b), which tends to 1 as b tends to infinity. Therefore, if in addition to (4.7) we require that

$$\frac{B}{\sqrt{B^2 + 1}} - \frac{\frac{B}{\sqrt{B^2+1}}(1 - \frac{B}{\sqrt{B^2+1}})}{e^{B\alpha} - 1} > q, \tag{4.13}$$

then property *(c)* in the lemma holds for all $s \in [0, 1]$ and $t \geq 1$. It remains to note that this property trivially holds for $s, t \in [0, 1]$ and that by symmetry, the case $s \in [0, 1]$, $t < 0$ is analogous to $1 - s \in [0, 1]$, $1 - t > 1$.

Thus, conditions *(a)*–*(c)* hold for g with $t_{q,\alpha} \in (1/2, 1/2 + s_0)$ and $L_{q,\alpha} = s_0 + s_1$. □

Remark 4.12. In addition to properties *(a)–(c)* of Lemma 4.11 we may assume that the curve $g_{q,\alpha}$ is a graph of Lipschitz piecewise smooth function $F_{q,\alpha}: \mathbb{R} \to \mathbb{R}$.

Proof. Let us analyze the tangent vector to $g_{q,\alpha}$ when the argument t changes from 1 to $1+s_0$ and from $1+s_0$ to $1+s_0+s_1$ (see the proof of Lemma 4.11).

The arc of $g_{q,\alpha}$ between $g_{q,\alpha}(1)$ and $g_{q,\alpha}(1+s_0)$ has the polar parametrization $z(\phi) = e^{b\phi}e^{-i\phi}$, ϕ increases from 0 to α. Then the x-coordinate $\operatorname{Re} dz/d\phi(\phi)$ of the tangent vector is equal to $e^{b\phi}(b\cos\phi - \sin\phi)$. This is positive provided $\tan\phi < b$. Thus, we impose the following additional restriction on B:

$$\tan\alpha < B. \tag{4.14}$$

Since $\operatorname{Re} dz/d\phi(\phi)$ is continuous, we conclude that its minimum on $\phi \in [0,\alpha]$ is positive. The y-coordinate of the tangent vector is continuous in t, therefore is bounded for $t \in [1, 1+s_0]$. Thus, the slope of the tangent vector is bounded. Hence $g_{q,\alpha}|_{[1,1+s_0]}$ is a graph of Lipschitz function.

The arc of $g_{q,\alpha}$ between $g_{q,\alpha}(1+s_0)$ and $g_{q,\alpha}(1+s_0+s_1)$ has the polar parametrization $z(\phi) = e^{2b\alpha+b\phi}e^{i\phi}$, where ϕ increases from $-\alpha$ to $-\beta$. Then $\operatorname{Re} dz/d\phi(\phi) = e^{2b\alpha+b\phi}(b\cos\phi - \sin\phi) > 0$ since $\cos\phi > 0$ and $\sin\phi < 0$ for $\phi \in (-\alpha, -\beta)$. In the same way this implies that $g_{q,\alpha}|_{[1+s_0,1+s_0+s_1]}$ is a graph of Lipschitz function. □

Proof. [Proof of Theorem 4.9 in the Euclidean case] Let $\alpha_n \to \pi/2$ and $q_n \to 1$, $n \geq 1$ be two increasing sequences of positive reals. For every pair (α_n, q_n) consider a Lipschitz function $F_n(x) = F_{q_n,\alpha_n}(x+1/2)$, where F_{q_n,α_n} is a Lipschitz piecewise smooth function described in Remark 4.12 (whose graph is the curve g_{q_n,α_n} from Lemma 4.11). The function F_n is even. Note that $F_n(x)$ is constant for $|x| \geq x_n = \operatorname{Re}(g_{q_n,\alpha_n}(1+L_{q_n,\alpha_n}))$. Denote by $\Gamma_{F_n}(x) = x + iF_n(x)$ the graph of F_n and for each $n \geq 1$ choose $L_n > x_n$ such that

$$L_n - x_n > n\mathcal{H}^1(\Gamma_{F_n}[-x_n, x_n]). \tag{4.15}$$

Now let $G_n(x) = (F_n(L_n x) - F_n(L_n))/L_n$. The function G_n has the following properties:

- G_n is a nonnegative even piecewise smooth Lipschitz function on $\mathbb{R}$,
- G_n is zero on $(-\infty,-1] \cup [1,\infty)$,
- $G_n(x)$ attains its maximum at $x = 0$, $G_n(0) < 1$ and $G_n^{-1}(G_n(0)) = [-1/L_n, 1/L_n]$ ($L_n > |F_n(x_n)|$ from (4.15)),
- If $\gamma_n = \Gamma_{G_n}$ is the graph of G_n, then $\mathcal{H}^1(\gamma_n[-a_n, a_n]) < 1/n$, where $a_n = \sup\{x\colon G_n(x) > 0\}$,
- There exists $t_n \in (0, a_n)$ such that the argument of $\gamma_n(t_n) - \gamma_n(0)$ is equal to $(-\alpha_n)$,
- The ratio $|\gamma_n(x) - \gamma_n(y)|/\mathcal{H}^1(\gamma_n[x,y])$ is bounded from below by q_n for all pairs of $x \neq y$ such that $|x| \leq 1/L_n$.

Denote by p_n the length of $\gamma_n[-a_n, a_n]$. Let $\theta_n \searrow 0$ ($n \geq 1$) be such that

$$\theta_{n+1}p_{n+1} < (\theta_n p_n)/4, \quad 2\theta_{n+1} < \theta_n/L_n, \quad \theta_1 < 1/2, \text{ and } \sum_{n\geq 1}\theta_n < 1.$$

Figure 1. A graph of $S_3(x)$

The first property of θ_n guarantees that for every $n \geq 1$

$$\sum_{j\geq 1} 2^{j-1}\theta_{n+j}p_{n+j} < 2\theta_{n+1}p_{n+1}.$$

Note that as $G_n(x)$ is a hat-like function on $[-1,1]$, and the graph of $G_n^{(\rho,\theta)}(x) = \theta G_n(\theta^{-1}(x-\rho))$ is the rescaled "hat" on $[\rho-\theta,\rho+\theta]$. For any closed interval $I = [\rho-\theta,\rho+\theta]$ denote by $I^{(L)}$ the interval $[\rho-\theta/L,\rho+\theta/L]$.

For $x \in [-1,1]$, let

$$h_1(x) = S_1(x) = \sum_{\rho\in\{-1+\theta_1,1-\theta_1\}} G_1^{(\rho,\theta_1)}(x).$$

Let $\mathcal{G}_1 = \{[-1,-1+2\theta_1],[1-2\theta_1,1]\}$ (since $\theta_1 < 1/2$, these intervals are disjoint). Now we define inductively two sequences of families of intervals as follows:

$$\mathcal{F}_n = \{I^{(L_n)} \text{ such that } I \in \mathcal{G}_n\};$$
$$\mathcal{G}_{n+1} = \{[a,a+2\theta_{n+1}],[b-2\theta_{n+1},b] \text{ such that } [a,b]\in\mathcal{F}_n\}.$$

For every $n \geq 1$, $x \in [-1,1]$, let

$$h_{n+1}(x) = \sum_{[a,b]\in\mathcal{F}_n} \sum_{\rho\in\{a+\theta_{n+1},b-\theta_{n+1}\}} G_{n+1}^{(\rho,\theta_{n+1})}(x); \tag{4.16}$$
$$S_{n+1}(x) = S_n(x) + h_{n+1}(x).$$

Figure 1 shows a possible graph of $S_3(x)$. Note that the definition of h_1 agrees with (4.16) if we let $\mathcal{F}_0 = \{[-1,1]\}$. For all n, $\mathcal{F}_n$ consists of 2^n disjoint closed intervals of the same length $2\theta_n/L_n$, whose union is equal to the preimage $S_n^{-1}(\max_x S_n(x))$. Since $4\theta_{n+1} \leq 2\theta_n/L_n$, intervals in $\mathcal{G}_{n+1}$ are disjoint.

For $x \in [-1,1]$, define $G(x) = \lim_n S_n(x)$. Note that each S_n is continuous and $|G - S_n| = |\sum_{k\geq n+1} h_k| \leq \sum_{k\geq n+1}\theta_k$ which tends to zero as $n\to\infty$. Therefore, G is continuous. Since the length ℓ of the graph of G is finite (it is bounded from above by $1+\sum_{n\geq 1} 2^n\theta_n p_n < 1+4\theta_1 p_1 < 5$), we conclude that the graph of G has an arc-length parametrization.

Let $\gamma = \Gamma_G\colon [-1,1]\to\mathbb{C}$ be the graph of G. The curve γ consists of points of two types: points in $A_1 = \bigcup \Gamma_{S_n}[-1,1]\cap\gamma[-1,1]$ and points in $A_2 = \gamma[-1,1]\setminus A_1$. The set A_2 is a Cantor-like set which will be described below.

For any $t \in \gamma^{-1}(A_1)$, the metric derivative of the normal parametrization of γ at t is clearly equal to 1, since the functions S_n are piecewise smooth. Consider $c \in C = \gamma^{-1}(A_2)$. Since $\gamma(c)$ does not belong to $\Gamma_{S_n}[-1,1]$ for any n, there is a sequence of intervals $I_n \in \mathcal{G}_n$ such that $c = \bigcap_{n\geq 1} I_n$. Then $\gamma(c)$ corresponds to a certain infinite sequence $\varepsilon \in \{0,1\}^\infty$: depending whether I_n has center at $\rho = a + \theta_n$ or at $\rho = b - \theta_n$ (see (4.16)), we let ε_n be equal to 0 or 1. Therefore, C is a Cantor set, and thus it is uncountable. We show that for any $c \in C$, the metric derivative of the normal parametrization of γ at c is equal to 1, but the normal parametrization of γ is not metrically differentiable at c.

For any point $c \in C$ there is a pair of sequences of points $y_n, z_n \to c$, $y_n < c < z_n$ such that $G(y_n) = S_n(y_n)$, $G(z_n) = S_n(z_n)$ and the points $\gamma(y_n)$, $\gamma(z_n)$ and $\Gamma_{S_n}((y_n+z_n)/2)$ form an isosceles triangle with vertex angle $\pi - 2\alpha_n$. This means that not only the ratio between the distance $|\gamma(y_n)-\gamma(z_n)|$ divided by the length of $\gamma[y_n, z_n]$ does not tend to 1, but moreover, it tends to 0. Therefore, the normal parametrization of γ is not metrically differentiable at c.

It remains to show that for any point $c \in C$ the metric derivative of the normal parametrization of γ at c is equal to 1. We will show that the ratio

$$\mathcal{H}^1(\gamma[c, c+t])/|\gamma(c+t) - \gamma(c)| \tag{4.17}$$

tends to 1 as $t \to 0$.

Assume $t > 0$ is small. Let $\varepsilon \in \{0,1\}^\infty$ be a sequence corresponding to $\gamma(c)$. Without any loss of generality we may assume $c + t \in \bigcup_{I \in \mathcal{G}_1} I$. Let $\delta \in \{0,1\}^\infty$ be a sequence corresponding to $\gamma(c+t)$. If $\gamma(c+t) \in A_1$, then δ is a finite sequence; otherwise, δ is infinite.

Since t is small, we may assume that $\delta_1 = \varepsilon_1$. Let $n \geq 1$ be such that $(\varepsilon_1, \dots, \varepsilon_n) = (\delta_1, \dots, \delta_n)$ and $\varepsilon_{n+1} \neq \delta_{n+1}$ (if such n does not exist, that is, if the sequence δ constitutes the beginning of the infinite sequence ε, we let n be equal to the length of δ). Note that when c is fixed and t tends to 0, then n tends to ∞.

In order to find an upper bound for (4.17), we will use the following estimate:

$$\begin{aligned}\frac{\mathcal{H}^1(\gamma[c, c+t])}{|\gamma(c+t) - \gamma(c)|} &\leq \frac{\mathcal{H}^1(\gamma[x, c+t]) + \mathcal{H}^1(\gamma[c, x])}{|\gamma(c+t) - \gamma(x)| - |\gamma(x) - \gamma(c)|} \\ &\leq \Big(\frac{\mathcal{H}^1(\gamma[x, c+t])}{|\gamma(c+t) - \gamma(x)|} + y\Big)/(1-y),\end{aligned} \tag{4.18}$$

for any $x \in (c, c+t)$, such that the expression $y = \mathcal{H}^1(\gamma[c,x])/|\gamma(c+t) - \gamma(x)|$ is strictly less than 1.

Consider first the case when δ coincides with $(\varepsilon_1, \dots, \varepsilon_n)$. In this case, $G(c+t) = S_n(c+t)$ and there is an interval $I_n \in \mathcal{G}_n$ of length $2\theta_n$ containing both c and $c+t$. Let $J_1, J_2 \subset I_n$ be disjoint intervals in $\mathcal{G}_{n+1}$ such that $c \in J_1 \cup J_2$. Since δ has length n, we get $c + t \notin J_1 \cup J_2$. Also note that $c \in J_1^{(L_{n+1})} \cup J_2^{(L_{n+1})}$ since $G(c) \neq S_{n+1}(c)$.

If $c \in J_i^{(L_{n+1})}$, then let $x = \sup\{z \in J_i \colon S_{n+1}(z) > S_n(z)\}$. Since $S_{n-1}|_{I_n}$ is constant and $G(x) = S_n(x)$, $G(c+t) = S_n(c+t)$, we may deduce that by the property of G_{n+1}, the expression $\mathcal{H}^1(\gamma[x, c+t])/|\gamma(c+t) - \gamma(x)|$ does not exceed q_{n+1}^{-1}.

Now we want to find an upper estimate for $y = \mathcal{H}^1(\gamma[c,x])/|\gamma(c+t) - \gamma(x)|$. The numerator is not greater than $\sum_{j\geq 1} 2^{j-1}\theta_{n+j}p_{n+j} < 2\theta_{n+1}p_{n+1}$, and the denominator is at least $\theta_{n+1}(n+1)p_{n+1}$ (this follows from the property of L_n, see (4.15)). Therefore, $y \leq 2/(n+1)$. Thus, the quantity (4.17) is at most $\psi_{n+1}(q_{n+1}^{-1})$, where $\psi_k(t) = (t+2/k)/(1-2/k)$.

Now consider the case when δ has length at least $n+1$. In the above notation this implies that $c \in J_1^{(L_{n+1})}$ and $c+t \in J_2$. Choose $x = \sup\{z \in J_1 \colon S_{n+1}(z) > S_n(z)\}$ as before. If $c+t \in J_2^{(L_{n+1})}$, then the same proof as in the previous paragraph shows that $\mathcal{H}^1(\gamma[x,c+t])/|\gamma(c+t)-\gamma(x)| \leq \psi_{n+1}(1)$ (in this case γ connects $\gamma(x)$ and $\gamma(x')$, where $x' = \inf\{z \in J_2 \colon S_{n+1}(z) > S_n(z)\}$, by a straight line interval). Thus, the quantity (4.17) is at most $\psi_{n+1}(\psi_{n+1}(1))$.

If $c+t \in J_2 \backslash J_2^{(L_{n+1})}$, then $\mathcal{H}^1(\gamma[x,c+t]) \leq |\gamma(c+t)-\gamma(x)| + \sum_{j\geq 1} 2^{j-1}\theta_{n+j}p_{n+j} < |\gamma(c+t)-\gamma(x)| + 2\theta_{n+1}p_{n+1}$, so together with $|\gamma(c+t)-\gamma(x)| > \theta_{n+1}(n+1)p_{n+1}$ we get that the quantity (4.17) is at most $\psi_{n+1}(1+2/(n+1))$.

It remains to observe that the length n of the initial part of sequences ε and δ tends to ∞ as $t \to 0$ and to note that $\psi_{n+1}(q_{n+1}^{-1})$, $\psi_{n+1}(\psi_{n+1}(1))$ and $\psi_{n+1}(1+2/(n+1))$ tend to 1 as n tends to infinity. □

Remark 4.13. Note that in fact we proved that the curve γ constructed above has the following property: for every $c \in C$ there exist $y_n < c < z_n$ such that $(y_n, z_n) \to (c,c)$ and

$$\frac{|\gamma(y_n) - \gamma(z_n)|}{\mathcal{H}^1(\gamma[y_n, z_n])} \to 0.$$

This means that this curve has uncountably many "spikes".

Remark 4.14. For a general norm $\|\cdot\|$ on the 2-dimensional plane, one can produce an analogue of the curve constructed in Lemma 4.11 in the following way.

We may assume the $\|\cdot\|$-norm of the point 1 on the complex plane is equal to 1. Define $g([0,1])$ to be a horizontal interval as in (4.8) ($g(t) = t$ for $t \in [0,1]$), then $\|g(t)\| = t$ for $0 \leq t \leq 1$. Next find a small $\varepsilon > 0$, such that if we define $g(t)|_{t>1}$ to be a ray with slope $-\varepsilon$, then the condition *(c)* in Lemma 4.11 with the norm $\|\cdot\|$ instead of Euclidean norm $|\cdot|$ holds for all $t > 1$. Next thing would be to note that the ratio $\|g(t) - g(s)\|/|t-s|$ tends to 1 as s remains in $[0,1]$ and t tends to infinity (we define $g(t) = 1 + (t-1)z_{-\varepsilon}$, where $\|z_{-\varepsilon}\| = 1$ and $\tan\arg z_{-\varepsilon} = -\varepsilon$). So we may choose a sufficiently large T_1 such that if we redefine $g(t)|_{t>T_1}$ to be a ray with slope -2ε, then we again have condition *(c)* in Lemma 4.11 still valid for $\|\cdot\|$. If we continue this way, the curve g would consist of straight intervals such that each new interval "turns" by less than $-\varepsilon$ with respect to the previous interval, and in the end point of each interval the ratio from condition *(c)* is very close to 1 (much closer to 1 than q is). Since $N\varepsilon \to \infty$, the angle between the horizontal axis and the subsequent intervals which form the curve g tends to $\pi/2$. So there will be a moment when this angle becomes bigger than α. At this moment, we stop the process, and start "rotating" intervals towards horizontal axis (making the slope less negative) in order to obtain a broken line satisfying the conditions *(a)*–*(c)* from Lemma 4.11.

One can check that since the arc-length parametrization of the boundary of a unit ball of arbitrary norm is uniformly continuous, the algorithm explained above can be implemented for every 2-dimensional norm (of course, ε would depend on the norm).

The curve g constructed above will in fact be an approximation of two logarithmic spirals (such as those used in the proof of Lemma 4.11). Then we prove Theorem 4.9 in the same way, each time putting two rescaled "hats" on top of the previous "hat". The curve obtained in this way will not be metrically differentiable at the points of the Cantor set, since if we consider a sequence of isosceles triangles $A_nB_nC_n$ with vertex angle $\angle B_n$ tending to 0, the ratio between $\|A_nC_n\|$ and $\|A_nB_n\| + \|B_nC_n\|$ will tend to zero as $n \to \infty$, for any norm $\|\cdot\|$.

Remark 4.15. If we work with the ℓ_1-norm, then for a fixed $\alpha \in (0, \pi/2)$ let $h = (1/2)\tan\alpha$ and

$$g(t) = \begin{cases} (t+h) - hi, & \text{if } t < -h, \\ ti, & \text{if } t \in [-h, 0], \\ t, & \text{if } t \in [0,1], \\ 1-(t-1)i, & \text{if } t \in [1, 1+h], \\ (t-h) - hi, & \text{if } t > 1+h. \end{cases}$$

The curve g satisfies conditions of Lemma 4.11 with any $q < 1$ (for the ℓ_1 norm), and although it cannot be made into a graph of a function in the usual sense, one can easily see that putting together such "boxes" (rescaling as necessary and taking $\alpha_n \to \pi/2$), we obtain the example of a planar curve with metric derivative 1 at every point, but with uncountable set of points where it is not metrically differentiable.

5 Metric regularity and metric differentiability

This section contains mainly auxiliary results. Let $f : [a,b] \to X$, $I = [a,b]$. We say that $x \in I$ is *bilaterally metrically regular point of the function* f, provided

$$\lim_{\substack{(y,z)\to(x,x) \\ a\le y\le x\le z\le b}} \frac{\|f(y) - f(z)\|}{\bigvee_y^z f} = 1.$$

See the beginning of section 4 for the definition of a metrically regular point. Note that every bilaterally metrically regular point of a function is also its metrically regular point.

Lemma 5.1. *Let X be a Banach space, $g : [a,b] \to X$, $x \in [a,b]$, g is metrically differentiable at x with $md(g,x) > 0$, and $md(g,\cdot)$ is continuous at x. Then x is bilaterally metrically regular point of the function g.*

Proof. Lemma 4.3 implies that x is a metrically regular point of g. Let $\varepsilon > 0$. By metric differentiability of g at x, by Lemma 4.2, and by continuity of $md(g,\cdot)$ at x find $\delta > 0$ such that $(1-\varepsilon)\, md(g,x)|z-y| \le \|g(z) - g(y)\|$, $\bigvee_y^z g = \int_y^z md(g,s)\, ds$, for

$x - \delta < y < x < z < x + \delta$, and $md(g, x + t) < (1 + \varepsilon) \cdot md(g, x)$ for $|t| < \delta$ with $x + t \in [a, b]$. Thus, for y, z with $x - \delta < y \leq x \leq z < x + \delta$ we have

$$\left(\frac{1-\varepsilon}{1+\varepsilon}\right) \bigvee_y^z g = \left(\frac{1-\varepsilon}{1+\varepsilon}\right) \int_y^z md(g, s)\, ds \leq (1 - \varepsilon) \cdot md(g, x)\, |z - y|$$
$$\leq \|g(z) - g(y)\| \leq \bigvee_y^z g.$$

If $y \neq z$, then by dividing by $\bigvee_y^z g$, we obtain $(1-\varepsilon)/(1+\varepsilon) \leq \|g(y) - g(z)\|/(\bigvee_y^z g) \leq 1$, and thus x is bilaterally metrically regular point of g. □

Lemma 5.2. *Let X be a Banach space, and let $f : [a, b] \to X$. If $md(f, \cdot)$ is continuous at $x \in [a, b]$, and x is a bilaterally metrically regular point of f, then f is metrically differentiable at x.*

Proof. If $md(f, x) = 0$, then the conclusion follows from Lemma 4.1*(i)*, and thus we can assume that $md(f, x) > 0$. Lemma 4.4 implies that the condition (1.1) holds provided $\operatorname{sign}(z - x) = \operatorname{sign}(y - x)$. Thus, we only need to treat the case $\operatorname{sign}(z - x) = -\operatorname{sign}(y - x)$ since the cases when either $y = x$ or $z = x$ follow easily from the existence of $md(f, x)$. Let $\varepsilon > 0$. Find $\delta > 0$ such that for all $x - \delta < y \leq x \leq z < x + \delta$ with $(y, z) \neq (x, x)$ we have that $\|f(y) - f(z)\| \geq (1 - \varepsilon) \bigvee_y^z f$, $\bigvee_y^z f = \int_y^z md(f, t)\, dt$, and $(1 - \varepsilon)\, md(f, t) \leq md(f, x) \leq (1 + \varepsilon)\, md(f, t)$ for $|x - t| < \delta$ with $t \in [a, b]$. Let $x - \delta < y \leq x \leq z < x + \delta$. Then

$$\|f(y) - f(z)\| \geq (1 - \varepsilon) \bigvee_y^z f = (1 - \varepsilon) \int_y^z md(f, t)\, dt$$
$$\geq (1 - \varepsilon)^2\, md(f, x)(z - y).$$

The other inequality follows from the same reasoning as in (4.2). □

Lemma 5.3. *Let X be a Banach space, let $f : [a, b] \to X$ be continuous, BV, and such that it is not constant on any subinterval of $[a, b]$. Let $x \in (a, b)$, $y = v_f(x)$, and $g = f \circ v_f^{-1}$. Then*

(i) if x is a metrically regular point of f, then $md(g, y) = 1$,

(ii) if x is a bilaterally metrically regular point of f, and there exists a neighbourhood U of x such that all $z \in U$ are metrically regular points of f, then g is metrically differentiable at y.

Proof. To prove *(i)*, note that

$$1 = \lim_{z \to x} \frac{\|f(z) - f(x)\|}{\left|\bigvee_x^z f\right|} = \lim_{z \to x} \frac{\|f(z) - f(x)\|}{|v_f(z) - v_f(x)|}$$
$$= \lim_{w \to y} \frac{\|f \circ v_f^{-1}(w) - f \circ v_f^{-1}(y)\|}{|w - y|} = md(g, y).$$

For *(ii)*, first note that $md(g, y) = 1$ by part *(i)*. Let U be the neighbourhood of x such that all $z \in U$ are metrically regular points of f. Then part *(i)* implies that $md(g, w) = 1$ for all $w = v_f(z)$, where $z \in U$. To apply Lemma 5.2, it is enough to show that y is a bilaterally metrically regular point of g, but

$$\lim_{\substack{(s,t)\to(y,y)\\ 0\le s\le y\le t\le v_f(b)}} \frac{\|g(t)-g(s)\|}{\bigvee_s^t g} = \lim_{\substack{(s,t)\to(y,y)\\ 0\le s\le y\le t\le v_f(b)}} \frac{\|f\circ v_f^{-1}(t) - f\circ v_f^{-1}(s)\|}{t-s}$$
$$= \lim_{\substack{(u,v)\to(x,x)\\ a\le u\le x\le v\le b}} \frac{\|f(v)-f(u)\|}{v_f(v)-v_f(u)} = 1,$$

where the last equality follows from the fact that x is a bilaterally metrically regular point of f, and $v_f(v) - v_f(u) = \bigvee_u^v f$ for any $u, v \in U$, $u < v$ by Lemma 4.2. Now, an application of Lemma 5.2 yields the conclusion. □

We will also need the following simple lemma.

Lemma 5.4. *Let $f : [a, b] \to X$, $x \in [a, b]$, be such that $md(f, x)$ exists, but f is not metrically differentiable at x. Then if h is a homeomorphism of $[a, b]$ onto itself such that $f \circ h$ is metrically differentiable at $h^{-1}(x)$, then $md(f \circ h, h^{-1}(x)) = 0$.*

Proof. Lemma 4.1 shows that $md(f, x) > 0$ (otherwise we have a contradiction with the fact that f is not metrically differentiable at x). Suppose that h is an (increasing) homeomorphism such that $f \circ h$ is metrically differentiable at $y = h^{-1}(x)$. For a contradiction, suppose that $md(f \circ h, y) > 0$. Note that

$$\frac{|h(y+t)-h(y)|}{|t|} = \frac{|h(y+t)-h(y)|}{\|f(h(y+t)) - f(h(y))\|} \cdot \frac{\|f(h(y+t)) - f(h(y))\|}{|t|},$$

and it follows that $h'(y) = md(f \circ h, y)/md(f, x) > 0$. Thus $h'(y)$ exists and is non-zero. This implies that $(h^{-1})'(x)$ exists. Because $f = (f \circ h) \circ h^{-1}$, Lemma 4.1 implies that f is metrically differentiable at x, a contradiction. We conclude that $md(f \circ h, y) = 0$. □

6 Continuous metric differentiability via homeomorphisms

Let $f : [a, b] \to X$. Let M_f be the set of all points $x \in [a, b]$ with the following property: there is no neighbourhood $U = (x - \delta, x + \delta)$ of x such that either $f|_U$ is constant or all points of U are metrically regular points of the function f. Obviously, M_f is closed, and $a, b \in M_f$.

Theorem 6.1. *Let X be Banach space, and let $f : [a,b] \to X$. Then the following are equivalent.*

(i) There exists a homeomorphism k of $[a,b]$ onto itself such that $md(f \circ k, \cdot)$ is continuous on $[a,b]$.

(ii) f is continuous, BV, and $\mathcal{H}^1(f(M_f)) = 0$.

Proof. To prove that *(i)* $\Longrightarrow$ *(ii)*, note that the existence of a continuous metric derivative implies continuity and boundedness of variation of the function, and these properties are preserved when the function is composed with a homeomorphism. Thus, it is enough to prove that $\mathcal{H}^1(f(M_f)) = 0$. Note that $M_f = k(M_{f\circ k})$, and thus it is enough to prove that $\mathcal{H}^1((f\circ k)(M_{f\circ k})) = 0$. Let $g = f \circ k$. We claim that

$$M_g \subset \{x \in [a,b] : md(g,x) = 0\}. \tag{6.1}$$

Indeed, Lemma 4.3 implies that every point $x \in (a,b)$, such that $md(g,x) > 0$, is metrically regular point of g. By continuity of $md(g,\cdot)$, there exists a neighbourhood U of x such that $md(g,y) > 0$ at all $y \in U$, and thus all points of U are metrically regular points of g. So we get (6.1), and then by Lemma 2.1, we see that $\mathcal{H}^1(g(M_g)) = 0$.

To prove that *(ii)* $\Longrightarrow$ *(i)*, let $(U_i)_i$ be the collection of all maximal open intervals inside $[a,b]$ such that $f|_{U_i}$ is constant, and put $U = \bigcup_i U_i$. Define $\varphi(t) = v_f(t) + \lambda(U \cap [a,t])$ for $t \in [a,b]$. Let (a_j,b_j) be the maximal open components of $[a,b]$ such that all points of (a_j,b_j) are metrically regular points of f. Let $\alpha_j = \varphi(a_j)$, $\beta_j = \varphi(b_j)$. Then $\varphi(b_j) - \varphi(a_j) = \bigvee_{a_j}^{b_j} f$. Note that

$$\begin{aligned}\varphi(b) &= \lambda(U) + \bigvee_a^b f = \lambda(U) + \sum_j \bigvee_{a_j}^{b_j} f \\ &= \lambda(U) + \sum_j (\beta_j - \alpha_j) = \lambda(\varphi([a,b] \setminus M_f)),\end{aligned} \tag{6.2}$$

by [8, Lemma 2.7], and thus $\lambda(\varphi(M_f)) = \lambda(M_{f\circ\varphi^{-1}}) = 0$ (the left-hand side of (6.2), $\varphi(b)$, is equal to $\lambda(\varphi[a,b])$, and φ is increasing). Let $g = f \circ \varphi^{-1}$. It is easy to see that g is Lipschitz (because φ is a homeomorphism). By Zahorski's lemma (see e.g. [11, p. 27]) there exists a continuously differentiable homeomorphism h of $[0,\varphi(b)]$ onto itself such that $h'(x) = 0$ if and only if $x \in h^{-1}(M_g)$. Now, by the equality

$$\frac{g(h(x+t)) - g(h(x))}{t} = \frac{g(h(x+t)) - g(h(x))}{h(x+t) - h(x)} \cdot \frac{h(x+t) - h(x)}{t}, \tag{6.3}$$

and by Lemma 5.3, we obtain that $md(g \circ h, x)$ exists and is continuous at all $x \in \varphi(U) \cup \bigcup_j (\alpha_j, \beta_j)$. By (6.3), by the choice of h and the fact that g is Lipschitz, we easily obtain that $md(g \circ h, x) = 0$ for all $x \in h^{-1}(M_g)$, and that $md(g \circ h, \cdot) = md(f \circ k, \cdot)$ is continuous at all such points (where $k = \varphi^{-1} \circ h$). □

Let M_f^{b} be the set of all points $x \in [a,b]$ with the following property: there is no neighbourhood $U = (x-\delta, x+\delta)$ of x such that either $f|_U$ is constant or all points of U are bilaterally metrically regular points of the function f. Obviously, M_f^{b} is closed and $a, b \in M_f^{\mathrm{b}}$.

Theorem 6.2. *Let X be Banach space, and let $f : [a,b] \to X$. Then the following are equivalent.*
(i) There exists a homeomorphism h of $[a,b]$ onto itself such that $f \circ h$ is metrically differentiable at every point of $[a,b]$, and $md(f \circ h, \cdot)$ is continuous.
(ii) f is continuous, BV, and $\mathcal{H}^1(f(M_f^{\mathrm{b}})) = 0$.

Proof. The proof is similar to the proof of Theorem 6.1, and thus we omit it. It uses Lemmas 5.1 and 5.3*(ii)*. □

The following example shows that the scopes of Theorems 6.1 and 6.2 are different (see also Remark 6.4).

Example 6.3. There exists 1-Lipschitz mapping $f : [0,1] \to \ell_2$ such that $md(f,x) = 1$ for all $x \in [0,1]$, but f is not metrically differentiable at a dense subset S of $[0,1]$.

Proof. Choose $t_n > 0$ with $\sum_n t_n^2 = 1$, and $q_n \in (0,1)$ such that $S = \{q_n : n \in \mathbb{N}\}$ is dense in $[0,1]$. Let $f_n : [0,1] \to \mathbb{R}^2$ be defined as

$$f_n(t) = \begin{cases} (t,0) & \text{for } 0 \le t \le q_n, \\ \frac{(t-q_n)}{\sqrt{2}} \cdot (1,1) + (q_n,0) & \text{for } q_n < t \le 1. \end{cases}$$

It is easy to see that $f_n(0) = 0$ and f_n is 1-Lipschitz for each $n \in \mathbb{N}$. Define $f : [0,1] \to \ell_2 = \sum \oplus_{\ell_2} \ell_2^2$ as $f(t) = (t_n \cdot f_n(t))_n$. It is easy to see that f is well defined, and 1-Lipschitz. First, we will show that $md(f,x) = 1$ for all $x \in [0,1]$. Choose $x \in [0,1]$ and $\varepsilon > 0$. Find $n_0 \in \mathbb{N}$ such that $\sum_{n \ge n_0} t_n^2 < \varepsilon^2$. Find $\delta > 0$ such that $(x-\delta, x+\delta) \cap \{q_j : j \le n_0\} \subset \{x\}$. Let $y \in (x-\delta, x+\delta)$ and notice that

$$\begin{aligned} |y-x| &\ge \|f(y) - f(x)\| \\ &= \Big(\sum_{n \le n_0} t_n^2 \, \|f_n(y) - f_n(x)\|_{\ell_2^2}^2 + \sum_{n > n_0} t_n^2 \, \|f_n(y) - f_n(x)\|_{\ell_2^2}^2 \Big)^{1/2} \\ &\ge \Big(\Big(\sum_{n \le n_0} t_n^2 \Big)^{1/2} - \varepsilon \Big) |y-x| \ge (1 - 2\varepsilon)|y-x|. \end{aligned}$$

Conclude by sending ε to 0.

Now we will show that f is not metrically differentiable at any $x \in S$. Fix $x = q_m \in S$ for some m, and let $\delta > 0$ be such that $0 \le x - \delta < x + \delta \le 1$. Then

$$\begin{aligned} \frac{\|f(x-\delta) - f(x+\delta)\|}{2\delta} &= \frac{1}{2\delta}\Big(t_m^2 \|f_m(x-\delta) - f_m(x+\delta)\|_{\ell_2^2}^2 \\ &\qquad + \sum_{n \neq m} t_n^2 \, \|f_n(x+\delta) - f_n(x-\delta)\|_{\ell_2^2}^2\Big)^{1/2} \\ &\le \frac{1}{2\delta}\Big(t_m^2 \delta^2 (2+\sqrt{2}) + \sum_{n\neq m} 4\delta^2 t_n^2\Big)^{1/2} \\ &= \Big(\frac{2+\sqrt{2}}{4} t_m^2 + \sum_{n \neq m} t_n^2\Big)^{1/2} = C_m < 1, \end{aligned}$$

and thus f is not metrically differentiable at x, as the condition (1.1) is violated. □

Remark 6.4. Lemma 5.4 implies that if h is a homeomorphism of $[0,1]$ onto itself such that $f \circ h$ is metrically differentiable at all $x \in [0,1]$, then $md(f \circ h, y) = 0$ for all $y \in h^{-1}(S)$, which is a dense subset of $[0,1]$. If h could be chosen to further make $md(f \circ h, \cdot)$ continuous, then f would have to be constant. Thus, there exists no homeomorphism h of $[0,1]$ onto itself such that $f \circ h$ is metrically differentiable at all points of $[0,1]$ while $md(f \circ h, \cdot)$ is continuous.

Acknowledgments. The authors would like to thank David Preiss and Luděk Zajíček for many valuable discussions.

References

[1] L. Ambrosio, *Metric space valued functions of bounded variation*, Ann. Scuola Norm. Sup. Pisa Cl. Sci. (4) **17** (1990), 439–478.

[2] L. Ambrosio, B. Kirchheim, *Rectifiable sets in metric and Banach spaces*, Math. Ann. **318** (2000), 527–555.

[3] Y. Benyamini, J. Lindenstrauss, *Geometric Nonlinear Functional Analysis, Vol. 1,* Colloquium Publications **48**, American Mathematical Society, Providence, 2000.

[4] A. Bruckner, *Differentiation of real functions*, Second edition, CRM Monograph Series, 5, American Mathematical Society, Providence, RI, 1994.

[5] G. De Cecco, G. Palmieri, *LIP manifolds: from metric to Finslerian structure*, Math. Z. **218** (1995), 223–237.

[6] J. Duda, *Absolutely continuous functions with values in metric spaces*, to appear in the Real Analysis Exchange.

[7] J. Duda, *Metric and w^*-differentiability of pointwise Lipschitz mappings*, to appear in the Journal of Analysis and its Applications.

[8] J. Duda, L. Zajíček, *Curves in Banach spaces – differentiability via homeomorphisms*, to appear in the Rocky Mountain J. of Math.

[9] M. J. Evans, P. D. Humke, *The equality of unilateral derivatives*, Proc. Amer. Math. Soc. **79** (1980), no. 4, 609–613.

[10] H. Federer, *Geometric Measure Theory,* Grundlehren der math. Wiss., vol. 153, Springer, New York, 1969.

[11] C. Goffman, T. Nishiura, D. Waterman, *Homeomorphisms in analysis,* Mathematical Surveys and Monographs, vol. 54, AMS, Providence, RI, 1997.

[12] R. Jeffery, *The theory of functions of a real variable*, Mathematical Expositions No. 6, University of Toronto Press, 1953.

[13] B. Kirchheim, *Rectifiable metric spaces: local structure and regularity of the Hausdorff measure*, Proc. Amer. Math. Soc. **121** (1994), 113–123.

[14] M. D. Kirszbraun, *Über die zusammenziehenden und Lipschitzschen Transformationen*, Fundam. Math. **22** (1934), 77–108.

[15] N. J. Korevaar, R. M. Schoen, *Sobolev spaces and harmonic maps for metric space targets*, Comm. Anal. Geom. **1** (1993), 561–659.

[16] L. Zajíček, *On σ-porous sets in abstract spaces*, Abstract and Applied Analysis **2005** (2005), 509–534.

Author information

Jakub Duda, ČEZ, a.s., Duhová 2/1444, 140 53 Praha 4, Czech Republic.
Email: jakub.duda@gmail.cz

Olga Maleva, Department of Pure Mathematics and Mathematical Statistics, Centre for Mathematical Sciences, University of Cambridge, Wilberforce Road, Cambridge CB3 0WB, UK.
Email: o.maleva@dpmms.cam.ac.uk

Banach Spaces and their Applications in Analysis, 331–338

Bohnenblust's theorem and norm-equivalent coordinates

Richard J. Fleming

Abstract. In 1940, F. Bohnenblust gave a set of conditions on a function $f(s,t)$ for which $f(s,t) = (|s|^p + |t|^p)^{1/p}$ for some p, $1 \leq p < \infty$, or $f(s,t) = \max\{|s|, |t|\}$. We will give an improvement of this theorem due to B. Randrianantoanina, and show how to use it to decompose certain sequence spaces into direct sums of ℓ^p-spaces. This can be used to describe the isometries of the spaces.

Key words. One-unconditional basis, norm-equivalent coordinates, direct sums of ℓ^p-spaces, isometries.

AMS classification. 46B45, 46B20.

Let X denote a Banach space with a normalized one-unconditional basis $\{e_j\}_{j\in N}$ where $N = \mathbb{N}$ or $\{1, 2, \dots, n\}$ for some n in $\mathbb{N}$. It is equivalent to think of X as a sequence space in which the coordinate vectors, also denoted by e_j, form a one-unconditional basis, with each basis element of norm one. Letting ν denote the norm on X, we will say that (X, ν) is an *admissable* sequence space. Note that an admissable sequence space has an *absolute norm*, that is, $\nu(x) = \nu(|x|)$, where $|x|$ is obtained from x by replacing its coordinates by their absolute values. It has been shown (in finite dimensions) by Schneider and Turner [6], by Kalton and Wood [4] and also in [2] that admissable spaces can by written as a direct sum $(\sum_{s\in S} \oplus X_s)_E$, where each X_s is a Hilbert space and E is itself an admissable sequence space. This is accomplished using the notion of equivalent coordinates, a notion introduced in [6]. Let us recall the definition here. Two coordinates j, k are equivalent with respect to ν if $\nu(x) = \nu(y)$ whenever $|x(j)|^2 + |x(k)|^2 = |y(j)|^2 + |y(k)|^2$ and $|x(m)| = |y(m)|$ for all $m \neq j, k$. If j and k are equivalent we will write

$$j \sim k.$$

We are going to generalize this idea slightly by introducing what we call *norm-equivalent* coordinates. As a result we will obtain decompositions which involve Banach spaces that are not necessarily Hilbert spaces.

The decomposition of an admissable sequence space as a direct sum of Banach spaces was earlier accomplished by Li and Randrianantoanina [5] and it is their paper that greatly influences what follows. The key idea in this approach is the introduction of the notion of a *fiber*.

Definition 1. Let (X, ν) be *an admissable* sequence space with one-unconditional basis $\{e_j\}_{j\in N}$. A non-empty proper subset S of N is said to be a *fiber* if for all finitely nonzero sequences $\{a_s\}_{s\in S}, \{a'_s\}_{s\in S}$ of scalars such that

$$\nu\Big(\sum_{s\in S} a_s e_s\Big) = \nu\Big(\sum_{s\in S} a'_s e_s\Big),$$

then

$$\nu\Big(\sum_{s\in S} a_s e_s + \sum_{j\notin S} b_j e_j\Big) = \nu\Big(\sum_{s\in S} a'_s e_s + \sum_{j\notin S} b_j e_j\Big),$$

for all finitely nonzero sequences of scalars $\{b_j\}_{j\in N\setminus S}$. The corresponding *fiber space* is given by

$$X_S = \overline{\text{span}}\{e_s : s \in S\}.$$

Clearly, any singleton set is a fiber, and every proper subset of N is a fiber for an ℓ^p space. For the space

$$X = \ell^1(2) \oplus_1 \ell^2(2) \oplus_1 \ell^\infty(2), \tag{1}$$

the fibers, in addition to the singletons, are

$$\{1,2\},\ \{3,4\},\ \{5,6\},\ \{3,4,5,6\},\\ \{1,2,3,4\},\ \{1,2,5,6\},\ \{1,3,4,5,6\},\ \{2,3,4,5,6\}.$$

We observe that the last four fibers listed above are maximal fibers, in that they are properly contained in no other fiber. Maximal fibers play an important role in the work of Li and Randrianantoanina [5], but not every space has maximal fibers. Consider the space $X_{p,q}$ with the norm defined inductively as follows:

$$\begin{aligned} \nu(x_1e_1) &= |x_1|, \\ \nu\Big(\sum_{j=1}^{n} x_j e_j\Big) &= \Big(\Big(\nu\Big(\sum_{j=1}^{n-1} x_j e_j\Big)\Big)^q + |x_n|^q\Big)^{1/q} \quad \text{if } n \ge 2 \text{ is odd} \\ &= \Big(\Big(\nu\Big(\sum_{j=1}^{n-1} x_j e_j\Big)\Big)^p + |x_n|^p\Big)^{1/p} \quad \text{if } n \ge 2 \text{ is even,} \end{aligned}$$

where $1 \le p, q < \infty$ and $p \ne q$. Here, the fibers are of the form $\{1, 2, \dots, m\}$ for $m \in \mathbb{N}$, and so in the infinite dimensional case, there are no maximal fibers. This example is given in [5].

Observe that if $j \sim k$ as defined above, then $\{j, k\}$ would be a fiber. We will refer to this as an ℓ^2-fiber. Now let us use this thought to generalize the notion of equivalent coordinates.

Definition 2. We will say that two coordinates j and k are *norm-equivalent*, written $j \sim_\nu k$, if $\{j, k\}$ is a fiber.

Clearly we should show that the above definition does, indeed, define an equivalence relation. The only item of difficulty is the transitivity. Before verifying that let us recall a lovely old theorem of Bohnenblust [1].

Theorem 3 (Bohnenblust, 1940). *Let f be a real valued function defined for $s \geq 0$, $t \geq 0$ so that it satisfies the following conditions:*

(i) $f(cs, ct) = cf(s,t)$ *for* $c, s, t \geq 0$*;*

(ii) $f(s,t) \leq f(s',t')$ *for* $0 \leq s \leq s'$ *and* $0 \leq t \leq t'$*;*

(iii) $f(s,t) = f(t,s)$*;*

(iv) $f(0,1) = 1$*;*

(v) $f(s, f(t,r)) = f(f(s,t), r)$ *for* $s, t, r \geq 0$.

Then there exists p, $0 < p \leq \infty$ such that

$$f(s,t) = (s^p + t^p)^{1/p},$$

where, in the case $p = \infty$, we mean $f(s,t) = \max\{s,t\}$.

This theorem is needed to show the above mentioned transitivity and was used also in the proof of [5, Proposition 2.2]. However, in the application there is generally no way to verify that the commutativity in item (iii) is satisfied. In a private communication, B. Randrianantoanina explained how that condition can be removed from the statement. We show that next, but omit the remainder of the proof which the reader can find in [1].

Theorem 4 (Randrianantoanina). *The conclusion of Bohnenblust's theorem remains true even with condition (iii) of the hypotheses removed.*

Proof. Let f be defined as given in the statement of Theorem 3. Define a sequence $\{a_n\}$ by $a_1 = 1$, and $a_n = f(1, a_{n-1})$ for $n > 1$. There are four steps in the proof of the original theorem:

1. $a_{n+m} = f(a_n, a_m)$;

2. $a_n a_m = a_{nm}$;

3. $a_2 = 1$ implies $f(s,t) = \max\{s,t\}$;

4. $a_2 > 1$ implies $a_n = n^{1/p}$ for some real $p > 0$ and $f(1, r^{1/p}) = (1+r)^{1/p}$.

Our first claim, is that $f(1, a_n) = f(a_n, 1)$ for each n. The statement is trivial for $n = 1$, and if it holds for $n-1$, we have

$$\begin{aligned} f(1, a_n) &= f(1, f(1, a_{n-1})) \\ &= f(1, f(a_{n-1}, 1)) \\ &= f(f(1, a_{n-1}), 1) \quad \text{(by (v) of Theorem 3)} \\ &= f(a_n, 1) \end{aligned}$$

and so the claim is established by induction. It was in proving this fact that Bohnenblust used the hypothesis that $f(s,t) = f(t,s)$. Now

$$\begin{aligned} f(a_n, a_m) &= f(f(1, a_{n-1}), a_m) \\ &= f(f(a_{n-1}, 1), a_m) = f(a_{n-1}, f(1, a_m)) \\ &= f(a_{n-1}, a_{m+1}) = f(a_{n-2}, a_{m+2}) \\ &= \cdots = f(a_1, a_{m+n-1}) = a_{m+n}, \end{aligned}$$

which completes step (i). We omit the proof of the remaining steps. □

Theorem 5. *The relation $\sim_\nu$ given in Definition 2 is an equivalence relation and an equivalence class is also a fiber. Furthermore, if N_j is an equivalence class with more than two elements, then the corresponding fiber space X_j is an ℓ^p-space for some p, $1 \leq p \leq \infty$.*

Proof. The only thing in question about equivalence is the transitive property. Hence, suppose that $j \sim_\nu k$ and $k \sim_\nu m$. If j, k, m are not all distinct, then there is nothing to prove. Hence, suppose they are three distinct positive integers. If s, t are non-negative scalars, then

$$\nu(se_j + te_k) = \nu(se_j + te_m) = \nu(se_k + te_m), \tag{2}$$

where the first equality is true because $k \sim_\nu m$ and the second follows from $j \sim_\nu k$. For any nonnegative numbers s, t, r, we note that

$$\begin{aligned} \nu(se_j + te_k + re_m) &= \nu(\nu(se_j + te_k)e_j + re_m) \ \text{(since } j \sim_\nu k) \\ &= \nu(\nu(se_j + te_k)e_j + re_k) \ \text{(by (2))}. \end{aligned} \tag{3}$$

Also,

$$\begin{aligned} \nu(se_j + te_k + re_m) &= \nu(se_j + \nu(te_k + re_m)e_k) \ \text{(since } k \sim_\nu m) \\ &= \nu(se_j + \nu(te_j + re_k)e_k) \ \text{(by (2))}. \end{aligned} \tag{4}$$

If we define $f(s,t) = \nu(se_j + te_k)$, then f clearly satisfies conditions (i), (ii), and (iv) of Bohnenblust's theorem. The fact that condition (v) holds is a consequence of (3) and (4). Hence, by the revised version of Bohnenblust's theorem, there exists p such that $\nu(se_j + te_k) = (|s|^p + |t|^p)^{1/p}$ or $\max\{|s|, |t|\}$ if $p = \infty$. It is therefore true that

$$\nu(se_j + te_k) = \nu(te_j + se_k). \tag{5}$$

Now suppose $\nu(se_j + re_m) = \nu(s'e_j + r'e_m)$. To show that $\{j, m\}$ is a fiber, it suffices to show that for any $t > 0$,and any u which is a finite linear combination of basis vectors with coordinates different from j, k, m that

$$\nu(se_j + te_k + re_m + u) = \nu(s'e_j + te_k + r'e_m + u). \tag{6}$$

For s, r, s', r' as given and for any $t > 0$ we obtain

$$\begin{aligned}
\nu(se_j + te_k + re_m + u) &= \nu(\nu(se_j + te_k)e_k + re_m + u) \quad (j \sim_\nu k)\\
&= \nu(\nu(te_j + se_k)e_k + re_m + u) \quad (\text{by } (5))\\
&= \nu(te_j + se_k + re_m + u) \quad (j \sim_\nu k)\\
&= \nu(te_j + \nu(se_k + re_m)e_m + u) \quad (k \sim_\nu m)\\
&= \nu(te_j + \nu(se_j + re_m)e_m + u) \quad (\text{from } (2))\\
&= \nu(te_j + \nu(s'e_j + r'e_m)e_m + u)\\
&= \nu(s'e_j + te_k + r'e_m + u),
\end{aligned}$$

where the last equality comes from reversing the previous steps.

We note that the argument as carried out above relied on the fact that there were three distinct integers in the resulting equivalence class, and in that case the norm restricted to the corresponding fiber space is an ℓ^p norm. In case there are only two integers in a class, we cannot make that conclusion nor can we know that for a class consisting of only $\{j, k\}$ is it true that equation (5) holds. □

The norm equivalence and corresponding equivalence classes give rise in a natural way to a decomposition of the space as indicated in the next theorem. Let $\mathcal{M} = \{N_j : j \in \Gamma\}$ denote the equivalence classes determined by the equivalence relation given in Definition 2. Here, Γ is of the form $\{1, 2, \ldots, k\}$, or $\Gamma = \mathbb{N}$. As in the statement above, X_j will mean the closed linear span of the basis vectors e_m for $m \in N_j$.

Theorem 6. *Let (X, ν) denote a sequence space with normalized absolute norm and suppose $\{N_j : j \in \Gamma\}$ is the collection of equivalence classes with respect to the norm equivalence. Then there exists a sequence space (E, μ) whose coordinate vectors form a normalized one-unconditional basis such that $X = \left(\sum_{j\in\Gamma} \oplus X_j\right)_E$, where X_j is the fiber space corresponding to N_j. Furthermore, for each X_j of dimension greater than 2, there is some p, $1 \le p \le \infty$, such that $X_j = \ell^p$ (where ℓ^p is of the appropriate dimension.)*

Proof. We let the elements of N_j be denoted by $p_{j1}, p_{j2}, \ldots$ and let E denote the space of sequences $\alpha = (\alpha(j))$ for which $\sum_j \alpha(j)e_{p_{j1}} \in X$. Let the norm μ be defined by $\mu(\alpha) = \nu(\sum_j \alpha(j)e_{p_{j1}})$. The proof is completed in much the same manner as the proof of the corresponding result for Hilbert space decompositions in [2]. In the direct sum notation for X we recall that the norm ν is given by

$$\nu\left(\sum x_j\right) = \mu((\|x_j\|))$$

where $\|\cdot\|$ is the norm on X_j. □

At this point let us consider some examples. Let ν_0 be a norm defined on $\mathbb{C}^2$ (or $\mathbb{R}^2$) by

$$\nu_0(s,t) = \begin{cases} |t|, & \text{if } 4|s| \le |t|;\\ |s| + \frac{3}{4}|t|, & \text{if } 4|s| > |t|. \end{cases} \tag{7}$$

Note that $\nu_0(1/4, 1) = 1 < \nu_0(1, 1/4)$. If we define ν_1 on $\mathbb{C}^3$ by

$$\nu_1(s,t,r) = \nu_0(|s| + \frac{4}{7}|t|, \frac{4}{7}|t| + |r|), \tag{8}$$

we get a three dimensional space with no equivalent coordinates and which does not satisfy the symmetry condition given by (5). On the other hand, the norm ν_2 defined on $\mathbb{C}^3$ by

$$\nu_2(s,t,r) = \nu_0(|s| + |t|, |r|) \tag{9}$$

gives a three-dimensional space in which the symmetry condition is not always satisfied, and there is one pair of equivalent coordinates, $1 \sim_\nu 2$. The corresponding fiber space is $\ell^1(2)$.

We are now ready to describe the isometries on the decompositions as given by Theorem 6. We find it necessary to assume our spaces are complex. We will see that the theorem does not carry over directly to the real case. By the *symmetry class* of a norm μ we mean the class of all permutations π for which $\mu((x(j))) = \mu((x(\pi(j))))$.

Theorem 7. *Let X be a complex sequence space with normalized one-unconditional basis $\{e_j : j \in N\}$ and suppose $X = \left(\sum_{j\in\Gamma} \oplus X_j\right)_E$ as given in Theorem 6. Then T is an isometry on X if and only if there exists a permutation π in the symmetry class of the norm μ and a family $\{T_j\}$ of surjective isometries with $T_j : X_{\pi(j)} \to X_j$ such that*

$$T(\sum_{j\in\Gamma} x_j) = \sum_{j\in\Gamma} T_j x_{\pi(j)}, \tag{10}$$

where we write $x = \sum_{j\in\Gamma} x_j$ for an element of the decomposition.

Proof. Let $\mathcal{M}_2 = \{J_\lambda : \lambda \in \Lambda\}$ denote the collection of maximal ℓ^2-fibers, that is, the collection of equivalence classes determined by the equivalence relation $\sim$ mentioned in the beginning. If T is an isometry, then by [3, Theorem 1] or [4, Theorem 6.1] there exists a permutation σ of Λ such that for all $\lambda \in \Lambda$,

$$\text{supp}(T(Y_\lambda)) = J_{\sigma(\lambda)} \tag{11}$$

where by supp we mean the indices of the basis elements for the given space and $Y_\lambda = \overline{\text{span}}\{e_j : j \in J_\lambda\}$. If $A \in \mathcal{M}$, then $A = \bigcup_{\lambda\in\Lambda_A} J_\lambda$, where $\Lambda_A \subset \Lambda$. This union consists of one element, A itself, if $A \in \mathcal{M}_2$, and a union of singletons otherwise. For each $A = \bigcup_{\lambda\in\Lambda_A} J_\lambda$, we define $\widetilde{T}$ by

$$\widetilde{T}(A) = \bigcup_{\lambda\in\Lambda_A} J_{\sigma(\lambda)}.$$

In case A is itself an ℓ_2-fiber, it is clear that $\widetilde{T}(A)$ is in $\mathcal{M}_2$ (also in $\mathcal{M}$) and of the same cardinality as A. If A is a union of singletons, then $\widetilde{T}(A) = \bigcup_{\lambda\in\Lambda_A} J_{\sigma(\lambda)}$ where each $J_{\sigma(\lambda)}$ is also a singleton.

Let $j, k \in \widetilde{T}(A)$, so that $j \in J_{\sigma(\lambda_1)}, k \in J_{\sigma(\lambda_2)}$ where $\lambda_1 \neq \lambda_2$. Let $a, b \in \overline{\text{span}}\{e_j, e_k\}$ with $\nu(a) = \nu(b)$. Further, suppose $c \in X$ with supp c contained in $N \setminus \{j, k\}$. Note that $T^{-1}(a)$ and $T^{-1}(b)$ have supports that are singletons and are in A, so they are norm-equivalent while supp $T^{-1}(c)$ is disjoint from both. Hence

$$\nu(a + c) = \nu(T^{-1}(a) + T^{-1}(c)) = \nu(T^{-1}(b) + T^{-1}(c)) = \nu(b + c),$$

from which we conclude that $j \sim_\nu k$ and $\widetilde{T}(A)$ is also in $\mathcal{M}$. Therefore, for each $N_j \in \mathcal{M}$, there is some $\pi(j) \in \Gamma$ such that $\widetilde{T}(N_{\pi(j)}) = N_j$. Moreover, this means that T must map $X_{\pi(j)}$ onto X_j and its restriction to $X_{\pi(j)}$ is an isometry. □

We see that the proof of the theorem really depends on the corresponding result for Hilbert space decompositions, and in general, there is not such a comprehensive theorem in the real case. In the following example [5], let X be the real space

$$X = \ell^1(3) \oplus_1 \ell^\infty(2). \tag{12}$$

The operator defined by

$$T(x(1), x(2), x(3), x(4), x(5)) = \big(U(x(4), x(5)), x(3), U^{-1}(x(1), x(2))\big),$$

where $U(s,t) = 2^{-1}(s+t, s-t)$, is an isometry which does not have the form of (11), since the norm equivalence classes are $\{1, 2, 3\}$ and $\{4, 5\}$. The problem here is that in the real case, $\ell^\infty(2)$ is isometric with $\ell^1(2)$ under the map U.

We want to mention here that Li and Randrianantoanina have shown that an admissable sequence space can be decomposed as in Theorem 6, but with the use of maximal fibers. This can lead to a slightly different decomposition. They have proved a theorem describing the isometries just as in Theorem 7 above. In fact, the proof above is modeled on their proof [5, Theorem 3.1]. Sometimes, the two decompositions can be used together to help in describing the isometries.

Let us consider the space $X_{p,q}(4)$, that is, the four dimensional version of the space $X_{p,q}$ that was defined earlier. The norm equivalence classes are $\{1, 2\}, \{3\}$, and $\{4\}$. The decomposition given by Theorem 6 would involve three spaces, $\ell^p(2)$ and two one-dimensional spaces and the space $E = X_{q,p}(3)$. An isometry T on the space would map $\ell^p(2)$ to itself and presumably could interchange the one-dimensional spaces. However $S_1 = \{1, 2, 3\}$ and $S_2 = \{4\}$ form a partition of maximal fibers so by [5, Theorem 3.1], the fiber space X_{S_1} is mapped to itself by T and X_{S_2} is mapped to itself, which shows that T must fix both of the one-dimensional spaces in the first decomposition.

References

[1] F. Bohnenblust, *Axiomatic characterization of L_p-spaces*, Duke Math J. 6 (1940), pp. 627–640.

[2] R. Fleming and J. Jamison, *Hermitian and adjoint abelian operators on certain Banach spaces*, Pacific J. Math. 52 (1974), pp. 67–84.

[3] ———, *Isometries on certain Banach spaces*, J. London Math. Soc. (2) 9 (1974), pp. 363–371.

[4] N. J. Kalton and G. Wood, *Orthonormal systems in Banach spaces and their applications*, Proc. Camb. Phil. Soc. 79 (1976), pp. 493–510.

[5] C. Li and B. Randrianantoanina, *Isometries of direct sums of sequence spaces*, Asian J. Math. 2 (1998), pp. 157–180.

[6] H. Schneider and R. Turner, *Matrices hermitian for an absolute norm*, Linear and Multilinear Algebra 1 (1973), pp. 9–31.

Author information

Richard J. Fleming, Department of Mathematics, Central Michigan University, Mt. Pleasant, Michigan 48859, USA.
Email: flemi1rj@cmich.edu

Banach Spaces and their Applications in Analysis, 339–342

An isometric form of a theorem of Lindenstrauss and Rosenthal on quotients of $\ell_1(\Gamma)$

Marcos González and Marek Wójtowicz

Abstract. Let X be an infinite dimensional Banach space and $\mathcal{Q}(X,\Gamma)$ denote the set of all quotient mappings Q_A from $\ell_1(\Gamma)$ onto X defined by the formula

$$Q_A((\xi_\gamma)_{\gamma\in\Gamma}) = \sum_{\gamma\in\Gamma} \xi_\gamma a_\gamma,$$

where the set $A = \{a_\gamma : \gamma \in \Gamma\}$ is dense in the unit ball of X, and $\mathrm{card}(A) = \mathrm{card}(\Gamma) = \chi(X)$ (the density character of X). For every pair $Q_A, Q_B \in \mathcal{Q}(X,\Gamma)$ there is a sequence (π_j) of autoisometries of $\ell_1(\Gamma)$ with

$$\lim_{j\to\infty} \|Q_A - Q_B\pi_j\| = 0.$$

This is an isometric form of the classical theorem of Lindenstrauss and Rosenthal about comparison of two general quotient mappings from $\ell_1(\Gamma)$ onto X.

A similar approximation property holds for a class of operators from $\mathrm{Lip}(V)$ into ℓ_∞, where V is an infinite compact metric space and $\mathrm{Lip}(V)$ denotes the space of all numerical functions on V fulfilling the Lipschitz condition.

Key words. Isometry, quotient mapping, space $\ell_1(\Gamma)$.

AMS classification. 46B25, 46B26.

1 Introduction

Throughout this paper X will denote an infinite dimensional Banach space, B_X its open unit ball, Γ an infinite set, and $\mathbb{N}$ the set of natural numbers. By $\chi(X)$ we denote the density character of X.

In 1969 Lindenstrauss and Rosenthal showed that if X is separable then the set of quotient mappings from ℓ_1 onto X is determined uniquely, up to automorphisms of ℓ_1: *for every two mappings T_1, T_2 of ℓ_1 onto X there is an automorphism τ of ℓ_1 such that*

$$T_1 = T_2\tau \tag{1.1}$$

(see [4, Theorem 2] and [5, p. 109]). In particular, this result holds for the class of natural quotient mappings $T = Q_A$, determined by dense sequences $A = (a_n)$ in B_X, of the form

$$Q_A((\xi_n)) = \sum_{n=1}^{\infty} \xi_n a_n\ , \quad (\xi_n) \in \ell_1\ . \tag{1.2}$$

A natural question to ask is whether τ in (1) can be replaced somehow by an autoisometry of ℓ_1. Although the question seems to be a little strange (as every such a τ is strictly connected with a permutation of the set $\mathbb{N}$), we provide a partial positive answer to it: for every pair Q_A, Q_B of mappings defined as in (1.2) there is a sequence (π_j) of autoisometries of ℓ_1 with

$$\lim_{j\to\infty} \|Q_A - Q_B\pi_j\| = 0 .$$

We show below that this type of approximation holds even in a more general case, i.e., for the set $\mathcal{Q}(X,\Gamma)$ of all quotient mappings Q_A acting from $\ell_1(\Gamma)$ onto X defined by the rule

$$Q_A((\xi_\gamma)_{\gamma\in\Gamma}) = \sum_{\gamma\in\Gamma} \xi_\gamma a_\gamma , \quad (\xi_\gamma)_{\gamma\in\Gamma} \in \ell_1(\Gamma) , \tag{1.3}$$

where the set $A = \{a_\gamma : \gamma \in \Gamma\}$ is dense in B_X, and $\mathrm{card}(\Gamma) = \chi(X) = \mathrm{card}(A)$. These mappings generalize (1.2), and are described in [6, Proposition 11.4.6]: every Q_A is open with $B_X \subset Q_A(\overline{B_{\ell_1(\Gamma)}}) \subset \overline{B_X}$.

Thus our main result reads as follows.

Theorem 1.1. *For every Q_A and Q_B in $\mathcal{Q}(X,\Gamma)$ and every $\varepsilon > 0$ there is an autoisometry π_ε of $\ell_1(\Gamma)$ such that*

$$\|Q_A - Q_B\pi_\varepsilon\| \le \varepsilon . \tag{1.4}$$

The method of proof of Theorem 1.1 can also be applied to yet another class of operators, acting this time into ℓ_∞, yielding a similar type of approximation as in (1.4). Let $V = (V,d)$ be an infinite compact metric space without isolating points and $\mathrm{Lip}(V)$ denote the Banach space of all (real or complex) functions on V fulfilling the Lipschitz condition with respect to d, endowed with the norm $\|f\| = \max\{\sup_{v\in V}|f(v)|, p(f)\}$, where $p(f) = \sup_{v\neq w}|f(v) - f(w)|/d(v,w)$. Let $\mathcal{T}(V)$ denote the set of all operators $T_A : \mathrm{Lip}(V) \to \ell_\infty$ of the form

$$T_A(f) = (f(a_n))_{n=1}^{\infty} ,$$

where $A = \{a_n : n \in \mathbb{N}\}$ is a dense set in V. Notice first that every element T_A of $\mathcal{T}(V)$ is injective, not a homeomorphism, and $T_A(\mathbf{1}_V) = \mathbf{1}_{\mathbb{N}}$, where the respective elements denote the constant-one functions in $\mathrm{Lip}(V)$ and ℓ_∞, respectively (hence, $\|T_A\| = 1$). Since for every pair T_A, T_B in $\mathcal{T}(V)$ their respective ranges are separable subspaces in ℓ_∞, part (i) of [4, Theorem 3] (see also [5, Theorem 2.f.12]) suggests that the pair might be comparable by means of an automorphism of ℓ_∞. Although we cannot confirm it exactly, we present an approximative result in the spirit of our Theorem 1.1.

Theorem 1.2. *For every T_A and T_B in $\mathcal{T}(V)$ and every $\varepsilon > 0$ there is an autoisometry ψ_ε of ℓ_∞ such that*

$$\|T_A - \psi_\varepsilon T_B\| \le \varepsilon . \tag{1.5}$$

2 The proofs

We shall present a full proof of Theorem 1.1 and only outline a proof for Theorem 1.2. The main tool in our proofs will be Banach's decomposition theorem ([3, Chapter 5,

§5, Theorem 2]), which is a useful form of the classical Cantor–Bernstein theorem and was already applied in topology and functional analysis (see [1, 2, 7, 8]):

Proposition 2.1. *Let U and W be two infinite sets, and let f and g be injective mappings from U into W and W into U, respectively. Then there exist decompositions of U and W: $U = U_1 \cup U_2$ and $W = W_1 \cup W_2$ such that*

$$f(U_1) = W_1 \quad \text{and} \quad g(W_2) = U_2. \tag{2.1}$$

Consequently, the mapping $h : U \to W$ defined by: $h(u) = f(u)$ for $u \in U_1$, and $h(u) = g^{-1}(u)$ for $u \in U_2$, is both injective and onto.

Proof. Let $\mathrm{m} = \chi(X) = \mathrm{card}(A) = \mathrm{card}(B) = \mathrm{card}(\Gamma)$, where $A = \{a_\gamma : \gamma \in \Gamma\}$ and $B = \{b_\gamma : \gamma \in \Gamma\}$ are dense subsets of B_X, and let $\varepsilon > 0$ be fixed. We shall prove first, using transfinite induction, that there exist two injections $f, g : \Gamma \to \Gamma$ with

$$\|b_{f(\gamma)} - a_\gamma\| \leq \varepsilon \text{ and } \|a_{g(\gamma)} - b_\gamma\| \leq \varepsilon \tag{2.2}$$

for all $\gamma \in \Gamma$. We identify Γ with the set of all ordinals $\{\gamma : \gamma < \mu\}$, where μ is the first ordinal of cardinality m. For $\gamma = 1$ there is evidently γ_1 with $\|b_{\gamma_1} - a_1\| \leq \varepsilon$; we put $f(1) = \gamma_1$. Let f injective, with property (2.2), be constructed for all γ with $\gamma < \beta < \mu$, and let us consider the set $B_\beta := B \setminus D_\beta$, where $D_\beta = \{b_{f(\gamma)} : \gamma < \beta\}$. Notice that $\mathrm{card}(D_\beta) < \mathrm{m}$.

We claim that the set B_β is dense in B_X. For this, it is enough to show that each element $b_{f(\gamma)}$, with $\gamma < \beta$, can be approximated by elements of B_β. To this end, let us fix $\gamma < \beta$ and an arbitrary number $r > 0$ such that the open ball $K_\gamma := B(b_{f(\gamma)}, r)$ is included in B_X. It is easy to check that the set $C_\gamma := K_\gamma \cap B$ is dense in K_γ, with $\mathrm{card}(C_\gamma) = \mathrm{m} > \mathrm{card}(D_\beta)$. The latter inequality implies that the set $C_\gamma \setminus D_\beta$ is a nonempty subset of B_β and contains an element b with $\|b - b_{f(\gamma)}\| < r$. This proves our claim.

Choose an element $b_{\gamma_\beta} \in B_\beta$ with $\|b_{\gamma_\beta} - a_\beta\| \leq \varepsilon$, both for β a successor ordinal and for β a limit ordinal. We put $f(\beta) = \gamma_\beta$, and we see that f defined in this way is both injective and fulfils (2.2).

The second function, g, is constructed similarly. □

By Proposition 2.1, there is a decomposition of Γ: $\Gamma = \Gamma_1 \cup \Gamma_2$ such that the mapping $h : \Gamma \to \Gamma$ of the form $h(\gamma) = f(\gamma)$ for $\gamma \in \Gamma_1$, and $h(\gamma) = g^{-1}(\gamma)$ for $\gamma \in \Gamma_2$, is both injective and onto, and fulfils, by (2.2), inequality

$$\|a_\gamma - b_{h(\gamma)}\| \leq \varepsilon,\ \gamma \in \Gamma. \tag{2.3}$$

Now we define a linear autoisometry π_ε of $\ell_1(\Gamma)$ by the formula

$$\pi_\varepsilon(e_\gamma) := e_{h(\gamma)}, \tag{2.4}$$

where e_γ denotes the γth unit vector of $\ell_1(\Gamma)$. Let $x = (\xi_\gamma) \in \ell_1(\Gamma)$ be fixed. By (2.3) and (2.4), we obtain $\|(Q_A - Q_B\pi_\varepsilon)(x)\| = \|\sum_{\gamma\in\Gamma} \xi_\gamma(a_\gamma - b_{h(\gamma)})\| \leq \varepsilon\|x\|$. Hence, $\|Q_A - Q_B\pi_\varepsilon\| \leq \varepsilon$, as claimed. The proof of Theorem 1.1 is complete. □

Concerning the proof of Theorem 1.2, we proceed similarly as above: we show first that if $A = (a_n)$ and $B = (b_n)$ are two dense sequences in V then, for every $\varepsilon > 0$ there exist strictly increasing sequences (k_n) and (l_n) in $\mathbb{N}$ such that $d(a_n, b_{k_n}) < \varepsilon$ and $d(b_n, a_{l_n}) < \varepsilon$ for all n's. By Proposition 2.1, there is a permutation h of $\mathbb{N}$ with $d(a_n, b_{h(n)}) < \varepsilon$ for all n's. It follows that for the autoisometry ψ_ε of ℓ_∞ of the form $\psi_\varepsilon(e_n) = e_{h(n)}$, where e_n is the nth unit vector of ℓ_∞, we have $\|(T_A - \psi_\varepsilon T_B)(f)\| \leq \varepsilon\|f\|$, $f \in \mathrm{Lip}(V)$. Hence, $\|T_A - \psi_\varepsilon T_B\| \leq \varepsilon$.

References

[1] L. Drewnowski, *A proof of Niederreiter's rearrangement theorem*, Arch. Math. 48 (1987), pp. 326–327.

[2] P. R. Halmos, *Permutations of sequences and the Schröder-Bernstein theorem*, Proc. Amer. Math. Soc. 19 (1968), pp. 509–510.

[3] K. Kuratowski and A. Mostowski, *Set Theory*, Polish Scientific Publishers, Warszawa 1971.

[4] J. Lindenstrauss and H. P. Rosenthal, *Automorphisms in c_0, ℓ_1 and m*, Israel J. Math. 7 (1969), pp. 227–239.

[5] J. Lindenstrauss and L. Tzafriri, *Classical Banach Spaces. I*, Springer-Verlag, Berlin 1977.

[6] Z. Semadeni, *Banach Spaces of Continuous Functions*, Polish Scientific Publishers, Warszawa 1971.

[7] M. Wójtowicz, *On the permutative equivalence of unconditional bases in F-spaces*, Funct. Approx., Comment. Math. 16 (1988), pp. 51–54.

[8] ______, *Universal Musielak-Orlicz sequence spaces*, in: Sequence Spaces and Applications (P. K. Jain and E. Malkovsky, eds), pp. 9–13, Narosa Publishing House, New Delhi 1999.

Author information

Marcos González, Departamento de Matemáticas, Universidad Simón Bolivar, Caracas 1080-A, Venezuela.
Email: mago@ma.usb.ve

Marek Wójtowicz, Instytut Matematyki, Uniwersytet Kazimierza Wielkiego (Casimir the Great University), Pl. Weyssenhoffa 11, 85-072 Bydgoszcz, Poland.
Email: mwojt@ukw.edu.pl

Banach Spaces and their Applications in Analysis, 343–355

Aspects of probabilistic Littlewood–Paley theory in Banach spaces

Tuomas P. Hytönen

Abstract. A probabilistic square function estimate, due to Varopoulos in the scalar case, is proved for functions with values in a UMD Banach space. As an application, close to optimal dimension-free norm bounds for certain spectral multipliers of the Laplacian in Bôchner spaces are obtained. The proof uses stochastic integration of vector-valued processes and especially decoupling estimates due to Garling.

Key words. Square function, stochastic integral, spectral multiplier, UMD space, martingale transform, decoupling, dimension-free estimate.

AMS classification. Primary 42B25; Secondary: 46B09, 60G46.

Dedicated to Professor Nigel Kalton on the occasion of his (quasi-)60th birthday.

1 Introduction

The notion of square-functions originating from Littlewood and Paley, and their use in the estimation of singular integrals, are among the central ideas in Harmonic Analysis. The theory around these topics is also an area where deep connections to Probability manifest themselves, and several authors (in particular, Stein [19], Meyer [14], and Varopoulos [21]) have exploited stochastic methods to reprove and improve results of the classical Littlewood–Paley theory. Besides offering insight into the original problems, the probabilistic approach has become an indispensable tool in developing Harmonic Analysis of Banach space-valued functions, a theory pioneered by Bourgain [1, 2], Burkholder [3, 4], and McConnell [12] in the early 80's. This program has further elaborated on the relationships of Harmonic and Stochastic Analysis by showing that the probabilistic UMD property of a Banach space X is equivalent to the validity of several estimates from Harmonic Analysis for X-valued functions.

Quadratic estimates, reformulated in a "randomized" way, have also played a crucial role in the vector-valued developments. This theory, however, was essentially restricted to discretely parameterized square functions, until Kalton and Weis's development of a more general framework, which has been a source of great inspiration to, I dare say, everyone who has had the chance to learn about the ideas contained in their still unpublished manuscript [11]. Using equivalents of these new square functions, although formulated in the different language of vector-valued stochastic integrals (cf. [15, 18]), I was recently [10] able to characterize the UMD property in terms of a vector-valued extension of Stein's [19] abstract Littlewood–Paley-type estimates.

The purpose of the present paper is to carry over (parts of) the Littlewood–Paley theory of Varopoulos [21] to the Banach space situation. Varopoulos's approach has a stronger probabilistic flavor than Stein's, and to reach the goal, I will need to employ the stochastic Itô integration of vector-valued *processes*, as opposed to the easier Wiener integration of (deterministic) functions used in [10]. The required integration theory, based on decoupling estimates due to Garling [7], has been developed by McConnell [13], with considerable recent elaboration by van Neerven, Veraar and Weis [16]. I wish to acknowledge that the application of this more advanced stochastic machinery to the vector-valued Harmonic Analysis was first suggested to me by Jan van Neerven, during the time of my finishing the work [10].

Compared to the Littlewood–Paley–Stein theory [19], which is valid for a general diffusion semigroup T^y on an abstract measure space, the methods of Varopoulos are more restricted (requiring, in particular, some topological structure on the underlying measure space), and I have here chosen to work in an even narrower setting, where T^y is the classical Gaussian semigroup on $\mathbb{R}^n$. One motivation for treating such a specialized case is the considerable simplification of the argument that is obtained here, even to the extent that I feel that the present proofs are the "right" ones for the theorems stated in this paper. There is also a more substantial reason: the new approach is very efficient in the use of the required martingale estimates in that fairly good constants in the various inequalities are obtained. In particular, as a corollary it is found that the norm of the imaginary powers of the Laplace operator, $(-\triangle)^{i\gamma}$, in the limit as $\gamma \to 0$, satisfies an upper bound involving the UMD constant, which differs from the known lower bound due to Guerre-Delabrière [9] only by a factor of 2. Thus, not more than this numerical factor is lost in the present estimates. Moreover, it is found that some of the Varopoulos-type square-function estimates only require "one half" of the UMD property (in the sense of the one-sided random martingale transform inequalities introduced by Garling [8]), so that more precise than before relations of the probabilistic and analytic estimates are achieved.

2 Preliminaries

The UMD condition and related notions

Let us recall the fundamental UMD condition of Burkholder [3]. A Banach space X has the property of *unconditional martingale differences*, for short UMD, provided that the best constant $\beta_{p,X}$ is finite for one (and then all; see [3]) $p \in (1,\infty)$ in the following inequality:

$$\Big(\mathsf{E}\Big|\sum_{j=1}^{N}\epsilon_j d_j\Big|^p\Big)^{1/p} \leq \beta_{p,X}\Big(\mathsf{E}\Big|\sum_{j=1}^{N} d_j\Big|^p\Big)^{1/p}, \tag{2.1}$$

which should hold for all $N \in \mathbb{N}$, all signs $(\epsilon_j)_{j=1}^N \in \{-1,+1\}^N$, any probability space $(\Omega,\mathscr{F},\mathsf{P})$ (with expectation $\mathsf{E} := \int_\Omega(\cdot)\,\mathrm{d}\mathsf{P}$) equipped with a filtration $\mathscr{F}_1 \subseteq \ldots \subseteq \mathscr{F}_N \subseteq \mathscr{F}$, and all martingale difference sequences $(d_j)_{j=1}^N \in L^p(\Omega,\mathscr{F},\mathsf{P};X)^N$ adapted to this filtration (i.e., $d_j \in L^p(\Omega,\mathscr{F}_j,\mathsf{P};X)$ and $\mathsf{E}[d_j|\mathscr{F}_{j-1}] = 0$ for all $j = 2,\ldots,N$).

We also consider some further estimates closely related to (2.1), where the quantities common with (2.1) have the same range as there. Let $\beta^{\mathbb{C}}_{p,X}$ be the best constant in

$$\Big(\mathsf{E}\Big|\sum_{j=1}^{N}\zeta_j d_j\Big|^p\Big)^{1/p} \leq \beta^{\mathbb{C}}_{p,X}\Big(\mathsf{E}\Big|\sum_{j=1}^{N} d_j\Big|^p\Big)^{1/p}, \tag{2.2}$$

where the ζ_j are arbitrary complex numbers of absolute value $|\zeta_j| = 1$. It is clear that $\beta_{p,X} \leq \beta^{\mathbb{C}}_{p,X}$ and a splitting into real and imaginary parts plus a standard convexity argument shows that $\beta^{\mathbb{C}}_{p,X} \leq 2\beta_{p,X}$. By more careful reasoning, using Lemma 4.11.5 of [17], this may be improved to $\beta^{\mathbb{C}}_{p,X} \leq \pi/2 \cdot \beta_{p,X}$.

There is a natural splitting of the inequality (2.1) into two parts. Let us denote by $\beta^{\pm}_{p,X}$ the best constants in the following estimates, where the ε_j designate independent *random* signs on another probability space $(\tilde{\Omega}, \tilde{\mathscr{F}}, \tilde{\mathsf{P}})$ with expectation $\tilde{\mathsf{E}}$, distributed by the symmetric law $\tilde{\mathsf{P}}(\varepsilon_j = -1) = \tilde{\mathsf{P}}(\varepsilon_j = +1) = 1/2$:

$$\frac{1}{\beta^{-}_{p,X}}\Big(\mathsf{E}\Big|\sum_{j=1}^{N} d_j\Big|^p\Big)^{1/p} \leq \Big(\mathsf{E}\tilde{\mathsf{E}}\Big|\sum_{j=1}^{N}\varepsilon_j d_j\Big|^p\Big)^{1/p} \leq \beta^{+}_{p,X}\Big(\mathsf{E}\Big|\sum_{j=1}^{N} d_j\Big|^p\Big)^{1/p}. \tag{2.3}$$

It is easy to see that $\max(\beta^{-}_{p,X}, \beta^{+}_{p,X}) \leq \beta_{p,X} \leq \beta^{-}_{p,X}\beta^{+}_{p,X}$. The two Banach space properties defined by the inequalities (2.3) have been studied by Garling [8] who showed, among other results, their independence from $p \in (1,\infty)$. He called the first and second inequality in (2.3) the *lower and upper estimate for random martingale transforms*, for short LERMT and UERMT, respectively. van Neerven, Veraar and Weis have proposed the alternative names UMD$^-$ and UMD$^+$, which are more suggestive of the close connection between (2.1) and (2.3).

We shall make use of a well-known theorem of Burkholder [3] which shows that the UMD property (2.1) (or (2.2)) self-improves into an estimate for more general *martingale transforms*. We say that $(v_j)_{j=1}^N \in L^\infty(\Omega, \mathscr{F}, \mathsf{P}; \mathbb{C})^N$ is a *predictable* sequence (adapted to the filtration $(\mathscr{F}_j)_{j=1}^N$) if $v_j \in L^\infty(\Omega, \mathscr{F}_{(j-1)\vee 1}, \mathsf{P}; \mathbb{C})$ for each $j = 1, \ldots, N$. If $(d_j)_{j=1}^N$ is a martingale difference sequence adapted to this same filtration, then so is $(v_j d_j)_{j=1}^N$. Only the case with $v_j \in L^\infty(\Omega, \mathscr{F}_{(j-1)\vee 1}, \mathsf{P}; \mathbb{R})$ is explicitly treated in [3], but the complex-valued result follows in the same way.

Theorem 2.1 ([3]). *Let X be a UMD space, $1 < p < \infty$, and $(\Omega, \mathscr{F}, \mathsf{P})$ be a probability space with filtration $(\mathscr{F}_j)_{j=1}^N$. For every martingale difference sequence $(d_j)_{j=1}^N$ and every predictable sequence $(v_j)_{j=1}^N$ adapted to this filtration, there holds*

$$\Big(\mathsf{E}\Big|\sum_{j=1}^{N} v_j d_j\Big|^p\Big)^{1/p} \leq \beta^{\mathbb{C}}_{p,X} \max_{1\leq j\leq N}\|v_j\|_\infty \Big(\mathsf{E}\Big|\sum_{j=1}^{N} d_j\Big|^p\Big)^{1/p}.$$

If the v_j are real-valued, $\beta^{\mathbb{C}}_{p,X}$ may be replaced by $\beta_{p,X}$ in the above estimate.

Stochastic integration

We next recall the elements of the theory of stochastic integration, which are relevant for the present purposes. Let $x(t)$ and $y(t)$, $t \geq 0$, be independent standard Brownian motions with continuous paths on $\mathbb{R}^n$ and $\mathbb{R}$, respectively, and $z(t) := (x(t), y(t))$. We write $\mathsf{P}_z \equiv \mathsf{P}_x \otimes \mathsf{P}_y$ for the probability on a measurable space $(\Omega, \mathscr{F})$ governing the motion of $z(t)$ with initial value $z(0) = z = (x, y) \in \mathbb{R}^n \times \mathbb{R}$; this is the Cartesian product of two independent processes as indicated by the notation. Similarly, we write $\mathsf{E}_z = \mathsf{E}_x \otimes \mathsf{E}_y$ for the related expectations. Let us finally introduce the Brownian motions $\tilde{z}(t) = (\tilde{x}(t), \tilde{y}(t))$, $t \geq 0$, which are independent copies of the above mentioned processes with starting point at 0. For convenience, we may take them to be defined on a different probability space $(\tilde{\Omega}, \tilde{\mathscr{F}}, \tilde{\mathsf{P}})$, with expectation $\tilde{\mathsf{E}}$. Let $(\mathscr{F}_t)_{t\geq 0}$ be any filtration (an increasing family of sub-σ-algebras of $\mathscr{F}$) such that $z(s)$ is $\mathscr{F}_t$-measurable for all $0 \leq s \leq t < \infty$, and let $(\tilde{\mathscr{F}}_t)_{t\geq 0}$ be similarly related to $\tilde{z}(t)$.

We then move to the vector-valued situation. Let X be a Banach space. In accordance with the stochastic integration theory of [16], we take the scalar field to be $\mathbb{R}$; in particular, by the dual space we understand $X^* := \mathscr{L}(X, \mathbb{R})$. Observe that a complex Banach space may always be viewed also as a real Banach space, where all the original operations are still defined, but we have to think of the mapping $X \in \xi \mapsto \lambda\xi \in X$ not as a scalar multiplication but as a bounded linear operator on X when $\lambda \in \mathbb{C} \setminus \mathbb{R}$.

Let us consider $\phi = (\phi_k)_{k=1}^{n+1} : [0, \infty) \times \Omega \to X^{n+1}$, where X is a Banach space. We identify $X^{n+1} \eqsim \mathscr{L}(\mathbb{R}^{n+1}, X)$, in particular when computing the norm in this space. Such a function is called *elementary adaptive* [16] if

$$\phi_k = \sum_{m=0}^{M} \sum_{\ell=1}^{L} 1_{(t_{m-1}, t_m] \times A_{\ell m}} \xi_{k\ell m}$$

for some $0 = t_{-1} \leq t_0 < t_1 < \ldots < t_M$, $A_{\ell m} \in \mathscr{F}_{t_{m-1}}$ and $\xi_{k\ell m} \in X$, where it is understood (with slight misuse of notation) that $(t_{-1}, t_0] := [0, t_0]$. A process $\phi : [0, \infty) \times \Omega \to X^{n+1}$ is called *adaptive* if $\omega \mapsto \phi(t, \omega)$ is (strongly) $\mathscr{F}_t$-measurable for every $t \in [0, \infty)$.

The stochastic integral of an elementary adaptive process is defined in the obvious way by

$$\int_0^\infty \phi(t)\, \mathrm{d}z(t) := \sum_{k,\ell,m} [z_k(t_m) - z_k(t_{m-1})] 1_{A_{\ell m}} \xi_{k\ell m}.$$

We may also define the stochastic integral where $z(t)$ is replaced by $\tilde{z}(t)$: both $\phi(t)$ and $\tilde{z}(t)$ may be extended to functions on $\Omega \times \tilde{\Omega}$ in a canonical way, and they satisfy the appropriate assumptions with respect to the product filtration $(\mathscr{F}_t \times \tilde{\mathscr{F}}_t)_{t\geq 0}$.

Definition 2.2 ([16]). A process $\phi : [0, \infty) \times \Omega \to X^{n+1}$ is called *L^p-stochastically integrable* with respect to $(z(t))_{t\geq 0}$ if $\langle \phi_k(\cdot, \cdot), \xi^* \rangle \in L^p(\Omega, L^2(0, \infty))$ for $k = 1, \ldots, n+1$, and there are elementary adaptive processes $\phi^{(j)} = (\phi_k^{(j)})_{k=1}^{n+1}$, $j \in \mathbb{N}$, such that

- $\lim_{j\to\infty} \left\langle \phi_k^{(j)}(\cdot, \cdot), \xi^* \right\rangle = \langle \phi_k(\cdot, \cdot), \xi^* \rangle$ in $L^p(\Omega, L^2(0, \infty))$ for every $\xi^* \in X^*$, and

- there exists $\Phi \in L^p(\Omega, X)$ such that

$$\Phi = \lim_{j\to\infty} \int_0^\infty \phi^{(j)}(t)\,\mathrm{d}z(t) \quad \text{in } L^p(\Omega, X).$$

In this case, we define $\int_0^\infty \phi(t)\,\mathrm{d}z(t) := \Phi$.

If $X_0 \subseteq X$ is a finite-dimensional subspace, an adaptive process $\phi : [0,\infty) \times \Omega \to X_0^{n+1}$ may be analyzed "component-wise", and the classical necessary and sufficient condition for its stochastic integrability is valid:

$$\mathsf{E}_z\Big(\int_0^\infty |\phi(t)|^2\,\mathrm{d}t\Big)^{p/2} < \infty. \tag{2.4}$$

For most of our purposes, this finite-dimensional situation would be enough as far as the existence of stochastic integrals is concerned; however, we want to make estimates of the integrals which are independent of the particular subspace X_0. For this purpose, we state the following result of Garling:

Theorem 2.3 ([7]). *Let X be a Banach space, and let the Brownian motion $z(t)$ be adapted to a filtration $(\mathscr{F}_t)_{t\geq 0}$. Let $v : [0,\infty[\, \times \Omega \to X_0^{n+1}$, where $X_0 \subseteq X$ is a finite-dimensional subspace, be adapted $(\mathscr{F}_t)_{t\geq 0}$ and satisfy* (2.4). *If X is UMD$^-$ resp. UMD^+, then there holds*

$$\begin{aligned} &\Big(\mathsf{E}_z\Big|\int_0^\infty v(t)\,\mathrm{d}z(t)\Big|^p\Big)^{1/p} \leq \beta^-_{p,X}\Big(\mathsf{E}_z\tilde{\mathsf{E}}\Big|\int_0^\infty v(t)\,\mathrm{d}\tilde{z}(t)\Big|^p\Big)^{1/p} \qquad \textit{resp.} \\ &\Big(\mathsf{E}_z\tilde{\mathsf{E}}\Big|\int_0^\infty v(t)\,\mathrm{d}\tilde{z}(t)\Big|^p\Big)^{1/p} \leq \beta^+_{p,X}\Big(\mathsf{E}_z\Big|\int_0^\infty v(t)\,\mathrm{d}z(t)\Big|^p\Big)^{1/p}. \end{aligned} \tag{2.5}$$

This concludes our general preliminaries, and we move on to matters more specific to the present paper.

Set-up for the theory of Varopoulos

Following Varopoulos [21], we introduce the operators

$$\mathsf{E}^\lambda(F) := \int_{\mathbb{R}^n} \mathsf{E}_{(x,\lambda)}(F)\,\mathrm{d}x, \qquad \mathsf{P}^\lambda(A) := \mathsf{E}^\lambda(1_A),$$

where the action of E^λ is defined for all measurable functions $F \geq 0$ on Ω. These represent the "expectation" and "probability" related to Brownian motion starting in a random point at height λ, except that P^λ is a σ-finite positive measure of infinite total mass, and hence not really a probability. However, as explained in [21], all the results from probability relevant to the present context remain valid in this extended setting, which may be seen either by inspection of the classical proofs or by splitting into countably many honest probability spaces and summing up.

An important role will be played by the stopping time

$$\tau := \inf\{t : y(t) \le 0\}, \tag{2.6}$$

where $y(t)$ is the vertical component of the Brownian motion $z(t) = (x(t), y(t))$.

We write $T^y := e^{y\triangle}$ and $P^y := e^{-y(-\triangle)^{1/2}}$ for the Gaussian and Poisson semigroups, where $\triangle$ is the Laplacian, $\triangle f = \sum_{k=1}^n \partial^2 f/\partial x_k^2$, on $\mathbb{R}^n$. For any reasonable enough, possibly vector-valued, function f on $\mathbb{R}^n$, we denote by u its harmonic extension to the upper half-space $\mathbb{R}^{n+1}_+ = \mathbb{R}^n \times [0,\infty)$, i.e., $u(x,y) = P^y f(x)$. Finally, let $\partial u/\partial x := (\partial u/\partial x_k)_{k=1}^n$, so that $\nabla u = (\partial u/\partial x, \partial u/\partial y)$.

3 Square function estimates

The main probabilistic Littlewood–Paley theorem of Varopoulos is the following:

Theorem 3.1 ([21]). *For all $p \in [2,\infty)$, there exists a constant K_p such that for all $\lambda \ge 0$ and all $f \in \mathscr{S}(\mathbb{R}^n)$, we have*

$$\mathsf{E}^\lambda\Big[\Big(\int_0^\tau \Big|\frac{\partial u}{\partial y}(z(s))\Big|^2 \,\mathrm{d}s\Big)^{p/2}\Big] \le K_p \|f\|_p^p, \qquad u(x,y) := P^y f(x). \tag{3.1}$$

In order to motivate our vector-valued extension, we rewrite the estimate (3.1) in a somewhat different but equivalent form. Recall that $(\tilde{y}(t))_{t\ge 0}$ is a Brownian motion independent of all the earlier-mentioned processes, defined on another probability space $(\tilde{\Omega}, \tilde{\mathscr{F}}, \tilde{\mathsf{P}})$, with expectation $\tilde{\mathsf{E}}$. Then the classical Itô isometry says that

$$\int_0^\tau \Big|\frac{\partial u}{\partial y}(z(s))\Big|^2 \,\mathrm{d}s = \tilde{\mathsf{E}}\Big|\int_0^\tau \frac{\partial u}{\partial y}(z(s))\,\mathrm{d}\tilde{y}(s)\Big|^2.$$

The stochastic integral $\int_0^\tau \partial u(z(s))/\partial y \,\mathrm{d}\tilde{y}(s)$ is a Gaussian random variable on $\tilde{\Omega}$, and thus all of its L^p norms a comparable, i.e.,

$$\Big(\tilde{\mathsf{E}}\Big|\int_0^\tau \frac{\partial u}{\partial y}(z(s))\,\mathrm{d}\tilde{y}(s)\Big|^2\Big)^{1/2} \eqsim_p \Big(\tilde{\mathsf{E}}\Big|\int_0^\tau \frac{\partial u}{\partial y}(z(s))\,\mathrm{d}\tilde{y}(s)\Big|^p\Big)^{1/p}.$$

In conclusion, we find that

$$\mathsf{E}^\lambda\Big[\Big(\int_0^\tau \Big|\frac{\partial u}{\partial y}(z(s))\Big|^2 \,\mathrm{d}s\Big)^{p/2}\Big] \eqsim_p \mathsf{E}^\lambda\tilde{\mathsf{E}}\Big[\Big|\int_0^\tau \frac{\partial u}{\partial y}(z(s))\,\mathrm{d}\tilde{y}(s)\Big|^p\Big],$$

where the left-hand side coincides with that of (3.1). Hence we see that the following result is indeed an extension of Theorem 3.1.

Theorem 3.2. *Let X be a UMD$^+$ Banach space and $p \in (1,\infty)$. Then for all $\lambda \ge 0$ and all $f \in \mathscr{S}(\mathbb{R}^n) \otimes X$, we have*

$$\Big\{\mathsf{E}^\lambda\tilde{\mathsf{E}}\Big|\int_0^\tau \frac{\partial u}{\partial y}(z(s))\,\mathrm{d}\tilde{y}(s)\Big|^p\Big\}^{1/p} \le \beta^+_{p,X}\big(\|f\|_{L^p_X} + \|P^\lambda f\|_{L^p_X}\big) \le 2\beta^+_{p,X}\|f\|_{L^p_X}. \tag{3.2}$$

Proof. To ensure the existence of the stochastic integrals, we may in the first place replace the stopping time τ by $\tau \wedge T$ and pass to the limit $T \to \infty$ in the end. However, we still write simply τ for convenience.

Let $(\tilde{x}(t))_{t\geq 0}$ be our independent copy of $(x(t))_{t\geq 0}$, also independent of the other processes. From the contraction principle and the decoupling inequality (2.5) we get

$$\begin{aligned}\Big\{\mathsf{E}^\lambda\tilde{\mathsf{E}}\Big|\int_0^\tau \frac{\partial u}{\partial y}\,\mathrm{d}\tilde{y}(s)\Big|^p\Big\}^{1/p} &\leq \Big\{\mathsf{E}^\lambda\tilde{\mathsf{E}}\Big|\int_0^\tau \frac{\partial u}{\partial x}\,\mathrm{d}\tilde{x}(s) + \int_0^\tau \frac{\partial u}{\partial y}\,\mathrm{d}\tilde{y}(s)\Big|^p\Big\}^{1/p} \\ &\leq \beta^+_{p,X}\Big\{\mathsf{E}^\lambda\Big|\int_0^\tau \frac{\partial u}{\partial x}\,\mathrm{d}x(s) + \int_0^\tau \frac{\partial u}{\partial y}\,\mathrm{d}y(s)\Big|^p\Big\}^{1/p}.\end{aligned} \tag{3.3}$$

By Itô's formula we have

$$u[z(\tau)] - u[z(0)] = \int_0^\tau \frac{\partial u}{\partial x}(z(s))\,\mathrm{d}x(s) + \int_0^\tau \frac{\partial u}{\partial y}(z(s))\,\mathrm{d}y(s),$$

and thus we have proved that

$$\Big\{\mathsf{E}^\lambda\tilde{\mathsf{E}}\Big|\int_0^\tau \frac{\partial u}{\partial y}\,\mathrm{d}\tilde{y}(s)\Big|^p\Big\}^{1/p} \leq \beta^+_{p,X}\big(\mathsf{E}^\lambda\left|u[(z(\tau))] - u[z(0)]\right|^p\big)^{1/p}.$$

The right-hand side may now be estimated by the triangle inequality to get the final upper bound

$$\beta^+_{p,X}\big\{\big(\mathsf{E}^\lambda\left|u[(z(\tau))]\right|^p\big)^{1/p} + \big(\mathsf{E}^\lambda\left|u[(z(0))]\right|^p\big)^{1/p}\big\} = \beta^+_{p,X}\big\{\,\|f\|_p + \big\|P^\lambda f\big\|_p\,\big\};$$

we refer to [21, Proposition 3.1] for the last equality, which is a result about non-negative functions (observe the norm signs) and does not involve anything new in the present setting. □

4 Multiplier theorems

We now consider applications of the square function estimates to the boundedness in $L^p(\mathbb{R}^n, X)$ of the Laplacian spectral multipliers

$$m(-\triangle)f = \int_0^\infty m(\lambda)\,\mathrm{d}E(\lambda)f, \tag{4.1}$$

where E is the spectral measure of $-\triangle$. The multipliers which we are going to treat are of the modified Laplace transform type,

$$m(\lambda) = 2\lambda\int_0^\infty tM(t)e^{-2\lambda^{1/2}t}\,\mathrm{d}t, \qquad M \in L^\infty(0,\infty;\mathbb{C}). \tag{4.2}$$

For such m, the operator $m(-\triangle)$ may also be written in terms of the Poisson semigroup P^t as

$$m(-\triangle)f = \int_0^\infty yM(y)\frac{1}{2}\frac{\partial^2}{\partial y^2}P^{2y}f\,\mathrm{d}y. \tag{4.3}$$

There is, of course, nothing new in the boundedness itself for these $m(-\Delta)$. It follows, for instance, from Bourgain's [2] or McConnell's [12] vector-valued Fourier multiplier theorems, and is also a special case of the Littlewood–Paley–Stein theory in [10]. However, the question we wish to address here concerns the actual size of the operator norms $\|m(-\Delta)\|_{\mathscr{L}(L^p(\mathbb{R}^n,X))}$.

Let us start with some preparations. Following Varopoulos [21], we introduce the stochastic multiplier transform, an obvious variant of (4.3),

$$U := \int_0^\tau M[z(s)]\frac{\partial u}{\partial y}[z(s)]\,\mathrm{d}y(s), \qquad u(x,y) = P^y f(x),$$

where $f \in \mathscr{S}(\mathbb{R}^n) \otimes X$, say. With $g_\lambda(x) := \mathsf{E}[U|z(\tau) = x]$, it is shown in [21], for the scalar-valued case, that $g_\lambda \to m(-\Delta)f$ weakly in $L^2(\mathbb{R}^n)$, and hence

$$\|m(-\Delta)f\|_p \leq \limsup_{\lambda\to\infty} \|g_\lambda\|_p \leq \limsup_{\lambda\to\infty} \left(\mathsf{E}^\lambda |U|^p\right)^{1/p}. \tag{4.4}$$

It is clear by linearity that (4.4) remains true for the vector-valued test functions $f \in \mathscr{S}(\mathbb{R}^n) \otimes X$.

The common way of exploiting square functions in the estimation of an operator T such as our $m(-\Delta)$ consists of three steps: the norm of Tf is controlled by a square function norm of the same quantity, this by a square function norm of f alone (which is often easy, if the square function and the operator are suitably related), and this by the norm of f. In the present setting, this approach yields the following easy consequence of Theorem 3.2.

Theorem 4.1. *Let X be a UMD space and $p \in (1,\infty)$. If $M \in L^\infty(0,\infty;\mathbb{C})$, then*

$$\|m(-\Delta)f\|_p \leq \frac{\pi}{2}\beta^-_{p,X}\beta^+_{p,X}\|M\|_\infty \|f\|_p, \qquad f \in L^p(\mathbb{R}^n, X).$$

If $M \in L^\infty(0,\infty;\mathbb{R})$, one may drop the factor $\pi/2$ above.

Proof. According to (4.4), it suffices to estimate $\limsup_{\lambda\to\infty}\left(\mathsf{E}^\lambda |U|^p\right)^{1/p}$. By the decoupling inequality (2.5) and the contraction principle (cf. [17, Lemma 4.11.5] for the complex-valued case), we find that

$$\begin{aligned}\left(\mathsf{E}^\lambda |U|^p\right)^{1/p} &\leq \beta^-_{p,X}\left\{\mathsf{E}^\lambda\tilde{\mathsf{E}}\left|\int_0^\tau M[z(s)]\frac{\partial u}{\partial y}[z(s)]\,\mathrm{d}\tilde{y}(s)\right|^p\right\}^{1/p} \\ &\leq \beta^-_{p,X}\cdot\frac{\pi}{2}\|M\|_\infty\left\{\mathsf{E}^\lambda\tilde{\mathsf{E}}\left|\int_0^\tau \frac{\partial u}{\partial y}[z(s)]\,\mathrm{d}\tilde{y}(s)\right|^p\right\}^{1/p},\end{aligned}$$

where the factor $\pi/2$ may be omitted if M is real-valued. Theorem 3.2 says that the L^p norm on the right is bounded by

$$\beta^+_{p,X}\left(\|f\|_p + \|P^\lambda f\|_p\right) \to \beta^+_{p,X}\|f\|_p \qquad \text{as } \lambda\to\infty,$$

and this completes the proof. □

When hunting for good constants, it turns out that the above three-step approach loses information. It is more desirable to combine all steps together, exploiting the full strength of the UMD property in one decisive strike, instead of dividing its power between the two inequalities (2.3) or their implications in (2.5). Our tool for implementing this strategy is Burkholder's Theorem 2.1, which involves the best constant. (This is, indeed, the beauty and strength of Burkholder's result. The analogous estimates with $\beta^-_{p,X}\beta^+_{p,X}$ and $\pi/2 \cdot \beta^-_{p,X}\beta^+_{p,X}$ in place of $\beta_{p,X}$ and $\beta^{\mathbb{C}}_{p,X}$, respectively, are rather direct consequences of the definitions and convexity.) More precisely, we shall use Theorem 2.1 via the following proposition for stochastic integrals.

Proposition 4.2. *Let X be a UMD space and $p \in (1, \infty)$. Let the process*

$$\Phi : \mathbb{R}_+ \times \Omega \to \mathscr{L}(\mathbb{R}^{n+1}, X)$$

be L^p-stochastically integrable with respect to an $(n+1)$-dimensional Brownian motion $(z(t))_{t\geq 0}$. Let $M_k \in L^\infty(\mathbb{R}_+ \times \Omega; \mathbb{C})$ be adaptive processes with $|M_k(t,\omega)| \leq 1$. Then

$$\Phi M(t,\omega) := (\Phi_k(t,\omega)M_k(t,\omega))_{k=1}^{n+1} \in \mathscr{L}(\mathbb{R}^{n+1}, X)$$

is also L^p-stochastically integrable, and there holds

$$\mathsf{E}\Big\| \int_0^\infty \Phi(t)M(t)\,\mathrm{d}z(t)\Big\|_X^p \leq (\beta^{\mathbb{C}}_{p,X})^p \mathsf{E}\Big\| \int_0^\infty \Phi(t)\,\mathrm{d}z(t)\Big\|_X^p.$$

If $M_k \in L^\infty(\mathbb{R}_+ \times \Omega; \mathbb{R})$, then $\beta^{\mathbb{C}}_{p,X}$ may be replaced by $\beta_{p,X}$.

Proof. Let first Φ be elementary adaptive,

$$\Phi_k(t,\omega) = \sum_{i,j} 1_{]t_{i-1},t_i]}(t)1_{A_{ij}}(\omega)\xi_{ijk}, \qquad A_{ij} \in \mathscr{F}_{t_{i-1}}, \qquad \xi_{ijk} \in X.$$

Then the existence of the asserted integral poses no problem, and

$$\begin{aligned}\mathsf{E}\Big| \int_0^\infty 1_{]t_{i-1},t_i]}1_{A_{ij}}M_k(t)\,\mathrm{d}z_k(t)\Big|^p &\eqsim_p \mathsf{E}\Big(\int_{t_{i-1}}^{t_i} |M_k(t)1_{A_{ij}}|^2\,\mathrm{d}t\Big)^{p/2} \\ &\leq \|M_k\|^p_{L^p(\Omega, L^2(0,T))},\end{aligned}$$

where $T := \max \operatorname{supp}[t \mapsto \Phi(t)] < \infty$. Since $\int_0^\infty \Phi(t)M(t)\,\mathrm{d}z(t)$ consists of a finite number of terms which can be estimated as above, we may assume, by density in $L^p(\Omega, L^2(0,T))$, that the M_k's are elementary adaptive processes. By making a common refinement if necessary, we may take their expansions to involve the same intervals $(t_{i-1}, t_i]$ as that of Φ. Thus

$$\int_0^\infty \Phi(t)M(t)\,\mathrm{d}z(t) = \sum_{i,k}\Big(\sum_\ell 1_{B_{i\ell}}\lambda_{i\ell k}\Big)\Big(\sum_j 1_{A_{ij}}\xi_{ijk}[z_k(t_i) - z_k(t_{i-1})]\Big).$$

The latter factors above, ordered primarily with increasing i and secondarily with increasing k, constitute a martingale difference sequence with respect to the filtration

$$\cdots \subset \mathscr{F}_{t_{i-1}} \subset \sigma(\mathscr{F}_{t_{i-1}}, \mathscr{Z}_{i-1,1}) \subset \cdots \subset \sigma(\mathscr{F}_{t_{i-1}}, \mathscr{Z}_{i-1,1}, \ldots, \mathscr{Z}_{i-1,n}) \subset \mathscr{F}_{t_i} \subset \cdots,$$

where $\mathscr{F}_t := \sigma[z(s) : 0 \leq s \leq t]$ and $\mathscr{Z}_{i-1,k} := \sigma[z_k(t_i) - z_k(t_{i-1})]$. The sum of this difference sequence is $\int_0^\infty \Phi(t)\,\mathrm{d}z(t)$. The first factors, on the other hand, constitute a bounded predictable transforming sequence. Hence the asserted estimate follows from the definition of the UMD constant $\beta_{p,X}$, and the proof is complete for elementary adaptive processes Φ.

If Φ be arbitrary then (cf. Definition 2.2) there is a sequence $(\Phi^{(\nu)})_{\nu=1}^\infty$ of elementary adaptive processes such that

$$\left\langle \Phi_k^{(\nu)}(\cdot,\cdot), \xi^* \right\rangle \to \langle \Phi_k(\cdot,\cdot), \xi^* \rangle \quad \text{in} \quad L^p(\Omega, L^2(0,\infty))$$

for all $k = 1, \ldots, n+1$ and $\xi^* \in X^*$, and

$$\int_0^\infty \Phi^{(\nu)}(t)\,\mathrm{d}z(t) \to \int_0^\infty \Phi(t)\,\mathrm{d}z(t) \quad \text{in} \quad L^p(\Omega, X)$$

as $\nu \to \infty$. Then also

$$\left\langle \Phi_k^{(\nu)} M_k(\cdot,\cdot), \xi^* \right\rangle \to \langle \Phi_k M_k(\cdot,\cdot), \xi^* \rangle \quad \text{in} \quad L^p(\Omega, L^2(0,\infty)),$$

and the fact that $\int_0^\infty \Phi^{(\nu)}(t)\,\mathrm{d}z(t)$ is Cauchy, together with the first part of the proof, implies that $\int_0^\infty \Phi^{(\nu)}(t)M(t)\,\mathrm{d}z(t)$ is Cauchy, hence convergent, in $L^p(\Omega, X)$. The functions $\Phi^{(\nu)}M$ are typically not elementary progressive but, as observed in the first part of the proof, we may, for each ν, find an elementary progressive $M^{(\nu)}$ so that

$$\mathsf{E}\left|\int_0^\infty \Phi^{(\nu)}(t)M^{(\nu)}(t)\,\mathrm{d}y(t) - \int_0^\infty \Phi^{(\nu)}(t)M(t)\,\mathrm{d}y(t)\right|^p < \epsilon_\nu^p,$$

and then also

$$\left\| \left\langle \Phi_k^{(\nu)} M_k^{(\nu)}(\cdot,\cdot), \xi^* \right\rangle - \left\langle \Phi_k^{(\nu)} M_k(\cdot,\cdot), \xi^* \right\rangle \right\|_{L^p(\Omega, L^2(0,\infty))} < K_p\, \epsilon_\nu\, |\xi^*|$$

for $k = 1, \ldots, n+1$ and $\xi^* \in X^*$. In this way, we can realize ΦM as the limit of the elementary adaptive functions $\Phi^{(\nu)}M^{(\nu)}$ in the appropriate norms so as to extend the required estimate to all L^p-stochastically integrable Φ. □

Let us point out that the conclusion of Proposition 4.2 with some constant $c_{p,X}$ in place of $\beta_{p,X}^{\mathbb{C}}$ and $\beta_{p,X}$ is an immediate consequence of results from [16], but this is not what we want here. With the help of Proposition 4.2, we reach the following improvement of Theorem 4.1 (recall that $2/\pi \cdot \beta_{p,X}^{\mathbb{C}} \leq \beta_{p,X} \leq \beta_{p,X}^- \beta_{p,X}^+$):

Theorem 4.3. *Let X be a UMD space and $p \in (1,\infty)$. If $M \in L^\infty(0,\infty;\mathbb{C})$, then*

$$\|m(-\Delta)f\|_p \leq \beta_{p,X}^{\mathbb{C}} \|M\|_\infty \|f\|_p, \qquad f \in L^p(\mathbb{R}^n, X).$$

If $M \in L^\infty(0,\infty;\mathbb{R})$, then one may replace $\beta_{p,X}^{\mathbb{C}}$ by $\beta_{p,X}$.

Proof. We start by observing that

$$U = \sum_{k=1}^{n} \int_0^T \frac{\partial u}{\partial x_k}[z(t)] \cdot 0 \cdot \mathrm{d}x_k(t) + \int_0^T \frac{\partial u}{\partial y}[z(t)] \cdot M[z(t)] \cdot \mathrm{d}y(t)$$

is an integral of the kind considered in Proposition 4.2 with $\Phi(t) = \nabla u[z(t)]$ and $(0, \dots, 0, M[z(t)])$ in place of $M(t)$. Thus, an application of Proposition 4.2 yields

$$\left(\mathsf{E}^\lambda |U|^p\right)^{1/p} \le \beta^{\mathbb{C}}_{p,X} \|M\|_\infty \left\{\mathsf{E}^\lambda \left| \int_0^T \frac{\partial u}{\partial x}[z(s)]\,\mathrm{d}x(s) + \int_0^T \frac{\partial u}{\partial y}[z(s)]\,\mathrm{d}y(s)\right|^p\right\}^{1/p},$$

or the similar estimate with $\beta_{p,X}$ under the reality assumption. The right-hand side is of the form already handled in the proof of Theorem 3.2. □

Corollary 4.4. *Let X be a UMD space and $p \in (1, \infty)$. Then*

$$\left\|(-\Delta)^{i\gamma} f\right\|_p \le \frac{2\beta^{\mathbb{C}}_{p,X}}{|\Gamma[2(1 - i\gamma)]|} \|f\|_p, \qquad f \in L^p(\mathbb{R}^n, X).$$

Proof. This is immediate from the identities $\lambda^{i\gamma} = 2\lambda \int_0^\infty tM(t)e^{-2\lambda^{1/2}t}\,\mathrm{d}t$, $M(t) = 2 \cdot (2t)^{-i2\gamma}/\Gamma[2(1 - i\gamma)]$ and Theorem 4.3. □

Corollary 4.4 implies in particular that

$$\limsup_{\gamma \to 0} \left\|(-\Delta)^{i\gamma}\right\|_{\mathscr{L}(L^p_X(\mathbb{R}))} \le 2\beta^{\mathbb{C}}_{p,X}. \tag{4.5}$$

A comparison of this with the following result of Guerre-Delabriere shows that this estimate, and therefore the ones from which it was derived, are close to optimal:

Theorem 4.5 ([9]). *Let X be a Banach space with $(-\Delta)^{i\gamma} \in \mathscr{L}(L^p(\mathbb{R}, X))$ for $\gamma \neq 0$. Then X is a UMD space, and more precisely*

$$\beta^{\mathbb{C}}_{p,X} \le \liminf_{\gamma \to 0} \left\|(-\Delta)^{i\gamma}\right\|_{\mathscr{L}(L^p(\mathbb{R},X))}. \tag{4.6}$$

The precise estimate above is not explicitly formulated in [9], but an investigation of the proof reveals that a version of (4.6) with $\beta_{p,X}$ in place of $\beta^{\mathbb{C}}_{p,X}$ is actually proved there, and Theorem 4.5 as stated follows *mutatis mutandis*.

5 Final comments

The results (4.5) and (4.6) show that the two different numerical indicators of the UMD property ($\beta_{p,X}$ and $\liminf_{\gamma\to 0} \left\|(-\Delta)^{i\gamma}\right\|_{\mathscr{L}(L^p(\mathbb{R},X))}$) are linearly bounded in terms of each other. This is in contrast to the best known relations between $\beta_{p,X}$ and $\alpha_{p,X} := \|H\|_{\mathscr{L}(L^p(\mathbb{R},X))}$, the norm of the Hilbert transform, which are

$$\beta_{p,X} \le \alpha^2_{p,X} \quad [1], \qquad \alpha_{p,X} \le \beta^-_{p,X}\beta^+_{p,X} \le \beta^2_{p,X} \quad [4, 7].$$

It is an interesting open problem whether one or both of these could be improved to linear bounds. The Hilbert transform, which is not a spectral multiplier of the Laplacian, is not in the scope of the present approach, although some of the proofs of its boundedness are not very far from the ones in this paper; still, two decoupling estimates seem always unavoidable to control it.

An obvious consequence of Theorem 4.3 (or Theorem 4.1) is the fact that the spectral multipliers of $-\triangle$ of the treated type (in particular, the imaginary powers $(-\triangle)^{i\gamma}$) satisfy norm bounds in $L^p(\mathbb{R}^n, X)$ which are independent of n. Such dimension-free estimates have been of interest to many authors, and the mentioned result is very classical for $X = \mathbb{C}$; cf. [20]. Since the exact value of the UMD constants of every Hilbert space H have been obtained by Burkholder (see [5]), the result being

$$\beta_{p,H} = \beta^{\mathbb{C}}_{p,H} = p^* - 1 := \max\{p, p'\} - 1 = \max\{p-1, \frac{1}{p-1}\},$$

Theorem 4.3 implies the following corollary. I do not know if these (in view of Theorem 4.5 close to optimal) explicit bounds have been proved before.

Corollary 5.1. *Let $p \in (1, \infty)$ and m be as in* (4.2). *Then* $\|m(-\triangle)\|_{\mathscr{L}(L^p(\mathbb{R}^n))} \leq \|M\|_\infty (p^* - 1)$, *and in particular* $\|(-\triangle)^{i\gamma}\|_{\mathscr{L}(L^p(\mathbb{R}^n))} \leq 2(p^* - 1)/\,|\Gamma[2(1 - i\gamma)]|$.

Let me finally mention that the reader may also find the recent paper of Dragičević and Volberg [6] quite interesting. While they work in finite-dimensional Hilbert spaces, proving a different square function estimate for controlling another family of operators, there is considerable ideological similarity between their methods and the present ones.

Acknowledgments. The results of this paper (in a slightly weaker form which I managed to improve since then) were first presented in the wonderful meeting "Banach spaces and their applications in analysis", a conference in honour of Nigel Kalton's 60th birthday. I wish to thank the organizers for the invitation and financial support to attend the meeting. Partial funding from the Finnish Academy of Science and Letters (Vilho, Yrjö and Kalle Väisälä Foundation) is gratefully acknowledged. Most of this research was carried out during my previous employment at the University of Turku.

References

[1] J. Bourgain, *Some remarks on Banach spaces in which martingale difference sequences are unconditional*, Ark. Mat. 21 (1983), pp. 163–168.

[2] ———, *Vector-valued singular integrals and the H^1-BMO duality*, Probability theory and harmonic analysis (J.-A. Chao and W. A. Woyczyński, eds.), pp. 1–19, Marcel Dekker, New York (1986).

[3] D. L. Burkholder, *A geometrical characterization of Banach spaces in which martingale difference sequences are unconditional*, Ann. Probab. 9 (1981), pp. 997–1011.

[4] ———, *A geometric condition that implies the existence of certain singular integrals of Banach-space-valued functions*, Conf. Harmonic Anal. in Honor of Antoni Zygmund (Chicago, 1981), pp. 270–886, Wadsworth, Belmont (1983).

[5] ———, *Martingales and singular integrals in Banach spaces*, Handbook of the geometry of Banach spaces, Vol. I, pp. 233–269, North-Holland, Amsterdam (2001).

[6] O. Dragičević and A. Volberg, *Bellman functions and dimensionless estimates of Littlewood-Paley type*, J. Operator Theory 56 (2006), pp. 167–198.

[7] D. J. H. Garling, *Brownian motion and UMD-spaces*, Probability and Banach spaces (Zaragoza, 1985), pp. 36–49, Lecture Notes in Math., 1221, Springer, Berlin (1986).

[8] ———, *Random martingale transform inequalities*, Probability in Banach spaces 6 (Sandbjerg, 1986), pp. 101–119, Progr. Probab., 20, Birkhäuser, Boston, MA (1990).

[9] S. Guerre-Delabriere, *Some remarks on complex powers of* $(-\Delta)$ *and UMD spaces*, Illinois J. Math. 35 (1991), pp. 401–407.

[10] T. P. Hytönen, *Littlewood–Paley–Stein theory for semigroups in UMD spaces*, Rev. Mat. Iberoamericana, to appear.

[11] N. J. Kalton and L. Weis, *The* H^∞*-functional calculus and square function estimates*, manuscript (2004).

[12] T. R. McConnell, *On Fourier multiplier transformations of Banach-valued functions*, Trans. Amer. Math. Soc. 285 (1984), pp. 739–757.

[13] ———, *Decoupling and stochastic integration in UMD Banach spaces*, Probab. Math. Statist. 10 (1989), pp. 283–295.

[14] P. A. Meyer, *Démonstration probabiliste de certaines inéqualités de Littlewood–Paley*, Sém. Probab. X, Lecture Notes in Math. 511, Springer-Verlag, Berlin, Heidelberg, New York (1976).

[15] J. M. A. M. van Neerven and L. Weis, *Stochastic integration of functions with values in a Banach space*, Studia Math. 166 (2005), pp. 131–170.

[16] J. M. A. M. van Neerven, M. C. Veraar, and L. Weis, *Stochastic integration in UMD Banach spaces*, Ann. Probab., to appear.

[17] A. Pietsch and J. Wenzel, *Orthonormal systems and Banach space geometry*. Encyclopedia of Mathematics and its Applications, 70. Cambridge University Press, Cambridge, 1998.

[18] J. Rosiński and Z. Suchanecki, *On the space of vector-valued functions integrable with respect to the white noise*, Colloq. Math. 43 (1980), pp. 183–201.

[19] E. M. Stein, *Topics in Harmonic Analysis Related to Littlewood–Paley Theory*, Ann. Math. Studies 63, Princeton Univ. Press, Princeton, N.J. (1970).

[20] ———, *Some results in harmonic analysis in* $\mathbb{R}^n$*, for* $n \to \infty$, Bull. Amer. Math. Soc. (N.S.) 9 (1983), pp. 71–73.

[21] N. Th. Varopoulos, *Aspects of probabilistic Littlewood–Paley theory*, J. Funct. Anal. 38 (1980), pp. 25–60.

Author information

Tuomas P. Hytönen, Department of Mathematics and Statistics, University of Helsinki, Gustaf Hällströmin katu 2b, FI-00014 Helsinki, Finland.
Email: `tuomas.hytonen@helsinki.fi`

Banach Spaces and their Applications in Analysis, 357–373

The q-concavity and q-convexity constants in Lorentz spaces

Anna Kamińska and Anca M. Parrish

Abstract. We study the q-convexity and q-concavity constants of quasi-Banach Lorentz spaces $\Lambda_{p,w}$, where $0 < p < \infty$ and w is a locally integrable positive weight function. We provide exact constants in the case when w is decreasing and their close estimates in the case when w is increasing. For the latter purpose we develop the notion of increasing rearrangement of a measurable function and prove several of its properties.

Key words. q-convexity constant, q-concavity constant, Lorentz space, increasing rearrangement.

AMS classification. 46B30, 46B20.

The purpose of this paper is to find exact q-convexity and q-concavity constants for quasi-Banach Lorentz spaces $\Lambda_{p,w}$, where $0 < p < \infty$ and w is a locally integrable positive weight function.

The criteria for Lorentz spaces $\Lambda_{p,w}$ to be q-concave or q-convex have been known already for some time and they have been developed by the effort of several authors [2, 3, 6, 7, 9, 10, 11, 12]. The most general and complete solution to this problem has been provided in [6]. However, in neither of these papers the exact convexity or concavity constants have been found. It was Jameson in [4] who first computed the exact constants in the case of sequence Lorentz spaces with a decreasing weight sequence. Inspired by his paper, we study here the best q-concavity and q-convexity constants of the Lorentz (function) spaces $\Lambda_{p,w}$ in the cases when the weight function w is either decreasing or increasing. In the first case we are able to provide exact constants, while in the second case we obtain their close estimates.

In the preliminary section 1, we set up notations, definitions and results needed further. In section 2, we introduce and investigate increasing rearrangement of a measurable function, showing several of its properties, which we shall further need to study the spaces induced by increasing weights. The main section 3 is devoted to the convexity and concavity constants of $\Lambda_{p,w}$. In the first part we provide the formulae for exact constants in the case when w is decreasing, that is when $\Lambda_{p,w}$ is a Banach space. While finding convexity constants in this case is rather an easy task, the concavity constants require more effort. We handle this problem by applying duality method, used also by Jameson in [4]. In the second part we investigate the spaces corresponding to increasing weights. In this case, the spaces $\Lambda_{p,w}$ are not necessarily normable and their duals could be trivial, and so we cannot apply the duality method used before. Instead we shall apply a different method, which leads us to some estimates of convexity constants. We close the paper by a corollary where we present the convexity and concavity constants in the classical Lorentz spaces $L_{q,p}$.

1 Preliminaries

Let $I = (0, \infty)$ and let $L^0 \equiv L^0(I, |\cdot|)$ be the set of all Lebesgue measurable functions $f : I \to \mathbb{R}$, where $|\cdot|$ is a Lebesgue measure on $\mathbb{R}$. The *distribution function* d_f of a function $f \in L^0$ is given by $d_f(\lambda) = |\{x \in I : |f(x)| > \lambda\}|$, for all $\lambda \geq 0$. We say that two functions $f, g \in L^0$ are *equimeasurable* and we denote it by $f \sim g$ if $d_f(\lambda) = d_g(\lambda)$, for all $\lambda \geq 0$. For $f \in L^0$ we define its *decreasing rearrangement* as $f^*(t) = \inf\{s > 0 : d_f(s) \leq t\}, t > 0$.

Definition 1.1. Given $0 < p < \infty$ and a measurable weight function $w : I \to (0, \infty)$, the Lorentz space $\Lambda_{p,w}$ is a subspace of L^0 such that

$$\|f\| = \|f\|_{p,w} := \left(\int_0^\infty f^{*p} w\right)^{1/p} = \left(\int_0^\infty f^{*p}(t) w(t) dt\right)^{1/p} < \infty.$$

We shall assume further that w satisfies the conditions

$$\int_0^\infty w = \infty \quad \text{and} \quad W(t) := \int_0^t w < \infty, \quad \text{for all} \quad t \in I.$$

We assume also that W satisfies the Δ_2-condition, i.e. $W(2t) \leq KW(t)$ for all $t > 0$ and some $K > 0$. Then $\|\cdot\|_{p,w}$ is a quasi-norm, and $(\Lambda_{p,w}, \|\cdot\|_{p,w})$ is a quasi-Banach space [6]. If w is decreasing, then $\|\cdot\|_{p,w}$ is a norm [1, 5] and

$$\|f\|_{p,w} = \left(\sup_{v \sim w} \int_0^\infty |f|^p v\right)^{1/p}. \tag{1.1}$$

In this case, $\Lambda_{1,w}$ is a separable Banach space and its dual is the Marcinkiewicz space M_W [5, Theorem 5.2] consisting of all functions $f \in L^0$ satisfying

$$\|f\|_{M_W} = \sup_{x > 0} \frac{\int_0^x f^*(t) dt}{W(x)} < \infty.$$

Observe that by the Monotone Convergence Theorem, the functional $\|\cdot\|_{p,w}$ has the *Fatou property* i.e. for any $f_n, f \in L^0$ with $0 \leq f_n \uparrow f$ a.e., we have that $\|f_n\|_{p,w} \uparrow \|f\|_{p,w}$. Recall that for $0 < p, q < \infty$, the classical Lorentz space $L_{q,p}$ is obtained from $\Lambda_{p,w}$ setting $w(t) = t^{p/q-1}$. Then for $f \in L_{q,p}$,

$$\|f\|_{q,p} := \|f\|_{p,w} = \left(\int_I (t^{1/q} f^*(t))^p \frac{dt}{t}\right)^{1/p} < \infty,$$

and $\|\cdot\|_{q,p}$ is a quasi-norm [1, 5].

Definition 1.2. For a quasi-normed lattice $(E, \|\cdot\|)$,

(i) the q-concavity constant $M_{(q)}(E)$ is the least constant M_1 such that

$$\left(\sum_{i=1}^n \|f_i\|^q\right)^{\frac{1}{q}} \leq M_1 \left\|\left(\sum_{i=1}^n |f_i|^q\right)^{\frac{1}{q}}\right\| \quad \text{for all } f_1, \ldots, f_n \in E;$$

(ii) the q-convexity constant $M^{(q)}(E)$ is the least constant M_2 such that

$$\Big\|\Big(\sum_{i=1}^{n}|f_i|^q\Big)^{\frac{1}{q}}\Big\| \le M_2\Big(\sum_{i=1}^{n}\|f_i\|^q\Big)^{\frac{1}{q}} \quad \text{for all } f_1,\dots,f_n \in E.$$

Clearly both constants $M_{(q)}(E)$ and $M^{(q)}(E)$ are bigger than or equal to one. Moreover they behave in a monotone way in the sense that for $0 < r < q$ we have

$$M_{(q)}(E) \le M_{(r)}(E) \quad \text{and} \quad M^{(q)}(E) \ge M^{(r)}(E).$$

It is also well known [8, Proposition 1.d.4] that if E is a Banach lattice then $M_{(q)}(E) = M^{(q^*)}(E^*)$, where q^* is such that $1/q + 1/q^* = 1$ and E^* is the dual of E. Given any quasi-normed lattice E, recall that its p-convexification is the space $E^{(p)} = \{x : |x|^p \in E\}$ equipped with the quasi-norm $\|x\|_{E^{(p)}} = \||x|^p\|^{1/p}$. Then we can easily calculate that

$$M_{(q)}(E^{(p)}) = M_{(q/p)}(E) \quad \text{and} \quad M^{(q)}(E^{(p)}) = M^{(q/p)}(E).$$

2 The increasing rearrangement of a function

In this section we shall consider the measure space (Ω, Σ, μ), where μ a σ-finite measure on Σ. By $L^0(\mu)$ we denote the space of all μ-measurable functions $f : \Omega \to \mathbb{R}$. Recall that $f^*(t) = \inf\{s > 0 : d_f(s) \le t\}$, $t > 0$, where $d_f(s) = \mu\{x \in \Omega : |f(x)| > s\}$, $s \ge 0$. We shall define an increasing rearrangement f_* of a μ-measurable function f and prove a number of its properties, analogous to those well known for decreasing rearrangement f^*. We shall apply the results of this section to study convexity constants of $\Lambda_{p,w}$ in the case when w is increasing.

Definition 2.1. Given $f \in L^0(\mu)$, for all $\lambda > 0$, define

$$\gamma_f(\lambda) = \mu\{x \in \operatorname{supp} f : |f(x)| < \lambda\}.$$

We say that two functions f and g are equivalent with respect to γ and we denote it by $f \sim_\gamma g$ if $\gamma_f = \gamma_g$.

Definition 2.2. Let $f \in L^0(\mu)$. We define the *increasing rearrangement* f_* of f as

$$f_*(t) = \begin{cases} \sup\{\lambda \ge 0 : \gamma_f(\lambda) \le t\}, & \text{if } t \in [0, \mu(\operatorname{supp} f)); \\ 0, & \text{if } t \ge \mu(\operatorname{supp} f). \end{cases} \tag{2.1}$$

Some of the properties of the increasing rearrangement are stated in the following proposition, which we leave without proof. The proofs are standard and could be modelled on analogous properties of f^* ([1]).

Proposition 2.3. *Suppose $f, g, f_n \in L^0(\mu)$ have supports of finite measure. The increasing rearrangement f_* of f is a function that is non-negative, increasing and right-continuous on $(0, \mu(\operatorname{supp} f))$. Furthermore:*

1. If $|g| \leq |f|$ *a.e. and* $\operatorname{supp} g = \operatorname{supp} f$*, then* $g_* \leq f_*$.
2. If $|f_n| \downarrow |f|$ *a.e. and* $\operatorname{supp} f_n = \operatorname{supp} f$*, then* $(f_n)_* \downarrow f_*$.
3. If $f > 0$ *a.e. and* $\gamma_f(\lambda) < \infty$ *for all* $\lambda > 0$*, then setting* $\Omega_n = \{x : f(x) < n\}$ *and* $f^{(n)} = f\chi_{\Omega_n}$, $n \in \mathbb{N}$*, we have* $(f^{(n)})_*(t) = f_*\chi_{(0,\mu\Omega_n)}(t)$*, for all* $n \in \mathbb{N}$.
4. If $r < 0$, $f > 0$ *a.e. and* $\gamma_f(\lambda) < \infty$ *for every* $\lambda > 0$*, then* $(f^r)^* = (f_*)^r$ *a.e.*

Lemma 2.4. *Let* $f \in L^0(\mu)$ *be such that* $f > 0$ *a.e. and* $\gamma_f(\lambda) < \infty$ *for every* $\lambda > 0$*. Then for any measurable set* E *we have*

$$\int_E f\, d\mu \geq \int_0^{\mu E} f_*(t)dt. \tag{2.2}$$

Proof. Let us first consider a non-negative simple function f with $\mu(\operatorname{supp} f) < \infty$. We can assume that $f(x) = \sum_{i=1}^n b_i\chi_{F_i}(x)$ for some $b_i > 0$ and increasing sequence of measurable sets F_i , $i = 1, \ldots, n$. It is clear that for any measurable set $E \subset \operatorname{supp} f$,

$$\int_E f\, d\mu = \int_E \Big(\sum_{i=1}^n b_i\chi_{F_i}\Big)d\mu = \sum_{i=1}^n b_i\mu(E \cap F_i).$$

On the other hand $f_* = \sum_{i=1}^n b_i\chi_{(\mu F_n - \mu F_i, \mu F_n)}$, and so

$$\int_0^{\mu E} f_*(t)dt = \sum_{i=1}^n b_i|(0, \mu E) \cap (\mu F_n - \mu F_i, \mu F_n)|.$$

But

$$(0, \mu E) \cap (\mu F_n - \mu F_i, \mu F_n) = \begin{cases} \emptyset, & \text{if } \mu E \leq \mu F_n - \mu F_i; \\ (\mu F_n - \mu F_i, \mu E), & \text{if } \mu E > \mu F_n - \mu F_i. \end{cases}$$

So

$$|(0, \mu E) \cap (\mu F_n - \mu F_i, \mu F_n)| = \begin{cases} 0, & \text{if } \mu E \leq \mu F_n - \mu F_i; \\ \mu E - \mu F_n + \mu F_i, & \text{if } \mu E > \mu F_n - \mu F_i. \end{cases}$$

Therefore in view of $E \subset \operatorname{supp} f$, we have that

$$\begin{aligned} |(0, \mu E) \cap (\mu F_n - \mu F_i, \mu F_n)| &\leq \mu E - \mu F_n + \mu F_i \\ &= \mu F_i - (\mu F_n - \mu E) \\ &= \mu(F_i \cap E) + \mu(F_i \setminus E) - \mu(F_n \setminus E) \\ &\leq \mu(F_i \cap E), \end{aligned}$$

because $\mu(F_i \setminus E) - \mu(F_n \setminus E) \leq 0$. Thus for $E \subset \operatorname{supp}\ f$,

$$\int_0^{\mu E} f_*(t)dt \leq \sum_{i=1}^n b_i\mu(E \cap F_i) = \int_E f\, d\mu.$$

Now consider f that is bounded with $\mu(\operatorname{supp} f) < \infty$. Let $E \subset \operatorname{supp} f$. Hence there exists a decreasing sequence of simple functions $f_n \downarrow f$ such that $\operatorname{supp} f_n = \operatorname{supp} f$, $n \in \mathbb{N}$. It follows by the Lebesgue dominated convergence theorem and the simple function case that,

$$\int_E f\,d\mu = \lim_{n\to\infty}\int_E f_n\,d\mu \geq \lim_{n\to\infty}\int_0^{\mu E}(f_n)_*(t)dt.$$

On the other hand, since $f_n \geq f$ and $\operatorname{supp} f_n = \operatorname{supp} f$, by Proposition 2.3 we have $(f_n)_* \geq f_*$ for all $n \in \mathbb{N}$, which implies that

$$\int_E f\,d\mu \geq \lim_{n\to\infty}\int_0^{\mu E}(f_n)_*(t)dt \geq \int_0^{\mu E} f_*(t)dt.$$

Let now $f > 0$ a.e., measurable and such that $\mu\Omega_n < \infty$, $n \in \mathbb{N}$, where $\Omega_n = \{x : f(x) < n\}$. Define the truncation $f^{(n)} = f\chi_{\Omega_n}$. By Proposition 2.3, we have $(f^{(n)})_* = f_*\chi_{(0,\mu\Omega_n)}$ a.e. Then for any measurable set E and $n \in \mathbb{N}$,

$$\begin{aligned}\int_E f\,d\mu &\geq \int_{E\cap\Omega_n} f^{(n)}\,d\mu \geq \int_0^{\mu(E\cap\Omega_n)}(f^{(n)})_*(t)\,dt\\ &= \int_0^{\mu E} f_*(t)\chi_{(0,\mu\Omega_n)\cap(0,\mu(E\cap\Omega_n))}(t)\,dt = \int_0^{\mu E} f_*(t)\chi_{(0,\mu(E\cap\Omega_n))}(t)\,dt.\end{aligned}$$

But $f_*\chi_{(0,\mu(E\cap\Omega_n))} \uparrow f_*\chi_{(0,\mu E)}$, and by limit argument we conclude that

$$\int_E f\,d\mu \geq \int_0^{\mu E} f_*(t)\,dt.$$

This completes the proof of the lemma. □

Theorem 2.5. *Let $f, g \in L^0(\mu)$ be non-negative functions such that $g > 0$ a.e. and $\gamma_g(\lambda) < \infty$ for every $\lambda > 0$. Then*

$$\int_\Omega fg\,d\mu \geq \int_0^\infty f^*(t)g_*(t)dt. \tag{2.3}$$

Proof. Clearly, it is enough to show this for simple functions. If f is a simple function, then there exist $a_1, \dots, a_n > 0$ and $E_1 \subset E_2 \subset \cdots \subset E_n$ such that $\mu E_n < \infty$,

$$f(x) = \sum_{i=1}^n a_i\chi_{E_i}(x) \quad\text{and}\quad f^*(t) = \sum_{i=1}^n a_i\chi_{[0,\mu E_i)}(t).$$

Applying Lemma 2.4 we get

$$\begin{aligned}\int_\Omega fg\,d\mu &= \sum_{i=1}^n a_i\int_{E_i} g\,d\mu \geq \sum_{i=1}^n a_i\int_0^{\mu E_i} g_*(t)dt\\ &= \sum_{i=1}^n a_i\int_0^\infty \chi_{(0,\mu E_i)}(t)g_*(t)dt = \int_0^\infty f^*(t)g_*(t)dt,\end{aligned}$$

as we wanted. □

Theorem 2.6. *Let μ be nonatomic and finite on Ω. Let $f \in L^0(\mu)$ be arbitrary and $g \in L^0(\mu)$ be bounded. Then there exists $\widetilde{g}$ such that $\widetilde{g} \sim_\gamma g$ and*

$$\int_0^\infty f^*(t) g_*(t) dt = \int_\Omega |f\widetilde{g}|\, d\mu.$$

If $g > 0$ a.e., then we can also find $\widetilde{g} > 0$ a.e.

Proof. Without loss of generality we may assume that f and g are non-negative. Let (g_n) be a sequence of simple functions such that $g_n \downarrow g$, g_1 is bounded and $\operatorname{supp} g_n = \operatorname{supp} g$, $n \in \mathbb{N}$. Then by Proposition 2.3 we have that $(g_n)_* \downarrow g_*$. We will construct $\widetilde{g}_n$ with the properties that $\widetilde{g}_n \sim_\gamma g_n$, $(\widetilde{g}_n)$ is decreasing and for all $n \in \mathbb{N}$,

$$\int_\Omega f\widetilde{g}_n\, d\mu = \int_0^\infty f^*(t)(\widetilde{g}_n)_*(t) dt.$$

Then defining $\widetilde{g}(x) = \lim_{n\to\infty} \widetilde{g}_n(x)$, by the Lebesgue convergence theorem,

$$\begin{aligned}\int_\Omega f\widetilde{g}\, d\mu &= \lim_{n\to\infty} \int_\Omega f\widetilde{g}_n\, d\mu = \lim_{n\to\infty} \int_0^\infty f^*(t)(\widetilde{g}_n)_*(t) dt \\ &= \lim_{n\to\infty} \int_0^\infty f^*(t)(g_n)_*(t)\, dt = \int_0^\infty f^*(t) g_*(t)\, dt,\end{aligned}$$

which proves the hypothesis.

Now fix $n = 1, 2, \ldots$, and for simplicity let $h = g_n$. Then h can be written as $h(x) = \sum_{j=1}^n b_j \chi_{F_j}(x)$, where $b_j > 0$, $j = 1, \ldots, n$, and $F_1 \subset F_2 \subset \cdots \subset F_n$. By [1, Lemma 2.5], there exist $E_1 \subset E_2 \subset \cdots \subset E_n$ such that $\mu E_j = \mu F_j$, for all $j = 1, \ldots, n$ and

$$\int_{E_j} f\, d\mu = \int_{\mu E_n - \mu E_j}^{\mu E_n} f^*(t) dt.$$

Let

$$\widetilde{h}(x) = \sum_{j=1}^n b_j \chi_{E_j}(x).$$

Then $\widetilde{h}_*(t) = \sum_{j=1}^n b_j \chi_{[\mu E_n - \mu E_j, \mu E_n)}(t) = h_*(t)$. Hence

$$\begin{aligned}\int_\Omega f\widetilde{h}\, d\mu &= \sum_{j=1}^n b_j \int_{E_j} f\, d\mu = \sum_{j=1}^n b_j \int_{\mu E_n - \mu E_j}^{\mu E_n} f^*(t) dt \\ &= \sum_{j=1}^n b_j \int_0^\infty f^*(t) \chi_{[\mu F_n - \mu F_j, \mu F_n]}(t) dt = \int_0^\infty f^*(t) h_*(t) dt.\end{aligned}$$

Since $h = g_n$ and $\widetilde{h} \sim_\gamma h$, we have $\widetilde{g}_n \sim_\gamma g_n$ such that

$$\int_\Omega f\widetilde{g}_n\, d\mu = \int_0^\infty f^*(t)(\widetilde{g}_n)_*(t) dt.$$

The proof is complete since by its construction the sequence $(\widetilde{g}_n)$ is decreasing. □

Theorem 2.7. *Let μ be nonatomic and let $g \in L^0(\mu)$ be such that for every $\lambda > 0$, $\gamma_g(\lambda) < \infty$ and $g > 0$ a.e. Then for every $f \in L^0(\mu)$ with $\mu(\operatorname{supp} f) < \infty$ and $\int_\Omega fg\, d\mu < \infty$, it holds*

$$\int_0^\infty f^*(t)g_*(t)dt = \inf\Big\{ \int_\Omega |f\widetilde{g}|\, d\mu : \widetilde{g} \sim_\gamma g, \widetilde{g} > 0 \text{ a.e.}\Big\}.$$

Proof. Let $\operatorname{supp} f = F$, $\mu F < \infty$ and let for $n \in \mathbb{N}$, $\Omega_n := \{x : g(x) < n\}$. Then $\Omega_1 \subset \Omega_2 \subset \dots$, $\mu\Omega_n < \infty$ for all $n \in \mathbb{N}$, and $\bigcup_{n=1}^\infty \Omega_n = \Omega$. Let us consider the sequence $(g\chi_{\Omega_n})$. By Proposition 2.3, for all $n \in \mathbb{N}$,

$$(g\chi_{\Omega_n})_* = g_*\chi_{(0,\mu\Omega_n)} \le g_*. \tag{2.4}$$

Assume now without loss of generality that f is non-negative. By Theorem 2.5, we have for all $\widetilde{g} \sim_\gamma g$,

$$\int_0^\infty f^*(t)g_*(t)dt \le \int_\Omega f|\widetilde{g}|\, d\mu.$$

In particular if we take $\widetilde{g} = g$,

$$\int_0^\infty f^*(t)g_*(t)dt \le \int_\Omega fg\, d\mu < \infty.$$

Now take any $0 < \alpha < \infty$ such that

$$\int_0^\infty f^*(t)g_*(t)dt < \alpha. \tag{2.5}$$

By the assumption that $\mu F < \infty$, $\lim_{n\to\infty} \mu(F\cap(\Omega\setminus\Omega_n)) = 0$, and by the integrability of fg, there exists $N \in \mathbb{N}$ such that

$$\int_{F\cap(\Omega\setminus\Omega_N)} fg\, d\mu < \varepsilon, \tag{2.6}$$

where $\varepsilon = \alpha - \int_0^\infty f^*(t)g_*(t)dt > 0$. We have by (2.4) that

$$\int_0^\infty (f\chi_{\Omega_N})^*(t)(g\chi_{\Omega_N})_*(t)dt \le \int_0^\infty f^*(t)g_*(t)dt < \alpha.$$

By Theorem 2.6 for finite measure applied to the measure space (Ω_N, μ), there exists $0 \le h \in L^0(\mu)$ with $\operatorname{supp} h = \Omega_N$ and such that $h \sim_\gamma g\chi_{\Omega_N}$ and

$$\int_0^\infty (f\chi_{\Omega_N})^*(t)(g\chi_{\Omega_N})_*(t)dt = \int_\Omega (f\chi_{\Omega_N})h\, d\mu = \int_{\Omega_N} fh\, d\mu = \int_\Omega fh\, d\mu.$$

Hence

$$\int_\Omega fh\, d\mu \le \int_0^\infty f^*(t)g_*(t)dt. \tag{2.7}$$

Let $\widetilde{g} = h\chi_{\Omega_N} + g\chi_{\Omega\setminus\Omega_N}$. It is clear that $\widetilde{g} > 0$ and $\gamma_{\widetilde{g}} = \gamma_g$, and so $\widetilde{g} \sim_\gamma g$. Finally by (2.6) and (2.7),

$$\int_\Omega f\widetilde{g}\,d\mu = \int_{F\cap\Omega_N} fh\,d\mu + \int_{F\cap(\Omega\setminus\Omega_N)} fg\,d\mu = \int_\Omega fh\,d\mu + \int_{F\cap(\Omega\setminus\Omega_N)} fg\,d\mu$$
$$< \int_0^\infty f^*(t)g_*(t)dt + \varepsilon = \int_0^\infty f^*(t)g_*(t)dt + \alpha - \int_0^\infty f^*(t)g_*(t)dt = \alpha,$$

and in view of Theorem 2.5, the proof is complete. □

3 The Lorentz function spaces

3.1 The case of w decreasing

In this case we are able to establish exact convexity and concavity constants of $\Lambda_{p,w}$. The first result is a consequence of (1.1) representing the norm $\|\cdot\|_{p,w}$.

Theorem 3.1. *Let $0 < q \le p$ and w be a decreasing weight function. Then*

$$M^{(q)}(\Lambda_{p,w}) = 1.$$

For $q > p$, the space $\Lambda_{p,w}$ is not q-convex.

Proof. It is enough to show that $\Lambda_{p,w}$ is p-convex with constant 1, since for $0 < q < p$, $1 \le M^{(q)}(\Lambda_{p,w}) \le M^{(p)}(\Lambda_{p,w}) = 1$. However by equation (1.1),

$$\Big\|\Big(\sum_{i=1}^n |f_i|^p\Big)^{1/p}\Big\|_{p,w} = \left[\sup_{v\sim w}\int_I \Big(\sum_{i=1}^n |f_i|^p\Big)v\right]^{1/p}$$
$$\le \left[\sum_{i=1}^n \sup_{v\sim w}\int_I |f_i|^p v\right]^{1/p} = \Big(\sum_{i=1}^n \|f_i\|_{p,w}^p\Big)^{1/p},$$

and so $\Lambda_{p,w}$ is p-convex with constant 1. Since $\Lambda_{p,w}$ contains an order isomorphic copy of l_p [6, Theorem 1] and l_p is not q-convex for $q > p$, $\Lambda_{p,w}$ is also not q-convex. □

Recall that for any Banach function space $(E, \|\cdot\|_E)$ over nonatomic measure space with rearrangement invariant norm, if $x, y \in E$ satisfy $\int_0^t y^* \le \int_0^t x^*$ for all $t \ge 0$, then $\|y\|_E \le \|x\|_E$ [1, Corollary 4.7].

Theorem 3.2. *Let $q > 1$ and w be a decreasing weight function. Then*

$$M_{(q)}(\Lambda_{1,w}) = \sup_{x>0} \frac{(\frac{1}{x}\int_0^x w^r)^{1/r}}{\frac{1}{x}\int_0^x w},$$

where $1/q + 1/r = 1$.

Proof. Let $\{\lambda_i\}_{i=1}^n$ be any decreasing nonnegative sequence. Fix $x > 0$ and define for all $t > 0$,

$$f(t) = \sum_{i=1}^{n} \lambda_i \chi_{(\frac{i-1}{n}x, \frac{i}{n}x]}(t).$$

It is clear that f is decreasing, and so $f^* = f$. Thus

$$\|f\|_{1,w} = \sum_{i=1}^{n} \lambda_i \left[W\Big(\frac{ix}{n}\Big) - W\Big(\frac{i-1}{n}x\Big) \right].$$

For $j = 1, \ldots, n$, and $t > 0$, let

$$f_j(t) = \sum_{i=1}^{n} \lambda_{(i+j) \pmod n} \chi_{(\frac{i-1}{n}x, \frac{i}{n}x]}(t).$$

Then $f_j^* = f^*$ and $\|f_j\|_{1,w} = \|f\|_{1,w}$. Hence

$$\Big(\sum_{i=1}^{n} \|f_j\|_{1,w}^q \Big)^{1/q} = n^{1/q} \left[\sum_{i=1}^{n} \lambda_i \Big(W\big(\frac{i}{n}x\big) - W\big(\frac{i-1}{n}x\big) \Big) \right]. \tag{3.1}$$

We also have that $(\sum_{i=1}^n |f_j|^q)^{1/q} = (\sum_{i=1}^n \lambda_i^q)^{1/q} \chi_{(0,x]}$, and thus

$$\Big\| \Big(\sum_{i=1}^{n} |f_i|^q \Big)^{1/q} \Big\|_{1,w} = \Big(\sum_{i=1}^{n} \lambda_i^q \Big)^{1/q} \int_0^x w(t)\, dt = \Big(\sum_{i=1}^{n} \lambda_i^q \Big)^{1/q} W(x). \tag{3.2}$$

Letting $M := M_{(q)}(\Lambda_{1,w}) < \infty$ we get

$$n^{1/q} \left[\sum_{i=1}^{n} \lambda_i \Big(W\big(\frac{i}{n}x\big) - W\big(\frac{i-1}{n}x\big) \Big) \right] \le M \Big(\sum_{i=1}^{n} \lambda_i^q \Big)^{1/q} W(x). \tag{3.3}$$

Since w is decreasing, the sequence $b_i := W(\frac{i}{n}x) - W(\frac{i-1}{n}x)$ is also decreasing. By Hölder's inequality, there exists a decreasing nonnegative sequence $(\lambda_i)_{i=1}^n$ such that

$$\sum_{i=1}^{n} \lambda_i^q = 1 \qquad \text{and} \qquad \sum_{i=1}^{n} \lambda_i b_i = \Big(\sum_{i=1}^{n} b_i^r \Big)^{1/r}, \qquad \text{where } \frac{1}{q} + \frac{1}{r} = 1.$$

Then from equation (3.3)

$$n^{1/q} \Big(\sum_{i=1}^{n} b_i^r \Big)^{1/r} \le M \Big(\sum_{i=1}^{n} \lambda_i^q \Big)^{1/q} W(x) = MW(x),$$

and so

$$n^{1/q} \left[\sum_{i=1}^{n} \Big(W\big(\frac{i}{n}x\big) - W\big(\frac{i-1}{n}x\big) \Big)^r \right]^{1/r} \le MW(x).$$

Since w is decreasing it follows that

$$W(\frac{i}{n}x) - W(\frac{i-1}{n}x) = \int_{\frac{i-1}{n}x}^{\frac{i}{n}x} w(t)dt \geq w(\frac{i}{n}x)\frac{x}{n}.$$

Thus

$$x^{1/q}\left[\sum_{i=1}^{n} w(\frac{i}{n}x)^r \frac{x}{n}\right]^{1/r} = n^{1/q}(\frac{x}{n})^{\frac{r-1}{r}}\left[\sum_{i=1}^{n} w(\frac{i}{n}x)^r \frac{x}{n}\right]^{1/r}$$

$$= n^{1/q}\left[\sum_{i=1}^{n} w(\frac{i}{n}x)^r (\frac{x}{n})(\frac{x}{n})^{r-1}\right]^{1/r} \leq MW(x).$$

Since $(in^{-1}x)_{i=1}^n$ is a partition of $[0, x]$,

$$\left(\int_0^x w(t)^r dt\right)^{1/r} = \left(\lim_{n\to\infty}\sum_{i=1}^{n} w(\frac{i}{n}x)^r \frac{x}{n}\right)^{1/r} \leq MW(x)x^{-1/q}.$$

Thus, for all $x > 0$,

$$M_{(q)}(\Lambda_{1,w}) \geq \frac{x^{1/q}}{W(x)}\left(\int_0^x w(t)^r\, dt\right)^{1/r}$$

and so

$$M_{(q)}(\Lambda_{1,w}) \geq \sup_{x>0}\frac{(\frac{1}{x}\int_0^x w^r)^{1/r}}{\frac{1}{x}\int_0^x w},$$

which proves half of the statement.

In order to establish an upper estimate we shall apply the dual relation between the best constants, namely the equality

$$M_{(q)}(\Lambda_{1,w}) = M^{(r)}(M_W),$$

where $1/r + 1/q = 1$, and M_W is the Marcinkiewicz space dual to $\Lambda_{1,w}$. We shall estimate the r-convexity constant of M_W. First, given $x > 0$, denote by

$$\|f\|_{r,x} = \left(\int_0^x f^*(t)^r dt\right)^{1/r},$$

and notice that $\|\cdot\|_{r,x}$ is r-convex with constant 1. Indeed, if $f = (\sum_{i=1}^n |f_i|^r)^{1/r}$,

$$\|f\|_{r,x} = \left(\int_0^\infty f^*(t)^r \chi_{(0,x]}(t)dt\right)^{1/r} = \left(\sup_{h\sim\chi_{(0,x]}}\int_0^\infty f(t)^r h(t)dt\right)^{1/r}$$

$$\leq \left(\sum_{i=1}^{n}\sup_{h\sim\chi_{(0,x]}}\int_0^\infty f_i(t)^r h(t)dt\right)^{1/r} = \left(\sum_{i=1}^{n}\|f_i\|_{r,x}^r\right)^{1/r}.$$

It is also clear that $\|\cdot\|_{r,x}$ is a rearrangement invariant norm. By the definition of the norm of M_W, for all $y>0$, $\int_0^y f^*(t)\,dt \leq \|f\|_{M_W}\int_0^y w(t)\,dt$, and so by the remark before the theorem we get

$$\left(\int_0^x f^*(t)^r\,dt\right)^{1/r} \leq \|f\|_{M_W}\left(\int_0^x w^r(t)\,dt\right)^{1/r}. \tag{3.4}$$

Now let $f_1,\dots,f_n \in M_W$, $f=(\sum_{i=1}^n |f_i|^r)^{1/r}$, and $\varepsilon>0$. Then there exists $x>0$ such that

$$(1-\varepsilon)\|f\|_{M_W} \leq \frac{\int_0^x f^*}{W(x)}.$$

In view of $1/r+1/q=1$ and Hölder's inequality,

$$\int_0^x f^*(t)\,dt \leq x^{1/q}\left(\int_0^x f^*(t)^r\,dt\right)^{1/r}.$$

Therefore from (3.4) and r-convexity of $\|\cdot\|_{r,x}$ with constant 1,

$$\begin{aligned}(1-\varepsilon)\|f\|_{M_W} &\leq \frac{x^{1/q}}{W(x)}\left(\int_0^x f^*(t)^r\,dt\right)^{1/r}\\ &\leq \frac{x^{1/q}}{W(x)}\Big(\sum_{i=1}^n \|f_i\|_{r,x}^r\Big)^{1/r}\\ &\leq \frac{x^{1/q}}{W(x)}\left(\sum_{i=1}^n \|f_i\|_{M_W}^r\|w\|_{r,x}^r\right)^{1/r}\\ &= \frac{x^{1/q}}{W(x)}\left(\int_0^x w(t)^r\,dt\right)^{1/r}\left(\sum_{i=1}^n \|f_i\|_{M_W}^r\right)^{1/r}.\end{aligned}$$

Since $\varepsilon>0$ is arbitrary we get that

$$\Big\|\Big(\sum_{i=1}^m |f_i|^r\Big)^{1/r}\Big\|_{M_W} \leq \frac{x^{1/q}}{W(x)}\left(\int_0^x w(t)^r\,dt\right)^{1/r}\left(\sum_{i=1}^m \|f_i\|_{M_W}^r\right)^{1/r},$$

and so for some $x>0$,

$$M_{(q)}(\Lambda_{1,w}) = M^{(r)}(M_W) \leq \frac{(\frac{1}{x}\int_0^x w^r)^{1/r}}{\frac{1}{x}\int_0^x w}.$$

Therefore,

$$M_{(q)}(\Lambda_{1,w}) \leq \sup_{x>0}\frac{(\frac{1}{x}\int_0^x w^r)^{1/r}}{\frac{1}{x}\int_0^x w},$$

and this concludes our proof. □

Corollary 3.3. *Let w be a decreasing weight function and $0 < p, q < \infty$. Then*
(1) If $q > p$ and r is such that $1/r + p/q = 1$, then

$$M_{(q)}(\Lambda_{p,w}) = \sup_{x>0} \frac{\left(\frac{1}{x}\int_0^x w^r\right)^{1/r}}{\frac{1}{x}\int_0^x w}.$$

(2) If $q < p$, then $\Lambda_{p,w}$ is not q-concave;
(3) If $q = p$ and $\lim_{t\to 0^+} w(t) = \infty$ or $\lim_{t\to\infty} w(t) = 0$, then $\Lambda_{p,w}$ is not q-concave.

Proof. (1) It follows from $M_{(q)}(\Lambda_{p,w}) = M_{(q/p)}(\Lambda_{1,w})$ and Theorem 3.2.

(2) Since $\Lambda_{p,w}$ contains an order isomorphic copy of l_p [6, Theorem 1] and l_p is not q-concave for $q < p$, $\Lambda_{p,w}$ is either not q-concave.

(3) Since w is decreasing, the function $W(t)/t$ is also decreasing. Moreover by the limit assumptions on w, $\lim_{t\to 0^+} W(t)/t = \infty$ or $\lim_{t\to\infty} W(t)/t = 0$, which shows that $W(t)/t$ can not be pseudo-increasing and by [6, Theorem 8], $\Lambda_{p,w}$ is not p-concave. □

Remark 3.4. The fact that $\sup_{x>0}\{(\frac{1}{x}\int_0^x w^r)^{1/r}/\frac{1}{x}\int_0^x w\}$ is finite, for $1/r + p/q = 1$ and $q > p$, is equivalent for the space $\Lambda_{p,w}$ being q-concave has already showed up in [7]. It appeared as one of several equivalent conditions stated in terms of the weight function w [7, Lemma 7]. Here we showed that the above constant is the best possible for the q-concavity property of $\Lambda_{p,w}$.

3.2 The case of w increasing

The first result on q-concavity constant is quickly obtained in the next theorem.

Theorem 3.5. *Let $q \geq p$ and w be an increasing weight such that $\lim_{x\to\infty} w(x) = \infty$. Then*

$$M_{(q)}(\Lambda_{p,w}) = 1.$$

For $0 < q < p$, $\Lambda_{p,w}$ is not q-concave.

Proof. It is enough to show that $\Lambda_{p,w}$ is p-concave with constant 1, since for $q \geq p$, $1 \leq M_{(q)}(\Lambda_{p,w}) \leq M_{(p)}(\Lambda_{p,w}) = 1$. By the assumptions on w and Theorem 2.7,

$$\|f\|_{p,w} = \left(\inf_{v\sim_\gamma w, v>0} \int_I |f|^p v\right)^{1/p},$$

for any f with $|\operatorname{supp} f| < \infty$. Then for $f_i \in \Lambda_{p,w}$ with $|\operatorname{supp} f_i| < \infty$, $i = 1, \ldots, n$, we get

$$\begin{aligned}\left(\sum_{i=1}^n \|f_i\|_{p,w}^p\right)^{1/p} &= \left(\sum_{i=1}^n \left(\inf_{v\sim_\gamma w,\ v>0} \int_I |f_i|^p v\right)\right)^{1/p}\\ &\leq \left(\inf_{v\sim_\gamma w,\ v>0}\left(\sum_{i=1}^n \int_I |f_i|^p v\right)\right)^{1/p} \leq \left\|\left(\sum_{i=1}^n |f_i|^p\right)^{1/p}\right\|_{p,w}.\end{aligned}$$

By the limit argument and the Fatou property of $\|\cdot\|_{p,w}$, we conclude that $\Lambda_{p,w}$ is p-concave with constant 1.

Since $\Lambda_{p,w}$ contains an order isomorphic copy of l_p and l_p [6, Theorem 1] is not q-concave for $0 < q < p$, $\Lambda_{p,w}$ is also not q-concave. □

For the increasing weight function w, our main result, Theorem 3.6, provides an estimate of $M^{(q)}(\Lambda_{1,w})$, and hence the estimate of $M^{(q)}(\Lambda_{p,w})$ in Corollary 3.7. In this case we cannot apply the duality method used before since the space $\Lambda_{p,w}$ is in general not normable, and its dual could be trivial.

Theorem 3.6. *If* $0 < q < 1$ *and* w *is increasing weight satisfying* $\lim_{x\to\infty} w(x) = \infty$, *then*

$$\sup_{x>0} \frac{\frac{1}{x}\int_0^x w}{\left(\frac{1}{x}\int_0^x w^r\right)^{1/r}} \le M^{(q)}(\Lambda_{1,w}) \le \sup_{x>0} \frac{w(x)}{\left(\frac{1}{x}\int_0^x w^r\right)^{1/r}},$$

where $1/q + 1/r = 1$.

Proof. First we shall show the lower estimate of the q-convexity constant. Let $(\lambda_i)_{i=1}^n$ be any decreasing nonnegative sequence. Fix $x > 0$ and define f and f_j in the same way as in the proof of Theorem 3.2. Then from the q-convexity for $K = M^{(q)}(\Lambda_{1,w})$, we get

$$\left(\sum_{i=1}^n \lambda_i^q\right)^{1/q} W(x) \le K n^{1/q} \left[\sum_{i=1}^n \lambda_i \left(W\left(\frac{i}{n}x\right) - W\left(\frac{i-1}{n}x\right)\right)\right]. \tag{3.5}$$

Since w is increasing, the sequence $b_i = W(in^{-1}x) - W((i-1)n^{-1}x)$ is also increasing. By the reverse Hölder inequality for $0 < q < 1$, we have

$$\inf\left\{\sum_{i=1}^n |a_i b_i| : \left(\sum_{i=1}^n |a_i|^q\right)^{1/q} = 1\right\} = \left(\sum_{i=1}^n |b_i|^r\right)^{1/r}.$$

Thus there exists a nonnegative and decreasing sequence $(\lambda_i)_{i=1}^n$ such that

$$\sum_{i=1}^n \lambda_i^q = 1 \qquad \text{and} \qquad \sum_{i=1}^n \lambda_i b_i = \left(\sum_{i=1}^n b_i^r\right)^{1/r}.$$

Then by (3.5) we get

$$W(x) \le K n^{1/q} \left[\sum_{i=1}^n \left(W\left(\frac{i}{n}x\right) - W\left(\frac{i-1}{n}x\right)\right)^r\right]^{1/r}, \tag{3.6}$$

which together with the inequality

$$W\left(\frac{i}{n}x\right) - W\left(\frac{i-1}{n}x\right) \le w\left(\frac{i}{n}x\right)\frac{x}{n}$$

implies that

$$Kx^{1/q}\left[\sum_{i=1}^{n} w\Big(\frac{i}{n}x\Big)^r \frac{x}{n}\right]^{1/r} \geq W(x).$$

It is clear that $(in^{-1}x)_{i=1}^n$ is a partition of $[0,x]$, and so

$$Kx^{1/q}\left(\int_0^x w(t)^r dt\right)^{1/r} = Kx^{1/q}\left(\lim_{n\to\infty}\sum_{i=1}^{n} w\Big(\frac{i}{n}x\Big)^r \frac{x}{n}\right)^{1/r} \geq W(x).$$

Therefore

$$\sup_{x>0}\frac{\frac{1}{x}\int_0^x w}{(\frac{1}{x}\int_0^x w^r)^{1/r}} \leq M^{(q)}(\Lambda_{1,w}),$$

which gives the lower estimate for $M^{(q)}(\Lambda_{1,w})$.

To prove the upper estimate, let

$$A = \sup_{x>0}\frac{w(x)}{(\frac{1}{x}\int_0^x w^r)^{1/r}} < \infty. \tag{3.7}$$

Again by the reverse Hölder inequality for any $(b_i)_{i=1}^n$, there exists a positive sequence $(a_i)_{i=1}^n$ such that

$$\sum_{i=1}^n a_i^r = 1 \qquad \text{and} \qquad \sum_{i=1}^n |a_i b_i| = \Big(\sum_{i=1}^n |b_i|^q\Big)^{1/q}.$$

Let $(f_i)_{i=1}^n \subset \Lambda_{w,1}$. In view of the Fatou property of $\|\cdot\|_{1,w}$, we can assume that each f_i is a simple function with compact support. By the above, there exists $(a_i)_{i=1}^n$, $a_i > 0$, $i = 1,\dots,n$, such that $\sum_{i=1}^n a_i^r = 1$ and $\sum_{i=1}^n a_i\|f_i\|_{1,w} = (\sum_{i=1}^n \|f_i\|_{1,w}^q)^{1/q}$. It is clear by Theorem 2.7 that

$$\|f_i\|_{1,w} = \inf\Big\{\int_0^\infty |f_i(t)h(t)|\,dt : \; h \sim_\gamma w,\; h > 0 \text{ a.e.}\Big\}.$$

Let $\varepsilon > 0$. Then for all $i = 1,\dots,n$, there exists h_i, $h_i \sim_\gamma w$ such that $h_i > 0$ and

$$\int_0^\infty |f_i(t)|h_i(t)dt - \frac{\varepsilon}{na_i} \leq \|f_i\|_{1,w} \leq \int_0^\infty |f_i(t)|h_i(t)\,dt.$$

Then by the Hölder inequality for $0 < q < 1$,

$$\begin{aligned}\Big(\sum_{i=1}^n \|f_i\|_{1,w}^q\Big)^{1/q} &\geq \sum_{i=1}^n a_i\Big(\int_0^\infty |f_i(t)|h_i(t)dt - \frac{\varepsilon}{na_i}\Big) \\ &= \int_0^\infty \sum_{i=1}^n a_i|f_i(t)|h_i(t)dt - \varepsilon \\ &\geq \int_0^\infty \Big(\sum_{i=1}^n |f_i(t)|^q\Big)^{1/q}\Big(\sum_{i=1}^n (a_ih_i(t))^r\Big)^{1/r} dt - \varepsilon.\end{aligned} \tag{3.8}$$

Let $H(t) = (\sum_{i=1}^n (a_ih_i(t))^r)^{1/r}$ and $S = \{s : H(s) < (1-\varepsilon)A^{-1}w(t)\}$, where

$A < \infty$ by assumption. We first show that $|S| \le t$. Suppose to the contrary that $|S| > t$. Then

$$\Big(\int_S H(s)^r ds\Big)^{1/r} = \Big(\int_S \sum_{i=1}^n (a_i h_i(s))^r ds\Big)^{1/r} \le \Big(\int_S (1-\varepsilon)^r A^{-r} w(t)^r ds\Big)^{1/r}$$
$$= (1-\varepsilon)A^{-1}w(t)|S|^{1/r} \le (1-\varepsilon)A^{-1}w(|S|)|S|^{1/r}.$$

On the other hand, in view of $r < 0$ and Proposition 2.3, for each $i = 1, \dots, n$,

$$\int_S h_i(s)^r ds \le \int_0^{|S|} (h_i(s)^r)^* ds = \int_0^{|S|} (h_i(s)_*)^r ds = \int_0^{|S|} w(s)^r ds.$$

Hence by (3.7) and Lemma 2.4,

$$\Big(\int_S H(s)^r ds\Big)^{1/r} \ge \Big(\sum_{i=1}^n a_i^r \int_0^{|S|} w(s)^r ds\Big)^{1/r} = \Big(\int_0^{|S|} w(s)^r ds\Big)^{1/r}$$
$$= \Big(\frac{1}{|S|}\int_0^{|S|} w(s)^r ds\Big)^{1/r} |S|^{1/r} \ge A^{-1}w(|S|)|S|^{1/r},$$

which is a contradiction. Thus we get $|S| \le t$. Therefore $H_*(t) \ge (1-\epsilon)A^{-1}w(t)$. So by (3.8) and Theorem 2.5,

$$\Big(\sum_{i=1}^n \|f_i\|^q\Big)^{1/q} \ge \int_0^\infty \Big(\sum_{i=1}^n |f_i(t)|^q\Big)^{1/q} H(t)dt - \varepsilon$$
$$\ge \int_0^\infty \Big(\big(\sum_{i=1}^n |f_i(t)|^q\big)^{1/q}\Big)^* H_*(t)dt - \varepsilon$$
$$\ge \int_0^\infty \Big(\big(\sum_{i=1}^n |f_i(t)|^q\big)^{1/q}\Big)^* (1-\varepsilon)A^{-1}w(t)dt - \varepsilon$$
$$= (1-\varepsilon)A^{-1}\Big\|\Big(\sum_{i=1}^n |f_i|^q\Big)^{1/q}\Big\| - \varepsilon.$$

It follows $M^{(q)}(\Lambda_{w,1}) \le A = \sup_{x>0} w(x)(\frac{1}{x}\int_0^x w(t)^r dt)^{-1/r}$, and this concludes our proof. □

Corollary 3.7. *Let w be an increasing weight function with $\lim_{x\to\infty} w(x) = \infty$. Then*
(1) If $0 < q < p$ and $1/r + p/q = 1$, we have

$$\sup_{x>0} \frac{\frac{1}{x}\int_0^x w}{(\frac{1}{x}\int_0^x w^r)^{1/r}} \le M^{(q)}(\Lambda_{p,w}) \le \sup_{x>0} \frac{w(x)}{(\frac{1}{x}\int_0^x w^r)^{1/r}}.$$

(2) If $q > p$, then $\Lambda_{p,w}$ is not q-convex.
(3) If $p = q$ and $\lim_{t\to 0^+} w(t) = 0$ or $\lim_{t\to\infty} w(t) = \infty$, then $\Lambda_{p,w}$ is not q-convex.

Proof. Part (1) is an immediate corollary of Theorem 3.6. (2) & (3) Applying [6, Theorems 1 and 4], and arguing similarly as in the proof of Corollary 3.3, we conclude that $\Lambda_{p,w}$ is not q-convex. □

Remark 3.8. Given r as in the above corollary, the constants

$$A := \sup_{x>0} \frac{\frac{1}{x}\int_0^x w}{\left(\frac{1}{x}\int_0^x w^r\right)^{1/r}} \quad \text{and} \quad B := \sup_{x>0} \frac{w(x)}{\left(\frac{1}{x}\int_0^x w^r\right)^{1/r}}$$

are equivalent, that is there exists $C \geq 1$ such that $A \leq B \leq CA$.

Indeed, assuming that $A < \infty$, for any $0 < y < x < \infty$, in view of the Hölder inequality for $r < 0$, we have

$$y^{1/r-1}\int_0^y w \geq \left(\int_0^x w^r\right)^{1/r} \geq A^{-1}x^{1/r-1}\int_0^x w.$$

Hence for any $b > 1$ and $y > 0$, $y^{1/r-1}\int_0^y w \geq A^{-1}b^{1/r-1}y^{1/r-1}\int_0^{by} w$. It follows that

$$(1 - A^{-1}b^{1/r-1})\int_0^y w \geq A^{-1}b^{1/r-1}\int_y^{by} w \geq A^{-1}b^{1/r-1}(b-1)yw(y),$$

and so for any $y > 0$, $w(y) \leq C\frac{1}{y}\int_0^y w$, where $C = (Ab^{1-1/r} - 1)/(b-1)$.

The last corollary summarizes our results on convexity and concavity constants in the case of the Lorentz spaces $L_{q,p}$, $0 < p, q < \infty$.

Corollary 3.9. *(1) If $p \leq q$, then*

(i) $M^{(s)}(L_{q,p}) = 1$ for $s \leq p$ and the space is not s-convex for $s > p$.

(ii) For $s > q$,

$$M_{(s)}(L_{q,p}) = \frac{p}{q\left[\left(\frac{p}{q} - 1\right)r + 1\right]^{1/r}},$$

where $1/r + p/s = 1$. For $s \leq q$, the space is not s-concave.

(2) If $p > q$, then

(i) $M_{(s)}(L_{q,p}) = 1$ for $s \geq p$ and the space is not s-concave for $s < p$.

(ii) For $0 < s < q$,

$$\frac{q\left[\left(\frac{p}{q} - 1\right)r + 1\right]^{1/r}}{p} \leq M^{(s)}(L_{q,p}) \leq \left[\left(\frac{p}{q} - 1\right)r + 1\right]^{1/r},$$

where $1/r + p/s = 1$. For $s \geq q$ the space is not s-convex.

References

[1] C. Bennett and R. Sharpley, *Interpolation of operators*, Pure and Applied Mathematics Series 129, Academic Press Inc., 1988.

[2] N. Carothers, *Symmetric structures in Lorents spaces*, Doctoral Dissertation, The Ohio State University (1982).

[3] J. Creekmore, *Type and cotype in Lorentz L_{pq} spaces*, Indag. Math. 43 (1982), pp. 145–152.

[4] G. J. O. Jameson, *The q-concavity constants of Lorentz sequence spaces and related inequalities,* Math. Z. 227 (1998), pp. 129–142.

[5] S. G. Kreĭn, Y. Ī. Petunīn and E. M. Semënov, *Interpolation of Linear Operators*, Translations of Mathematical Monographs Series 54, AMS (1982).

[6] A. Kamińska and L. Maligranda, *Order convexity and concavity of Lorentz spaces*, Studia Math. 160 (3) (2004), pp. 267–286.

[7] A. Kamińska, L. Maligranda, and L. E. Persson,*Convexity, concavity, type and cotype of Lorentz spaces*, Indagationes Mathematicae. New series 9 (1998), pp. 367–382.

[8] J. Lindenstrauss and L. Tzafriri, *Classical Banach spaces II*, Springer-Verlag, Berlin, 1979.

[9] S. J. Montgomery-Smith, *Boyd indices of Orlicz-Lorentz spaces*. Function spaces (Edwardsville, IL, 1994), pp. 321–334, Lecture Notes in Pure and Appl. Math., 172, Dekker, New York, 1995.

[10] Y. Raynaud, *On Lorentz-Sharpley spaces*, Interpolation Spaces and Related Topics (M. Cwikel, M. Milmann and R. Rochberg, eds); Israel Math. Conf. Proc. 5, Bar-Ilan University (1992), pp. 207–228.

[11] S. Reisner, *A factorization theorem in Banach lattices and its application to Lorentz spaces*, Ann. Inst. Fourier 31 (1981), pp. 239–255.

[12] C. Schütt, *Lorentz spaces that are isomorphic to subspaces of L_1*, Trans. Amer. Math. Soc. 314 (1989), pp. 583–595.

Author information

Anna Kamińska, Department of Mathematical Sciences, The University of Memphis, Memphis, TN 38152, USA.
Email: `kaminska@memphis.edu`

Anca M. Parrish, Department of Mathematical Sciences, The University of Memphis, Memphis, TN 38152, USA.
Email: `abuican1@memphis.edu`

Banach Spaces and their Applications in Analysis, 375–389

On properties of Chalmers–Metcalf operators

Grzegorz Lewicki and Lesław Skrzypek

Abstract. We investigate Chalmers–Metcalf operators. This class of operators is a main tool for finding and describing minimal projections. We show that Chalmers–Metcalf operators can be inscribed in the general approximation scheme. We also point out that we can treat these operators as functionals. This allows to obtain a series of interesting properties. Finally, a connection between the Chalmers–Metcalf operators and the Rudin theorem is presented.

Key words. Minimal projections.

AMS classification. 41A65, 46A22, 41A50.

1 Introduction

By a projection we mean any bounded linear operator P from a Banach space X onto its linear subspace V that acts as an identity on V. The set of all projections from X onto V will be denoted by $\mathcal{P}(X, V)$.

It is well known that not every subspace of a given Banach space is the range of a bounded projection. For example, there is no projection from $\mathcal{B}[0, 1]$ onto $\mathcal{C}[0, 1]$. Subspaces which are ranges of bounded projections are called complemented subspaces and are crucial in the study of Banach spaces. Projections also play an important role in numerical analysis. The error of approximation of an element x by Px, i.e., the quantity $\|x - Px\|$ can be estimated by means of elementary inequalities

$$\|x - Px\| \leq \|Id - P\| \cdot \operatorname{dist}(x, V) \leq (1 + \|P\|) \cdot \operatorname{dist}(x, V), \tag{1.1}$$

where $\operatorname{dist}(x, V)$ denotes the infimum of $\|x - v\|$ over all $v \in V$. This motivates the consideration of a related problem of minimizing $\|P\|$ and led to the following definition: a projection $P_0 \in \mathcal{P}(X, V)$ is called *minimal* if

$$\|P_0\| = \lambda(V, X) := \inf\{\|P\| : P \in \mathcal{P}(X, V)\}. \tag{1.2}$$

The constant $\lambda(V, X)$ is called the *relative projection constant.*

There is also a significant connection between projections and general functional analysis. We may consider a minimal projection as an extension of the Hahn–Banach theorem. Indeed, using a particular minimal projection we can linearly extend any functional $v^* \in V^*$ to X^* by setting $x^* = v^* \circ P$, or equivalently, we can speak of a linear extension of the operator $Id_V : V \to V$ to the operator $P : X \to V$. Moreover, uniqueness of minimal projections results in uniqueness of such extensions. Minimal projections and their properties have been studied by several authors. See, for example, [5, 8, 2, 17, 14, 19] and the references contained therein.

One of the main tools for the study of minimal projections is the class of operators called *Chalmers–Metcalf operators* which we define below. Throughout, we assume that X is a Banach space and V is a finite-dimensional subspace of X. We denote by $S(X)$ the unit sphere of X and by $\text{ext}S(X)$ the set of its extreme points.

Definition 1.1 ([5]). A pair $(x, y) \in S(X^{**}) \times S(X^*)$ is called an *extremal pair* for $P \in \mathcal{P}(X, V)$ if $y(P^{**}x) = \|P\|$.

Let $\mathcal{E}(P)$ be the set of all extremal pairs for P. To each $(x, y) \in \mathcal{E}(P)$, we associate the rank-one operator $y \otimes x$ from X into X^{**} given by $(y \otimes x)(z) = y(z) \cdot x$ for $z \in X$.

Theorem 1.2 (Chalmers–Metcalf, [5]). *A projection $P \in \mathcal{P}(X, V)$ is minimal if and only if the closed convex hull of $\{y \otimes x : (x, y) \in \mathcal{E}(P)\}$ contains an operator E_P for which V is an invariant subspace.*

The operator $E_P : X \to X^{**}$ in Theorem 1.2 is called *Chalmers–Metcalf operator* and is given by the formula:

$$E_P = \int_{\mathcal{E}(P)} y \otimes x \, d\mu(x, y), \tag{1.3}$$

where μ is a Borel probability measure on $\mathcal{E}(P)$.

Theorem 1.2 and the formal definition of E_P were first formulated in [5] but have their origin in [3] and [4]. In fact, in [3] and [4] Chalmers and Metcalf constructed this operator to solve two long standing open problems. Precisely, they found formulae for minimal projections from $C[-1, 1]$ and $L_1[-1, 1]$ onto the quadratics. Since then, this operator found many applications in finding formulae for minimal projections, mainly for the case $X = L_1$, (see e.g., [1], [2], [14]). Also the invertibility of this operator (when restricted to the finite-dimensional subspace V) plays an essential role in the uniqueness of minimal projections (see [14], [21], [22]).

The aim of this paper is to present some properties (some of which may be known to specialists) of Chalmers–Metcalf operators. Our main purpose is to show that these operators can be treated as linear functionals on $\mathcal{L}(X, V)$ (Theorem 2.13). This will provide another proof of Theorem 1.2. Also a connection of the Chalmers–Metcalf operator with trace duality (Theorem 3.5) and the Rudin theorem (Theorem 3.20) will be presented.

2 Definitions and preliminary results

In this section, we recall some definitions and basic known theorems without proof. Here, and in the following sections, we consider projections from a Banach space X, onto a finite-dimensional subspace V.

Definition 2.1. Let $x \in X \setminus V$. An element $v_0 \in V$ is called a *best approximation* to x in V if $\|x - v_0\| = \inf\{\|x - v\| : v \in V\}$.

We denote by $P_V(x)$ the set of all elements of best approximation to x in V. To view minimal projections as best approximation elements, we need to define the appropriate subspace that will provide this correspondence. Let $\mathcal{L}(X,V)$ be the set of all continuous linear operators from X to V. We will denote by $\mathcal{L}_V(X,V)$ the subspace of $\mathcal{L}(X,V)$ given by:

$$\mathcal{L}_V(X,V) := \{L \in \mathcal{L}(X,V) : L|_V \equiv 0\}. \tag{2.1}$$

Let us formulate the relevant theorems.

Theorem 2.2. *A projection $P_0 \in \mathcal{P}(X,V)$ is minimal if and only if 0 is a best approximation to P_0 in $\mathcal{L}_V(X,V)$, i.e., $0 \in P_{\mathcal{L}_V(X,V)}(P_0)$.*

Theorem 2.3. *Let Y be a subspace of X and $x \in X \setminus \overline{Y}$. Then $y \in P_Y(x)$ if and only if there is a functional $f \in S(X^*)$ such that $f|_Y = 0$ and $f(x-y) = \operatorname{dist}(x,Y) = \|x-y\|$.*

Theorem 2.2 implies

Theorem 2.4. *A projection $P_0 \in \mathcal{P}(X,V)$ is minimal if and only if there is a functional $f \in S(\mathcal{L}(X,V)^*)$ such that $f|_{\mathcal{L}_V(X,V)} = 0$ and $f(P_0) = \|P_0\|$.*

Since V is finite dimensional, any projection $P \in \mathcal{P}(X,V)$ is compact. Therefore, there is $y \in S(V^*)$ such that $\|y \circ P\| = \|P\|$ (see [17, Theorem III.2.1]). If in addition X is reflexive, then by James's theorem, there is $x \in S(X)$ such that $yPx = \|y \circ P\|$. Therefore, there is an extremal pair for P i.e. there is $(x,y) \in S(X) \times S(V^*)$ such that $yPx = \|P\|$ (see Definition 1.1). If X is not reflexive then in general, this is not true. For example, the classical n^{th} Fourier projection from $C_0(2\pi)$ onto the subspace of trigonometric polynomials of degree less than n does not attain its norm for $n \geq 1$. But since any functional attains its norm in X^{**}, there always exists an extremal pair for P in $S(X^{**}) \times S(V^*)$.

By the above observation, it is better to consider the Chalmers–Metcalf operators as linear operators from V to X^{**} (instead of from X to X^{**}, as in the original definition). Note that once we have a Chalmers–Metcalf operator from V into X^{**} then, by Hahn–Banach theorem, it induces a Chalmers–Metcalf operator from X to X^{**}. We will modify the original definition of extremal pairs to:

Definition 2.5. A pair $(x,y) \in S(X^{**}) \times S(V^*)$ will be called an *extremal pair* for $P \in \mathcal{P}(X,V)$ if $y(P^{**}x) = \|P\|$.

As above, we also denote by $\mathcal{E}(P)$ the set of all extremal pairs for P in this new setting.

We now state two results which are necessary to prove the main theorem. The first one is the Choquet representation theorem (see e.g. [18, Section 3]).

Theorem 2.6 (Choquet). *Let K be a metrizable compact convex subset of a locally convex space E and $x_0 \in K$. Then there exists a regular Borel probability measure μ on K supported by $\operatorname{ext}(K)$ such that for very continuous linear functional f on E,*

$$f(x_0) = \int_{\operatorname{ext} K} f(z)\, d\mu(z).$$

The next theorem provides characterization of the extreme points of the unit sphere in of the dual of the space of compact operators $\mathcal{K}(X, V)$.

Theorem 2.7 (Collins–Ruess [6, 13]). *Let $\mathcal{K}(X, V)$ denote the ideal of all compact operators from X to V. Then*

$$\text{ext}S(\mathcal{K}(X, V)^*) = \text{ext}S(X^{**}) \otimes \text{ext}S(V^*), \tag{2.2}$$

where $x^{**} \otimes v^* \in \text{ext}S(X^{**}) \otimes \text{ext}S(V^*)$ is considered as a functional in $\mathcal{K}(X, V)^*$ given by the duality:

$$\langle x^{**} \otimes v^*, L\rangle = \langle v^*, L^{**}x^{**}\rangle, \quad \textit{for all } L \in \mathcal{K}(X, V).$$

If $x \in X^{**}$ and $y \in V^*$ then $x \otimes y$ may be interpreted as a functional on $\mathcal{K}(X, V)$ in the sense of Theorem 2.7, but also may be interpreted as a rank one operator from V onto X^{**} as follows $x \otimes y(v) = y(v) \cdot x$. As a result, in the proof of Theorem 3.1 below, the same notation will have two different meanings. We hope that this will not confuse the reader, but actually will help in understanding the correspondence.

3 Results

Using the Choquet representation theorem, we relate Chalmers–Metcalf operators to functionals.

Theorem 3.1. *Assume that X^* is separable and V is finite-dimensional. Let $P_{\min}$ be a minimal projection from X onto V. There is a one-to-one correspondence between functionals corresponding to $P_{\min}$ from Theorem 2.4 and Chalmers–Metcalf operators corresponding to $P_{\min}$. Namely, each such functional can be written as*

$$f = \int_{\mathcal{E}(P_{\min})} x \otimes y \, d\mu(x, y) \in (\mathcal{L}(X, V))^*, \tag{3.1}$$

where $(x \otimes y)(L) = y(L^{**}x)$. While, the corresponding Chalmers–Metcalf operator can be written as

$$E_{P_{\min}} = \int_{\mathcal{E}(P_{\min})} x \otimes y \, d\mu(x, y) \in \mathcal{L}(V, X^{**}), \tag{3.2}$$

where $(x \otimes y)(z) = y(z) \cdot x$. In particular, each Chalmers–Metcalf operator can be treated as a functional and vice versa.

Proof. First we will prove the existence of a functional of the form (3.1) and that each such functional can be written in the form of Chalmers–Metcalf operator (3.2).

Let P_0 be a minimal projection in $\mathcal{P}(X, V)$. By Theorem 2.4, there is a functional $f \in S(\mathcal{L}(X, V)^*)$ such that $f|_{\mathcal{L}_V(X,V)} = 0$ and $f(P_0) = \|P_0\|$.

Observe that since X^* is separable, so is $\mathcal{L}(X, V) = \mathcal{K}(X, V)$. Thus $S(\mathcal{K}(X, V)^*)$ is compact and metrizable when endowed with the weak*-topology. By Choquet's representation theorem (see Theorem 2.6) with $K = B(\mathcal{K}(X,V)^*)$ and $E = \mathcal{K}(X,V)^*$ with the weak*-topology, there exists a probability measure μ on $\text{ext}S(\mathcal{K}(X,V)^*)$ such that for every $L \in \mathcal{K}(X,V)$,

$$f(L) = \int_{\text{ext}S(\mathcal{K}(X,V)^*)} t(L)\, d\mu(t).$$

Using Theorem 2.7, we deduce that for every $L \in \mathcal{L}(X,V)$,

$$f(L) = \int_{\text{ext}S(X^{**})\otimes\text{ext}S(V^*)} \langle x \otimes y, L\rangle\, d\mu(x \otimes y).$$

Thus, f can be expressed as:

$$f = \int_{\text{ext}S(X^{**})\otimes\text{ext}S(V^*)} x \otimes y\, d\mu(x \otimes y). \tag{3.3}$$

Since $f(P_0) = \|P_0\|$ we have

$$\begin{aligned}\|P_0\| = f(P_0) &= \int_{\text{ext}S(X^{**})\otimes\text{ext}S(V^*)} \langle x \otimes y, P_0\rangle\, d\mu(x \otimes y)\\ &= \int_{\text{ext}S(X^{**})\otimes\text{ext}S(V^*)} y(P_0^{**}x)\, d\mu(x \otimes y)\\ &\leq \int_{\text{ext}S(X^{**})\otimes\text{ext}S(V^*)} \|P_0\|\, d\mu(x \otimes y) = \|P_0\|.\end{aligned}$$

This means that $y(P_0^{**}x) = \|P_0\|$ for μ-almost every $x \otimes y \in \text{ext}S(X^{**}) \otimes \text{ext}S(V^*)$. Therefore, $(x,y) \in \mathcal{E}(P_0)$ μ-almost everywhere. Hence, (3.3) can be written as

$$f = \int_{\mathcal{E}(P_0)} x \otimes y\, d\mu(x,y). \tag{3.4}$$

We now define an operator where y, x are the same as in (3.4) but $x \otimes y$'s are viewed as operators from V into X^{**}

$$E_{P_0} := \int_{\mathcal{E}(P_0)} x \otimes y\, d\mu(x,y) : V \to X^{**}. \tag{3.5}$$

Since $f|_{\mathcal{L}_V(X,V)} = 0$, we have $f(v^* \otimes v) = 0$ whenever $v \in V$ and $v^* \in V^\perp$. Thus

$$\begin{aligned}0 = f(v^* \otimes v) &= \int_{\mathcal{E}(P_0)} \langle x \otimes y, v^* \otimes v\rangle\, d\mu(x,y)\\ &= \int_{\mathcal{E}(P_0)} v^*(x) \cdot y(v)\, d\mu(x,y)\\ &= \langle v^*, \int_{\mathcal{E}(P_0)} y(v) \cdot x\, d\mu(x,y)\rangle\\ &= v^*(E_{P_0}(v)).\end{aligned}$$

Therefore $E_{P_0}(V) \subset V$. This shows that E_{P_0} is indeed a Chalmers–Metcalf operator.

Conversely, we will prove that every Chalmers–Metcalf operator corresponds to a functional. Assume that for a projection $P \in \mathcal{P}(X, V)$ we have a Chalmers–Metcalf operator E_P ($E_P(V) \subset V$) given by the formula (1.3):

$$E_P = \int_{\mathcal{E}(P)} x \otimes y \, d\mu(x, y) : V \to X^{**}, \tag{3.6}$$

where μ is a Borel probability measure on $\mathcal{E}(P)$. We define f as follows ($x \otimes y$ are the same as in (3.6) but viewed as functionals on $\mathcal{L}(X, V)$):

$$f := \int_{\mathcal{E}(P)} x \otimes y \, d\mu(x, y). \tag{3.7}$$

It is clear that $f(P) = \|P\|$. Also for every $L \in \mathcal{L}(X, V)$, we have

$$|f(L)| = |\int_{\mathcal{E}(P)} \langle x \otimes y, L \rangle \, d\mu(x, y)| = |\int_{\mathcal{E}(P)} yL^{**}x \, d\mu(x, y)| \leq \|L^{**}\| = \|L\|.$$

Thus f is a norm one functional.

To finish the proof we will verify that $f|_{\mathcal{L}_V(X,V)} = 0$. For this, observe first that since $E_P(V) \subset V$, we have $v^*(E_p(v)) = 0$ whenever $v \in V$ and $v^* \in V^\perp$. A similar computation as above leads to $f(v^* \otimes v) = 0$ whenever $v \in V$ and $v^* \in V^\perp$. Since $\{P - Q : Q \in \mathcal{P}(X, V)\} = \operatorname{span}\{v^* \otimes v : v \in V, v^* \in V^\perp\}$, we deduce that

$$f(Q) = f(P), \quad \text{for all } Q \in \mathcal{P}(X, V). \tag{3.8}$$

Observe that for any $L \in \mathcal{L}_V(X, V)$, both P and $P - L$ are projections from X onto V. Hence, by (3.8), $f(L) = f(P) - f(P - L) = 0$, as claimed. □

As an immediate corollary of Theorem 3.1, we get:

Corollary 3.2. *For any projection P, the set of Chalmers–Metcalf operators E_P is convex.*

If X is finite-dimensional then the measure μ giving f or E_P has finite support.

Corollary 3.3. *If X is finite-dimensional, then any Chalmers–Metcalf operator is determined by a measure μ supported by at most* $(\dim X \cdot \dim V + 1)$ *points.*

Proof. Let $f \in S(\mathcal{L}(X, V)^*)$. Since $\mathcal{L}(X, V)^*$ is finite-dimensional by Caratheodory's theorem, f can be written as a convex combination of $\dim(\mathcal{L}(X, V)^*) + 1$ extreme points of $S(\mathcal{L}(X, V)^*)$. Now we can follow the proof of Theorem 2.13 to show that all such f's correspond to Chalmers–Metcalf operators. □

When X is finite-dimensional the correspondence between the functional from Theorem 2.4 and the Chalmers–Metcalf operator as described in Theorem 3.1 can be obtained using trace duality. Trace duality has been used in [10, 11, 12] to estimate absolute projection constants of finite-dimensional spaces.

We will need the following facts concerning trace duality (see [7, 9] for details).

Let X, V be finite-dimensional Banach spaces. For $L \in \mathcal{L}(X,V)$, its nuclear norm is defined by:

$$\nu(L) = \inf\left\{ \sum_{i=1}^{n} \|x_i^*\| \cdot \|y_i\| \,:\, L = \sum_{i=1}^{n} x_i^*(\cdot)\, y_i \right\}.$$

The space $\mathcal{L}(V,X)$ endowed with the nuclear norm forms a Banach operator ideal. Moreover, the trace duality identifies $\mathcal{L}(X,V)^*$ with $(\mathcal{L}(V,X),\nu)$ isometrically by:

$$L(V,X) \ni L \mapsto L^{\mathrm{tr}} = (K \to \mathrm{tr}(L \circ K)) \in L(X,V)^*. \tag{3.9}$$

Theorem 3.4. *Assume that X is finite dimensional. Let P be a minimal projection from X onto V. Then for each Chalmers–Metcalf operator $E : V \to X$ we have:*

(i) $E(V) \subset V$ *and* $\nu(E) = 1$;

(ii) *for any projection* $Q \in \mathcal{P}(X,V)$, $\mathrm{tr}(E \circ Q) = \mathrm{tr}(E) = \|P\|$.

Conversely, every operator E fulfilling the above conditions generates, by trace duality, a linear functional as in Theorem 2.4.

Proof. Let the Chalmers–Metcalf operator E_P ($E_P(V) \subset V$) be represented by

$$E_P = \sum_{\mathcal{E}(P)} \alpha_i x_i \otimes y_i : V \to X, \tag{3.10}$$

where $\alpha_i > 0$ and $\sum_i \alpha_i = 1$ (by Corollary 3.3, we can assume that this sum is finite).

Since (x,y) is an extremal pair for P, and the trace of an operator does not depend on its particular representation we have

$$\mathrm{tr}(E_P \circ P) = \mathrm{tr}\Big(\sum_{\mathcal{E}(P)} \alpha_i x_i \otimes (y_i \circ P) \Big) = \sum_{\mathcal{E}(P)} \alpha_i y_i(P x_i) = \sum_{\mathcal{E}(P)} \alpha_i \|P\| = \|P\|.$$

Let $e_1, \dots, e_n$ be a basis for V. Since $E_P(V) \subset V$, we can represent E_P by $E_P = \sum_{k=1}^{n} e_k^* \otimes e_k$ (for some vectors $e_1^*, \dots, e_n^*$ from V^*). Again, since the trace does not depend on a particular representation, it follows that for any projection $Q : X \to V$,

$$\mathrm{tr}\,(E_P \circ Q) = \sum_{k=1}^{n} e_k^*(Q e_k) = \sum_{k=1}^{n} e_k^*(e_k) = \mathrm{tr}\,(E_P).$$

This proves that for any projection $Q : X \to V$, we have $\mathrm{tr}(E_P \circ Q) = \mathrm{tr}(E_P) = \|P\|$ and thus *(ii)* is verified.

Now it is easy to see that $\nu(E_p) = 1$. Indeed, from (3.10) and the definition of $\nu(\cdot)$ we have $\nu(E_P) \leq 1$. On the other hand, $\nu(E_P) = \|K \mapsto \mathrm{tr}(E_P \circ K)\|$ and $\|K \mapsto \mathrm{tr}(E_P \circ K)\| \geq \mathrm{tr}(E_P \circ P)/\|P\| = 1$, so *(i)* holds.

For the converse, let $(E_P)^{\mathrm{tr}}$ denote the functional related to E_P by the trace duality. By *(i)*, we have $\|(E_P)^{\mathrm{tr}}\| = 1$. For any $L \in \mathcal{L}_V(X, V)$, we observe that $P + L \in \mathcal{P}(X, V)$. Thus by *(ii)*, we have $(E_P)^{\mathrm{tr}}(L) = (E_P)^{\mathrm{tr}}(P + L) - (E_P)^{\mathrm{tr}}(P) = 0$. □

Combining Theorem 3.1, Corollary 3.3, and Theorem 3.4, the finite dimensional case can be summarized as follows:

Theorem 3.5. *Assume that X is finite dimensional. There is a one-to-one correspondence between functionals from Theorem 2.4 and Chalmers–Metcalf operators. This correspondence is given by trace duality. Namely, each functional can be written as*

$$f = \sum_{\mathcal{E}(P)} \alpha_i \cdot x_i \otimes y_i \in (\mathcal{L}(X, V))^*,$$

where $\alpha_i > 0, \sum \alpha_i = 1$ and $(x \otimes y)(L) = y(Lx)$. While, the corresponding Chalmers–Metcalf operator can be written as

$$E_P = \sum_{\mathcal{E}(P)} \alpha_i \cdot x_i \otimes y_i : V \to X,$$

where $\alpha_i > 0, \sum \alpha_i = 1$ and $(x \otimes y)(z) = y(z)x$.

Additionally, $(E_P)^{\mathrm{tr}} = 1$.

We now consider other properties of Chalmers–Metcalf operators. We begin with the following lemma.

Lemma 3.6. *Let P and Q be two minimal projections from X onto V. Let E_P be a Chalmers–Metcalf operator for P and μ be its representing measure. Then*

$$\mathcal{E}(P) \cap \operatorname{supp} \mu \subset \mathcal{E}(Q) \quad \mu\text{-almost everywhere.}$$

Proof. Since $\|Q\| = \|P\|$, it follows from (3.8) that

$$\begin{aligned} \|Q\| = E_P(Q) &= \int_{\mathcal{E}(P)} \langle x \otimes y, Q \rangle \, d\mu(x, y) \\ &= \int_{\mathcal{E}(P)} y(Q^{**}x) \, d\mu(x, y) \qquad (3.11) \\ &\leq \int_{\mathcal{E}(P)} \|Q\| \, d\mu(x, y) = \|Q\|. \end{aligned}$$

Hence for μ-almost every $(x, y) \in \mathcal{E}(P)$, we have $y(Q^{**}x) = \|Q\|$ and therefore $\mathcal{E}(P) \cap \operatorname{supp} \mu \subset \mathcal{E}(Q)$ μ-almost everywhere. □

As an immediate consequence of Lemma 3.6, we can state:

Theorem 3.7. *Let P and Q be two minimal projections from X onto V. Let E_P be a Chalmers–Metcalf operator for P. Then E_P is also a Chalmers–Metcalf operator for Q, i.e, any Chalmers–Metcalf operator is "good" for all minimal projections.*

Remark 3.8. In general Lemma 3.6 cannot be improved in the sense that $\mathcal{E}(P)$ and $\mathcal{E}(Q)$ may not be equal.

As an example, consider $X = l_1^2(\mathbb{R})$ and $V = \{(0,t) : t \in \mathbb{R}\}$. Then the norm of a minimal projection from $l_1^2(\mathbb{R})$ onto V is equal to one. The following two projections are minimal

$$P_0(a,b) = (0, b-a),$$
$$P_1(a,b) = (0, b+a).$$

It is easy to check that every minimal projection from $l_1^2(\mathbb{R})$ onto V is a convex combination of P_0 and P_1. One may also check that

$$\begin{aligned}\mathcal{E}(P_0) =&\{(y,x) : y = (b,1), x = (a, a+1), \text{ where } a \in [-1,0], b \in [-1,1]\}\\ &\cup \{(y,x) : y = (b,-1), x = (a, a-1), \text{ where } a \in [0,1], b \in [-1,1]\},\end{aligned}$$

while

$$\begin{aligned}\mathcal{E}(P_1) =&\{(y,x) : y = (b,1), x = (a, -a+1), \text{ where } a \in [0,1], b \in [-1,1]\}\\ &\cup \{(y,x) : y = (b,-1), x = (a, -a-1), \text{ where } a \in [-1,0], b \in [-1,1]\}.\end{aligned}$$

Thus neither $\mathcal{E}(P_0) \subset \mathcal{E}(P_1)$ nor $\mathcal{E}(P_1) \subset \mathcal{E}(P_0)$.

However,

Theorem 3.9. *Assume that X is finite-dimensional. Then the norming pairs in support of the measure of any Chalmers–Metcalf operator are norming pairs for any minimal projection.*

Proof. We proceed as in Lemma 3.6. Since the measure μ has finite support, by examining (3.11), we see that if for some $(x,y) \in \mathcal{E}(P)$ we would have $y(Qx) < \|Q\|$ then we would have also strict inequality in (3.11). □

The above theorem is quite interesting. It shows that minimal projections share many norming pairs and as a result they have many common norming points. Until now it was known (and widely used) that minimal projections have at least one common norming point (this is easy to see if we take two minimal projections P and Q and consider the minimal projection $(P+Q)/2$). It would be interesting to determine whether all common norming pairs must appear in the Chalmers–Metcalf operator.

Theorem 3.1 with a Kolmogorov criterion for best approximation leads to another characterization of minimal projections. First we recall this criterion (see e.g [13]).

Theorem 3.10 (Kolmogorov type criterion). *A projection $P_0 \in \mathcal{P}(X,V)$ is minimal if and only if for any $L \in \mathcal{L}_V(X,V)$ there exists $y \in \operatorname{ext}S(\mathcal{L}(X,V)^*)$ and $y(P_0) = \|P_0\|$ such that* $(\operatorname{Re} y)(L) \leq 0$.

Theorem 3.10 leads to the following result.

Theorem 3.11 ([17, Theorem III.2.2]). *A projection $P_0 \in \mathcal{P}(X,V)$ is minimal if and only if for any $P \in \mathcal{P}(X,V)$ there exists $y \in \operatorname{ext}S(V^*)$ and $\|y \circ P_0\| = \|P_0\|$ such that*

$$\|\operatorname{Re}(y \circ P)\| \geq \|P_0\|.$$

Our next result is a slight improvement of Theorem 3.11.

Theorem 3.12. *Assume that X is finite dimensional and E is a Chalmers–Metcalf operator represented by a measure μ_E. Define*

$$B := \{y \in \operatorname{ext}S(V^*) : (y,x) \in \operatorname{supp}(\mu_E) \text{ for some } x \in \operatorname{ext}S(X)\}.$$

Then a projection $P_0 \in \mathcal{P}(X,V)$ is minimal if and only if for any $P \in \mathcal{P}(X,V)$ there exists $y \in B$ such that

$$\|y \circ P\| \geq \|P_0\|.$$

Proof. Let $P \in \mathcal{P}(X,V)$. Then $\|P\| \geq \|y \circ P\| \geq \|P_0\|$ and therefore, P_0 is minimal. To prove the converse, assume to the contrary that P_0 is a minimal projection and there is a projection P such that for any $y \in B$ we have $\|y \circ P\| < \|P_0\|$. Using Theorem 3.4, we have $\|P_0\| = \operatorname{tr}(E \circ P) = \sum_{\mathcal{E}(P)} \alpha_i y(Px) < \sum_{\mathcal{E}(P)} \alpha_i \|P_0\| = \|P_0\|$, which is a contradiction. □

Following the proof of Theorem 3.1, we know that the representing measure μ of any Chalmers–Metcalf operator is supported by the set $\operatorname{ext}S(X^{**}) \times \operatorname{ext}S(V^*)$. In general, we may find that (x_1, y) and (x_2, y) (or (x, y_1) and (x, y_2)) are both in the support of μ. We can avoid this problem by extending the measure to either $S(X^{**}) \times \operatorname{ext}S(V^*)$ or $\operatorname{ext}S(X^{**}) \times S(V^*)$. For this, we collect first some elementary facts in the next lemma. For simplicity, we assume that X is finite dimensional.

Lemma 3.13. *Let $\alpha, \beta > 0$. Assume that $(x_1, y) \in \mathcal{E}(P)$ and $(x_2, y) \in \mathcal{E}(P)$. Then*

(i) $((\alpha x_1 + \beta x_2)/(\alpha + \beta), y) \in \mathcal{E}(P)$;

(ii) $\alpha y(Px_1) + \beta y(Px_2) = \langle(\alpha + \beta) y, P((\alpha x_1 + \beta x_2)/(\alpha + \beta))\rangle$;

(iii) $\alpha y \otimes x_1 + \beta y \otimes x_2 = (\alpha + \beta) y \otimes (\alpha x_1 + \beta x_2)/(\alpha + \beta)))$.

Similarly, if $(x, y_1) \in \mathcal{E}(P)$ and $(x, y_2) \in \mathcal{E}(P)$, then

(iv) $(x, (\alpha y_1 + \beta y_2)/(\alpha + \beta)) \in \mathcal{E}(P)$;

(v) $\alpha y_1(Px) + \beta y_2(Px) = (\alpha + \beta)\langle(\alpha y_1 + \beta y_2)/(\alpha + \beta), Px\rangle$;

(vi) $\alpha y_1 \otimes x + \beta y_2 \otimes x = (\alpha + \beta)[(\alpha y_1 + \beta y_2)/(\alpha + \beta) \otimes x]$.

Proof. Let $s = \alpha/(\alpha+\beta)$ then $s \in (0,1)$, $(\alpha x_1 + \beta x_2)/(\alpha+\beta)) = sx_1 + (1-s)x_2$, and $\|sx_1 + (1-s)x_2\| \le 1$. Moreover, it follows from the assumption that

$$\langle y, P(sx_1 + (1-s)x_2)\rangle = s\langle y, Px_1\rangle + (1-s)\langle y, Px_2\rangle = \|P\|.$$

Since $y \otimes (sx_1 + (1-s)x_2)$ is norming for P, we also have $\|sx_1 + (1-s)x_2\| = 1$ and therefore $(sx_1 + (1-s)x_2, y) \in \mathcal{E}(P)$. This proves *(i)*. The identities *(ii)* and *(iii)* are straightforward.

The proof of the second part of the lemma goes the same way. □

Definition 3.14. We say that a Banach space X is strictly convex (or rotund) if for any $x_1, x_2 \in S(X)$ $(x_1 \neq x_2)$ and any $\alpha \in (0,1)$ we have $\|\alpha x_1 + (1-\alpha)x_2\| < 1$.

Remark 3.15. It follows from Lemma 3.13 that if X is strictly convex then all functionals y's in a Chalmers–Metcalf operator are different. Similarly if V^* is strictly convex then all x's in a Chalmers–Metcalf operator are different.

In fact, Lemma 3.13 allows us to group together norming pairs which have the same norming functional or norming point. The following remark summarizes this.

Remark 3.16. Assume that X is finite dimensional and P is a minimal projection from X onto V. Then any Chalmers–Metcalf operator for P can be written either as

$$E_P = \sum_{\mathcal{E}(P)} \alpha_i x_i \otimes y_i$$

where $y_i \in \mathrm{ext}S(V^*)$, $x_i \in S(X)$ and all y_i are different, or $y_i \in S(V^*)$, $x_i \in \mathrm{ext}S(V^*)$ and all x_i are different. In particular if $X = \ell_1^n$ then any Chalmers–Metcalf operator can be written as

$$E_P = \sum_{\mathcal{E}(P)} \alpha_i e_i \otimes y_i,$$

where $y_i \in S(V^*)$ and $\alpha_i \ge 0$.

As an application, one can easily check the existence of the Chalmers–Metcalf operator for $X = \ell_1^n$. Another application is to show that the set of critical functionals for minimal projection (i.e., $\{y : y \in \mathrm{ext}S(X^*), \|y \circ P\| = \|P\|\}$) is linearly dependent over V. This was first observed by Cheney and Morris and has many important applications.

Theorem 3.17 ([16, 17]). *Let P be a minimal projection of a real normed space X onto a finite-dimensional subspace V. Then either $\|P\| = 1$, or the set $F = \{y : y \in \mathrm{ext}S(X^*), \|y \circ P\| = \|P\|\}$ can be written as $K \cup -K$ with $K \cap -K = \emptyset$ and K linearly dependent over V.*

Our next result is an analogue of Theorem 3.17 for Chalmers–Metcalf operators.

Theorem 3.18. *Let E be a Chalmers–Metcalf operator for a set of minimal projections from a finite-dimensional space X onto its subspace V. Then either $\|P\| = 1$, or the set $F = \{y : y \in \text{ext}S(X^*), y$ appears in Chalmers–Metcalf operator$\}$ can be written as $K \cup -K$ with $K \cap -K = \emptyset$ and K linearly dependent over V.*

Proof. Observe that if $(x, y) \in \mathcal{E}(P)$ then also $(-x, -y) \in \mathcal{E}(P)$ and $x \otimes y = (-x) \otimes (-y)$. Therefore we can assume that the Chalmers–Metcalf operator contains only functionals from the set K. By Remark 3.16, we can write $E_P = \sum_{\mathcal{E}(P)} \alpha_i x_i \otimes y_i$, where $y_i \in K$ are all different. Since $E_P(V) \subset V$ we have $\sum_{\mathcal{E}(P)} \alpha_i y_i(v) x_i \in V$, for any $v \in V$. If we assume that $\lambda(V, X) > 1$ then $x_i \notin V$ and we may choose $x^* \in V^\perp$ such that $x^*(x_i) \neq 0$. We deduce that for any $v \in V$,

$$0 = x^*(\sum_{\mathcal{E}(P)} \alpha_i y_i(v) x_i) = \sum_{\mathcal{E}(P)} \alpha_i x^*(x_i) y_i(v).$$

That is, $\sum_{\mathcal{E}(P)} \alpha_i x^*(x_i)(y_i|_V) = 0$ where $\alpha_i x^*(x_i) \neq 0$. □

It is worth mentioning that all of the above results apply to the case of minimal extensions (see [5, 4] for definition and properties of minimal extensions).

Now we connect Chalmers–Metcalf operator with another classical tool in the theory of minimal projections namely the Rudin theorem (see e.g. [20, 17, 21]).

Theorem 3.19 (The Rudin theorem). *Let X be a Banach space and let $V \subset X$ be a closed subspace (not necessarily finite-dimensional). Assume that $\mathcal{P}(X, V) \neq \emptyset$. Let G be a compact topological group which acts as a group of linear operators on X such that*

(i) $T_g(x)$ is a continuous function of g, for every $x \in X$;

(ii) $T_g(V) \subset V$, for all $g \in G$;

(iii) T_g's are isometries, for all $g \in G$.

Furthermore, assume that there exists a unique projection $P : X \to V$ that commutes with G, in the sense that $T_g P = P T_g$ for all $g \in G$. Then P is minimal and for any given projection $Q \in \mathcal{P}(X, V)$,

$$Px = Bochner - \int_G T_{g^{-1}} Q T_g(x)\, d\mu(g), \quad for\ x \in X, \tag{3.12}$$

where μ is the normalized Haar measure on G.

The statement of Theorem 3.19 has been motivated by a proof of the minimality of the classical n-th Fourier projection from $C_o(2\pi)$-the space of 2π-periodic, real-valued continuous functions onto the space of trigonometric polynomials of degree $\leq n$ (see e.g. [17]). Theorem 3.19 found numerous applications in searching for minimal projections (see e.g. [15, 14, 23, 24]). Now we provide a connection between the Chalmers–Metcalf operators and the Rudin theorem.

Theorem 3.20. *Let X be a reflexive Banach space and $V \subset X$ be a finite-dimensional subspace. Assume that $P \in \mathcal{P}(X,V)$ is minimal determined by Theorem 3.19. Fix $x_0^* \in \text{ext}S(V^*)$ and $x_0 \in \text{ext}S(X)$ such that $\langle x_0^* \otimes x_0, P\rangle = \|P\|$. Define $E_p : V \to X$ by:*

$$E_P(v) = \int_G ((x_0^* \circ T_{g^{-1}}) \otimes T_g(x_0))v \, d\mu, \quad \textit{for } v \in V.$$

Then $E_P(V) \subset V$. Moreover, there exists a regular Borel probability measure ν on $\mathcal{E}(P)$ such that

$$E_P = \int_{\mathcal{E}(P)} (x^* \otimes x) \, d\nu(x, x^*).$$

Proof. For simplicity, we write g instead of T_g. Fix $v \in V$ and $f \in X^*$ such that $f|_V = 0$. Define $L \in \mathcal{L}_V(X,V)$ by $Lz = f(z)v$ for $z \in X$. Fix $Q \in \mathcal{P}(X,V)$ and let $Q_1 = Q - L$. By (3.12)

$$\int_G g^{-1} \circ Q \circ g(x_0) \, d\mu(g) = \int_G g^{-1} \circ Q_1 \circ g(x_0) \, d\mu(g).$$

Hence $\int_G g^{-1} \circ L \circ g(x_0) \, d\mu(g) = 0$. A fortiori, $\int_G x_0^*(g^{-1} \circ L \circ g(x_0)) \, d\mu(g) = 0$. Observe that

$$\begin{aligned} 0 &= \int_G x_0^*(g^{-1} \circ L \circ g(x_0)) \, d\mu(g) = \int_G (x_0^* \circ g^{-1})(f(gx_0)v) \, d\mu(g) \\ &= f\Big(\int_G x_0^* \circ g^{-1}(v) g x_0 \, d\mu\Big) = f(E_p(v)). \end{aligned}$$

Since the above equality holds for any $f \in V^\perp$, for any $v \in V$, $E_P(v) \in V$, as required.

Now we show that for any $g \in G$, $(gx_0, x_0^* \circ g^{-1}) \in \mathcal{E}(P)$. Indeed, since any $g \in G$ is a linear isometry, $gx_0 \in \text{ext}S(x)$ and $x_0^* \circ g^{-1} \in \text{ext}S(X^*)$. Moreover, since P commutes with G and $x_0^* P x_0 = \|P\|$, we deduce that

$$(x_0^* \circ g^{-1})P(gx) = (x_0^* \circ g^{-1} \circ g)(Px) = x_0^* P x_0 = \|P\|$$

for any $g \in G$, which shows our claim.

To end the proof, we will construct the measure ν. To do this, define $F : G \to B(X) \times B(X^*)$ by

$$F(g) = (gx_0, x_0^* \circ g^{-1}).$$

By our assumptions on G, the map F is continuous when $B(X) \times B(X^*)$ is endowed with the product of the weak topology and the weak*-topology. Put $G_1 = F(G)$. By the assumptions on G, and the above reasoning, $G_1 \subset \mathcal{E}(P)$ and is a closed subset of $B(X) \times B(X^*)$. The Haar measure μ on G induces a regular Borel probability measure ν on $B(X) \times B(X^*)$ by setting:

$$\nu(B) = \mu(F^{-1}(B \cap G_1)),$$

for any Borel subset of $B(X) \times B(X^*)$. It is clear that $\mathrm{supp}(\nu) \subset \mathcal{E}(P)$ and

$$E_P = \int_{\mathcal{E}(P)} (x^* \otimes x)\, d\nu(x, x^*).$$

The proof is complete. □

We hope that the sequence of results above will convince the reader that Chalmers–Metcalf operators bear very important information, and many interesting properties are included in or can be derived from them.

Acknowledgments. The research presented in this paper was partially supported by by the KBN Grant 1 P03A 10 26 (Polish State Committee for Scientific Research).

References

[1] B. L. Chalmers and G. Lewicki, *Symmetric subspaces of l_1 with large projection constants*, Studia Math. 134 (1999), pp. 119–133.

[2] ———, *Symmetric spaces with maximal projection constants*, J. Funct. Anal. 200 (2003), pp. 1–22.

[3] B. L. Chalmers and F. T. Metcalf, *Determination of a minimal projection from $C[-1,1]$ onto the quadratics*, Numer. Funct. Anal. Optim. 11 (1990), pp. 1–10.

[4] ———, *The determination of minimal projections and extensions in L^1*, Trans. Amer. Math. Soc. 329 (1992), pp. 289–305.

[5] ———, *A characterization and equations for minimal projections and extensions*, J. Operator Theory 32 (1994), pp. 31–46.

[6] H. S. Collins and W. Ruess, *Weak compactness in spaces of compact operators and of vector-valued functions*, Pacific J. Math. 106 (1983), pp. 45–71.

[7] J. Diestel, H. Jarchow, and A. Tonge, *Absolutely summing operators*, Cambridge Studies in Advanced Mathematics. Cambridge University Press, 43, Cambridge, 1995.

[8] B. Grünbaum, *Projection constants*, Trans. Amer. Math. Soc. 95 (1960), pp. 451–465.

[9] G. J. O. Jameson, *Summing and nuclear norms in Banach space theory*, London Mathematical Society Student Texts. Cambridge University Press, 8, Cambridge, 1987.

[10] H. König, *Spaces with large projection constants*, Israel J. Math. 50 (1985), pp. 181–188.

[11] H. König, C. Schütt, and N. Tomczak-Jaegermann, *Projection constants of symmetric spaces and variants of Khintchine's inequality*, J. Reine Angew. Math. 511 (1999), pp. 1–42.

[12] H. König and N. Tomczak-Jaegermann, *Bounds for projection constants and 1-summing norms*, Trans. Amer. Math. Soc. 320 (1990), pp. 799–823.

[13] G. Lewicki, *Best approximation in spaces of bounded linear operators*, Dissertationes Math. (Rozprawy Mat.) 330 (1994), p. 103.

[14] G. Lewicki and L. Skrzypek, *Chalmers-Metcalf operator and uniqueness of minimal projections*, J. Approx. Theory (to appear).

[15] W. A. Light and E. W. Cheney, *Approximation theory in tensor product spaces*, Lecture Notes in Mathematics 1169. Springer-Verlag, Berlin, 1985.

[16] P. D. Morris and E. W. Cheney, *On the existence and characterization of minimal projections*, J. Reine Angew. Math. 270 (1974), pp. 61–76.

[17] W. Odyniec and G. Lewicki, *Minimal projections in Banach spaces*, Lecture Notes in Mathematics 1449. Springer-Verlag, Berlin, 1990, Problems of existence and uniqueness and their application.

[18] R. R. Phelps, *Lectures on Choquet's theorem*. D. Van Nostrand Co., Inc., Princeton, N.J.-Toronto, Ont.-London, 1966.

[19] B. Randrianantoanina, *Norm-one projections in Banach spaces*, Taiwanese J. Math. 5 (2001), pp. 35–95. International Conference on Mathematical Analysis and its Applications (Kaohsiung, 2000).

[20] W. Rudin, *Projections on invariant subspaces*, Proc. Amer. Math. Soc. 13 (1962), pp. 429–432.

[21] B. Shekhtman and L. Skrzypek, *Uniqueness of minimal projections onto two-dimensional subspaces*, Studia Math. 168 (2005), pp. 273–284.

[22] L. Skrzypek, *Uniqueness of minimal projections in smooth matrix spaces*, J. Approx. Theory 107 (2000), pp. 315–336.

[23] ———, *Minimal projections in spaces of functions of N variables*, J. Approx. Theory 123 (2003), pp. 214–231.

[24] P. Wojtaszczyk, *Banach spaces for analysts*, Cambridge Studies in Advanced Mathematics, 25. Cambridge University Press, Cambridge, 1991.

Author information

Grzegorz Lewicki, Department of Mathematics, Jagiellonian University, Reymonta 4, 30-059 Cracow, Poland.
Email: `lewicki@im.uj.edu.pl`

Lesław Skrzypek, Department of Mathematics, University of South Florida, 4202 E. Fowler Ave., PHY 114, Tampa, FL 33620-5700, USA.
Email: `skrzypek@math.usf.edu`

Banach Spaces and their Applications in Analysis, 391–398 © de Gruyter 2007

Approximating fixed points of asymptotically Φ-hemicontractive type mappings

Renxing Ni

Abstract. A new class of asymptotically Φ-hemicontractive type mapping is introduced, which includes the classes of asymptotically pseudocontractive mappings, uniformly Φ-hemicontractive mappings, asymptotically Φ-hemicontractive mappings, asymptotically nonexpansive mappings and monotone mappings as special cases. By the norm inequality and the concept of generalized uniformly L-Lipschitzian mapping, the convergence of the iterative approximation of fixed points for generalized uniformly L-Lipschitzian asymptotically Φ-hemicontractive type mapping by the modified Mann type iterative sequence with errors in real q-uniformly smooth Banach spaces is proved. The results presented extend and improve some recent results due to Goebel and Kirk, Yao, Chen and Zhou, Schu, Ofoedu, and others.

Key words. Asymptotically Φ-hemicontractive type mapping, generalized L-Lipschitzian mapping, modified Mann type iterative sequence with errors, q-uniformly smooth Banach spaces, fixed points.

AMS classification. 47H09, 47H10, 47J99.

1 Introduction and preliminaries

Let $(E, \|\cdot\|)$ be a real Banach space and K be a nonempty convex subset of E. For $1 < q < \infty$, the duality mapping $J_q : E \to 2^{E^*}$ is defined by:

$$J_q(x) = \{u^* \in E^*; \langle x, u^*\rangle = \|x\| \cdot \|u^*\|, \|u^*\| = \|x\|^{q-1}\}, \quad x \in E.$$

It is well known (see e.g. [1]) that for $1 < q < \infty$, $J_q(\lambda x) = \lambda^{q-1} J_q(x)$ and $J_q(x) = \|x\|^{q-2} J(x)$ for $x \in E \setminus \{0\}$ and $\lambda \geq 0$, where $J(\cdot)$ is the normalized duality mapping. In the sequel we shall denote the single-valued duality mapping by $j_q(\cdot)$ and denote the set of fixed points of self-mapping T by $F(T) := \{x \in K; Tx = x\}$. In this paper we consider the following classes of mappings:

Definition 1.1. A mapping $T : K \to K$ is said to be *asymptotically nonexpansive* if there exists a sequence $\{r_n\} \subset [1, \infty)$ with $\lim_{n\to\infty} r_n = 1$ and for every $n \geq 1$,

$$\|T^n x - T^n y\| \leq r_n \|x - y\|, \quad x, y \in K. \tag{1.1}$$

Definition 1.2 ([10, 11]). A mapping $T : K \to K$ is said to be *asymptotically pseudocontractive*, if there exists a sequence $\{r_n\} \subset [1, \infty)$ with $\lim_{n\to\infty} r_n = 1$, and for all $x, y \in K$, there exists $j_q(x - y) \in J_q(x - y)$ such that for every $n \geq 1$,

$$\langle T^n x - T^n y, j_q(x - y)\rangle \leq r_n \|x - y\|^q. \tag{1.2}$$

Supported by the National Natural Science Foundation of China (Grant No.10271025), the Provincial Natural Science Foundation of Zhejiang (Grant No.Y606717), and the Key Scientific Research Foundation from Zhejiang Province Education Committee (Grant No.20061154).

Definition 1.3 ([15]). Let $T : K \to K$ and $\Phi : [0, \infty) \to [0, \infty)$ be a strictly increasing function with $\Phi(0) = 0$. We say that T is *uniformly Φ-hemicontractive* if $F(T) \neq \emptyset$ and for every $x \in K$, there exists $j_q(x-p) \in J_q(x-p)$ such that for every $n \geq 1$,

$$\langle T^n x - p, j_q(x-p)\rangle \leq \|x-p\|^q - \Phi(\|x-p\|)\|x-p\|^{q-1}. \tag{1.3}$$

Definition 1.4. Let $T : K \to K$ and $\Phi : [0, \infty) \to [0, \infty)$ be a strictly increasing function with $\Phi(0) = 0$. We say that T is *asymptotically Φ-hemicontractive type* if there exists a sequence $\{r_n\} \subset [1, \infty)$ with $\lim_{n\to\infty} r_n = 1$, $F(T) \neq \emptyset$, and for every $x \in K$, there exists $j_q(x-p) \in J_q(x-p)$ satisfying

$$\langle T^{m_n} x - p, j_q(x-p)\rangle \leq r_n\|x-p\|^q - \Phi(\|x-p\|)\|x-p\|^{q-1}, \quad n \geq 1, \tag{1.4}$$

where $\{m_n\}$ is any positive integer sequence satisfying $\lim_{n\to\infty} m_n = \infty$.

Definition 1.5 ([9, 15]). A mapping $T : K \to K$ is said to be *uniformly L-Lipschitzian*, if there exists a constant $L > 0$ such that for every $n \geq 1$,

$$\|T^n x - T^n y\| \leq L\|x-y\|, \quad x, y \in K. \tag{1.5}$$

Definition 1.6 ([5, 14]). A mapping $T : K \to K$ is said to be *generalized uniformly L-Lipschitzian*, if there exists some constant $L > 0$ such that for every $n \geq 1$,

$$\|T^n x - T^n y\| \leq L(1 + \|x-y\|), \quad x, y \in K. \tag{1.6}$$

If (1.1) and (1.2) hold for all $x \in K$ and $y \in F(T)$, then the mapping T is said to be *asymptotically quasi-nonexpansive* [11] and *asymptotically hemicontractive* [4, 8] respectively. When $n = 1$ and $q = 2$ in (1.3), we get the usual definition of a Φ-hemicontractive mapping [3]. If $m_n = n$ for all $n \geq 1$ in (1.4), then the mapping T is said to be *asymptotically Φ-hemicontractive* [12]. It should be pointed out that the new class of mappings introduced in Definition 1.4 includes numbers of known classes of nonlinear Lipschitzian mappings and non-Lipschitzian mappings in Banach spaces as special cases. For example, the classes of (asymptotically) nonexpansive mappings, (asymptotically) pseudocontractive mappings, monotone mappings, (uniformly) Φ-hemicontractive mappings and (asymptotically) Φ-hemicontractive mappings.

Remark 1.7. If T is uniformly L-Lipschitzian, then T is generalized uniformly L-Lipschitzian. The converse does not hold in general. In fact, if T is a generalized uniformly L-Lipschitzian, it may fail to be continuous (see [5]).

The concept of asymptotically nonexpansive mappings was introduced by Goebel and Kirk ([2]) and is closely related to the theory of fixed points of mappings in Banach spaces. A fundamental result proved in [2] is that if E is a uniformly convex Banach space, D is a nonempty bounded closed convex subset of E and $T : D \to D$ is an asymptotically nonexpansive mapping, then T has a fixed point in D. In [11], Schu introduced the concept of asymptotically pseudocontractive mapping. It is well known that the class of asymptotically pseudocontractive mappings is wider than the class of asymptotically nonexpansive mappings.

The iterative approximation problems for (asymptotically) nonexpansive, asymptotically pseudocontractive and uniformly Φ-hemicontractive mappings were studied extensively by several authors (see for example [2–10, 13–15]).

Recall that the modulus $\rho_E(\cdot)$ of smoothness of a Banach space E is defined by

$$\rho_E(\tau) = \sup\left\{ \frac{\|x+y\| + \|x-y\|}{2} - 1; x, y \in E, \|x\| = 1, \|y\| \le \tau \right\}, \quad \tau > 0,$$

and that E is said to be *uniformly smooth* if $\lim_{\tau\to 0^+} \rho_E(\tau)/\tau = 0$. It is known ([12]) that if E is uniformly smooth, then E is smooth and reflexive. Moreover, $J_q(\cdot)$ is single-valued, and uniformly continuous on any bounded subset of E. Recall also that for a real number $1 < q < \infty$, a Banach space E is said to be *q-uniformly smooth* if there is a constant $d > 0$ such that for every $\tau > 0$, $\rho_E(\tau) \le d\tau^q$. It is clear that a real Hilbert space H is 2-uniformly smooth and L_q-spaces are $\min(q, 2)$-uniformly smooth ([12]).

The next two lemmas will be used repeatedly in the sequel.

Lemma 1.8 ([12]). *Let $q > 1$ be a fixed real number and E be a real q-uniformly smooth Banach space. Then there exists a constant $c_q > 0$ such that*

$$\|x+y\|^q \le \|x\|^q + q\langle y, J_q(x)\rangle + c_q\|y\|^q, \quad x, y \in E.$$

Lemma 1.9 ([5, 8]). *Let $\{a_n\}$, $\{b_n\}$, and $\{\lambda_n\}$ be three sequences of nonnegative real numbers. Assume that there exists $n_0 \in \mathbb{N}$ such that for every $n \ge n_0$,*

$$a_{n+1} \le (1+\lambda_n)a_n + b_n.$$

If $\sum_{n=1}^{\infty} \lambda_n < \infty$ and $\sum_{n=1}^{\infty} b_n < \infty$, then $\lim_{n\to\infty} a_n$ exists.

2 Main results

Theorem 2.1. *Let E be a real q-uniformly smooth Banach space for some $1 < q < \infty$ and K be a nonempty convex subset of E with $K + K \subset K$. Assume that $T : K \to K$ is a (not necessarily continuous) generalized uniformly L-Lipschitzian asymptotically Φ-type mapping with the sequence $\{r_n\} \subset [1, \infty)$ and $\lim_{n\to\infty} r_n = 1$. Let $\{\alpha_n\}_{n=1}^{\infty}$ be a real sequence in $[0, 1]$ satisfying the following conditions:*

(1) $\sum_{n=1}^{\infty} \alpha_n^q < \infty$;
(2) $\sum_{n=1}^{\infty} \alpha_n(1-\alpha_n)^{q-1} = \infty$;
(3) $\sum_{n=1}^{\infty} \alpha_n(1-\alpha_n)^{q-1}(r_n - 1) < \infty$,

$\{m_n\}$ be a sequence of positive integers satisfying $\lim_{n\to\infty} m_n = \infty$, $\{u_n\}$ be a sequence in K satisfying the condition $\|u_n\| = o(\alpha_n)$. Let $\{x_n\} \subset K$ be the modified Mann type iterative sequence with errors defined by:

$$x_{n+1} = (1-\alpha_n)x_n + \alpha_n T^{m_n} x_n + u_n, \quad n \ge 1.$$

If $F(T) \ne \emptyset$, then the sequence $\{x_n\}$ converges to the unique fixed point p of T.

Proof. In the sequel, we assume that the generalized uniformly L-Lipschitzian constant L is greater than 1. Let $\|u_n\| = d_n\alpha_n$ where $\lim_{n\to\infty} d_n = 0$. Without loss of generality, we may assume that $d_n \in [0,1)$. If $p_1, p_2 \in F(T)$, then we have

$$\langle T^{m_n}(p_1 - p_2), j_q(p_1 - p_2)\rangle \le r_n\|p_1 - p_2\|^q - \Phi(\|p_1 - p_2\|)\|p_1 - p_2\|^{q-1}.$$

Since $\lim_{n\to\infty} r_n = 1$, we deduce that $p_1 = p_2$. That is, T has a unique fixed point p.
Using the definition of the sequence $\{x_n\}$ and Lemma 1.8, we obtain that

$$\begin{aligned}
&\|x_{n+1} - p\|^q = \|(1-\alpha_n)(x_n - p) + \alpha_n(T^{m_n}x_n - p) + u_n\|^q \\
&\le (1-\alpha_n)^q\|x_n - p\|^q + q(1-\alpha_n)^{q-1}\langle \alpha_n(T^{m_n}x_n - p) + u_n, J_q(x_n - p)\rangle \\
&\quad + c_q(\alpha_n\|T^{m_n}x_n - T^{m_n}p\| + \|u_n\|)^q \\
&\le (1-\alpha_n)^q\|x_n - p\|^q + q\alpha_n(1-\alpha_n)^{q-1}\langle T^{m_n}x_n - p, J_q(x_n - p)\rangle \\
&\quad + q(1-\alpha_n)^{q-1}\langle u_n, J_q(x_n - p)\rangle + c_q(\alpha_n\|T^{m_n}x_n - T^{m_n}p\| + \|u_n\|)^q.
\end{aligned}$$

From the assumption on the map T, we have

$$\begin{aligned}
&\|x_{n+1} - p\|^q \le (1-\alpha_n)^q\|x_n - p\|^q \\
&\quad + q\alpha_n(1-\alpha_n)^{q-1}[r_n\|x_n - p\|^q - \Phi(\|x_n - p\|)\|x_n - p\|^{q-1}] \\
&\quad + q(1-\alpha_n)^{q-1}\|u_n\|\|x_n - p\|^{q-1} + \alpha_n^q c_q[L(1 + \|x_n - p\|) + d_n]^q \\
&\le [(1-\alpha_n)^q + q\alpha_n(1-\alpha_n)^{q-1} + q\alpha_n(1-\alpha_n)^{q-1}(r_n - 1)]\|x_n - p\|^q \\
&\quad - q\alpha_n(1-\alpha_n)^{q-1}[\Phi(\|x_n - p\|) - d_n]\|x_n - p\|^{q-1} \qquad (2.1)\\
&\quad + \alpha_n^q L^q c_q(1 + \|x_n - p\| + d_n)^q \\
&\le [1 + q\alpha_n(1-\alpha_n)^{q-1}(r_n - 1)]\|x_n - p\|^q \\
&\quad - q\alpha_n(1-\alpha_n)^{q-1}[\Phi(\|x_n - p\|) - d_n]\|x_n - p\|^{q-1} \\
&\quad + \alpha_n^q L^q c_q(1 + \|x_n - p\| + d_n)^q.
\end{aligned}$$

We claim that $\liminf_{n\to\infty} \|x_n - p\|/(1 + \|x_n - p\| + d_n) := 2\delta = 0$. Indeed, if $\delta > 0$ then we may assume that for every $n \ge 1$, $\|x_n - p\|/(1 + \|x_n - p\| + d_n) \ge \delta > 0$. A fortiori, $\|x_n - p\| \ge \delta$. It follows that

$$\Phi(\|x_n - p\|) \ge \Phi(\delta) > 0. \qquad (2.2)$$

Observe that since $\lim_{n\to\infty} d_n = 0$, there exists $N_1 > 0$ such that for every $n \ge N_1$, we have $d_n < \Phi(\delta)$. Thus by (2.2), $d_n < \Phi(\delta) \le \Phi(\|x_n - p\|)$ and therefore,

$$\begin{aligned}
\|x_{n+1} - p\|^q &\le [1 + q\alpha_n(1-\alpha_n)^{q-1}(r_n - 1)]\|x_n - p\|^q + M\alpha_n^q\|x_n - p\|^q \\
&= [1 + q\alpha_n(1-\alpha_n)^{q-1}(r_n - 1) + M\alpha_n^q]\|x_n - p\|^q,
\end{aligned}$$

where $M = L^q c_q \delta^{-q} > 0$.

Since $\sum_{n=1}^{\infty} \alpha_n(1-\alpha_n)^{q-1}(r_n-1) < \infty$, and $\sum_{n=1}^{\infty} \alpha_n^q < \infty$, it follows from Lemma 1.9 that $\lim_{n\to\infty} \|x_n - p\|$ exists. In particular, the sequence $\{\|x_n - p\|\}$ is bounded. Let $M_1 > 0$ be a constant satisfying

$$\|x_n - p\|^q \le M_1, \quad n \ge 1. \tag{2.3}$$

Now, we consider the following two cases:

Case 1. Assume that $\inf\{\|x_n - p\|; n \ge 1\} = 0$. Then it is clear that

$$\liminf_{n\to\infty} \|x_n - p\|/(1 + \|x_n - p\| + d_n) = 0.$$

But this contradicts the assumption. So Case 1 does not hold.

Case 2. If $\inf\{\|x_n - p\|; n \ge 1\} = r > 0$, then for every $n \ge 1$, $\|x_n - p\| \ge r$. Since Φ is strictly increasing, we have

$$\Phi(\|x_n - p\|)\|x_n - p\|^{q-1} \ge \Phi(r)r^{q-1}.$$

Choose a positive integer $N_2(> N_1)$ such that for every $n \ge N_2$,

$$d_n < \frac{1}{2}\Phi(r).$$

It follows from (2.1) and (2.3) that

$$\begin{aligned}\|x_{n+1} - p\|^q &\le \|x_n - p\|^q + q\alpha_n(1-\alpha_n)^{q-1}(r_n - 1)\|x_n - p\|^q \\ &\quad - q\alpha_n(1-\alpha_n)^{q-1}[\Phi(\|x_n - p\|) - d_n]\|x_n - p\|^{q-1} + M\alpha_n^q\|x_n - p\|^q \\ &\le \|x_n - p\|^q - \frac{q}{2}\alpha_n(1-\alpha_n)^{q-1}\Phi(r)r^{q-1} \\ &\quad + M_1[q\alpha_n(1-\alpha_n)^{q-1}(r_n - 1) + M\alpha_n^q].\end{aligned}$$

Hence $2^{-1}q\Phi(r)r^{q-1}\sum_{n=N_2}^{\infty} \alpha_n(1-\alpha_n)^{q-1} \le \|x_{N_2} - p\|^q + M_1\sum_{n=N_2}^{\infty}[q\alpha_n(1-\alpha_n)^{q-1}(r_n-1) + M\alpha_n^q] < \infty$. This contradicts the condition $\sum_{n=1}^{\infty} \alpha_n(1-\alpha_n)^{q-1} = \infty$. Thus Case 2 does not hold and our claim is verified.

To conclude the proof of Theorem 2.1, fix a subsequence $\{\|x_{n_j} - p\|\}$ of $\{\|x_n - p\|\}$ such that $x_{n_j} \to p$ $(j \to \infty)$. For $\varepsilon \in (0,1)$, there exists a positive integer $N_3(\ge N_2)$ such that for $n_j \ge N_3$,

$$\|x_{n_j} - p\| < \varepsilon. \tag{2.4}$$

Next we prove that $\|x_{n_j+1} - p\| \le \varepsilon$. To see this, assume by contradiction that $\|x_{n_j+1} - p\| > \varepsilon$. Since $\lim_{j\to\infty} \alpha_{n_j} = \lim_{j\to\infty} d_{n_j} = \lim_{j\to\infty}(2^{q-1}\varepsilon(r_{n_j} - 1) + 2^{q-1}d_{n_j}) = \lim_{j\to\infty} \alpha_{n_j}^{q-1}/(q(1-\alpha_{n_j})^{q-1}) = 0$, there exists a positive integer $N_4(> N_3)$ such that for every $n_j \geq N_4$, the following inequalities hold:

$$\begin{aligned}
&\alpha_{n_j} < \frac{\varepsilon}{4(1+L)\varepsilon + 2L},\\
&2^{q-1}\varepsilon(r_{n_j} - 1) + 2^{q-1}d_{n_j} < \frac{1}{2}\Phi(\frac{\varepsilon}{2}),\\
&\frac{L^q c_q (3+L)^q \alpha_{n_j}^{q-1}}{q(1-\alpha_{n_j})^{q-1}} < \frac{1}{2}\Phi(\frac{\varepsilon}{2})(\frac{\varepsilon}{2})^{q-1},\\
&d_{n_j} < (L+1)\varepsilon.
\end{aligned} \tag{2.5}$$

By (2.4), (2.5), and the definition of the modified Mann type iterative sequence with errors, we obtain that for $n_j \geq N_4$,

$$\begin{aligned}
\|T^{m_{n_j}} x_{n_j} - x_{n_j}\| + d_{n_j} &\leq \|T^{m_{n_j}} x_{n_j} - T^{m_{n_j}} p\| + \|x_{n_j} - p\| + (L+1)\varepsilon\\
&\leq L(1 + \|x_{n_j} - p\|) + \|x_{n_j} - p\| + (L+1)\varepsilon\\
&\leq L(1+\varepsilon) + \varepsilon + (L+1)\varepsilon = 2(1+L)\varepsilon + L.
\end{aligned}$$

We deduce that

$$\begin{aligned}
\|x_{n_j} - p\| &\geq \|x_{n_j+1} - p\| - \alpha_{n_j}\|T^{m_{n_j}} x_{n_j} - x_{n_j}\| - \|u_{n_j}\|\\
&\geq \|x_{n_j+1} - p\| - \alpha_{n_j}[\|T^{m_{n_j}} x_{n_j} - x_{n_j}\| + d_{n_j}]\\
&> \varepsilon - \alpha_{n_j}(2(1+L)\varepsilon + L) \geq \varepsilon - \frac{2(1+L)\varepsilon + L}{4(1+L)\varepsilon + 2L}\varepsilon\\
&= \frac{\varepsilon}{2},
\end{aligned} \tag{2.6}$$

and therefore $\Phi(\|x_{n_j} - p\|) > \Phi(\varepsilon/2)$. It follows from (2.1), (2.5), and (2.6) that

$$\begin{aligned}
\|x_{n_j+1} - p\|^q &\leq \|x_{n_j} - p\|^q + q\alpha_{n_j}(1-\alpha_{n_j})^{q-1}(r_{n_j} - 1)\varepsilon^q\\
&\quad - q\alpha_{n_j}(1-\alpha_{n_j})^{q-1}\Phi(\frac{\varepsilon}{2})(\frac{\varepsilon}{2})^{q-1}\\
&\quad + q\alpha_{n_j}(1-\alpha_{n_j})^{q-1} d_{n_j}\varepsilon^{q-1} + c_q\alpha_{n_j}^q L^q (1 + \|x_{n_j} - p\| + d_{n_j})^q\\
&\leq \|x_{n_j} - p\|^q\\
&\quad - q\alpha_{n_j}(1-\alpha_{n_j})^{q-1}(\frac{\varepsilon}{2})^{q-1}[\frac{1}{2}\Phi(\frac{\varepsilon}{2}) - 2^{q-1}\varepsilon(r_{n_j} - 1) - 2^{q-1}d_{n_j}]\\
&\quad - \alpha_{n_j}[q(1-\alpha_{n_j})^{q-1}(\frac{\varepsilon}{2})^{q-1}\frac{1}{2}\Phi(\frac{\varepsilon}{2}) - \alpha_{n_j}^{q-1}L^q c_q(1 + \varepsilon + \varepsilon + L\varepsilon)^q]\\
&\leq \|x_{n_j} - p\|^q - q\alpha_{n_j}(1-\alpha_{n_j})^{q-1}(\frac{\varepsilon}{2})^{q-1}0\\
&\quad - \alpha_{n_j}[q(1-\alpha_{n_j})^{q-1}(\frac{\varepsilon}{2})^{q-1}\frac{1}{2}\Phi(\frac{\varepsilon}{2}) - \alpha_{n_j}^{q-1}L^q c_q(3+L)^q]\\
&\leq \|x_{n_j} - p\|^q \leq \varepsilon^q.
\end{aligned}$$

Thus we have $\|x_{n_j+1}-p\| \leq \|x_{n_j}-p\| \leq \varepsilon$ which contradicts the assumption $\|x_{n_j+1}-p\| > \varepsilon$. We have proved that $\|x_{n_j+1}-p\| \leq \varepsilon$.

One can repeat the argument above inductively to deduce that for every $m \geq 1$, $\|x_{n_j+m}-p\| \leq \varepsilon$. Since ε is arbitrary, we conclude that $x_n \to p$ $(n \to \infty)$. □

Remark 2.2. Most of previously known results on the convergence of the iterative approximation of fixed points for asymptotically Φ-hemicontractive mappings by the modified Mann (or Ishikawa) type iteration procedure with errors impose one or more of the following assumptions (see, [2–8, 11, 13–15]):

1. continuity of the mapping T (e.g. uniformly L-Lipschitzian continuity or uniformly continuity);
2. boundedness of the subset K of the Banach space E;
3. boundedness of range $R(T)$ of the mapping T;
4. boundedness of the iterative sequence $\{x_n\}$(or $\{Tx_n\}$ and $\{Ty_n\}$);
5. compulsory condition on the function Φ such as $\liminf_{x\to\infty} \Phi(x)/x > 0$.

Theorem 2.1 shows that all the of the extra assumptions above can be removed. In particular, Theorem 2.1 generalizes [14, Theorem 2.2].

Remark 2.3. Let $r_n = m_n = n = 1$ in Theorem 2.1, and the generalized uniformly L-Lipschitzian asymptotically Φ-hemicontractive type mapping be replaced by Lipschitzian (or uniform) continuity Φ-hemicontractive mapping, then Theorem 2.1 is also held. Thus, Theorem 2.1 also extends and improves the corresponding results in [3, 7, 15].

Remark 2.4. Rhoades and Soltue [9] proved the equivalence between the convergence of Ishikawa and Mann iteration for an asymptotically pseudocontractive and Lipschitzian (with Lipschitz constant $L \geq 1$) self-map of K. Recently, Özdemir and Akbulut [8] showed that the convergence of the Mann, Ishikawa and Noor iteration procedures are equivalent for Lipschitzian self maps with Lispschitz constant $L \leq 1$. Thus, to some extent, it is without loss of generality that we only give the results on the convergence of the modified Mann type iterative sequence with errors.

Remark 2.5. For the parameters of our Theorem 2.1, one can make the following choices: $\alpha_n = 1/n$, $r_n = 1 + 1/n$, $n = 1, 2, \ldots$.

References

[1] E. Asplund, *Positivity of duality mappings*, Bull. Amer. Math. Soc. 73 (1967), pp. 200–203.

[2] K. Goebel and W. A. Kirk, *A fixed point theorem for asymptotically nonexpansive mappings*, Proc. Amer. Math. Soc. 35 (1972), pp. 171–174.

[3] N. Hirano and Z. Y. Huang, *Convergence theorems for multi-valued Φ-hemicontractive operators and Φ-strongly accretive operators*, Computers and Math. Appli. 46 (2003), pp. 1461–1471.

[4] C. Moore and B. V. C. Nnoli, *Iterative sequence for asymptotically demicontractive maps in Banach spaces*, J. Math. Anal. Appl. 302 (2005), pp. 557–562.

[5] R. Ni, *Ishikawa iteration procedures with errors for certain generalized Lipschitzian nonlinear operators*, Acta Math. Sinica (in Chinese), 44 (2001), pp. 701–712.

[6] E. U. Ofoedu, *Strong convergence theorem for uniformly L-Lipschitzian asymptotically pseudocontractive mapping in real Banach spaces*, J. Math. Anal. Appl. 321 (2006), pp. 722–728.

[7] M. O. Osilike, *Iterative solution of nonlinear equations of the Φ-strongly accretive type*, J. Math. Anal. Appl. 200 (1996), pp. 259–271.

[8] M. O. Osilike, S. C. Aniagbosor, and B. G. Akuchu, *Fixed points of asymptotically demicontractive mappings in arbitrary Banach spaces*, PanAmer. Math. J. 12 (2002), pp. 77–88.

[9] M. Özdemir and S. Akbulut, *On the equivalence of some fixed point iterations*, Kyungpook Math. J. 46 (2006), pp. 211–217.

[10] B. E. Rhoades and S. M. Soltuz, *The equivalence between the convergences of Ishikawa and Mann iterations for an asymptotically pseudocontractive map*, J. Math. Anal. Appl. 283 (2003), pp. 681–688.

[11] J. Schu, *Iterative construction of fixed points of asymptotically nonexpansive mappings*, J. Math. Anal. Anal. 158 (1991), pp. 407–413.

[12] H. K. Xu, *Inequalities in Banach spaces with applications*, Nonlinear Anal. T.M.A. 16 (1991), pp. 1127–1138.

[13] L. P. Yang, *The strong convergence of three-step iterative set sequence for a multivalued asymptotically Φ-hemicontractive type mappings* (in Chinese), Acta Math. Applicatae Sinica 29 (2006), pp. 555–566.

[14] Y. H. Yao, R. D. Chen, and H. Y. Zhou, *Iterative process for certain nonlinear mapping in uniformly smooth Banach spaces*, Nonlinear Funct. Anal. and Appl. 10 (2005), pp. 651–664.

[15] H. Y. Zhou and Y. J. Cha, *Ishikawa and Mann iterative processes with errors for nonlinear Φ-strong quasi-accretive mappings in normal linear spaces*, J. Korean Math. Soc. 36 (1999), pp. 1061–1073.

Author information

Renxing Ni, Department of Mathematics, Shaoxing College of Arts and Sciences, Zhejiang 312000, P. R. China.
Email: nrx64@yahoo.com.cn

Banach Spaces and their Applications in Analysis, 399–401

Some properties related to the Daugavet property

Timur Oikhberg

Abstract. A recent paper of Kadets, Martín, and Merí introduced some properties related to the well-known Daugavet property of Banach spaces. In this short note, we further investigate the spaces possessing such properties.

Key words. Daugavet property, slices of unit balls.

AMS classification. 46B04, 46B20, 46B22.

Suppose X is a real Banach space, and ε stands for $+$ or $-$. We say that X has the *εSquare Daugavet Property* (*εSDP*, for short) if, for any rank one $T \in B(X)$, $\|I + \varepsilon T^2\| = 1 + \|T^2\|$. These properties were introduced in [3]. This short note complements some results of that paper.

The authors of [3] were investigating various generalizations of the Daugavet Property. Recall that a Banach space X is said to have the *Daugavet Property* (*DP*, for short) if, for every rank one $T \in B(X)$, we have

$$\|I + T\| = 1 + \|T\|. \tag{1}$$

Banach spaces with the DP have been studied extensively in the last 10–15 years (see e.g. Chapter 11 of [1] for information). For instance, if X is a Banach space with the DP, and $T \in B(X)$ doesn't fix a copy of ℓ_1, then (1) holds (Theorem 11.62 of [1]). Moreover, the unit balls of Banach spaces with the DP have no strongly exposed points (Lemma 11.46 of [1]).

The DP can be reformulated in terms of slices of the unit ball (see e.g. [4]). Similarly, slices are important for studying $\pm$SDP. For a Banach space X, we denote by S_X and B_X the unit sphere and the (closed) unit ball of X, respectively. For $x^* \in S_{X^*}$ and $\varepsilon \in (0, 1)$, we define the *slice*

$$S(x^*, \varepsilon) = \{x \in B_X \mid \langle x^*, x \rangle > 1 - \varepsilon\}.$$

Below we establish some geometric properties of Banach spaces with $\pm$SDP. Our main result is:

Theorem 1. *(a) Suppose X is a Banach space with the $-$Square Daugavet Property. Then any slice of B_X has diameter 2. Consequently, B_X has no strongly exposed points. Therefore, X fails the Radon–Nikodým Property.*

(b) If a Banach space X has the $+$Square Daugavet Property, then B_X has no strongly exposed points.

The author was partially supported by the NSF grant DMS-0500957.

Proof of Theorem 1(a). Consider $x^* \in S_{X^*}$, $\varepsilon \in (0,1)$, and $x \in S_X \cap S(x^*,\varepsilon)$. By Theorem 4.9(b) of [3], there exists $y \in S_X \cap S(x^*,\varepsilon)$ s.t. $\|(-x)+y\| > 2-\varepsilon$. This shows that diam $S(x^*,\varepsilon) = 2$, and therefore, B_X has no strongly exposed points. The last statement now follows from Section VII.3 of [2]. □

To prove Theorem 1(b), we need

Lemma 2. *Suppose a Banach space X has the +Square Daugavet Property, and x is a strongly exposed point of B_X, with an exposing functional x^*. Then X is isometric to* $\ker x^* \oplus_1 \mathbb{R}$.

Proof. We have to show that, for any $y \in \ker x^*$, $\|x+y\| = 1 + \|y\|$. First suppose $\|y\| = 1$. Then, by Theorem 4.9(a) of [3], for every $n \in \mathbb{N}$ there exists $x_n \in S(x^*, 1/n)$, s.t. $\|x_n + y\| > 2 - 1/n$. But x is a strongly exposed point, hence $\lim_n x_n = x$. Therefore, $\|x+y\| = 2$.

For the general case, consider $y \in S_X \cap \ker x^*$, and a function $\phi : [0,\infty) \to [0,\infty) : t \mapsto \|x+ty\|$. Clearly, $\phi(0) = 1$, ϕ is convex, and $\phi(t) \leqslant 1+t$ for any t. On the other hand, $\phi(t) \geqslant 1+t$ for $t \in [0,1]$. Indeed,

$$\phi(t) \geqslant \|x+y\| - (1-t)\|y\| = 2 - (1-t) = 1+t.$$

Taken together, these facts show that $\phi(t) = 1+t$, which is what we need. □

Proof of of Theorem 1(b). Suppose, for the sake of contradiction, that B_X has a strongly exposed point x. Let $x^* \in S_{X^*}$ be an exposing functional, and set $X_0 = \ker x^*$. By Lemma 2, X is isometric to $X_0 \oplus_1 \mathbb{R}$. Find $x_0^* \in X_0^*$ and $x_0 \in X_0$ s.t. $\|x_0\| = 2/3$, $\|x_0^*\| = 1/2$, and $\langle x_0^*, x_0\rangle = 1/3$. Then, for $x = x_0 \oplus (-1/3) \in S_X$ and $x^* = x_0^* \oplus 1 \in S_{X^*}$, $\langle x^*, x\rangle = 0$.

If X has the +SDP, then, by Theorem 4.9(a) of [3], there exists $y = y_0 \oplus \gamma \in S_X$, such that

$$\langle x^*, y\rangle = \langle x_0^*, y_0\rangle + \gamma > 11/12, \tag{2}$$

and

$$\|x+y\| = \|x_0 + y_0\| + |\gamma - 1/3| > 23/12. \tag{3}$$

We shall show that these two inequalities cannot hold simultaneously.

Recall that $\|y_0\| = 1 - |\gamma|$, hence $|\langle x_0^*, y_0\rangle| \leqslant \|y_0\|/2 \leqslant (1-|\gamma|)/2$. By (2), $(1-|\gamma|)/2 + \gamma > 11/12$, and therefore, $\gamma > 5/6$. Then

$$\|x_0 + y_0\| + |\gamma - 1/3| \leqslant \|x_0\| + \|y_0\| + \gamma - 1/3 = \|x_0\| + 2/3 \leqslant 5/3,$$

which contradicts (3). □

Acknowledgments. We would like to thank M. Martín for valuable conversations concerning the Daugavet Property and related phenomena.

References

[1] A. Abramovich and C. Aliprantis, *An invitation to operator theory*, American Mathematical Society, Providence, RI, 2002.

[2] J. Diestel and J. J. Uhl, jr, *Vector Measures*, American Mathematical Society, Providence, RI, 1977.

[3] V. Kadets, M. Martín, and J. Merí, *Norm inequalities for operators on Banach spaces*, Indiana Univ. Math. J., to appear.

[4] V. Kadets, R. Shvidkoy, G. Sirotkin, and D. Werner, *Banach spaces with the Daugavet property*, Trans. Amer. Math. Soc. 352 (2000), pp. 855–873.

Author information

Timur Oikhberg, Department of Mathematics, University of California - Irvine, Irvine CA 92697, USA.
Email: toikhber@math.uci.edu

Banach Spaces and their Applications in Analysis, 403–411

Pathological hypercyclic operators II

Héctor N. Salas

Abstract. We exhibit an unbounded operator which satisfies the Hypercyclicity Criterion but is not hypercyclic. We also exhibit an unbounded Hilbert space operator whose non-zero vectors are hypercyclic but doesn't have an infinite dimensional closed subspace on which a sequence of powers of the operator goes to zero. Finally, we show examples of (bounded) weighted shifts having supercyclic subspaces but not satisfying the spectral condition.

Key words. Hypercyclicity Criterion, Supercyclicity Criterion, unbounded operators, Hamel basis, transfinite induction.

AMS classification. Primary 47A16; Secondary 47B37, 47B99.

1 Introduction

Let E be an infinite dimensional separable complex Banach space. An operator (linear transformation) T acting on E is *hypercyclic* if there exists an $x \in E$ such that the sequence $(T^n(x))_n$ is dense in E. The operator T is *supercyclic* if the normalization of the above sequence in the unit sphere of E is dense in the sphere.

A good place to learn not only the fundamentals of hypercyclicity but its roots is the paper by Godefroy and Shapiro, [7]. For a good sample of recent activity in the area the Oberwolfach report, [11], is recommended. Usually, the hypercyclic and supercyclic operators studied are continuous, and many of their properties have been discovered. In [16], we showed that several of these properties no longer hold for unbounded operators; for instance, the celebrated Ansari's result, [1], which says that powers of hypercyclic operators are also hypercyclic. We continue this study of unbounded operators in the present paper.

Since every result in this paper refers to one of the following criteria, the referee suggested to have them stated explicitly. As usual, $\mathcal{L}(E)$ denotes the bounded operators on E.

Definition A. Let $T \in \mathcal{L}(E)$ and let $(\lambda_n)_n \subset \mathbb{C} \setminus \{0\}$ be a sequence of scalars. We say that T satisfies the *Supercyclicity Criterion* with respect to the subsequence $(\lambda_{n_k})_k$ provided there exist dense subsets D_1 and D_2 of E and (possibly discontinuous) mappings $S_k : D_2 \to E$ $(k = 1, 2, \dots)$ so that

- $\lambda_{n_k} T^{n_k} \to 0$ on D_1,
- $S_{n_k} \to 0$ pointwise on D_2,
- $\lambda_{n_k} T^{n_k} S_{n_k} \to I_{D_2}$ (Identity on D_2).

We say that the operator T above satisfies the *Hypercyclicity Criterion*, with respect to $(n_k)_k$ provided $\lambda_{n_k} = 1$ for all k. Finally, we say that the operator T satisfies the

Kitai Criterion provided it satisfies the Hypercyclicity Criterion with respect to the full sequence $(n)_n$.

The importance of the Hypercyclicity Criterion is, as it was first shown for the full sequence by Kitai in [10] and by Gethner and Shapiro in [6], that if an operator satisfies it then it is hypercyclic. The analogous result for the Supercyclicity Criterion is in [13]; a different but equivalent criterion is in [15]. All versions of the Hypercyclicity Criterion (for subsequences) are equivalent, see Bermúdez, Bonilla and Peris, [2]. But also some unbounded hypercyclic operators satisfy a version of it, see for instance the paper by Bès, Chan and Seubert, [3].

Very recently De la Rosa and Read [5] have shown the striking fact that there exists a bounded hypercyclic operator which doesn't satisfy the Hypercyclicity Criterion. This answers negatively a long-standing question of Herrero, [9].

In Section 2, we construct an unbounded non-hypercyclic operator, defined on the whole space, which satisfies the Hypercyclicity Criterion. This result, although not unexpected, is quite curious.

Recall that a *hypercyclic (supercyclic) subspace* for the operator T on E is an infinite dimensional closed subspace such that any nonzero vector is hypercyclic (supercyclic) for T. The following two theorems were proved in [13], pp. 87 and 90 respectively.

Theorem B. *Let $T \in \mathcal{L}(E)$. Suppose that T satisfies:*

1. The Supercyclicity Criterion for a sequence $(\lambda_{n_k})_k \subset \mathbb{C} \setminus \{0\}$.

2. There exists an infinite dimensional closed subspace B_0 of E such that for every $x \in B_0$, $\lambda_{n_k} T^{n_k}(x) \to 0$.

Then T has a supercyclic subspace.

The second condition in Theorem B is called *the B_0 condition.* Montes-Rodríguez discovered it in the context of hypercyclicity, actually universality, see [12, Theorem 2.2, Remark 1]. The converse of Theorem B is true in the classes of unilateral and bilateral backward weighted shifts. Moreover, for these operators the existence of supercyclic subspaces was also identified in terms of their weights, see [13, Theorem 5.1, Theorem 6.1]. It is not known whether the converse of Theorem B is true, see [13, p. 133]. In Section 3 we show that the converse is not true if T is allowed to be unbounded. As usual, $\sigma_e(T)$ denotes the essential spectrum of T.

Theorem C. *Let $T \in \mathcal{L}(E)$. Suppose that T satisfies:*

1. The Supercyclicity Criterion for $(\lambda_{n_k})_k$.

2. $\sup_k\{|\lambda^{n_k} \lambda_{n_k}|\} < \infty$ for some $\lambda \in \sigma_e(T)$.

Then T has a supercyclic subspace.

Evidently, it suffices to consider a $\lambda \in \sigma_e(T)$ with minimum modulus. The case $0 \in \sigma_e(T)$ was proved in [15]. The second condition in Theorem C is called *the spectral condition.* It is known, [13, p. 130], that the converse of Theorem C is false. In

Section 4 we show that the converse doesn't hold even for the classes of unilateral and bilateral backward shifts.

2 A non-hypercyclic operator satisfying the Hypercyclicity criterion

Definition D. Let T be an operator, bounded or not, on the Banach space E.

- T is *topologically transitive* if given open sets U and V there exists n such that $U \cap T^{-n}(V) \neq \emptyset$.
- T is *topologically mixing* if given open sets U and V there exists n such that $U \cap T^{-k}(V) \neq \emptyset$ for $n \leq k$.
- T is *weakly mixing* if $T \oplus T$ is hypercyclic.

It is easy to show that if an operator T, bounded or not, satisfies the Hypercyclicity (Kitai) Criterion, then T is topologically transitive (topologically mixing). Indeed, an argument in [6] works also for unbounded operators. Let U and V be open sets and $y \in U \cap D_1$ and $z \in V \cap D_2$. (With the notation in Definition A.) If k is large enough, then $x = y + S_{n_k} z$ is in U and $T^{n_k}(y) + T^{n_k} S_{n_k} z$ is in V.

The following theorem highlights the contrast that we will find in this section between the bounded and the unbounded case. Equivalence (1) is part of Theorem 1.2 of [7], whereas equivalence (2) is part of Theorem 2.3 of [4].

Theorem E. *Let E be an infinite dimensional separable complex Banach space and $T \in \mathcal{L}(E)$. Then:*

1. *T is hypercyclic if and only if it is topologically transitive.*
2. *T is weakly mixing if and only if it satisfies the Hypercyclicity Criterion.*

In what follows the set $X_k = \{x_{k,m} : m \in \mathbb{N}\}$ is, for each $k \in \mathbb{Z}$, a "transversal" of $X = \{x_{n,m} : (n, m) \in \mathbb{Z} \times \mathbb{N}\}$ because the action of T_c is "perpendicular"to X_k.

$$\begin{array}{ccccccccc} \dots & \longrightarrow & x_{-1,1} & \longrightarrow & x_{0,1} & \longrightarrow & x_{1,1} & \longrightarrow & \dots \\ \dots & \longrightarrow & x_{-1,2} & \longrightarrow & x_{0,2} & \longrightarrow & x_{1,2} & \longrightarrow & \dots \\ \dots & \longrightarrow & x_{-1,3} & \longrightarrow & x_{0,3} & \longrightarrow & x_{1,3} & \longrightarrow & \dots \\ & & \vdots & & \vdots & & \vdots & & \end{array}$$

Proposition 2.1. *Let E be an infinite dimensional separable Banach space. Then there is an unbounded operator T which satisfies the Hypercyclicity Criterion, is supercyclic but is not hypercyclic.*

Proof. Let $X = \{x_{n,m} : (n, m) \in \mathbb{Z} \times \mathbb{N}\}$ be a linearly independent set in E such that:

1. $\{x_{0,m} : m \in \mathbb{N}\}$ is dense in E.

2. For $n \neq 0$, $||x_{n,m}|| = 2^{-|n|}$.

3. $\{\lambda x_{n,1} : n \in \mathbb{N}$ and $\lambda \in \mathbb{C}\}$ is dense in E.

Let A be a linearly independent set such that $X \cap A = \emptyset$ and $X \cup A$ is a Hamel basis for E. For $c \in \mathbb{C}$, we define T_c on this basis, and then extend it linearly to E, in this way:

$$T_c(v) = \begin{cases} x_{n+1,m}, & \text{if } v = x_{n,m}; \\ cv, & \text{if } v \in A. \end{cases}$$

T_c satisfies the Hypercyclicity Criterion since T_c is bijective when restricted to X and we can define $S = T_c^{-1}|_X$. The set X is dense because of (1). For $x \in X$ condition (2) implies that $T_c^n(x)$ and $S^n(x)$ go to zero when n goes to infinity.

T_c is supercyclic because of (3). For $z = \sum_{j=1}^{p} \alpha_j z_j$ with $z_j \in X$ condition (2) implies that $T^n(z)$ goes to zero. Assume further that $v_k \in A$ and $w = \sum_{k=1}^{r} \beta_k v_k$. Then for $x = z + w$, we have that $||T_c^n(x)||$ goes to zero if $|c| < 1$, goes to $||w||$ if $|c| = 1$, and goes to infinity if $|c| > 1$. Since any vector in E can be written as $z + w$, with $z \in \text{span}X$ and $w \in \text{span}A$, it follows that T_c is not hypercyclic. □

Proposition 2.2. *Let E be an infinite dimensional separable Banach space. Then there is an unbounded operator T which satisfies the Hypercyclicity Criterion but is not even supercyclic.*

Proof. Let $X = \{x_{n,m} : (n,m) \in \mathbb{Z} \times \mathbb{N}\}$ be a linearly independent set in E such that:

1. $\{x_{0,m} : m \in \mathbb{N}\}$ is dense in E.

2. For $n \neq 0$, $||x_{n,m}|| = 2^{-|n|}$.

3. $B = \overline{\text{span}}\{x_{n,m} : (n,m) \in \mathbb{Z} \setminus \{0\} \times \mathbb{N}\}$ is an infinite dimensional closed subspace with infinite codimension.

Let T_c be defined as in Proposition 2.1. For the same reasons given earlier, T_c satisfies the Hypercyclicity Criterion.

Let $e = \sum_{j=1}^{p} \alpha_j z_j + \sum_{k=1}^{r} \beta_k v_k$ with each $z_j \in X$ and each $v_k \in A$. Let

$$F = \text{span}\{x_{0,m} : \text{ if } z_j = x_{u,m} \text{ for some } j\} \cup \{v_k : 1 \leq k \leq r\}.$$

Then $\{T_c^l(e) : l = 0, 1, 2, \ldots\}$ is contained in $B \oplus F$. Since $\dim F \leq p + r$, condition (3) implies that $B \oplus F$ is properly contained in E. Thus T_c is not supercyclic. Indeed, the argument shows that the family $\{\lambda T_c^l : \lambda, c \in \mathbb{C}, l = 0, 1, 2, \ldots\}$ is not universal. □

Remark 2.3. In both examples T is bijective in span$X$ and therefore $S = T^{-1}$ can be defined there. Notice also that in both examples T_c actually satisfies the Kitai Criterion. Moreover, T_c is topologically mixing but not weakly mixing.

3 The $\mathcal{B}_0$ condition is not necessary for having hypercyclic subspaces

The following theorem of González, León-Saavedra and Montes-Rodríguez was a precursor of Theorems B and Theorem C, see [8, Theorem 1.2].

Theorem F. *Let $T \in \mathcal{L}(E)$ be a weakly mixing operator. Then the following are equivalent:*

1. T has a hypercyclic subspace.

2. Some subsequence $(T^{n_k})_k$ converges pointwise to zero on some infinite dimensional closed subspace of E.

3. $\sigma_e(T)$ meets the unit disk.

The operator T below, which has a hypercyclic subspace, answers affirmatively the (restated) Question 2 in [17]. Its proof consists of a combination of a key idea in Shields' paper, [18] (also in [14, p. 96]), for getting an unbounded Hilbert space operator without invariant subspaces, with the approach used in [16] for obtaining an operator whose whole space is a hypercyclic subspace.

Theorem 3.1. *Let E be an infinite dimensional separable complex Banach space. There exists an unbounded operator T defined on the whole space E which satisfies the following properties:*

1. Every nonzero vector is hypercyclic.

2. For any infinite dimensional closed subspace M and any sequence $(n_k)_k$ there exists $x_0 \in M$ with $T^{n_k}(x_0)$ not going to 0.

3. T is not weakly mixing; i.e., $T \oplus T$ is not hypercyclic.

4. $T \oplus T$ is topologically mixing.

Proof. Let Ω be the first uncountable ordinal; we assume the continuum hypothesis and so $\operatorname{card}(\Omega) = c = \operatorname{card}(\mathbb{R})$. Since $c = \operatorname{card}(\mathcal{E})$, where

$$\mathcal{E} = \{M : M \text{ is an infinite dimensional closed subspace of } E\},$$

it follows that $\mathcal{E} = \{E_w : w \in \Omega\}$. By transfinite induction we can find for each E_w a countable infinite dense subset B_w such that they are pairwise disjoint; i.e., $B_w \cap B_\eta = \emptyset$ if $w \neq \eta$, and $B = \bigcup_{w\in\Omega} B_w$ is a linearly independent set. Let S be a Hamel basis of E which contains B. (Thus $\operatorname{card}(S) = \operatorname{card}(B) = c$.)

We will also need the following: Let $\{V_k : k \in \mathbb{N}\}$ be a basis of open sets for the topology on E given by the norm. (Thus $\{V_j \times V_k : (j,k) \in \mathbb{N}^2\}$ is a basis for $E \oplus E$.) Let $\varphi : \mathbb{N}^5 \longrightarrow \mathbb{N}$ be a bijective function.

By using a similar argument to that used in the proof of Theorem 3.1 (and its remark) in [16] we can have:

1. $S = \bigcup_{w\in\Omega} A_w$ such that each A_w is dense in E and $A_w \cap A_\zeta = \emptyset$ if $w \neq \zeta$.

2. There are c sequences $(n_{j,w})_j$, one for each $w \in \Omega$, such that any two of these sequences have only a finite number of terms in common. Therefore the same is true for the lacunary sequence $(2^{n_{j,w}})_j$.
3. $(x_{2^{n_{j,w}},w})_j$ is dense in E for each $w \in \Omega$.
4. $(x_{n,w})_n$ tends to $x_{1,w}/||x_{1,w}||$ if $n \to \infty$ and $n \neq 2^{n_{j,w}}$.
5. If $\varphi(i,j,k,l,n) = p$, then $x_{2,2p} \in V_i$ and $x_{2,2p+1} \in V_j$ and also $x_{2+n,2p} \in V_k$ and $x_{2+n,2p+1} \in V_l$.

The operator T is now defined as $T(x_{n,w}) = x_{n+1,w}$ and extended linearly to the whole space. The proof that every nonzero vector is hypercyclic for T is similar to the proof of Theorem 3.1 in [16]. Choose a sequence $(n_k)_k$ and an infinite dimensional closed subspace M. Thus $M = E_\beta$ for some β. Let $x_{j,w}$ and $x_{g,\mu}$ be two different vectors in $B_\beta \subset E_\beta = M$. The only way that $T^{n_k}(x_{j,w}) = x_{j+n_k,w}$ can go to zero is that $(j+n_k)_k$ be a subsequence (except for a finite number of terms) of $(2^{n_{s,w}})_s$. We now consider two cases:

First case. If $w = \mu$, then $j \neq g$; therefore $(g+n_k)_k$ has at most a finite number of terms in common with $(2^{n_{s,w}})_s$ which means, according to (4), that $T^{n_k}(x_{g,w}) = x_{g+n_k,w}$ goes to $x_{1,w}/||x_{1,w}||$.

Second case. If $w \neq \mu$, then $(g+n_k,\mu)_k$ cannot have more than a finite number of terms in common with $(2^{n_{s,\mu}})_s$. Therefore $T^{n_k}(x_{g,\mu}) = x_{g+n_k,\mu}$ goes to $x_{1,\mu}/||x_{1,\mu}||$.

In any event we have found $x_0 \in M$ such that $T^{n_k}(x_0)$ doesn't go to zero.

We now see that T is not weakly mixing.

Let $p < q$ and let $a \oplus b = \sum_{i=1}^{p} \alpha_i x_{j_i,w_i} \oplus \sum_{i=p+1}^{q} \alpha_i x_{j_i,w_i}$. The only vectors that are limits of $T^n \oplus T^n(a \oplus b)$ for $n \to \infty$ are of the form $z \oplus \sum_{i=p+1}^{q} \alpha_i x_{1,w_i}/||x_{1,w_i}||$ and $\sum_{i=1}^{p} \alpha_i x_{1,w_i}/||x_{1,w_i}|| \oplus z$ for $z \in E$.

It remains to show that $T \oplus T$ is topologically mixing. But for $\varphi(i,j,k,l,n) = p$ condition (5) implies that $(V_i \times V_j) \cap (T \oplus T)^{-n}(V_k \times V_l) \neq \emptyset$. □

Remark 3.2. (a) In the original statement of Theorem 3.1, T satisfied only properties (1) and (2), but the referee asked if T could also be weakly mixing or $T \oplus T$ be topologically transitive (both conditions being equivalent in the bounded case; this is the Bès and Peris result stated as the second part of Theorem E) since this would imply that, for Theorem F, (1) $\Rightarrow$ (2) no longer holds for the unbounded case.

(b) It would be interesting to know whether there exists a weakly mixing unbounded operator T which also satisfies properties (1) and (2) of the above theorem.

(c) The proof also shows that the operator T in [16, Theorem 3.1] is not weakly mixing.

4 For weighted shifts the spectral condition is not necessary for having supercyclic subspaces

The following two operators are, of course, bounded. They may be considered pathological in so far that in the class of weighted shifts we would have expected that the spectral condition be necessary for having supercyclic subspaces. Both propositions are stated, without proof, in [11]. Information on the spectra of weighted shifts can be found in [19, Section 5].

Proposition 4.1. *Let T be the bilateral backward shift on $\ell^2(\mathbb{Z})$ with weights*

$$w_i = \begin{cases} 1, & \text{if } 0 \le i; \\ \frac{1}{2}, & \text{if } i = -1; \\ \frac{j}{j+1}, & \text{if } i = -\frac{n(n+1)}{2} - j \text{ and } j = 1, \dots, n+1. \end{cases}$$

Then T has a supercyclic subspace but doesn't satisfy the spectral condition.

Proof. The weights are displayed graphically below.

$$\dots, 1/2, 3/4, 2/3, 1/2, 2/3, 1/2, 1/2, \overbrace{1}^{0}, 1, 1, 1, 1, 1, 1, 1, \dots$$

To show that $\sigma(T) = \sigma_e(T) = \{z : |z| = 1\}$, we observe that the spectral radius $r(T) = 1$ and also $r(T^{-1}) = 1$ since $||T^{-\frac{n(n+1)}{2}}|| = (n+1)!$ and therefore Stirling's formula implies that

$$\lim_{n\to\infty} ||T^{-\frac{n(n+1)}{2}}||^{\frac{2}{n(n+1)}} = \lim_{n\to\infty} ((n+1)!)^{\frac{2}{n(n+1)}} = 1.$$

Therefore $\sigma(T)$ and $\sigma(T^{-1})$ are contained in the unit disc and the spectral mapping theorem ([14, p. 34]), implies that $\sigma(T)$ is contained in the unit circle. It is also well known that the spectrum of a weighted shift has circular symmetry and consequently $\sigma(T) = \{z : |z| = 1\}$, but then also $\sigma(T) = \sigma_e(T)$ since each $\lambda \in \sigma(T)$ is an approximate eigenvalue but not an eigenvalue of T.

Since $\prod_{i=1}^{n} w_i$ is bounded, [13, Proposition 2.8] asserts that any sequence $(\lambda_{n_k})_k$ which satisfies the Supercyclicity Criterion for T must be unbounded. Therefore T doesn't satisfy the spectral condition.

To show that T has a supercyclic subspace we check that T satisfies condition (b) in [13, Theorem 6.1]. Let

$$J = \{-1\} \cup \{m_j = -\frac{j(j+1)}{2} - 1 : j = 1, 2, 3, \dots\}.$$

Then for each fixed q,

$$\lim_{n\to\infty} \sup_{m_j \in J} \left\{ \frac{\prod_{i=0}^{n-1} w_{m_j - i}}{\prod_{i=1}^{n} w_{q+i}} \right\} = 0.$$

□

Proposition 4.2. *Let a_k be a positive sequence increasing to 1 and let ϵ_k be a positive sequence decreasing to 0. Let n_k be a rapidly increasing sequence, with $n_0 = 0$, such that*

$$a_k^{n_k - n_{k-1}} < \epsilon_k.$$

Let T be the unilateral backward shift on $\ell^2(\mathbb{N})$ with weights, for $1 \leq k$ and $0 \leq s \leq k-1$ and where $A = 3\sum_{j<k} n_j$

$$w_i = \begin{cases} 1, & \textit{if } A+1 \leq i \leq A+n_k; \\ a_{k-s}, & \textit{if } A+2n_k-n_{k-s}+1 \leq i \leq A+2n_k-n_{k-s-1}; \\ a_{s+1}^{-1}, & \textit{if } A+2n_k+n_s+1 \leq i \leq A+2n_k+n_{s+1}. \end{cases}$$

Then T has a supercyclic subspace but doesn't satisfy the spectral condition.

Proof. The definition of the weights may look cumbersome, but we can visualize them as follows:

$$\overbrace{1,1,\ldots,1}^{n_1},\overbrace{a_1,a_1,\ldots,a_1}^{n_1},\overbrace{a_1^{-1},a_1^{-1},\ldots,a_1^{-1}}^{n_1},\overbrace{1,1,\ldots,1}^{n_2},\overbrace{a_2,a_2,\ldots,a_2}^{n_2-n_1},$$

$$\overbrace{a_1,a_1,\ldots,a_1}^{n_1},\overbrace{a_1^{-1},a_1^{-1},\ldots,a_1^{-1}}^{n_1},\overbrace{a_2^{-1},a_2^{-1},\ldots,a_2^{-1}}^{n_2-n_1},\overbrace{1,1,\ldots,1}^{n_3},$$

and so forth.

Note that $\sigma_e(T)$ is the unit circle. We prove this by showing that $\sigma_e(T^\star)$ is the unit circle. Indeed $T^\star$ can be seen as the restriction to an invariant subspace of a bilateral forward shift L whose weights corresponding to negative indexes are 1. With essentially the same argument as in the preceding proposition, $\sigma(L) = \{z : |z| = 1\}$ and therefore $L - wI$ is bounded below for $|w| < 1$. Thus $T^\star - wI$ is also bounded below. On the other hand, the spectral radius $r(T^\star) = 1$ and $\sigma(T^\star) = \{z : |z| \leq 1\}$; therefore to see that $T^\star - wI$ is invertible in the Calkin algebra it suffices to see that it has finite codimension. This is true, one can check easily that dim $\mathrm{Ker}(T - \overline{w}I) \leq 1$.

Since $\prod_{i=1}^{n} w_i$ is bounded, [13, Proposition 2.7] asserts that any sequence $(\lambda_{n_k})_k$ which satisfies the Supercyclicity Criterion for T must be unbounded. Therefore T doesn't satisfy the spectral condition.

To show that T has a supercyclic subspace we check that T satisfies condition (b) in [13, Theorem 5.1]. Let $J = \{m_1 = 2n_1\} \cup \{m_j = 3\sum_{i=1}^{j-1} n_i + 2n_j : 2 \leq j\}$. Then for $k \to \infty$,

$$\sup\left\{\max_{0\leq q\leq k-1} \frac{\prod_{i=0}^{n_k-1} w_{m_j-i}}{\prod_{i=1}^{n_k} w_{q+i}} : n_k < m_j\right\} \leq a_k^{n_k-n_{k-1}} < \epsilon_k \to 0$$

since

$$\prod_{i=0}^{n_k-1} w_{m_j-i} = \prod_{i=1}^{k} a_i^{n_i-n_{i-1}}$$

and the denominator is bounded below by $\prod_{i=1}^{k-1} a_i^{n_i-n_{i-1}}$. □

In both propositions the underlying space may be ℓ^p with $1 \leq p < \infty$.

Acknowledgments. I wish to thank Professor Juan P. Bès for helpful comments. I also wish to thank the referee whose comments helped to improve the presentation of the paper and put some of the results in a wider perspective.

References

[1] S. I. Ansari, *Hypercyclic and cyclic vectors*, J. Funct. Anal. 128 (1995), pp. 374–383.

[2] T. Bermúdez, A. Bonilla and A. Peris, *On hypercyclicity and supercyclicity criteria*, Bull. Austral. Math. Soc. 70 (2004), pp. 45–54.

[3] J. Bès, K. C. Chan and S. M. Seubert, *Chaotic unbounded differentiation operators*, Integral Equations Operator Theory 40 (2001), pp. 257–267.

[4] J. Bès and A. Peris, *Hereditarily hypercyclic operators*, J. Funct. Anal. 167 (1999), pp. 94–112.

[5] M. De la Rosa and C. J. Read, *A hypercyclic operator whose direct sum $T \oplus T$ is not hypercyclic*, Preprint 2006.

[6] R. M. Gethner and J. H. Shapiro, *Universal vectors for operators on spaces of holomorphic functions*, Proc. Amer. Math. Soc. 100 (1987), pp. 281–288.

[7] G. Godefroy and J. H. Shapiro, *Operators with dense, invariant, cyclic vector manifolds*, J. Funct. Anal. 98 (1991), pp. 229–269.

[8] M. González, F. León-Saavedra, and A. Montes-Rodríguez, *Semi-Fredholm theory: hypercyclic and supercyclic subspaces*, Proc. London Math. Soc. 81 (2000), pp. 169–189.

[9] D. A. Herrero, *Hypercyclic operators and chaos*, J. Operator Theory 28 (1992), pp. 93–103.

[10] C. Kitai, *Invariant closed sets for linear operators*, Ph.D. Thesis, Univ. of Toronto, 1982.

[11] Mini-workshop: Hypercyclicity and Linear Chaos. Abstracts from the mini-workshop held August 13–19, 2006. Organized by T. Bermúdez, G. Godefroy, K. G. Grosse-Erdmann and A. Peris. Mathematisches Forschungsinstitut Oberwolfach Report No. 37/2006.

[12] A. Montes-Rodríguez, *Banach spaces of hypercyclic vectors*, Michigan Math. J. 43 (1996), pp. 419–436.

[13] A. Montes-Rodríguez and H. N. Salas, *Supercyclic subspaces: spectral theory and weighted shifts*, Adv. Math. 163 (2001), pp. 74–134.

[14] H. Radjavi and P. Rosenthal, *Invariant subspaces*, Second edition. Dover Publications, Inc., Mineola, NY, 2003. xii+248 pp.

[15] H. N. Salas, *Supercyclicity and weighted shifts*, Studia Math. 135 (1999), pp. 55–74

[16] ———, *Pathological hypercyclic operators*, Arch. Math. (Basel) 86 (2006), pp. 241–250.

[17] ———, *Questions on supercyclic and hypercyclic subspaces.* Mathematisches Forschungsinstitut Oberwolfach Report No. 37/2006.

[18] A. L. Shields, *A note on invariant subspaces*, Michigan Math. J. 17 (1970), pp. 231–233.

[19] ———, *Weighted shift operators and analytic function theory.* Topics in operator theory, pp. 49–128. Math. Surveys, No. 13, Amer. Math. Soc., Providence, R.I., 1974.

Author information

Héctor N. Salas, Department of Mathematics, University of Puerto Rico, Mayagüez 00681, Puerto Rico.
Email: `salas@math.uprm.edu`

Banach Spaces and their Applications in Analysis, 413–421

On perturbations of ideal complements

Boris Shekhtman

Abstract. Let $\mathbb{F}[x]$ be the space of polynomials in d variables, let $\mathfrak{G}^N$ be the Grassmannian of N-dimensional subspaces of $\mathbb{F}[x]$ and let $\mathfrak{J}_N$ stand for the family of all ideals in $\mathbb{F}[x]$ of codimension N. For a given $G \in \mathfrak{G}^N$ we let

$$\mathfrak{J}_G := \{J \in \mathfrak{J}_N : J \cap G = \{0\}\}.$$

Is it true, that (with appropriate topology on $\mathfrak{J}_N$) the set $\mathfrak{J}_G$ is dense in $\mathfrak{J}_N$? In general the answer is "No". What is even more surprising, is that there are "good ideals" $J \in \mathfrak{J}_N$ such that every "neighborhood" $\mathcal{U}(J) \subset \mathfrak{J}_N$ has a non-empty intersection with $\mathfrak{J}_G$ for any $G \in \mathfrak{G}^N$ and there are "bad" ideals $J \in \mathfrak{J}_N$ (for $d \geq 3$) such that some "neighborhoods" $\mathcal{U}(J) \subset \mathfrak{J}_N$ have an empty intersection with $\mathfrak{J}_G$ for some $G \in \mathfrak{G}^N$. This contrast illuminates the non-homogeneous nature of $\mathfrak{J}_N$.

Key words. Ideal complements, ideal projections.

AMS classification. Primary 46H10, 46J05, 46J20; Secondary 14C05.

To Nigel Kalton on occasion of his 60th birthday.

1 Preliminaries

Let F be a normed linear space, let $\mathfrak{G}^N(F)$ be the Grassmannian of N-dimensional subspaces of F and let $\mathfrak{G}_N(F)$ denotes the Grassmannian of all subspaces of F of codimension N. For a given $G \in \mathfrak{G}_N(F)$ let

$$\mathfrak{G}_G(F) := \{J \in \mathfrak{G}_N(F) : J \cap G = \{0\}\},$$

i.e., $\mathfrak{G}_G(F)$ is a family of all subspaces J in F that are "missing" G or, equivalently, the family of all subspaces of F that complement G.

It is well-known and easy to see that (with appropriate topology on $\mathfrak{G}_N(F)$) for any $G \in \mathfrak{G}^N(F)$, the set $\mathfrak{G}_G(F)$ is an open and dense subset of $\mathfrak{G}_N(F)$.

The main focus of this article is an investigation of the following "ideal" version of this statement.

Let $F = \mathbb{F}[x] = [x_1, \ldots, x_d]$ be the space of polynomials in d variables over the field $\mathbb{F}$ of real or complex numbers. Let $\mathfrak{J}_N$ stand for the family of all ideals in $\mathbb{F}[x]$ of codimension N. Let $\mathfrak{G}^N := \mathfrak{G}^N(\mathbb{F}[x])$. For a given $G \in \mathfrak{G}^N$ we let

$$\mathfrak{J}_G := \{J \in \mathfrak{J}_N : J \cap G = \{0\}\}$$

Question. *Is it still true, that $\mathfrak{J}_G$ is dense in $\mathfrak{J}_N$?*

In general the answer is "No". What is even more surprising, is that there are "good ideals" $J \in \mathfrak{J}_N$ such that every "neighborhood" $\mathcal{U}(J) \subset \mathfrak{J}_N$ has a non-empty intersection with $\mathfrak{J}_G$ for any $G \in \mathfrak{G}^N$ and there are "bad" ideals $J \in \mathfrak{J}_N$ (for $d \geq 3$) such that some "neighborhoods" $\mathcal{U}(J) \subset \mathfrak{J}_N$ have an empty intersection with $\mathfrak{J}_G$ for some $G \in \mathfrak{G}^N$. This contrast illuminates the non-homogeneous nature of $\mathfrak{J}$ (as oppose to $\mathfrak{G}_N$).

We will use the rest of this section to describe the topology on $\mathfrak{J}_N$. Actually, any norm on $\mathbb{F}[\mathbf{x}]$ and any "reasonable" notion of convergence of sequences of ideals in $\mathfrak{J}_N$ will lead to the same results. The underlined reason is the existence of intrinsic (Zariski) topology on the Hilbert scheme $\mathfrak{F}^N(\mathbb{F}^d)$ parametrizing $\mathfrak{J}_N$, which is formally weaker then any reasonable topology (cf. [6]).

For the sake of specificity, define the norm of $f = \sum \hat{f}(\mathbf{k})\mathbf{x}^{\mathbf{k}} \in \mathbb{F}[\mathbf{x}]$ by

$$\|f\| := \sum_{\mathbf{k}} |\hat{f}(\mathbf{k})|,$$

which turns $\mathbb{F}[\mathbf{x}]$ into a normed space with continuous multiplication. Let $(\mathbb{F}[\mathbf{x}])'$ be the dual of $\mathbb{F}[\mathbf{x}]$. An ideal $J \in \mathfrak{J}_N$ induces an N-dimensional subspace $J^{\perp} \subset (\mathbb{F}[\mathbf{x}])'$ defined as

$$J^{\perp} := \{\lambda \in (\mathbb{F}[\mathbf{x}])' : \lambda(f) = 0, \ \forall f \in J\}$$

that uniquely identifies thc ideal J via $(J^{\perp})^{\top} = J$. We will adopt the following definition of convergence:

Definition 1.1. Let $(J_m, m \in \mathbb{N})$ be a sequence of ideals in $\mathfrak{J}_N$ and let $J \in \mathfrak{J}_N$. We say that $J_m \to J$ if for every $\lambda \in J^{\perp}$ there exists $\lambda_m \in J_m^{\perp}$ such that

$$\lambda_m(f) \longrightarrow \lambda(f) \tag{1.1}$$

for every $f \in \mathbb{F}[\mathbf{x}]$.

A simple perturbation argument yields the following:

Theorem 1.2 ([8]). *Let $(J, J_m, m \in \mathbb{N})$ be a sequence of ideals in $\mathfrak{J}_N$ such that $J_m \to J$. Let $E \in \mathfrak{G}^N$ complement J. Then the space E complements J_m for sufficiently large m and*

$$P_m f \to Pf \tag{1.2}$$

for all $f \in \mathbb{F}[\mathbf{x}]$, where P_m and P are projections onto E with $\ker P_m = J_m$, $\ker P = J$.

Conversely, let P_m and P be projections onto a space E with ideal kernels, such that (1.2) *holds. Then* $\ker P_m \to \ker P$.

Definition 1.3 (Birkhoff [1]). A linear idempotent operator P on $\mathbb{F}[\mathbf{x}]$ is called an ideal projection if $\ker P$ is an ideal in $\mathbb{F}[\mathbf{x}]$.

The symbol $\mathfrak{P}_N$ will stand for all ideal projections with the N-dimensional range. For a $G \in \mathfrak{G}^N$ we let $\mathfrak{P}_G$ be the family of all ideal projections onto G. Thus $\mathfrak{J}_G$ is in one-to-one correspondence with $\mathfrak{P}_G$.

A nice characterization of ideal projections is due to de Boor [2]:

Proposition 1.4. *Let P be a linear mapping on $\mathbb{F}[\mathbf{x}]$. Then P is an ideal projection if and only if*

$$P(fg) = P(f \cdot Pg) \tag{1.3}$$

for all $f, g \in \mathbb{F}[\mathbf{x}]$.

In one variable, the space $F_{<N}[x]$ of polynomials of degree less than N complements every ideal $J \in \mathfrak{J}_N$ (cf. proof of Proposition 2.1 below). In two or more variables there does not exist a subspace $G \in \mathfrak{G}^N$ with this property. However:

Theorem 1.5. *For every N and d there exists a fixed finite family $\mathcal{E}^N$ of N-dimensional (translation-invariant and spanned by monomials) subspaces of $\mathbb{F}[\mathbf{x}]$ such that every $J \in \mathfrak{J}_N$ complements at least one $E \in \mathcal{E}^N$.*

Let $\mathbb{F}_{<M}[\mathbf{x}]$ be the space of polynomials of degree less than M and $\mathbb{F}_{\leq M}[\mathbf{x}]$ be the space of polynomials of degree at most M. It follows from Theorem 1.5 that for every N there exists M such that every $E \in \mathcal{E}^N$ is a subspace of $\mathbb{F}_{<M}[\mathbf{x}]$, i.e., $\mathcal{E}^N \subset \mathfrak{G}^M(\mathbb{F}_{<M}[\mathbf{x}])$. Let $M = M(N)$ be the least integer such that $\mathcal{E}^N \subset \mathbb{F}_{<M}[\mathbf{x}]$.

The next theorem is a consequence of Theorem 1.5 and of de Boor's formula.

Theorem 1.6. *Let $(J, J_m,\ m \in \mathbb{N})$ be a sequence of ideals in $\mathfrak{J}_N$. Then $J_m \to J$ if and only if there exists a subspace $E \in \mathbb{F}_{<M}[\mathbf{x}]$ and a sequence of ideal projections $(P, P_m, m > M_0)$ such that* $\ker P = J$, $\ker P_m = J_m$ *and*

$$\|P - P_m\|_{\mathbb{F}_{\leq M}[\mathbf{x}]} \to 0. \tag{1.4}$$

Proof. If $J \in \mathfrak{J}_N$ then by Theorem 1.5 there exists a subspace $E \subset \mathbb{F}_{<M}[\mathbf{x}]$ that complements J and since $\mathbb{F}_{\leq M}[\mathbf{x}]$ is finite-dimensional, by Theorem 1.2, (1.3) implies (1.4).

Conversely, suppose that $f \in \mathbb{F}_{\leq K}[\mathbf{x}]$ for some $K \in \mathbb{N}$. If $K \leq N$ then (1.2) follows from (1.4). Assuming (1.2) for all monomials $f \in \mathbb{F}_{\leq K}[\mathbf{x}]$, let g be a monomial of degree $K+1$. Then $g = x_j f$ for some $j = 1, \dots, d$ and by (1.3):

$$\|Pg - P_m g\|_{\mathbb{F}_{\leq N}[\mathbf{x}]} = \|P(x_j Pf) - P_m(x_j P_m f)\|_{\mathbb{F}_{\leq N}[\mathbf{x}]} \to 0$$

since $x_j Pf, x_j P_m f \in \mathbb{F}_{\leq N}[\mathbf{x}]$ and $x_j P_m f \to x_j Pf$ by inductive assumption. Hence (1.2) holds for all monomials. □

This theorem allows us to define an ε-neighborhood of an ideal $J \in \mathfrak{J}_N$: let $E \in \mathcal{E}^N$ be such that $J \in \mathfrak{J}_E$. Let P be an ideal projection onto E with $\ker P = J$. Define

$$\mathcal{U}(E, J, \varepsilon) = \left\{ \ker Q : Q \in \mathfrak{P}_E, \|P - Q\|_{\mathbb{F}_{\leq N}[\mathbf{x}]} < \varepsilon \right\}$$

and

$$\mathcal{U}(J, \varepsilon) = \cup U(E, J, \varepsilon),$$

where the union is taken over all subspaces $E \subset \mathbb{F}_{<M}[\mathbf{x}]$ that complement J.

2 "Good ideals"

We will now examine closely one special space $E \in \mathcal{E}^N$:

$$E := \mathrm{span}\{1, x_1, x_1^2, \ldots, x_1^{N-1}\} \tag{2.1}$$

viewed as an N-dimensional subspace of $\mathbb{F}[\mathbf{x}] = \mathbb{F}[x_1, x_2, \ldots, x_d]$. Let P be an ideal projection onto E. Then

$$\begin{aligned} P(x_1^N) &= p_1 = \textstyle\sum_{j=0}^{N-1} b_{1,j} x_1^j, \\ P(x_2) &= p_2 = \textstyle\sum_{j=0}^{N-1} b_{2,j} x_1^j, \\ &\vdots \\ P(x_d) &= p_d = \textstyle\sum_{j=0}^{N-1} b_{d,j} x_1^j. \end{aligned} \tag{2.2}$$

Let B be the collection of coefficients

$$B = (b_{k,j}, k = 1, \ldots, d; j = 0, \ldots, N-1). \tag{2.3}$$

Then, by de Boor's formula (1.3), we have

$$\begin{aligned} P(x_1^{N+1}) &= P(x_1 P x_1^N) = P(x_1(\sum_{j=0}^{N-1} b_{1,j} x_1^j)) \\ &= P(\sum_{j=0}^{N-1} b_{1,j} x_1^{j+1}) = b_{1,N-1}(\sum_{j=0}^{N-1} b_{1,j} x_1^j) + \sum_{j=0}^{N-2} b_{1,j} x_1^{j+1} \\ &= b_{1,N-1} b_{1,0} + \sum_{j=1}^{N-1} b_{1,N-1}(b_{1,j} + b_{1,j-1}) x_1^j. \end{aligned}$$

Inductively, we conclude that

$$P(x_1^k) = \sum_{j=0}^{N-1} q_{k,j}(B) x_1^j$$

for all k, where $q_{k,j} \in \mathbb{F}[B]$ are polynomials in dN variables B. Now, using $P(x_j f) = P(f P x_j)$ for every $f \in E$ we conclude that

$$Pf = \sum_{j=0}^{N-1} q_{f,j}(B) x_1^j \tag{2.4}$$

for $f \in x_j E$, where $q_{f,j} \in \mathbb{F}[B]$. Inductively, (2.4) holds for all $f \in \mathbf{x}^{\mathbf{k}} E$ and therefore for all $f \in \mathbb{F}[\mathbf{x}]$ with $q_{f,j} \in \mathbb{F}[B]$.

Hence a sequence of d polynomials $(p_1, \ldots, p_d)$ given by (1.4) (or equivalently the sequence of dN coefficients B) completely determines the ideal projector P onto E. What is so special about this particular space E is that the converse also holds:

Proposition 2.1. *Every sequence $(p_1, \ldots, p_d)$ of polynomials in E defines an ideal $J = \langle x_1^N - p_1, x_2 - p_2, \ldots, x_d - p_d\rangle$ that complements E. Hence every sequence B of dN scalars defines an ideal projection P_B onto E by (2.4) and every ideal projection P onto E defines a sequence B_P by (2.2). Clearly*

$$P_{B_P} = P \text{ and } B_{P_B} = B.$$

Proof. It is clear from the construction of E that $E \cap J = \{0\}$. Let $f \in \mathbb{F}[x_1]$ be a polynomial in only one variable. Using the division algorithm in $\mathbb{F}[x_1]$ we have $f = q(x_1^N - p_1) + r$ with $\deg r < N$. Thus the ideal $\langle x_1^N - p_1\rangle$ generated by $x_1^N - p_1$ complements the space $\mathbb{F}_{<N}[x_1]$ of polynomials of degree less than N in $\mathbb{F}[x_1]$ and $E + J \supset \mathbb{F}[x_1]$. Inductively, we assume that $E + J \supset \mathbb{F}[x_1, \ldots, x_k]$, $k < d$ and prove that $E + J \supset \mathbb{F}[x_1, \ldots, x_k, x_{k+1}]$, i.e., we need to show that $x_{k+1}^n \mathbb{F}[x_1, \ldots, x_k] \subset E + J$ for all n. For $f \in \mathbb{F}[x_1, \ldots, x_k]$ we have

$$x_{k+1} f = (x_{k+1} - p_{k+1}) f + p_{k+1} f \in E + J \tag{2.5}$$

since the first term is in the ideal J and the second belongs to $E + J$ by inductive assumption. Using induction on n, assume that $f \in x_{k+1}^n \mathbb{F}[x_1, \ldots, x_k]$ and conclude that $x_{k+1} f \in x_{k+1}^{n+1} \mathbb{F}[x_1, \ldots, x_k]$ has a representation (2.5). □

Corollary 2.2. $B_m \to B$ *if and only if* $\ker P_{B_m} \to \ker P_B$.

Proof. Since the functions $q_{f,j}(B)$ in (2.4) are polynomials (hence continuous function of B) $B_m \to B$ implies $P_{B_m} f \to P_B f$ for every f. Conversely, if $P_{B_m} f \to P_B f$ then $P_{B_m} x_1^N \to P_B x_1^N$ and $P_{B_m} x_j \to P_B x_j$ for all $j = 2, \ldots, d$. Now $B_m \to B$ follows from (2.2). □

We are now ready to prove that every ideal $J \in \mathfrak{J}_E$ is a "good ideal".

Theorem 2.3. *Let $J \in \mathfrak{J}_E$ and $G \in \mathfrak{G}^N$. Then every neighborhood $\mathcal{U}(J)$ has a non-empty intersection with $\mathfrak{J}_G$.*

Proof. First, we establish that $\mathfrak{J}_E \cap \mathfrak{J}_G \neq \emptyset$. Let $g_1, \ldots, g_N$ be a bases in G and $e_1, \ldots, e_N$ be a bases in E. There exists a sequence $\{\mathbf{z}_1^*, \ldots, \mathbf{z}_N^*\}$ of points in $\mathbb{F}^d$ such that $g_j(\mathbf{z}_k^*) = \delta_{j,k}$ and hence the polynomial

$$\det(g_j(\mathbf{z}_k), j = 1, \ldots, N)$$

in dN variables (coordinates of the points $\mathbf{z}_k$) is not identically zero. Therefore the set

$$\mathcal{Z}_G := \{(\mathbf{z}_1, \ldots, \mathbf{z}_N) : \det(g_j(\mathbf{z}_k)) \neq 0\}$$

is an open and dense set in $(\mathbb{F}^d)^N$. Similarly the set

$$\mathcal{Z}_E := \{(\mathbf{z}_1, \ldots, \mathbf{z}_N) : \det(e_j(\mathbf{z}_k)) \neq 0\}$$

is an open and dense set in $(\mathbb{F}^d)^N$. Thus $\mathcal{Z}_G \cap \mathcal{Z}_E \neq \emptyset$ and for any $(\mathbf{z}_1, \ldots, \mathbf{z}_N) \in \mathcal{Z}_G \cap \mathcal{Z}_E$, the ideal

$$J := \{f \in \mathbb{F}[\mathbf{x}] : f(\mathbf{z}_j) = 0,\ j = 1, \ldots, N\}$$

complements E and G. Therefore, there exists a sequence $B^* \in \mathbb{F}^{dN}$ such that P_{B^*} is an ideal projection onto E and $\ker P_{B^*} \in \mathfrak{J}_G$. Let

$$\mathcal{B}_G := \{B \in \mathbb{F}^{dN} : \ker P_B \in \mathfrak{J}_G\}.$$

It follows that $\mathcal{B}_G \neq \emptyset$. Suppose that $\ker P_B \in \mathfrak{J}_G$, i.e., $\ker P_B \cap G = \{0\}$. Then

$$P_B(\sum_{k=1}^{N} \alpha_k g_k) = \sum_{k=1}^{N} \alpha_k P_B g_k = 0$$

implies $\alpha_k = 0$ for all $k = 1, \ldots, N$. Hence $\ker P_B \in \mathfrak{J}_G$ is the same as linear independency of the sequence of polynomials $(P_B g_k, k = 1, \ldots, N)$. Since, by (2.4), $P_B f = \sum_{j=0}^{N-1} q_{g_k,j}(B) x_1^j$ this is equivalent to

$$\det(q_{g_k,j}(B)) \neq 0.$$

Since $\mathcal{B}_G \neq \emptyset$, this determinant is a non-zero polynomial in $\mathbb{F}[B]$, hence there exists an open and dense set of $B \subset \mathbb{F}^{dN}$ such that $\ker P_B \in \mathfrak{J}_G$. □

3 "Bad Ideals"

In this section we will use a beautiful construction of Iarrobino and modified reasoning of [7] to show that for $d \geq 3$ and for sufficiently large N there exists an ideal $J \in \mathfrak{J}_N$ such that J is not the limit of ideals in $\mathfrak{J}_E$, where $E := \operatorname{span}\{1, x_1, x_1^2, \ldots, x_1^{N-1}\}$ is the space considered in the previous section.

Let $W := M^d_{<n}[\mathbf{x}]$ be the set of monomials of degree less than n. Let $U \cup V = M^d_n[\mathbf{x}]$ be a non-trivial partition of the set $M^d_n[\mathbf{x}]$ of all monomials of degree n in $\mathbb{F}[\mathbf{x}]$. Let H be the subspace of $\mathbb{F}[\mathbf{x}]$ spanned by $\mathbb{F}^d_{<n}[\mathbf{x}]$ and V and let

$$N = \dim H = \#V + \#W \tag{3.1}$$

where # is the cardinality of the set. For any choice of matrix $C \in \mathbb{F}^{U \times V}$, the space J_C spanned by monomials of degree greater than n and the specific polynomials

$$p_u := u - \sum_{v \in V} C(u,v)v,\ u \in U, \tag{3.2}$$

complements H and is an ideal. The latter is so because each p_u is homogeneous, thus every product of a monomial with p_u is in J_C. Furthermore, as is easy to see

$$\mathbb{F}[\mathbf{x}] = H \oplus J_C \tag{3.3}$$

for each C.

Theorem 3.1. *For any $d \geq 3$ and for sufficiently large n, there exists a partition $U \cup V = M_n^d[\mathbf{x}]$ and a matrix $C \in \mathbb{F}^{U \times V}$ such that the ideal J_C can not be perturbed to complement E.*

Proof. Assume that J_C complements H and that there is a small perturbation of J_C that complements E. In other words, assume the existence of a sequence of ideals $\{J_m\}$ such that each J_m complements G and $J_m \to J_C$. This is the same as the existence of a sequence of ideal projections P_m onto H such that $\ker P_n$ complements E and H at the same time and $P_m f \to Pf$ for every $f \in \mathbb{F}^d[\mathbf{x}]$. In particular

$$u - P_m u \to u - \sum_{v \in V} C(u,v)v, \forall\, u \in U.$$

Since P_m is a projection onto H it follows that

$$P_m u = \sum_{v \in V} C_m(u,v)v + \sum_{w \in W} C_m(u,w)w \tag{3.4}$$

and

$$C_m(u,v) \to C(u,v), \forall u \in U \text{ and } C_m(u,w) \to 0, \forall w \in W. \tag{3.5}$$

Since $\ker P_m$ complements E, we let Q_m be the ideal projection onto E with $\ker Q_m = \ker P_m$. Then $u - P_m u \in \ker P_m = \ker Q_m$ and

$$0 = Q_m(u - P_m u) = Q_m u - (\sum_{v \in V} C_m(u,v) Q_m v + \sum_{w \in W} C_m(u,w) Q_m w)$$

or

$$\sum_{v \in V} C_m(u,v) Q_m v + \sum_{w \in W} C_m(u,w) Q_m w = Q_m u. \tag{3.6}$$

At this point it is important to notice that, as projections onto E, the operators Q_m depend polynomially on $d \times N$ parameters B as in (2.4):

$$Q_m f = \sum_{k=0}^{N-1} q_{k,f}^{(m)}(B) x_1^k \tag{3.7}$$

where $q_{k,f}^{(n)} \in \mathbb{F}^{dN}[B]$. Rewriting (3.6) we have:

$$\sum_{k=0}^{N-1} (\sum_{v \in V} C_m(u,v) q_{k,v}^{(m)}(B) + \sum_{w \in W} C_m(u,W) q_{k,w}^{(m)}(B)) x_1^k = \sum_{k=0}^{N-1} q_{k,u}^{(m)}(B) x_1^k \tag{3.8}$$

or equivalently

$$\sum_{v \in V} C_m(u,v) q_{k,v}^{(m)}(B) + \sum_{w \in W} C_m(u,w) q_{k,w}^{(m)}(B) = q_{k,u}^{(m)}(B), u \in U, k = 0, \dots, N-1. \tag{3.9}$$

Since $\#V + \#W = N$, this is the system of $N \times \#U$ equations with the same number of unknowns $\{C_m(u,v), u \in U\}$ and $\{C_m(u,w), u \in U\}$. By Cramer's rule

$$C_m(u,v) = \frac{\det V_{m,u,v}(B)}{\det V_m(B)}, \tag{3.10}$$

where $\det V_m(B)$ is the determinant of the matrix on the left-hand side of (3.9), and $\det V_{m,u,v}(A)$ is the determinant of the same matrix with the (u,k)-th column replaced by the column $(q_{k,u}^{(m)}(B))$. Notice that it makes the $C_m(u,v)$ rational functions of $d \times N$ parameters B with common denominator. Thinking of $1/\det V_m(B)$ as just another variable, say z, we conclude that the set

$$\mathfrak{C} := \{z \det V_{n,u,v}(B),\ B \in \mathbb{F}^{dN}, z \in \mathbb{F}\backslash\{0\}\} \subset \mathbb{F}^{\#U \times \#V}$$

is a polynomial image of $\mathbb{F}^{dN+1}$ where the first dN parameters are the parameters of B. As such $\mathfrak{C}$ forms an affine subvariety of $\mathbb{F}^{\#U \times \#V}$ with $\dim \mathfrak{C} \leq dN + 1$ (cf. [4, Theorem 2, p. 446]. Hence if

$$\#U \times \#V > dN + 1 \tag{3.11}$$

then there exists a collection $\{C(u,v) \in \mathbb{F}^{\#U \times \#V}\}$ which is not in the closure of $\mathfrak{C}$, contradicting (3.5).

Now we only need to count. As was pointed out in [5], we have $\#M_n^d[\mathbf{x}] = \binom{n+d-1}{d-1} \approx (\frac{n^{d-1}}{(d-1)!})$, choosing the partition $U \cup V = M_n^d[\mathbf{x}]$ such that $\#U = \left\lfloor \frac{1}{2}\binom{n+d-1}{d-1} \right\rfloor$ we have

$$\#U \times \#V = \left\lfloor \frac{1}{2}\binom{n+d-1}{d-1} \right\rfloor \left\lceil \frac{1}{2}\binom{n+d-1}{d-1} \right\rceil \approx \frac{1}{4}\left(\frac{n^{d-1}}{(d-1)!}\right)^2 \approx n^{2d-2}$$

while $N = \dim H = \dim \mathbb{F}_{\leq n}^d[\mathbf{x}] - \#U = \binom{n+d}{d} - \left\lfloor \frac{1}{2}\binom{n+d-1}{d-1} \right\rfloor \approx \frac{n^d}{d!} \approx n^d$. Hence for sufficiently large n, (3.11) holds. Direct computation yields (3.11) with $n = 7$, for $d = 3$; $n = 3$ for $d = 4$ or 5 and $n = 2$ for $d > 5$. □

Remark 3.2. The "bad" ideals do not exist in $\mathbb{F}[\mathbf{x}]$ for $d = 1$, as follows from Theorem 2.3, since, for $d = 1$, every ideal in $\mathfrak{J}_N$ is complemented to

$$E := \text{span}\{1, x_1, x_1^2, \ldots, x_1^{N-1}\}.$$

"Bad" ideals also do not exist in $\mathbb{C}[\mathbf{x}]$ for $d = 2$ as noted in [8]. The existence of "bad" ideals in $\mathbb{R}[\mathbf{x}]$ for $d = 2$ is an open problem.

References

[1] G. Birkhoff, *The algebra of multivariate interpolation*, in Constructive approaches to mathematical models (C. V. Coffman and G. J. Fix, eds.), pp. 345–363, Academic Press, New York, 1979.

[2] C. de Boor, *Ideal interpolation*, in Approximation Theory XI: Gatlinburg 2004 (C. K. Chui, M. Neamtu, and L. L. Schumaker, eds.), Nashboro Press (2005), pp. 59–91.

[3] D. Cox, J. Little, and D. O'Shea, *Using algebraic geometry*, Graduate Texts in Mathematics, Springer-Verlag, New York, Berlin, Heidelberg, 1997.

[4] ———, *Ideals, varieties, and algorithms*, second edition, Springer-Verlag, New York, Berlin, Heidelberg, 1997.

[5] A. Iarrobino, *Reducibility of the families of* 0*-dimensional schemes on a variety*, Invent. Math. 15 (1972), pp. 72–77.

[6] D. Mumford, *The red book of varieties and schemes*, Lecture Notes in Mathematics 1358, Springer-Verlag, 1988.

[7] B. Shekhtman, *On a conjecture of Carl de Boor regarding the limits of Lagrange interpolants*, Constr. Approx. 24 (2006), pp. 365–370.

[8] ———, *On bivariate ideal projectors and their perturbations*, Advances in Comput. Math. (to appear).

Author information

Boris Shekhtman, Department of Mathematics, University of South Florida, Tampa, FL 33620, USA.
Email: `boris@math.usf.edu`

Banach Spaces and their Applications in Analysis, 423–438

Asymptotically transitive Banach spaces

Jarno Talponen

Abstract. We introduce a flexible almost isometric version of the almost transitivity property of Banach spaces. With the help of this new notion we generalize to several directions a strong recent rotational characterization of Hilbert spaces due to Randrianantoanina. This characterization is a partial answer to the classical Banach–Mazur rotation problem.

Key words. Asymptotically transitive, almost transitive, almost isometric, norm geometry, subspaces of L^p.

AMS classification. 46C15, 46B04.

1 Introduction

A Banach space $\mathbf{X}$ is called *transitive* if for each $x \in \mathbf{S}_{\mathbf{X}}$ the orbit

$$\mathcal{G}_{\mathbf{X}}(x) = \{T(x) |\, T\colon \mathbf{X} \to \mathbf{X} \text{ is an isometric automorphism}\} = \mathbf{S}_{\mathbf{X}}.$$

If $\overline{\mathcal{G}_{\mathbf{X}}(x)} = \mathbf{S}_{\mathbf{X}}$ for all $x \in \mathbf{S}_{\mathbf{X}}$ then $\mathbf{X}$ is called *almost transitive*. The following classical *Banach–Mazur rotation problem*, which already appears in Banach's book [3, p. 242], remains unsolved: *Is every separable transitive Banach space* $\mathbf{X}$ *in fact isometrically a Hilbert space?*

Various partial answers to this problem are known and together with related results they form a substantial theory of rotations in Banach spaces. A recent comprehensive survey of this field is found in [5]. The following recent partial answer to the problem is due to Randrianantoanina:

Theorem 1.1 ([19]). *If* $\mathbf{X}$ *is an almost transitive Banach space, which contains a* 1*-codimensional* 1*-complemented subspace* $\mathbf{Z} \subset \mathbf{X}$*, then* $\mathbf{X}$ *is isometric to a Hilbert space.*

Theorem 1.1 is very general among 1-codimensional characterizations of Hilbert spaces in terms of rotations. Its proof implicitly applies 1-dimensional linear closest point selections. One should note that the assumptions of the above result are isometric in nature. In this paper we show that the rigidity provided by the isometric conditions can be relaxed and that the corresponding almost isometric conditions are actually sufficient for this type of characterization (see Theorem 2.2 below). As an application we also obtain some additional information about the structure of projections onto 1-codimensional and other subspaces.

In Section 2 we will introduce and study asymptotic transitivity property for Banach spaces which implicitly appeared in [17], [7], and [18] in connection with universal

This work has been supported financially by the Academy of Finland projects # 53893 and # 210970 during the years 2003–2005 and by the Finnish Cultural Foundation in 2006.

disposition property (see [11]). In the framework of rotations, asymptotic transitivity property seems to be a natural generalization of almost transitivity. Thus the asymptotic transitive setting also yields some new information about classical almost transitive spaces.

We now introduce notations. Throughout $\mathbf{X}$, $\mathbf{Y}$, and $\mathbf{Z}$ denote real Banach spaces. The unit ball and the unit sphere of $\mathbf{X}$ are denoted by $\mathbf{B_X}$ and $\mathbf{S_X}$ respectively. The space of continuous linear operators $T : \mathbf{X} \to \mathbf{Y}$ is denoted by $L(\mathbf{X}, \mathbf{Y})$ (abbreviated to $L(\mathbf{X})$ if $\mathbf{X} = \mathbf{Y}$). Denote

$$\mathrm{Aut}(\mathbf{X}) = \{T \in L(\mathbf{X}) : T \text{ is an isomorphism}\}.$$

For two mutually isomorphic Banach spaces $\mathbf{X}$ and $\mathbf{Y}$, we recall that their *Banach–Mazur distance* is given by

$$d(\mathbf{X}, \mathbf{Y}) = \inf\{\|T\| \cdot \|T^{-1}\| : \ T \in L(\mathbf{X}, \mathbf{Y}) \text{ isomorphism}\}.$$

Spaces $\mathbf{X}$ and $\mathbf{Y}$ are said to be *almost isometric* if $d(\mathbf{X}, \mathbf{Y}) = 1$. We denote the *group of rotations* of $\mathbf{X}$ by

$$\mathcal{G}_{\mathbf{X}} = \{T \in \mathrm{Aut}(\mathbf{X}) : \max(\|T\|, \|T^{-1}\|) = 1\}$$

and the *orbit* of $x \in \mathbf{S_X}$ by $\mathcal{G}_{\mathbf{X}}(x) = \{T(x) : T \in \mathcal{G}_{\mathbf{X}}\}$. If $\overline{\mathrm{conv}}(\mathcal{G}_{\mathbf{X}}(x)) = \mathbf{B_X}$ for all $x \in \mathbf{S_X}$ then $\mathbf{X}$ is called *convex-transitive.*

We say that $\mathbf{Y}, \mathbf{Z} \subset \mathbf{X}$ are L^p-summands of $\mathbf{X}$ if $\mathbf{X} = \mathbf{Y} \oplus_p \mathbf{Z}$, where as usual, for $(y, z) \in \mathbf{Y} \times \mathbf{Z}$, $\|(y, z)\| = (\|y\|_{\mathbf{Y}}^p + \|z\|_{\mathbf{Z}}^p)^{1/p}$ (respectively, $\max(\|y\|_{\mathbf{Y}}, \|z\|_{\mathbf{Z}})$ for $p = \infty$).

For $f \in \mathbf{X}^*$ and $x \in \mathbf{X}$ we let $f \otimes x \colon \mathbf{X} \to [x]$ be the map $y \mapsto f(y)x$.

Next we recall basic concepts concerning the geometry of the norm which we refer the reader to [13] for more details. A norm $\|\cdot\|$ is said to be *Gâteaux differentiable* at $x \in \mathbf{S_X}$ if there is $l \in \mathbf{X}^*$ such that

$$\lim_{t \to 0} \frac{\|x + th\| - 1}{t} = l(h)$$

for every $h \in \mathbf{X}$. If $\|\cdot\|$ is Gâteaux differentiable at $x \in \mathbf{S_X}$ and the derivative above satisfies

$$\lim_{\|h\| \to 0} \frac{\|x + h\| - 1}{\|h\|} - l(h) = 0$$

then $\|\cdot\|$ is said to be *Fréchet differentiable* at $x \in \mathbf{S_X}$. Alternatively, the point x above is called *(Gâteaux-)smooth*, respectively *Fréchet-smooth*.

We end this section with classical results that we will use in the sequel (see e.g. [12, p. 92, p. 300]).

Lemma 1.2 (Šmulyan). *For a Banach space $(\mathbf{X}, \|\cdot\|)$, the following conditions are equivalent:*

(1) The norm $\|\cdot\|$ is Fréchet differentiable at $x \in \mathbf{S_X}$.
(2) Let $(f_n), (g_n) \subset \mathbf{S_{X^*}}$. If $\lim_{n\to\infty} f_n(x) = \lim_{n\to\infty} g_n(x) = 1$ then $\lim_{n\to\infty} \|f_n - g_n\| = 0$.

Theorem 1.3 (Bishop–Phelps). *The norm-attaining functionals of $\mathbf{S_{X^*}}$ are dense in $\mathbf{S_{X^*}}$.*

2 Asymptotic transitivity and orthogonal structures

Definition 2.1. We say that a Banach space $\mathbf{X}$ is *asymptotically transitive* if for every $x \in \mathbf{S_X}$ the generalized orbit $\mathcal{O}_\mathbf{X}(x)$ of x defined by

$$\mathcal{O}(x) = \mathcal{O}_\mathbf{X}(x) = \bigcap_{\varepsilon>0} \left\{T(x) : T \in \mathrm{Aut}(\mathbf{X}),\ \max(\|T\|, \|T^{-1}\|) \leq 1+\varepsilon\right\}$$

is equal to $\mathbf{S_X}$.

It is immediate that:

(i) $\mathcal{O}(x)$ is norm closed for all $x \in \mathbf{S_X}$.
(ii) Almost transitive spaces are asymptotically transitive (see e.g. [17]).
(iii) If $\mathbf{X}$ and $\mathbf{Y}$ are almost isometric and $\mathbf{X}$ is almost transitive, then $\mathbf{Y}$ is asymptotically transitive.
(iv) The convex-transitive space $L^\infty(0,1)$ is not asymptotically transitive.

For (iv), we observe that if $T \in \mathrm{Aut}(L^\infty)$ then $\|T(x)\|_\infty \geq \|T^{-1}\|^{-1}\mathrm{ess\ inf}_t|x(t)|$ for all $x \in L^\infty(0,1)$. Hence $\chi_{[0,\frac{1}{2}]} \notin \mathcal{O}_{L^\infty}(\chi_{[0,1]})$ even though L^∞ is convex-transitive (see [5]).

The generalized orbit $\mathcal{O}(x)$ and the closure $\overline{\mathcal{G}(x)}$ need not coincide in general (see Example 5.3), but unfortunately we do not know at the moment any examples of asymptotically transitive spaces, which are not almost transitive.

Our work in this section is aimed at showing the following main result:

Theorem 2.2. *Let $\mathbf{X}$ be an asymptotically transitive Banach space such that for each $\varepsilon > 0$ there is 1-codimensional $(1+\varepsilon)$-complemented subspace $Z_\varepsilon \subset \mathbf{X}$. Then $\mathbf{X}$ is isometrically a Hilbert space.*

We will first establish some crucial observations. Define the local projectional indices $\hat{\eta}_\mathbf{X}, \check{\eta}_\mathbf{X} : \mathbf{S_X} \to \mathbb{R}$ by setting for $y \in \mathbf{S_X}$

$$\hat{\eta}(y) = \hat{\eta}_\mathbf{X}(y) = \inf\{\|\mathbf{I} - f \otimes y\| :\ f \in \mathbf{X}^*,\ f(y) = 1\},$$

and

$$\check{\eta}(y) = \check{\eta}_\mathbf{X}(y) = \inf_{\varepsilon>0} \sup\{\|\mathbf{I} - f \otimes y\| :\ f \in (1+\varepsilon)\mathbf{B}_{\mathbf{X}^*},\ f(y) = 1\}.$$

Clearly $1 \leq \hat{\eta}_{\mathbf{X}} \leq \check{\eta}_{\mathbf{X}} \leq 3$, $\hat{\eta}_{\mathbf{Y}} \leq \hat{\eta}_{\mathbf{X}}$, and $\check{\eta}_{\mathbf{Y}} \leq \check{\eta}_{\mathbf{X}}$ pointwise for $\mathbf{Y} \subset \mathbf{X}$.

Lemma 2.3. *For any Banach space $\mathbf{X}$ the following holds: For each $x \in \mathbf{S_X}$ there is $f \in \mathbf{X}^*$ such that $f(x) = 1$ and $\|\mathbf{I}_X - f \otimes x\| = \hat{\eta}(x)$.*

Proof. Let $x \in \mathbf{S_X}$ and $(f_n) \subset \mathbf{X}^*$ be a sequence such that $f_n(x) = 1$ and $\|\mathbf{I_X} - f_n \otimes x\| \leq \hat{\eta}(x) + n^{-1}$ for each $n \in \mathbb{N}$. Note that

$$\|f_n\| = \|f_n \otimes x\| \leq 1 + \|\mathbf{I_X} - f_n \otimes x\| \leq 1 + \hat{\eta}(x) + n^{-1},$$

so that $(f_n) \subset 5\mathbf{B_{X^*}}$. Thus, (f_n) has a weak*-cluster point $f \in 5\mathbf{B_{X^*}}$. Clearly $f(x) = 1$ and it can be easily checked that $\|\mathbf{I_X} - f \otimes x\| \leq \limsup_{n\to\infty} \|\mathbf{I_X} - f_n \otimes x\| = \hat{\eta}(x)$. □

Proposition 2.4. *Suppose that $\mathbf{X}$ is an asymptotically transitive Banach space. Then $\hat{\eta}$ and $\check{\eta}$ are constant functions.*

Proof. Let $x, y \in \mathbf{S_X}$. By the asymptotic transitivity of $\mathbf{X}$ there is a sequence $(T_n) \subset \operatorname{Aut}(\mathbf{X})$ such that $T_n(x) = y$ and $\max(\|T_n\|, \|T_n^{-1}\|) \leq 1 + n^{-1}$ for each $n \in \mathbb{N}$. By Lemma 2.3, there is $f \in \mathbf{X}^*$, $f(x) = 1$ such that $\|\mathbf{I} - f \otimes x\| = \hat{\eta}(x)$. Define $g_n = f \circ T_n^{-1}$ for $n \in \mathbb{N}$. Then $g_n(y) = 1$ for $n \in \mathbb{N}$, so that

$$\hat{\eta}(y) \leq \limsup_{n\to\infty} \|\mathbf{I} - g_n \otimes y\| \leq \limsup_{n\to\infty} (\|T_n\| \cdot \|\mathbf{I} - f \otimes x\| \cdot \|T_n^{-1}\|) = \hat{\eta}(x).$$

Reversing the role of $\hat{\eta}(x)$ and $\hat{\eta}(y)$ shows that $\hat{\eta}(x) = \hat{\eta}(y)$.

In the case of $\check{\eta}$ suppose similarly as above that $(f_n) \subset 2\mathbf{B_{X^*}}$ is a sequence so that $f_n(x) = 1$ for each $n \in \mathbb{N}$, $\|f_n\| \to 1$ and $\|\mathbf{I} - f_n \otimes x\| \to \check{\eta}(x)$ as $n \to \infty$. Then putting $g_n = f_n \circ T_n^{-1}$, $n \in \mathbb{N}$, as above yields the claim. □

Proof of Theorem 2.2. If $\dim(\mathbf{X}) \leq 2$ then the theorem is clear. Assume that $\dim(\mathbf{X}) \geq 3$. Recall that a combination of [1, Lemma 13.1] and [1, Criterion 13.4'] yields the following characterization of Hilbert spaces:
A Banach space E, $\dim(E) \geq 3$, is isometrically a Hilbert space if and only if for all $e \in \mathbf{S}_E$ there is $f \in E^*$ such that $f(e) = 1$ and $\|\mathbf{I} - f \otimes e\| = 1$.

By Proposition 2.4, $\hat{\eta}(\cdot)$ is a constant function. For $\varepsilon > 0$, there exist $x \notin Z_\varepsilon$ and $g \in \mathbf{X}^*$ such that $\|\mathbf{I} - g \otimes x\| \leq 1 + \varepsilon$. Thus $\hat{\eta}(x) \leq 1 + \varepsilon$. Since ε is arbitrary, we have $\hat{\eta}(y) = 1$ for all $y \in \mathbf{S_X}$. Lemma 2.3 insures that for $e \in \mathbf{S_X}$, there is $f \in \mathbf{X}^*$ such that $\|\mathbf{I} - f \otimes e\| = 1$. Thus the proof is complete. □

One should note that the above result was anticipated (in the almost transitive setting) in F. Cabello's Ph.D. thesis [6]. See also [2, Proposition 1.5] for a related result.

Remark 2.5. The referee suggested an alternative route to Theorem 2.2 using Theorem 1.1 as follows:

(i) If $\mathbf{X}$ is asymptotically transitive, then each non-trivial ultrapower $\mathbf{X}_\mathcal{U}$ is transitive.

(ii) If for each $\varepsilon > 0$, $\mathbf{X}$ has $(1 + \varepsilon)$-complemented hyperplanes, then each non-trivial ultrapower $\mathbf{X}_\mathcal{U}$ has 1-complemented hyperplanes.

(iii) $\mathbf{X}$ is an inner product space if and only if $\mathbf{X}_{\mathcal{U}}$ is.

Proposition 2.6. *Assume that* $\mathbf{X}$ *is asymptotically transitive and Fréchet-smooth Banach space. Then* $\check{\eta} = C$ *is a constant function and* $\|\mathbf{I} - P\| = C$ *for any* 1*-dimensional projection* $P \in L(\mathbf{X})$ *such that* $\|P\| = 1$.

Proof. Let $y \in \mathbf{S_X}$. Suppose that a sequence $(g_n) \subset 2\mathbf{B_{X^*}}$ is chosen as in the proof of Proposition 2.4 for $\check{\eta}$. Then an application of the Fréchet-smoothness together with Lemma 1.2 gives that $\lim_{n\to\infty} \|g_n - g\| = 0$ where g is the unique element of $\mathbf{S_{X^*}}$ satisfying $g(y) = 1$ and $g \otimes y = P$. We obtain that $||\mathbf{I} - g_n \otimes y|| \to ||\mathbf{I} - g \otimes y||$ as $n \to \infty$. Thus $||\mathbf{I} - g \otimes y|| = \check{\eta}(y)$. This yields the claim, since $\check{\eta}$ is a constant function by Proposition 2.4. □

It turns out (see Theorem 3.1 below) that in the asymptotically transitive setting rather mild geometric conditions, for example the Fréchet-differentiability of the norm at *some* point $x \in \mathbf{S_X}$, already imply the Fréchet-differentiability of the norm at *all* $y \in \mathbf{S_X}$.

3 Asymptotic transitivity and the geometry of the norm

We will show that in the case of asymptotically transitive spaces some suitable control of the convexity or of the smoothness of the unit ball actually guarantees simultaneously *uniform convexity* and *uniform smoothness*. See [8] and [5] for similar results in the almost transitive setting. We will use some concepts and results about the geometry of the norm, for a discussion of which we refer to [13, Chapter 1]. Denote *closed slices* by $S(\mathbf{B_X}, f, \alpha) = \{x \in \mathbf{B_X} : f(x) \geq \alpha\}$ for $f \in \mathbf{S_{X^*}}$, $0 \leq \alpha \leq 1$. It is said that $\mathbf{B_X}$ is *dentable* (respectively $\mathbf{B_{X^*}}$ is ω^**-dentable*) if

$$\inf_{\substack{f\in \mathbf{S_{X^*}},\\ 0<\alpha<1}} \operatorname{diam}(S(\mathbf{B_X}, f, \alpha)) = 0 \ \ (\text{resp.} \inf_{\substack{x\in \mathbf{S_X},\\ 0<\alpha<1}} \operatorname{diam}(S(\mathbf{B_{X^*}}, x, \alpha)) = 0).$$

Recall that the local modulus of convexity $\Delta_\mathbf{X} \colon \mathbf{S_X} \times [0,1) \to \mathbb{R}$ is defined by:

$$\Delta_\mathbf{X}(z, \varepsilon) = \inf\{1 - \lambda : \lambda \leq 1, \text{ there is } v \in \mathbf{X} \text{ such that } \|v\| \geq \varepsilon,\ \|\lambda z \pm v\| \leq 1\}$$

for $z \in \mathbf{S_X}$ and $0 \leq \varepsilon < 1$. Note that $\Delta_\mathbf{X}(z, 0) = 0$ for each $z \in \mathbf{S_X}$.

Theorem 3.1. *Let* $\mathbf{X}$ *be an asymptotically transitive Banach space. If* $\mathbf{B_X}$ *is dentable or* $\mathbf{B_{X^*}}$ *is* ω^**-dentable, then* $\mathbf{X}^*$ *is also asymptotically transitive, and both* $\mathbf{X}$ *and* $\mathbf{X}^*$ *are uniformly convex and uniformly smooth.*

Proof. We will first prove the following auxiliary technical claim:
Claim. Suppose that $\mathbf{X}$ *and* $\mathbf{Y}$ *are Banach spaces,* $x \in \mathbf{S_X}$*, and* $y \in \mathbf{S_Y}$*. If* $T \colon \mathbf{X} \to \mathbf{Y}$ *is a linear isomorphism such that* $T(x) = y$*, then*

$$\Delta_\mathbf{Y}\Big(y, \frac{\varepsilon \cdot \Delta_\mathbf{X}(x,\varepsilon)}{||T^{-1}||(||T|| + \Delta_\mathbf{X}(x,\varepsilon) - 1)}\Big) \leq \Delta_\mathbf{X}(x, \varepsilon). \tag{3.1}$$

With the convention that $0/0 = 0$.

To prove (3.1), observe first that $\min(\|T\|, \|T^{-1}\|) \geq 1$. Note that (3.1) holds trivially if $\varepsilon\Delta_{\mathbf{X}}(x,\varepsilon) = 0$. Suppose that $0 < \varepsilon < 1$, $\Delta_{\mathbf{X}}(x,\varepsilon) > 0$, and write $1 - \lambda = \Delta_{\mathbf{X}}(x,\varepsilon)$. Thus $0 \leq \lambda \leq 1$. Let (v_n) be a sequence such that for $n \in \mathbb{N}$, $\|\lambda x \pm v_n\| \leq 1$ and $\liminf_{n\to\infty} \|v_n\|_{\mathbf{X}} \geq \varepsilon$.

Define an auxiliary mapping $\phi\colon \mathbb{R} \to (0,\infty)$ by $\phi(t) = (1-\lambda)/(t-\lambda)$ if $t > 1$ and $\phi(t) = 1$ otherwise. Write $T(v_n) = w_n,\ n \in \mathbb{N}$.

We aim to prove that for $n \in \mathbb{N}$,

$$\Delta_{\mathbf{Y}}(y, \alpha_n\|w_n\|) \leq \Delta_{\mathbf{X}}(x,\varepsilon), \tag{3.2}$$

where $\alpha_n = \min_{\theta=\pm1} \phi(\|\lambda y + \theta w_n\|)$. Indeed, since $h \mapsto (\|\lambda y + hw\| - \lambda)/h$ is a non-decreasing map for each w, we obtain by substituting $h = (1-\lambda)/(\|\lambda y + w_n\| - \lambda)$ once $\|\lambda y + w_n\| > 1$ that

$$\frac{\left\|\lambda y + \frac{1-\lambda}{\|\lambda y+w_n\|-\lambda} w_n\right\| - \lambda}{\frac{1-\lambda}{\|\lambda y+w_n\|-\lambda}} \leq \|\lambda y + w_n\| - \lambda.$$

Consequently

$$\frac{\left\|\lambda y + \frac{1-\lambda}{\|\lambda y+w_n\|-\lambda} w_n\right\| - \lambda}{1-\lambda} \leq 1,$$

and hence

$$\left\|\lambda y + \frac{1-\lambda}{\|\lambda y + w_n\| - \lambda} w_n\right\| \leq 1.$$

Therefore we obtain for w_n that in general

$$\|\lambda y + \phi(\|\lambda y + w_n\|)w_n\| \leq 1$$

and similarly

$$\|\lambda y - \phi(\|\lambda y - w_n\|)w_n\| \leq 1.$$

Note that the set $\{h \in \mathbb{R} : \|\lambda y + hw_n\| \leq 1\} \subset \mathbb{R}$ is a compact interval that contains 0. Moreover, the previous estimates give that

$$[-\phi(\|\lambda y - w_n\|), \phi(\|\lambda y + w_n\|)] \subset \{h \in \mathbb{R} : \|\lambda y + hw_n\| \leq 1\}.$$

In particular we obtain that $[-\alpha_n, \alpha_n] \subset \{h \in \mathbb{R} : \|\lambda y + hw_n\| \leq 1\}$ and $\|\lambda y \pm \alpha_n w_n\| \leq 1$. Hence we have proved (3.2).

Next, we observe that since by assumption $\|\lambda x \pm v_n\| \leq 1$ and $\Delta_{\mathbf{X}}(x,\varepsilon) = 1 - \lambda$,

$$\|\lambda y \pm w_n\| - \lambda = \|T(\lambda x \pm v_n)\| - \lambda \leq \|T\| - \lambda = \|T\| + \Delta_{\mathbf{X}}(x,\varepsilon) - 1.$$

Thus

$$\frac{\Delta_{\mathbf{X}}(x,\varepsilon)}{\|T\| + \Delta_{\mathbf{X}}(x,\varepsilon) - 1} \leq \min_{\theta=\pm1}\left(\frac{1-\lambda}{\|\lambda y + \theta w_n\| - \lambda}\right) = \alpha_n. \tag{3.3}$$

Note that since $||T|| - 1 \geq 0$ we have $\Delta_{\mathbf{X}}(x,\varepsilon)/(\|T\| + \Delta_{\mathbf{X}}(x,\varepsilon) - 1) \leq 1$. Moreover, as $\|w_n\| \geq \|v_n\|/\|T^{-1}\|$, $\liminf_{n\to\infty} \|v_n\|/\|T^{-1}\| \geq \varepsilon/\|T^{-1}\|$, and $\Delta_{\mathbf{Y}}(y,\cdot)$ is non-decreasing, (3.1) follows from (3.2) and (3.3).

Let $x, y \in \mathbf{S}_{\mathbf{X}}$. Then $\Delta_{\mathbf{X}}(x,\cdot), \Delta_{\mathbf{X}}(y,\cdot) \colon (0,1) \to \mathbb{R}$ are non-decreasing maps. Thus they are continuous a.e. with respect to the Lebesgue measure m on $(0,1)$. Let $t_0 \in (0,1)$ be a point of continuity for both $\Delta_{\mathbf{X}}(x,\cdot)$ and $\Delta_{\mathbf{X}}(y,\cdot)$. Then by the asymptotic transitivity of $\mathbf{X}$ there is a sequence $(T_n) \subset \mathrm{Aut}(\mathbf{X})$ such that $T_n(x) = y$, $n \in \mathbb{N}$, and $||T_n|| \cdot ||T_n^{-1}|| \to 1$ as $n \to \infty$. It follows from (3.1) and symmetry that $\Delta_{\mathbf{X}}(x,t_0) = \Delta_{\mathbf{X}}(y,t_0)$. We conclude that $\Delta_{\mathbf{X}}(x,\cdot)$ and $\Delta_{\mathbf{X}}(y,\cdot)$ coincide m-a.e. Hence

$$\lim_{t\to s^-} \Delta_{\mathbf{X}}(x,t) = \lim_{t\to s^-} \Delta_{\mathbf{X}}(y,t) \tag{3.4}$$

for all $s \in (0,1)$ and $(x,y) \in \mathbf{S}_{\mathbf{X}} \times \mathbf{S}_{\mathbf{X}}$.

• Consider first the case where $\mathbf{B}_{\mathbf{X}}$ is dentable. Fix sequences $(f_n) \subset \mathbf{S}_{X^*}$ and $(\alpha_n) \subset (0,1)$ such that $\alpha_n \to 1$ and $\mathrm{diam}(S(\mathbf{B}_{\mathbf{X}}, f_n, \alpha_n)) \to 0$ as $n \to \infty$. Since norm-attaining functionals of $\mathbf{S}_{\mathbf{X}^*}$ are dense subset by Theorem 1.3, we can find norm-attaining functionals $(g_n) \subset \mathbf{S}_{\mathbf{X}^*}$ such that $||f_n - g_n|| < (1-\alpha_n)/2$ for $n \in \mathbb{N}$. Then $S(\mathbf{B}_{\mathbf{X}}, g_n, \alpha_n + (1-\alpha_n)/2) \subset S(\mathbf{B}_{\mathbf{X}}, f_n, \alpha_n)$ for $n \in \mathbb{N}$. Let $(x_n) \subset \mathbf{S}_{\mathbf{X}}$ be such that $g_n(x_n) = 1$ for $n \in \mathbb{N}$. Then for every $0 < \varepsilon < 1$,

$$\Delta_{\mathbf{X}}(x_n,\varepsilon) \geq \inf\{1-\lambda : \mathrm{diam}(S(\mathbf{B}_{\mathbf{X}}, g_n, \lambda)) \geq \varepsilon\}. \tag{3.5}$$

This is a consequence of the following observation: if $\lambda x_n \pm v \in \mathbf{B}_{\mathbf{X}}$, $||v|| \geq \varepsilon$, then at least one of $\lambda x_n \pm v$ is contained in $S(\mathbf{B}_{\mathbf{X}}, g_n, \lambda)$, so that $\mathrm{diam}(S(\mathbf{B}_{\mathbf{X}}, g_n, \lambda)) \geq \varepsilon$. Hence, choosing $\varepsilon = 2\mathrm{diam}(S(\mathbf{B}_{\mathbf{X}}, g_n, \alpha_n + (1-\alpha_n)/2))$ in (3.5) we get that for $n \in \mathbb{N}$,

$$\begin{aligned}
&\Delta_{\mathbf{X}}\Big(x_n, 2\mathrm{diam}\big(S(\mathbf{B}_{\mathbf{X}}, g_n, \alpha_n + \frac{1-\alpha_n}{2})\big)\Big)\\
&\quad \geq \inf\Big\{1-\lambda :\ \mathrm{diam}(S(\mathbf{B}_{\mathbf{X}}, g_n, \lambda)) \geq 2\mathrm{diam}\big(S(\mathbf{B}_{\mathbf{X}}, g_n, \alpha_n + \frac{1-\alpha_n}{2})\big)\Big\}\\
&\quad > 1 - \big(\alpha_n + \frac{1-\alpha_n}{2}\big),
\end{aligned}$$

where we used the fact that $\mathrm{diam}(S(\mathbf{B}_{\mathbf{X}}, g_n, \lambda)) \geq 2\mathrm{diam}(S(\mathbf{B}_{\mathbf{X}}, g_n, \alpha_n + (1-\alpha_n)/2))$ implies $\lambda < \alpha_n + (1-\alpha_n)/2$. Let us summarize the facts established so far: $\Delta_{\mathbf{X}}(x,\cdot) = \Delta_{\mathbf{X}}(y,\cdot)$ holds m-a.e. for each pair $(x,y) \in \mathbf{S}_{\mathbf{X}} \times \mathbf{S}_{\mathbf{X}}$, these maps are non-decreasing and $\Delta_{\mathbf{X}}(x_n, \varepsilon_n) > 0$ for $n \in \mathbb{N}$, where $\varepsilon_n \doteq 2\mathrm{diam}(S(\mathbf{B}_{\mathbf{X}}, g_n, \alpha_n + (1-\alpha_n)/2)) \to 0$ as $n \to \infty$. By (3.4) we get that $\lim_{t\to\varepsilon_n^-} \Delta_{\mathbf{X}}(x,t) = \lim_{t\to\varepsilon_n^-} \Delta_{\mathbf{X}}(y,t) > 0$ for any $x, y \in \mathbf{S}_{\mathbf{X}}$, $n \in \mathbb{N}$. Hence we conclude that $\inf_{x\in\mathbf{S}_{\mathbf{X}}} \Delta_{\mathbf{X}}(x,t) > 0$ for all $t > 0$. It follows easily that $\mathbf{X}$ is uniformly convex.

• Consider next the case where $\mathbf{B}_{\mathbf{X}^*}$ is ω^*-dentable. We claim that $\mathbf{X}^*$ is asymptotically transitive. Let $(z_n) \subset \mathbf{S}_{\mathbf{X}} \subset \mathbf{S}_{\mathbf{X}^{**}}$ be a sequence such that

$$\lim_{n\to\infty} \mathrm{diam}\big(S(\mathbf{B}_{\mathbf{X}^*}, z_n, \frac{1}{1+2^{-n}})\big) = 0.$$

Suppose that $f, g \in \mathbf{S}_{\mathbf{X}^*}$ are norm-attaining functionals and $x, y \in \mathbf{S}_{\mathbf{X}}$ are such that $f(x) - g(y) = 1$. Since $\mathbf{X}$ is asymptotically transitive, there exist sequences $(T_n), (S_n) \subset \mathrm{Aut}(\mathbf{X})$ such that for every $n \in \mathbb{N}$:

(i) $T_n^{-1}(x) = S_n^{-1}(y) = z_n$;

(ii) $\max(||T_n||, ||T_n^{-1}||, ||S_n||, ||S_n^{-1}||) < 1 + 2^{-n}$.

Since $T_n^* f/\|T_n^* f\|$ and $S_n^* f/\|S_n^* f\|$ belong to $S(\mathbf{B}_{\mathbf{X}^*}, z_n, 1/1 + 2^{-n})$ for $n \in \mathbb{N}$, we obtain that $\|T_n^* f - S_n^* g\| \to 0$ as $n \to \infty$. Hence $\|(T_n^*)^{-1} S_n^* g - f\| \to 0$ and by assumption $\max(\|(S_n^*)^{-1} T_n^*\|, \|(T_n^*)^{-1} S_n^*\|) \to 1$ as $n \to \infty$. By Theorem 1.3 the norm-attaining functionals are dense in $\mathbf{S}_{\mathbf{X}^*}$, and since $\mathcal{O}(h)$ is norm closed for all $h \in \mathbf{S}_{\mathbf{X}^*}$, we get that $\mathbf{X}^*$ is asymptotically transitive. Thus an application of the first part of the proof gives that $\mathbf{X}^*$ is uniformly convex.

We conclude that in both cases $\mathbf{X}$ and $\mathbf{X}^*$ are reflexive. It follows by applying the RNP together with the previous arguments that in fact both $\mathbf{X}$ and $\mathbf{X}^*$ are asymptotically transitive and uniformly convex. Hence $\mathbf{X}$ and $\mathbf{X}^*$ are also uniformly smooth. □

Using Theorem 3.1 and [5, Theorem 6.8] one can deduce the interesting fact that not every Banach space admits an equivalent asymptotically transitive norm. The following fact was inspired by a comment from the referee and a question of P. Hájek.

Remark 3.2. Let $\mathbf{X}$ be a non-superreflexive Banach space, which is an Asplund space or has the RNP. Then $\mathbf{X}$ is not asymptotically transitive or convex-transitive under any equivalent renorming.

4 Projections onto subspaces of L^p

In this section we obtain, perhaps surprisingly, some information about the classical L^p spaces that appears to be new. Note that the L^p spaces are in particular asymptotically transitive as they are almost transitive (see e.g. [5, p. 8]). Hence the main idea here is to apply the rotational structure of L^p to study its other types of structures.

Let us fix some notations and recall some results, which are applied in this section. For $1 < p < \infty$ we denote $L^p = L^p(0,1)$, by 1 the unit function. Write $M^p = \ker(1) \subset L^p$, where we consider $1 \in L^{p^*}$. Similarly we consider $2^{1/p^*}\chi_{[0,2^{-1}]}$ as a functional in L^{p^*}.

Recall also that if $\mathbf{Y} \subset \mathbf{X}$, then a continuous linear projection $P : \mathbf{X} \to \mathbf{Y}$ is called *minimal* if $\|P\| = \inf\{\|Q\| : Q : \mathbf{X} \to \mathbf{Y}$ is a linear projection$\}$.

Theorem 4.1. *For $1 \leq p \leq \infty$ denote*

$$\alpha_p = \sup_{t \in (0,1)} (t^{p-1} + (1-t)^{p-1})^{\frac{1}{p}} (t^{p^*-1} + (1-t)^{p^*-1})^{\frac{1}{p^*}} - 1, \quad \frac{1}{p} + \frac{1}{p^*} = 1.$$

Suppose that $P\colon L^p \to [x]$, $x \in \mathbf{S}_{L^p}$, is a linear projection. Then the following conditions are equivalent:

(1) $\|P\| = 1$.

(2) $\|\mathbf{I} - P\| = 1 + \alpha_p$.

(3) $\mathbf{I} - P : L^p \to \ker(P)$ *is a minimal projection.*

Proof. The case $p = 2$ is clear. Let us consider the case $1 < p < \infty$, $p \neq 2$. By Proposition 2.6 we have that if $\|P\| = 1$ then $\|\mathbf{I} - P\| = \|I - 1 \otimes 1\|$. On the other hand, by [20, Theorem 6] every linear projection Q onto a 1-codimensional subspace $Z \subset L^p$ satisfies $\|Q\| \geq \|\mathbf{I} - 1 \otimes 1\|$. The exact value $\|\mathbf{I} - 1 \otimes 1\| = \alpha_p$ was calculated by Franchetti in [10, Theorem 3]. Combining the Riesz Lemma, the Hahn–Banach theorem, and the weak compactness of $\mathbf{B}_{L^p}$, we observe that for each 1-codimensional subspace $Z \subset L^p$, there is a point $y \in S_{L^p}$ and a norm-one projection $P : L^p \to [y]$ such that $\ker(P) = Z$. Finally, recall that in a uniformly convex space each minimal projection Q, $\|Q\| > 1$, onto a 1-codimensional subspace is unique (see [15, p. 28]).

The case $p \in \{1, \infty\}$ follows directly from the fact that L^1 and L^∞ have the Daugavet property: whenever $T : L^p \to L^p$, $p \in \{1, \infty\}$, is a compact operator, then $\|\mathbf{I} + T\| = 1 + \|T\|$ (see e.g. [22, p. 78]). On the other hand $\alpha_p = 1$ for $p \in \{1, \infty\}$. Since $\|P\| \geq 1$, it follows that $\|\mathbf{I} - P\| = 1 + \|P\| \geq 2$ and equality holds if and only if $\mathbf{I} - P$ is minimal. □

The following result which is obtained using classical facts about rotations of L^p, has also some nice consequences (see Theorem 4.3 below).

Proposition 4.2. *Let $Z_1, Z_2 \subset L^p$ be 1-codimensional subspaces, where $1 < p < \infty$. Then $Z_1 \oplus_p L^p = Z_2 \oplus_p L^p$ isometrically. Moreover,* exactly *one of the following holds isometrically:*

(1) $Z_1 = Z_2$.

(2) Either (2a) $Z_1 = Z_1 \oplus_p L^p$ or (2b) $Z_2 = Z_2 \oplus_p L^p$ (but not both).

We note that the second condition above does occur, see Example 4.6 below.

Proof. The case $p = 2$ is clear, so let us consider the case $p \neq 2$. Since L^p is reflexive, there exists by a standard application of the Riesz lemma and the weak compactness of $\mathbf{B}_{L^p}$ points $x, y \in \mathbf{S}_{L^p}$ such that $\mathrm{dist}(x, Z_1) = \mathrm{dist}(y, Z_2) = 1$. By the Hahn–Banach theorem there are $f, g \in \mathbf{S}_{L^{p*}}$ such that $\ker(f) = Z_1$, $\ker(g) = Z_2$, and $f(x) = g(y) = 1$.

We use the fact that there are exactly two disjoint orbits in $\mathbf{S}_{L^p}$ (see [3, p. 178]):

$$\{x \in \mathbf{S}_{L^p} : m(\mathrm{supp}(x)) = 1\} \text{ and } \{x \in \mathbf{S}_{L^p} : m(\mathrm{supp}(x)) < 1\}. \tag{4.1}$$

Hence there exists a rotation $T \in \mathcal{G}_{L^p(0,2)}$ for which $T((x,0)) = (y,0)$ under the identification that $(x,0), (y,0) \in L^p \oplus_p L^p = L^p(0,2)$. The corresponding support functionals are $\tilde{f} = (f,0)$ and $\tilde{g} = (g,0)$ in $L^{p^*} \oplus_{p^*} L^{p^*} = L^{p^*}(0,2)$. Since $L^p(0,2)$ is smooth, it follows that $\tilde{f} = \tilde{g} \circ T$. Hence

$$Z_1 \oplus_p L^p = \ker(\tilde{f}) = \ker(\tilde{g} \circ T) = \ker(\tilde{g}) = Z_2 \oplus_p L^p$$

isometrically, which is the first part of the proposition. This also means that the conditions (2a) and (2b) are mutually exclusive in the case where Z_1 and Z_2 are non-isometric.

If $T \in \mathcal{G}_{L^p}$ is such that $T(x) = y$ then from the smoothness of L^p we have as above that $f = g \circ T$, so that

$$Z_1 = \ker(f) = \ker(g \circ T) = \ker(g) = Z_2 \tag{4.2}$$

isometrically.

Suppose now that Z_1 and Z_2 are non-isometric. By relabelling we may assume without loss of generality that $m(\mathrm{supp}(x)) = 1$ and $m(\mathrm{supp}(y)) < 1$. This means that $2^{1/p}\chi_{[0,2^{-1}]} \in \mathcal{G}_{L^p}(y)$. Hence we obtain that $\ker(g)$ is isometric to $M^p \oplus_p L^p$. Clearly $(M^p \oplus_p L^p) \oplus_p L^p$ and $M^p \oplus_p L^p$ are isometric. Thus (2b) holds. □

Note the difference between the L^p-summands appearing in the proposition above and the 1-complemented subspaces appearing in the next theorem.

Theorem 4.3. *Let $1 < p < \infty$ and $Z \subset L^p$ be a finite codimensional subspace. Then there is a subspace $N \subset Z$ isometric to L^p such that $N \subset L^p$ is 1-complemented.*

We first verify the following special case.

Lemma 4.4. *The space M^p contains an isometric copy of L^p which is 1-complemented in L^p.*

Proof. The space $N := \{f \in M^p : f(t) = -f(2^{-1} + t) \text{ for a.e. } t \in [0, 2^{-1}]\}$ is such a subspace. Indeed, $T : N \to L^p(0, 2^{-1})$ given by $T(f) = 2^{1/p} f_{|[0,2^{-1}]}$ is an isometric isomorphism. The required projection $P : M^p \to N$ is given by $P(f)(t) = 2^{-1}(f(t) - f(2^{-1} + t))$ for $t \in [0, 2^{-1}]$ and $P(f)(t) = -2^{-1}(f(t - 2^{-1}) - f(t))$ for $t \in [2^{-1}, 1]$. Clearly P is a linear projection. To verify that $\|P\| = 1$ we use the estimate $|a - b| \leq 2^{1/p^*}(|a|^p + |b|^p)^{1/p}$ to obtain that

$$\begin{aligned}
\|Pf\|_p^p &= \int_0^{2^{-1}} |P(f)(t)|^p \, dt + \int_{2^{-1}}^1 |P(f)(t)|^p \, dt \\
&= 2^{1-p} \int_0^{2^{-1}} |f(t) - f(2^{-1} + t)|^p \, dt \\
&\leq 2^{1-p} 2^{\frac{p}{p^*}} \int_0^{2^{-1}} |f(t)|^p + |f(2^{-1} + t)|^p \, dt = \|f\|_p^p.
\end{aligned}$$

Hence $\|P\| = 1$ and we have the lemma. □

Proof of Theorem 4.3. Suppose that $Z_1 \subset Z_2 \subset \cdots \subset Z_n = L^p$ are subspaces such that $\dim(Z_i/Z_{i-1}) = 1$ for all $i - 1, i \in \{1, \ldots, n\}$. We proceed inductively.

Step 1. We apply Proposition 4.2 and its proof. We know that Z_{n-1} is isometric to either M^p or $M^p \oplus_p L^p$. According to (4.1) and (4.2), by applying a suitable rotation

$T\colon L^p \to L^p$, we may assume without loss of generality that $Z_{n-1} \subset L^p$ is either M^p or $\ker(2^{\frac{1}{p^*}}\chi_{[0,2^{-1}]})$.

If $Z_{n-1} = M^p$ the the previous lemma gives a norm-1 projection $P_{n-1}\colon L^p \to N_{n-1}$, where $N_{n-1} \subset Z_{n-1}$ is an isometric copy of L^p.

If $Z_{n-1} = \ker(2^{\frac{1}{p^*}}\chi_{[0,2^{-1}]})$ then the operator $P_{n-1} : L^p \to \{f \in L^p | f_{|[0,2^{-1}]} \equiv 0 \text{ a.e.}\}$ defined by $P_{n-1}(f) = \chi_{[2^{-1},1]}f$ is also a norm-1 onto projection. Hence, in both the cases there exists a linear norm-1 projection $P_{n-1}\colon L^p \to N_{n-1}$, where N_{n-1} is an isometric copy of L^p.

Step 2. Observe that $\dim(N_{n-1}/N_{n-1} \cap Z_{n-2}) \leq 1$. Since N_{n-1} is isometric to L^p we may apply Step 1 to conclude that there exists a subspace $N_{n-2} \subseteq N_{n-1} \cap Z_{n-2}$, which is isometric to L^p and 1-complemented in N_{n-1}. Denote the corresponding norm-1 projection by $P_{n-2}\colon N_{n-1} \to N_{n-2}$.

We continue in this manner to define subspaces $N_{n-1} \supset N_{n-2} \supset \cdots \supset N_1$ which are isometric to L^p together with the norm-1 projections $P_{i-1}\colon N_i \to N_{i-1}$ for $i-1, i \in \{1,\dots,n\}$. Hence $N_1 \subset Z_1$ is the required subspace isometric to L^p and $P_1 \circ P_2 \circ \cdots \circ P_{n-1}\colon L^p \to N_1$ is the corresponding norm-1 projection. □

Recall that L^p is primary, that is, if $L^p = M \oplus N$, then either $M = L^p$ or $N = L^p$ isomorphically (see e.g. [16, 2.d.11]). Hence we know that in the following result $\ker(P_1)$ is isomorphic to $\ker(P_2)$. The crux of the following result is that $\ker(P_1)$ and $\ker(P_2)$ below are almost isometric. Thus we obtain examples of subspaces of L^p (hence both uniformly convex and uniformly smooth), which are mutually non-isometric but still almost isometric (see Example 4.7).

Theorem 4.5. *Let $n \in \mathbb{N}$, $1 < p < \infty$, $p \neq 2$ and suppose that $Y_1, Y_2 \subset L^p$ are isometric copies of ℓ^p_n or of ℓ^p. Then there exist unique linear projections $P_1\colon L^p \to Y_1$, $P_2\colon L^p \to Y_2$ such that $||P_1|| = ||P_2|| = 1$ and* $\ker(P_1)$ *and* $\ker(P_2)$ *are* almost isometric.

Proof. We consider only the case where Y_1 and Y_2 are isometric copies of ℓ^p, since the argument for the other cases is similar. Let $S\colon \ell^p \to Y_1$ be an isometry and (e_k) be the unit vector basis of ℓ^p. For $k \in \mathbb{N}$, put $\hat{e}_k = S(e_k)$. Since S is an isometry we obtain from Lamperti's well-known result (see e.g. [14, p. 163]) that the sequence $(\hat{e}_k)$ have pairwise disjoint supports. It is well-known that there is a normalized sequence $(\hat{e}_k^*) \subset L^{p^*}$ with $\operatorname{supp}(\hat{e}_k^*) = \operatorname{supp}(\hat{e}_k)$ for $k \in \mathbb{N}$ and such that

$$P(x) = \sum_{k\in\mathbb{N}} \hat{e}_k^*(x)\hat{e}_k$$

defines a norm-1 projection $L^p \to \overline{\operatorname{span}}(\{\hat{e}_k | k \in \mathbb{N}\})$. By [4, Proposition 5] this projection P is unique since L^p is smooth.

We may partition $[0,1] = \bigcup_{k\in\mathbb{N}} A_k$, where $A_1 = [0,1] \setminus \bigcup_{k\geq 2} \operatorname{supp}(\hat{e}_k)$, $A_2 = \operatorname{supp}(\hat{e}_2)$, $A_3 = \operatorname{supp}(\hat{e}_3), \dots$ For each $\hat{e}^*_{k\,|A_k} \in L^{p^*}(A_k)$, $\ker\left(\hat{e}^*_{k\,|A_k}\right) \subset L^p(A_k)$ is a 1-codimentional subspace. Under the above notations one can write

$$\ker(P) = \bigoplus_{k\in\mathbb{N}} \ker\left(\hat{e}^*_{k\,|A_k}\right) \subset L^p,$$

where the direct sum is understood in the ℓ^p-sense. It suffices to show that for every $k \in \mathbb{N}$,

$$d\left(\ker\left(\hat{e}^*_{k\,|A_k}\right), \ker\left(1_{|A_k}\right)\right) = 1,$$

where we consider $1_{|A_k} \in L^{p^*}(A_k)$. Indeed, if $T_k\colon \ker\left(1_{|A_k}\right) \to \ker\left(\hat{e}^*_{k\,|A_k}\right)$ are isomorphisms such that $\|T_k\| \leq 1 + C2^{-k}$, $C \geq 0$, and $\|T_k^{-1}\| = 1$, then

$$\boldsymbol{T} = \bigoplus_{k\in\mathbb{N}} T_k\colon \bigoplus_{k\in\mathbb{N}} \ker\left(1_{|A_k}\right) \to \bigoplus_{k\in\mathbb{N}} \ker\left(\hat{e}^*_{k\;A_k}\right),$$

(where the direct sums are in ℓ^p-sense) is such that $\|\boldsymbol{T}^{-1}\| = 1$ and $\|\boldsymbol{T}\| \leq 1 + C$.

Fix k_0. As $L^p(A_{k_0})$ and $L^p(0,1)$ are isometric, we may assume without loss of generality that $A_{k_0} = [0,1]$. Since L^p is asymptotically transitive, there is a sequence $(T_k) \subset \mathrm{Aut}(L^p)$ with $T_k(\hat{e}_{k_0}) = 1$ for all $k \in \mathbb{N}$ and $\lim_{k\to\infty} \max(||T_k||, ||T_k^{-1}||) = 1$. Observe that $(\mathbf{I} - 1 \otimes 1)_{|\ker(1)} = \mathbf{I}_{|\ker(1)}$. For $k \in \mathbb{N}$, define $S_k \in L(L^p, \ker(1))$ by $S_k = (\mathbf{I} - 1 \otimes 1) \circ T_k$. Since $\hat{e}^*_{k_0} \circ T_k^{-1}(1) = 1$ for $k \in \mathbb{N}$ and $||\hat{e}^*_{k_0} \circ T_k^{-1}|| \to 1$ as $k \to \infty$ we obtain by the Fréchet-smoothness of L^p and Lemma 1.2 that $\lim_{k\to\infty} \|1 - \hat{e}^*_{k_0} \circ T_k^{-1}\| = 0$. As

$$\begin{aligned} T_k \circ (\mathbf{I} - \hat{e}^*_{k_0} \otimes \hat{e}_{k_0}) - S_k &= (\mathbf{I} - (\hat{e}^*_{k_0} \circ T_k^{-1}) \otimes T_k(\hat{e}_{k_0})) \circ T_k - S_k \\ &= (\mathbf{I} - (\hat{e}^*_{k_0} \circ T_k^{-1}) \otimes 1) \circ T_k - (\mathbf{I} - 1 \otimes 1) \circ T_k, \end{aligned}$$

We deduce that $\lim_{k\to\infty} \|T_k \circ (\mathbf{I} - \hat{e}^*_{k_0} \otimes \hat{e}_{k_0}) - S_k\| = 0$. This justifies that if k is large enough then

$$\mathrm{codim}\left(S_k\left(\ker\left(\hat{e}^*_{k_0}\right)\right)\right) = \mathrm{codim}\left(\left(\mathbf{I} - \hat{e}^*_{k_0} \otimes \hat{e}_{k_0}\right)\left(\ker\left(\hat{e}^*_{k_0}\right)\right)\right) = 1.$$

Since $\mathrm{codim}(S_k(L^p)) = 1$ for $k \in \mathbb{N}$, we deduce that for sufficiently large k,

$$S_k\left(\ker\left(\hat{e}^*_{k_0}\right)\right) = S_k(L^p).$$

Therefore the map $S_{k\,|\ker\left(\hat{e}^*_{k_0}\right)}\colon \ker\left(\hat{e}^*_{k_0}\right) \to \ker(1)$ is onto. Observe that

$$\lim_{k\to\infty} \max\left(\|T_k \circ \left(\mathbf{I} - \hat{e}^*_{k_0} \otimes \hat{e}_{k_0}\right)_{|\ker\left(\hat{e}^*_{k_0}\right)}\|, \|(T_k \circ \left(\mathbf{I} - \hat{e}^*_{k_0} \otimes \hat{e}_{k_0}\right)_{|\ker\left(\hat{e}^*_{k_0}\right)})^{-1}\|\right) = 1$$

and

$$\lim_{k\to\infty} \|(T_k \circ \left(\mathbf{I} - \hat{e}^*_{k_0} \otimes \hat{e}_{k_0}\right))_{|\ker\left(\hat{e}^*_{k_0}\right)} - S_{k\,|\ker\left(\hat{e}^*_{k_0}\right)}\| = 0.$$

Thus the restriction $S_{k\,|\ker\left(\hat{e}^*_{k_0}\right)}\colon \ker\left(\hat{e}^*_{k_0}\right) \to \ker(1)$ is an isomorphism for sufficiently large k, and moreover

$$\lim_{k\to\infty} \max(\|S_{k\,|\ker\left(\hat{e}^*_{k_0}\right)}\|, \|(S_{k\,|\ker\left(\hat{e}^*_{k_0}\right)})^{-1}\|) = 1.$$

This proves that $d\left(\ker\left(\hat{e}^*_{k_0}\right), \ker(1)\right) = 1$. □

Example 4.6. The subspace $M^p \subset L^p$ for $1 \leq p < \infty$, $p \neq 2$, is not isometrically of the form $N \oplus_p L^p$ for any closed subspace $N \neq \{0\}$.

Assume to the contrary that $M^p = N \oplus_p K$ isometrically for some non-trivial subspaces $N, K \subset L^p$, where K is isometric to L^p. From Lamperti's disjointness condition ([14]), there exists a measurable decomposition $[0,1] = A \cup B$ such that for any functions $f \in N$ and $g \in K$, $\mathrm{supp}(f) \subset A$ and $\mathrm{supp}(g) \subset B$.

Hence $\int_A f\,dt = \int_B g\,dt = 0$. This means that

$$M^p = \left\{ f \in L^p : \int_A f\,dt = \int_B f\,dt = 0 \right\}.$$

But this is a contradiction since there exist $a, b \in (0, \infty)$ such that $\int_{[0,1]} a\chi_A - b\chi_B\,dt = 0$ and hence $a\chi_A - b\chi_B \in M^p$. □

Example 4.7. The subspaces $\ker(1)$ and $\ker(2^{\frac{1}{p^*}} \chi_{[0,2^{-1}]})$ of L^p are almost isometric but not isometric.

We may use Theorem 4.5 for 1-dimensional subspaces to obtain that the subspaces above are almost isometric. Observe that $\ker(2^{\frac{1}{p^*}} \chi_{[0,2^{-1}]}) = M^p \oplus_p L^p$ isometrically. On the other hand $\ker(1)$ does not have such an isometric decomposition according to Example 4.6. Hence $\ker(1)$ and $\ker(2^{\frac{1}{p^*}} \chi_{[0,2^{-1}]})$ are not isometric. □

5 Concluding remarks

Next we will give an asymptotic analogue of the following known comparison principle (see [5, p. 16]):

For a convex-transitive space $(\mathbf{X}, \|\cdot\|)$ the condition $\mathcal{G}_{\|\cdot\|} \subset \mathcal{G}_{|||\cdot|||}$ for some equivalent norm $|||\cdot||| \sim \|\cdot\|$ implies that $|||\cdot||| = c\|\cdot\|$ for some constant $c > 0$.

First we introduce asymptotic analogues of the expressions '$\mathcal{G}_{\|\cdot\|}$' and '$\mathcal{G}_{\|\cdot\|} \subset \mathcal{G}_{|||\cdot|||}$'. For a Banach space $(\mathbf{X}, \|\cdot\|)$ and $\delta > 0$, let

$$\mathcal{F}_{\|\cdot\|}(\delta) := \{T \in \mathrm{Aut}(\mathbf{X}, \|\cdot\|) : \max(\|T\|, \|T^{-1}\|) \leq 1 + \delta\}.$$

The increasing family $\{\mathcal{F}_{\|\cdot\|}(\delta)\}_{\delta \geq 0}$ is denoted by $\mathcal{F}_{\|\cdot\|}$. If $\|\cdot\|$ and $|||\cdot|||$ are norms on $\mathbf{X}$, then the condition that for each $\varepsilon > 0$ there exists $\delta > 0$ such that $\mathcal{F}_{\|\cdot\|}(\delta) \subset \mathcal{F}_{|||\cdot|||}(\varepsilon)$ will be denoted by $\mathcal{F}_{\|\cdot\|} << \mathcal{F}_{|||\cdot|||}$.

Proposition 5.1. *Let $(\mathbf{X}, \|\cdot\|)$ be an asymptotically transitive normed space and let $|||\cdot|||$ be* any *norm on $\mathbf{X}$. If $\mathcal{F}_{\|\cdot\|} << \mathcal{F}_{|||\cdot|||}$, then $|||\cdot||| = c\|\cdot\|$ for some $c > 0$.*

Proof. Let $x, y \in \mathbf{S}_{\|\cdot\|}$ be such that $|||x||| \leq |||y|||$. The assumptions yield that for each $\varepsilon > 0$ there exists $\delta \in (0, \varepsilon)$ such that $\mathcal{F}_{\|\cdot\|}(\delta) \subset \mathcal{F}_{|||\cdot|||}(\varepsilon)$. The asymptotic transitivity of $(\mathbf{X}, \|\cdot\|)$ implies that for such $\delta > 0$ there is for each $\delta > 0$ an automorphism $T \in \mathcal{F}_{\|\cdot\|}(\delta)$ such that $T(x) = y$. This means that

$$|||y||| - |||x||| \leq (|||T||| - 1)|||x||| \leq \varepsilon |||x|||.$$

Since ε was arbitrary, we obtain that $|||x||| = |||y|||$. Since $x, y \in \mathbf{S}_{\|\cdot\|}$ were arbitrary, we conclude that $||| \cdot ||| = c\| \cdot \|$ for some $c > 0$. □

We would like to stress the significance of the following problem.

Problem 5.2. *Does there exist an asymptotically transitive Banach space* $\mathbf{X}$*, which is not almost transitive?*

In particular it is not known whether there exists an asymptotically transitive Banach space whose only rotations are $\pm\mathbf{I}$.

The following example shows that for a given $x \in \mathbf{S_X}$ the generalized orbit $\mathcal{O}(x)$ does not necessarily coincide with the closure of the regular orbit $\overline{\mathcal{G}(x)}$.

Example 5.3. Let $p \in (1, \infty)$, $p \neq 2$, and $M, N \subset L^p$ be the 1-codimensional spaces appearing in Example 4.7. Recall that M and N are almost isometric and non-isometric. Put

$$\mathbf{X} = (M \oplus_p \mathbb{R}) \oplus_1 (N \oplus_p \mathbb{R}).$$

We apply notation $(f, a, g, b) \in \mathbf{X}$, where $f \in M$, $g \in N$ and $a, b \in \mathbb{R}$. Since M and N are almost isometric there is for each $\varepsilon > 0$ an automorphism $S\colon \mathbf{X} \to \mathbf{X}$ such that $\max(\|S\|, \|S^{-1}\|) \leq 1+\varepsilon$ and $S((0,1,0,0)) = (0,0,0,1)$. Fix $T \in \mathcal{G}_{\mathbf{X}}$. We claim that

$$T((M \oplus_p \mathbb{R}) \oplus_1 (\{0\} \oplus_p \{0\})) = (M \oplus_p \mathbb{R}) \oplus_1 (\{0\} \oplus_p \{0\}).$$

Observe that $(f, a, 0, 0), (0, 0, g, b) \in \mathbf{S_X}$ are exactly all the extreme points of $\mathbf{B_X}$. Clearly T preserves extreme points. Hence $T((f, a, 0, 0))$ is either of the form $(f_1, a_1, 0, 0)$ or $(0, 0, g_1, b_1)$. For any $(f_1, a_1), (f_2, a_2) \subset M \oplus_p \mathbb{R} \setminus \{(0,0)\}$ such that $(f_1, a_1) \notin [(f_2, a_2)]$ it holds that $||(f_1, a_1) + (f_2, a_2)||_{M \oplus_p \mathbb{R}} < ||(f_1, a_1)||_{M \oplus_p \mathbb{R}} + ||(f_2, a_2)||_{M \oplus_p \mathbb{R}}$. Hence $T((M \oplus_p \mathbb{R}) \oplus_1 (\{0\} \oplus_p \{0\}))$ is either $(M \oplus_p \mathbb{R}) \oplus_1 (\{0\} \oplus_p \{0\})$ or $(\{0\} \oplus_p \{0\}) \oplus_1 (N \oplus_p \mathbb{R})$. Observe that the set $\{\pm(0,1,0,0), \pm(0,0,0,1)\}$ can be defined in purely metric terms. Indeed, for $z \in \{\pm(0,1,0,0), \pm(0,0,0,1)\}$ there does not exist $x, y \in \mathbf{X} \setminus \{0\}$, $x \notin [y]$, such that $x + y = z$ and $||z||_{\mathbf{X}}^r = ||x||_{\mathbf{X}}^r + ||Y||_{\mathbf{X}}^r$ for any $r \in \{1, p\}$. It is easy to see that there do not exist such atoms in $\mathbf{S_X}$ apart from $\{\pm(0,1,0,0), \pm(0,0,0,1)\}$. Thus, if $T((M \oplus_p \mathbb{R}) \oplus_1 (\{0\} \oplus_p \{0\})) = (\{0\} \oplus_p \{0\}) \oplus_1 (N \oplus_p \mathbb{R})$ then $T((0,1,0,0)) = \pm(0,0,0,1)$. But this gives that $T((M \oplus_p \{0\}) \oplus_1 (\{0\} \oplus_p \{0\})) = (\{0\} \oplus_p \{0\}) \oplus_1 (N \oplus_p \{0\})$, which is impossible since N and M are non-isometric. □

The fact that M^p and $M^p \oplus_p L^p$ are almost isometric suggest the question whether M^p could be somehow exhausted by successive almost isometric embeddings of L^p. This leads to the following problem.

Problem 5.4. *What is* $d(M^p, L^p)$*?*

Acknowledgments. This article is part of the writer's ongoing Ph.D. work. It is supervised by H.-O. Tylli to whom I am grateful for careful suggestions about the presentation. I am indebted to B. Randrianantoanina for suggesting the present formulation of Theorem 2.2 in response to the author's Licentiate Thesis [21]. I am grateful to G. Lewicki for helpful comments regarding minimal projections. I also thank the referee for very useful remarks.

References

[1] D. Amir, *Characterizations of inner product spaces*, Operator Theory: Advances and Applications, Vol 20, 1986.

[2] D. Amir and C. Franchetti, *The radius ratio and convexity properties in normed linear spaces*, Trans. Amer. Math. Soc. 282 (1984), pp. 275–291.

[3] S. Banach, *Théorie des opérations linéaires*, Warsaw 1932.

[4] B. Beauzamy and B. Maurey, *Points minimaux et ensembles optimaux dans les espaces de Banach*, J. Funct. Anal. 24 (1977), pp. 107–139.

[5] J. Becerra Guerrero and A. Rodriguez-Palacios, *Transitivity of the norm on Banach spaces*, Extracta Math. 17 (2002), pp. 1–58.

[6] F. Cabello Sánchez, *10 Variaciones sobre un Tema de Mazur*, Thèse, Universidad de Extremadura, Badajoz, 1996.

[7] ______, *Regards sur le probléme des rotations de Mazur*, Extracta Math. 12 (1997), pp. 95–116.

[8] ______, *Maximal symmetric norms on Banach spaces*, Math. Proc. R. Ir. Acad. 98A (1998), pp. 121–130.

[9] R. Fleming and J. Jamison, *Isometries on Banach Spaces*, function spaces, Monographs and Surveys in Pure and Applied Mathematics, Vol. 129, Chapman & Hall, 2006.

[10] C. Franchetti, *The norm of the minimal projection onto hyperplanes in $L^p(0,1)$ and the radial constant*, Boll. Un. Mat. Ital B(7) 4 (1990), pp. 803–821.

[11] V. I. Gurarij, *Space of universal disposition, isotropic spaces and the Mazur problem on rotations of Banach spaces*, Sibirskij Mat. Zhurnal 7 (1966), pp. 1002–1013.

[12] P. Habala, P. Hajek, and V. Zizler, *Introduction to Banach Spaces*, Vol. I–II, Matfyzpress, 1996.

[13] W. B. Johnson and J. Lindenstrauss, *Basic concepts in the geometry of Banach spaces*, Handbook of the Geometry of Banach Spaces, Vol I, pp. 1–84, North Holland, 2001.

[14] H. E. Lacey, *The isometric theory of classical Banach spaces*, Die Grundlehren der mathematischen Wissenschaften, Band 208, Springer-Verlag, 1974.

[15] G. Lewicki and W. Odyniec, *Minimal Projections in Banach Spaces*, Lecture Notes in Mathematics, Vol. 1449, Springer-Verlag, 1990.

[16] J. Lindenstrauss and L. Tzafriri, *Classical Banach spaces II*, Function spaces, Vol. 97, Springer-Verlag, 1979.

[17] W. Lusky, *A note on rotations in separable Banach spaces*, Studia Math. 65 (1979), pp. 239–242.

[18] E. D. Positselkij and E. V. Tokarev, *The amalgamation property in classical Lebesgue-Riesz spaces, Banach spaces with almost transitive norm and projection constants*, Preprint, arXiv:math.FA/0206182, 2002.

[19] B. Randrianantoanina, *A Note on the Banach-Mazur problem*, Glasgow J. Math. 44, (2002), pp. 159–165.

[20] S. Rolewicz, *On projections on subspaces of codimension one*, Studia Math. 96 (1990), pp. 17–19.

[21] J. Talponen, *On the Banach-Mazur rotation problem*, Ph.Lic. thesis, Helsinki, 2005.

[22] D. Werner, *Recent progress on the Daugavet property*, Irish Math. Soc. Bull. 46 (2001), pp. 77–97.

Author information

Jarno Talponen, University of Helsinki, Department of Mathematics and Statistics, Box 68 (Gustaf Hällströminkatu 2b), FI-00014 University of Helsinki, Finland.
Email: talponen@cc.helsinki.fi

Participants of the conference

TROND ABRAHAMSEN, Agder University College, Kristiansand, Norway
LIPI ACHARYA, Indian Institute of Technology Kanpur, Kanpur, India
MARIA ACOSTA, University of Granada, Granada, Spain
YAKOV ALBER, Technion, Israel Institute of Technology, Haifa, Israel
FERNANDO ALBIAC, University of Missouri, Columbia, MO
DALE ALSPACH, Oklahoma State University, Stillwater, OK
GEORGE ANDROULAKIS, University of South Carolina, Columbia, SC
RAZVAN ANISCA, Lakehead University, Thunder Bay, Canada
RAJAPPA K. ASTHAGIRI, Miami University, Middletown, OH
HOLLY ATTENBOROUGH, Miami University, Oxford, OH
ANTONIO AVILÉS, University of Murcia, Espinardo, Spain
PRADIPTA BANDYOPADHYAY, Indian Statistical Institute, Kolkata, India
SUDESHNA BASU, Morgan State University, Baltimore, MD
ELIZABETH BATOR, University of North Texas, Denton, TX
KEVIN BEANLAND, University of South Carolina, Columbia, SC
JOEL BEIL, Kent State University, Kent, OH
OSCAR BLASCO, University of Valencia, Burjasot, Spain
SARITA BONDRE, The M.S. University of Baroda, Vadodara, India
OLGA BREZHNEVA, Miami University, Oxford, OH
BO BRINKMAN, Miami University, Oxford, OH
QINGYING BU, University of Mississippi, Oxford, MS
DENNIS BURKE, Miami University, Oxford, OH
FELIX CABELLO SANCHEZ, University of Extremadura, Badajoz, Spain
JESÚS M. F. CASTILLO, University of Extremadura, Badajoz, Spain
CONSTANTIN DAN CAZACU, ING, West Chester, PA
SIMON COWELL, University of Missouri, Columbia, MO
MARIANNA CSÖRNYEI, University College, London, UK
MICHAEL CWIKEL, Technion, Israel Institute of Technology, Haifa, Israel
STEFAN CZERWIK, Silesian University of Technology, Gliwice, Poland
JERRY DAY, University of Pittsburgh, Pittsburgh, PA
MIENIE DE KOCK, Kent State University, Kent, OH
GEOFF DIESTEL, University of Missouri, Columbia, MO

JOE DIESTEL, Kent State University, Kent, OH
STEPHEN DILWORTH, University of South Carolina, Columbia, SC
PAWEŁ DOMAŃSKI, Adam Mickiewicz University, Poznań, Poland
DETELIN DOSEV, Texas A&M University, College Station, TX
IAN DOUST, University of New South Wales, Sydney, Australia
PADDY DOWLING, Miami University, Oxford, OH
JAKUB DUDA, Weizmann Institute of Science, Rehovot, Israel
YVES DUTRIEUX, Université de Franche-Comté, Besançon, France
SUDIPTA DUTTA, Ben-Gurion University of the Negev, Beer-Sheva, Israel
MAHMOUD M. EL-BORAI, Alexandria University, Alexandria, Egypt
MARIAN FABIAN, Czech Academy of Sciences, Prague, Czech Republic
VASSILIKI FARMAKI, University of Athens, Athens, Greece
ADERAW FENTA, Kent State University, Kent, OH
VALENTIN FERENCZI, Université Paris VI, Paris, France
RICHARD FLEMING, Central Michigan University, Mt. Pleasant, MI
JULIO FLORES, Universidad Rey Juan Carlos, Mostoles, Spain
JAN FOURIE, North-West University, Potchefstroom, South Africa
MARIA FRAGOULOPOULOU, University of Athens, Athens, Greece
DANIEL FREEMAN, Texas A&M University, College Station, TX
NADIA GAL, The University of Memphis, Memphis, TN
MIKHAIL GANICHEV, University of Missouri, Columbia, MO
PANDO GEORGIEV, University of Cincinnati, Cincinnati, OH
IOANA GHENCIU, University of Wisconsin-River Falls, River Falls, WI
GILLES GODEFROY, Université Paris VI, Paris, France
STANISŁAW GOLDSTEIN, University of Łódź, Łódź, Poland
MANUEL GONZÁLEZ, Universidad de Cantabria, Santander, Spain
KATHERINE GRIMES, Miami University, Oxford, OH
KARL GROSSE-ERDMANN, FernUniversität Hagen, Düsseldorf, Germany
MANJUL GUPTA, Indian Institute of Technology Kanpur, Kanpur, India
ALEXANDER HELEMSKII, Moscow State University, Moscow, Russia
FRANCISCO L. HERNANDEZ, Universidad Complutense de Madrid, Madrid, Spain
TUOMAS HYTÖNEN, University of Turku, Turku, Finland
MAR JIMENEZ-SEVILLA, Universidad Complutense de Madrid, Madrid, Spain
WILLIAM B. JOHNSON, Texas A&M University, College Station, TX

MATTHEW JORDAN, Miami University, Oxford, OH

STEN KAIJSER, Uppsala University, Uppsala, Sweden

NIGEL J. KALTON, University of Missouri, Columbia, MO

ANNA KAMIŃSKA, The University of Memphis, Memphis, TN

ROBERT KAUFMAN, The University of Illinois, Urbana-Champain, IL

RICHARD KETCHERSID, Miami University, Oxford, OH

LEONID KOVALEV, Texas A&M University, College Station, TX

TAMARA KUCHERENKO, University of California at Los Angeles, Los Angeles, CA

DENKA KUTZAROVA, The University of Illinois, Urbana, IL

COENRAAD LABUSCHAGNE, University of the Witwatersrand, Johannesburg, South Africa

FLORENCE LANCIEN, Université de Franche-Comté, Besançon, France

GILLES LANCIEN, Université de Franche-Comté, Besançon, France

PAUL B. LARSON, Miami University, Oxford, OH

CHRISTIAN LE MERDY, Université de Franche-Comté, Besançon, France

BAS LEMMENS, University of Warwick, Coventry, UK

CHRIS LENNARD, University of Pittsburgh, Pittsburgh, PA

CAMINO LERANOZ, Universidad Pública de Navarra, Pamplona, Spain

DENNY H. LEUNG, National University of Singapore, Singapore

GRZEGORZ LEWICKI, Jagiellonian University, Kraków, Poland

CHONG LI, Zhejiang University, Hangzhou, China

JINLU LI, Shawnee State University, Portsmouth, OH

VEGARD LIMA, Agder University College, Kristiansand, Norway

PEI-KEE LIN, The University of Memphis, Memphis, TN

JORAM LINDENSTRAUSS, Hebrew University, Jerusalem, Israel

ALEXANDER LITVAK, University of Alberta, Edmonton, Canada

VICTOR LOMONOSOV, Kent State University, Kent, OH

OLGA MALEVA, University of Cambridge, Cambridge, UK

KRISTEL MIKKOR, University of Tartu, Tartu, Estonia

LUIZA AMALIA MORAES, Universidade Federal do Rio de Janeiro, Rio de Janeiro, Brazil

DOUGLAS MUPASIRI, University of Northern Iowa, Cedar Falls, IA

ASSAF NAOR, Microsoft Research, Redmond, WA

JAN VAN NEERVEN, Technical University Delft, Delft, The Netherlands

PIOTR NOWAK, Vanderbilt University, Nashville, TN

OLAV NYGAARD, Agder University College, Kristiansand, Norway

EDWARD ODELL, University of Texas, Austin, TX

TIMUR OIKHBERG, University of California-Irvine, Irvine, CA

EVE OJA, University of Tartu, Tartu, Estonia

ELENA OURNYCHEVA, Kent State University, Kent, OH

MIKHAIL OSTROVSKII, St. John's University, Queens, NY

IMRE PATYI, Georgia State University, Atlanta, GA

ANNA PELCZAR, Jagiellonian University, Kraków, Poland

JAVIER PELLO, Universidad Rey Juan Carlos, Mostoles, Spain

HENRIK PETERSSON, Chalmers, Goteborg, Sweden

PIERRE PORTAL, Australian National University, Canberra, Australia

JOSE ANTONIO PRADO-BASSAS, Universidad Autonoma de Madrid, Madrid, Spain

DANIEL RADELET, University of Pittsburgh, Pittsburgh, PA

BEATA RANDRIANANTOANINA, Miami University, Oxford, OH

NARCISSE RANDRIANANTOANINA, Miami University, Oxford, OH

NIRINA LOVA RANDRIANARIVONY, University of Missouri, Columbia, MO

YVES RAYNAUD, Institut de Mathematiques de Jussieu, Paris, France

MARIA DEL CARMEN REGUERA RODRIGUEZ, University of Missouri, Columbia, MO

JAMES ROBERTS, University of South Carolina, Columbia, SC

CHRISTIAN ROSENDAL, University of Illinois, Urbana-Champaign, IL

HÉCTOR SALAS, University of Puerto Rico at Mayaguez, Mayaguez, Puerto Rico

FRANK SANACORY, University of South Carolina, Columbia, SC

BUNYAMIN SARI, University of North Texas, Denton, TX

GIDEON SCHECHTMAN, Weizmann Institute of Science, Rehovot, Israel

JEFFREY SCHLAERTH, Kent State University, Kent, OH

THOMAS SCHLUMPRECHT, Texas A&M University, College Station, TX

MORTEZA SEDDIGHIN, Indiana University East, Richmond, IN

EVGENY SEMENOV, Voronezh State University, Voronezh, Russia

PETER SEMRL, University of Ljubljana, Ljubljana, Slovenia

VELI SHAKHMUROV, Istanbul University, Istanbul, Turkey

BORIS SHEKHTMAN, University of South Florida, Tampa, FL

SCOTT SIMON, Purdue University, West Lafayette, IN

GIERI SIMONETT, Vanderbilt University, Nashville, TN

LESLAW SKRZYPEK, University of South Florida, Tampa, FL

MARK SMITH, Miami University, Oxford, OH

EMILY SPRAGUE, University of Wisconsin-Richland, Richland, WI

ABEBAW TADESSE, University of Pittsburgh, Pittsburgh, PA

JARNO TALPONEN, University of Helsinki, Helsinki, Finland

WEE-KEE TANG, Nanyang Technological University, National Institute of Education, Singapore

ADI TCACIUC, University of Alberta, Edmonton, Canada

EDUARDO TEIXEIRA, Rutgers University, Piscataway, NJ

VLADIMIR TEMLYAKOV, University of South Carolina, Columbia, SC

CASEY TRAIL, Miami University, Oxford, OH

DAVE TRAUTMAN, The Citadel, Charleston, SC

DRAGAN TRNINIC, Miami University, Oxford, OH

VLADIMIR TROITSKY, University of Alberta, Edmonton, Canada

STANIMIR TROYANSKI, University of Murcia, Espinardo, Spain

BARRY TURETT, Oakland University, Rochester, MI

HANS-OLAV TYLLI, University of Helsinki, Helsinki, Finland

ONNO VAN GAANS, Leiden University, Leiden, The Netherlands

ROMAN VERSHYNIN, University of California-Davis, Davis, CA

IGNACIO VILLANUEVA, Universidad Complutense de Madrid, Madrid, Spain

MIRCEA VOISEI, The University of Texas-Pan American, Edinburg, TX

GRIFFITH WARE, Australian National University, Canberra, Australia

MEGAN WAWRO, Miami University, Oxford, OH

LUTZ WEIS, Universität Karlsruhe, Karlsruhe, Germany

PRZEMYSŁAW WOJTASZCZYK, Warsaw University, Warsaw, Poland

MAREK WÓJTOWICZ, Casimir the Great University, Bydgoszcz, Poland

GEOFF WOOD, University of Wales Swansea, Swansea, UK

VLAD YASKIN, University of Missouri, Columbia, MO

MARYNA YASKINA, University of Missouri, Columbia, MO

JINDRICH ZAPLETAL, University of Florida, Gainesville, FL

BENTUO ZHENG, Texas A&M University, College Station, TX

ARTEM ZVAVITCH, Kent State University, Kent, OH

Plenary talks

YURI BRUDNYI, *Technion, Israel Institute of Technology, Haifa, Israel.*
Multivariate functions of bounded variation[1]

JESÚS M. F. CASTILLO, *University of Extremadura, Badajoz, Spain.*
Limit Banach spaces and the extension of operators

MARIANNA CSÖRNYEI, *University College, London, UK.*
Structure of null sets, differentiability of Lipschitz functions, and other problems

JESÚS M. F. CASTILLO, *University of Extremadura, Badajoz, Spain.*
Limit Banach spaces and the extension of operators

MARIANNA CSÖRNYEI, *University College, London, UK.*
Structure of null sets, differentiability of Lipschitz functions, and other problems

STEPHEN DILWORTH, *University of South Carolina, Columbia, SC.*
Coefficient quantization in Banach spaces

VALENTIN FERENCZI, *Université Paris VI, Paris, France.*
Uniqueness of complex structure and real hereditarily indecomposable Banach spaces

GILLES GODEFROY, *Université Paris VI, Paris, France.*
Non-linear isomorphisms between Banach spaces

WILLIAM B. JOHNSON, *Texas A&M University, College Station, TX.*
Non linear factorization of linear operators

JORAM LINDENSTRAUSS, *Hebrew University, Jerusalem, Israel.*
Porous sets, Fréchet differentiability and Γ_n null sets

ASSAF NAOR, *Microsoft Research, Redmond, WA.*
Dvoretzky's theorem in metric spaces

EDWARD ODELL, *University of Texas, Austin, TX.*
On the structure of asymptotic ℓ_p spaces

GIDEON SCHECHTMAN, *Weizmann Institute of Science, Rehovot, Israel.*
Fine estimates in Dvoretzky's theorem

THOMAS SCHLUMPRECHT, *Texas A&M University, College Station, TX.*
Embeddings into Banach spaces with finite dimensional decompositions

VLADIMIR TEMLYAKOV, *University of South Carolina, Columbia, SC.*
Greedy approximations in Banach spaces

NICOLE TOMCZAK-JAEGERMANN, *University of Alberta, Edmonton, Canada.*
Saturating normed spaces[1]

---

[1] Talk canceled due to last minute medical reasons.

ROMAN VERSHYNIN, *University of California-Davis, Davis, CA.*
Functional analysis approach to linear programming?

LUTZ WEIS, *Universität Karlsruhe, Karlsruhe, Germany.*
R-bounded sets of operators in analysis

PRZEMYSŁAW WOJTASZCZYK, *Warsaw University, Warsaw, Poland.*
Results and problems on wavelet greedy approximation of functions of bounded variation

Talks in special sessions

LIPI RANI ACHARYA, *Indian Institute of Technology, Kanpur, India.*
Diagonal operators between vector-valued sequence spaces and measure of compactness

MARÍA D. ACOSTA, *Universidad de Granada, Granada, Spain.*
An "isomorphic" version of James's theorem

YAKOV ALBER, *Technion, Israel Institute of Technology, Haifa, Israel.*
Generalized projection operators in Banach spaces

FERNANDO ALBIAC, *University of Missouri, Columbia, MO.*
Lipschitz maps between quasi-Banach spaces

DALE ALSPACH, *Oklahoma State University, Stillwater, OK.*
Partition and weight spaces with the KP property

GEORGE ANDROULAKIS, *University of South Carolina, Columbia, SC.*
The invariant subspace problems for Banach spaces

RAZVAN ANISCA, *Lakehead University, Thunder Bay, Canada.*
Theorem of Komorowski and Tomczak-Jaegermann, revisitcd

ANTONIO AVILÉS, *University of Murcia, Murcia, Spain.*
Renormings of the duals of James tree spaces

PRADIPTA BANDYOPADHYAY, *Indian Statistical Institute, Kolkata, India.*
Almost constrained subspaces of $C(K)$-II

SUDESHNA BASU, *Morgan State University, Baltimore, MD.*
Ball intersection properties in Banach spaces

ELIZABETH BATOR, *University of North Texas, Denton, TX.*
c_0, ℓ^∞, and complemented subspaces of $L(X,Y)$

KEVIN BEANLAND, *University of South Carolina, Columbia, SC.*
A hereditarily indecomposable asymptotic ℓ_2 Banach space

OSCAR BLASCO, *University of Valencia, Burjasot, Spain.*
A Coiffman–Weiss transference method for bilinear and maximal bisublinear operators and applications

SARITA BONDRE, *The M.S. University of Baroda, Vadodara, India.*
Structural properties of spaces of fuzzy number valued functions

OLGA BREZHNEVA, *Miami University, Oxford, OH.*
Implicit function theorems for nonregular mappings in Banach spaces

BO BRINKMAN, *Miami University, Oxford, OH.*
A technique for bounding the dimensionality of ℓ_1 spaces

QINGYING BU, *University of Mississippi, University, MS.*
Radon–Nikodym property for Fremlin projective tensor product of Banach lattices

FÉLIX CABELLO SÁNCHEZ, *Universidad de Extremadura, Badajoz, Spain.*
Some weak forms of injectivity for Banach spaces, with applications to ultraproducts

MICHAEL CWIKEL, *Technion, Israel Institute of Technology, Haifa, Israel.*
K-divisibility constants for some special Banach and Hilbert couples

STEFAN CZERWIK, *Silesian University of Technology, Gliwice, Poland.*
On the Ulam–Hyers stability of the functional equations in Banach spaces

JERRY DAY, *University of Pittsburgh, Pittsburgh, PA.*
Minimal invariant sets of Alspach's map

GEOFF DIESTEL, *University of Missouri, Columbia, MO.*
Sobolev spaces with only trivial isometries

PAWEŁ DOMAŃSKI, *Adam Mickiewicz University, Poznań, Poland.*
Interpolation inequalities between norms in spaces of distributions and real analytic functions

IAN DOUST, *University of New South Wales, Sydney, Australia.*
Norm bounds for families of projections

JAKUB DUDA, *Weizmann Institute of Science, Rehovot, Israel.*
Metric derived numbers and continuous metric differentiability via homeomorphisms

YVES DUTRIEUX, *Université de Franche-Comté, Besançon, France.*
Lipschitz free space of $C(K)$-spaces

SUDIPTA DUTTA, *Ben-Gurion University of the Negev, Beer-Sheva, Israel.*
On tree characterizations of some Banach spaces

MAHMOUD MOHAMMED MOSTAFA EL-BORAI, *Alexandria University, Alexandria, Egypt.*
On some stochastic fractional integrodifferential equations

MARIÁN FABIAN, *Czech Academy of Sciences, Prague, Czech Republic.*
Weakly compactly generated spaces and their relatives

VASILIKI FARMAKI, *Athens University, Athens, Greece.*
Schreier families of variable located words

RICHARD J. FLEMING, *Central Michigan University, Mt. Pleasant, MI.*
Bohnenblust's Theorem and norm-equivalent coordinates

JULIO FLORES, *Universidad Rey Juan Carlos, Madrid, Spain.*
Domination by positive Banach–Saks operators

JAN FOURIE, *North-West University, Potchefstroom, South Africa.*
On operator valued sequences of multipliers and R-boundedness

MARIA FRAGOULOPOULOU, *University of Athens, Athens, Greece.*
When an enveloping pro-C^*-algebra is a genuine C^*-algebra?

ONNO VAN GAANS, *Leiden University, Leiden, The Netherlands.*
One-complemented subspaces of finite dimensional vector spaces with smooth lattice norms

NADIA J. GAL, *The University of Memphis, Memphis, TN.*
Isometries and isometric equivalence of Hermitian operators on $A^{1,p}(X)$

PANDO GEORGIEV, *University of Cincinnati, Cincinnati, OH.*
Variational principles in Banach spaces – parametric versions

IOANA GHENCIU, *University of Wisconsin, River Falls, WI.*
Complemented spaces of operators

STANISŁAW GOLDSTEIN, *University of Łódź, Łódź, Poland.*
Twisting noncommutative L^p-spaces

MANUEL GONZÁLEZ, *Universidad de Cantabria, Santander, Spain.*
Local duality for Banach spaces: examples and characterizations

KARL GROSSE-ERDMANN, *Fachbereich Mathematik, Hagen, Germany.*
Frequently hypercyclic operators

MANJUL GUPTA, *Indian Institute of Technology, Kanpur, India.*
Banach spaces of entire sequences and their Köthe duals

ALEXANDER YA. HELEMSKII, *Moscow State University, Moscow, Russia.*
Highlights of homology theory in classical and quantum functional analysis

FRANCISCO L. HERNANDEZ, *Universidad Complutense de Madrid, Madrid, Spain.*
Strict singularity and disjoint strict singularity in rearrangement invariant spaces

TUOMAS HYTÖNEN, *University of Turku, Turku, Finland.*
Probabilistic Littlewood–Paley theory in Banach spaces

MAR JIMENEZ-SEVILLA, *Universidad Complutense de Madrid, Madrid, Spain.*
Approximation by smooth functions with no critical points in separable Banach spaces

STEN KAIJSER, *Uppsala University, Uppsala, Sweden.*
More on K-divisibility

ANNA KAMIŃSKA, *The University of Memphis, Memphis, TN.*
On the extensions of homogeneous polynomials

TAMARA KUCHERENKO, *University of California Los Angeles, Los Angeles, CA.*
Absolute functional calculus for sectorial operators

DENKA KUTZAROVA, *University of Illinois, Urbana-Champain, IL.*
On strongly asymptotic ℓ_p spaces and minimality

FLORENCE LANCIEN, *Université de Franche-Comté, Besançon, France.*
Square functions and H^∞ functional calculus for sectorial operators on subspaces of L^p

GILLES LANCIEN, *Université de Franche-Comté, Besançon, France.*
Spectral theory for linear operators on L^1 or $C(K)$ spaces

CHRISTIAN LE MERDY, *Université de Franche-Comté, Besançon, France.*
H^p-Maximal regularity and operator valued multipliers on Hardy spaces

BAS LEMMENS, *University of Warwick, Coventry, UK.*
Dynamics of 1-Lipschitz maps

CHRIS LENNARD, *University of Pittsburgh, Pittsburgh, PA.*
L^p_+ is uniformly concave for $0 < p < 1$

CAMINO LERÁNOZ, *Universidad Pública de Navarra, Pamplona, Spain.*
Geometric properties of quasi-Banach spaces

DENNY H. LEUNG, *National University of Singapore, Singapore.*
Comparing mixed Tsirelson spaces and their modified versions

GRZEGORZ LEWICKI, *Jagiellonian University, Kraków, Poland.*
Minimal multi-convex projections

CHONG LI, *Zhejiang University, Hangzhou, China.*
Existence and well-posedness in approximation theory in Banach spaces

JINLU LI, *Shawnee State University, Postsmouth, OH.*
Relationship between metric and generalized projections in Banach spaces

VEGARD LIMA, *Agder University College, Kristiansand, Norway.*
Ideals of operators and the weak metric approximation property

PEI-KEE LIN, *The University of Memphis, Memphis, TN.*
There is an equivalent norm of ℓ_1 that has the fixed point property

ALEXANDER LITVAK, *University of Alberta, Edmonton, Canada.*
A covering lemma and its applications

OLGA MALEVA, *University of Cambridge, Cambridge, UK.*
Unavoidable sigma-porous sets

KRISTEL MIKKOR, *Tartu University, Tartu, Estonia.*
Uniform factorization for compact sets of operators acting from a Banach space to its dual space

LUIZA A. MORAES, *Universidade Federal do Rio de Janeiro, Rio de Janeiro, Brazil.*
On the extension of polynomials on Banach spaces

DOUGLAS MUPASIRI, *University of Northern Iowa, Cedar Falls, IA.*
On the difficulty of preserving monotonicity via projections and related results

JAN VAN NEERVEN, *Technical University Delft, Delft, The Netherlands.*
Uniformly γ-radonifying families of operators

PIOTR NOWAK, *Vanderbilt University, Nashville, TN.*
Coarse embeddings into Banach spaces and the Novikov conjecture

TIMUR OIKHBERG, *University of California-Irvine, Irvine, CA.*
Hyperreflexivity with respect to operator ideals

EVE OJA, *Tartu University, Tartu, Estonia.*
Grothendieck's theorem on nuclear operators revisited

MIKHAIL I. OSTROVSKII, *St.-John's University, New York, NY.*
Compositions of projections in Banach spaces and relations between approximation properties

ELENA OURNYCHEVA, *Kent State University, Kent, OH.*
Composite cosine transforms on Stiefel manifolds

IMRE PATYI, *Georgia State University, Atlanta, GA.*
Lifting and right invertibility of holomorphic operator functions

ANNA PELCZAR, *Jagiellonian University, Kraków, Poland.*
Stabilization of Tsirelson-type norms on ℓ_p spaces

HENRIK PETERSSON, *Chalmers, Goteborg, Sweden.*
Hypercyclic operators with hypercyclic adjoints

PIERRE PORTAL, *Australian National University, Canberra, Australia.*
Sectorial operators and H^∞-functional calculus in a discrete setting

DANIEL P. RADELET, *University of Pittsburgh, Pittsburgh, PA.*
Reconstruction using Markuschevich and Cesàro bases

NIRINA LOVA RANDRIANARIVONY, *University of Missouri, Columbia, MO.*
L_p is not uniformly homeomorphic to $\ell_p + \ell_2$ when $p < 2$

YVES RAYNAUD, *Institut de Mathématiques de Jussieu, Paris, France.*
Positive contractive projections in $L_p(L_q)$, with applications to axiomatizability theory

CHRISTIAN ROSENDAL, *University of Illinois, Urbana-Champaign, IL.*
Infinite asymptotic games

HÉCTOR SALAS, *University of Puerto Rico, Mayagüez, Puerto Rico.*
Pathological hypercyclic operators

FRANK SANACORY, *University of South Carolina, Columbia, SC.*
On the compact plus multiple of the inclusion problem

BÜNYAMIN SARI, *University of North Texas, Denton, TX.*
Lattice structures and spreading models

MORTEZA SEDDIGHIN, *Indiana University East, Richmond, IN.*
Algorithms for computation of antieigenvalues

BORIS SHEKHTMAN, *University of South Florida, Tampa, FL.*
Perturbation of ideal complements

LESLAW SKRZYPEK, *University of South Florida, Tampa, FL.*
Minimal projections in L_p spaces

SCOTT SIMON, *Purdue University, West Lafayette, IN.*
Inhomogeneous Cauchy–Riemann equations in Banach spaces

GIERI SIMONETT, *Vanderbilt University, Nashville, TN.*
Operator-valued symbols for elliptic and parabolic problems on wedges

EMILY H. SPRAGUE, *University of Wisconsin, Richland, WI.*
An application of uniform integrability to the characterization of Walsh Fourier series

ABEBAW TADESSE, *University of Pittsburgh, Pittsburgh, PA.*
Compact composition operators on the Hardy and Bergman spaces

JARNO TALPONEN, *University of Helsinki, Helsinki, Finland.*
On the Banach–Mazur rotation problem

WEE-KEE TANG, *Nanyang Technological University, National Institute of Education, Singapore.*
Extension of functions with small oscillation

ADI TCACIUC, *University of Alberta, Edmonton, Canada.*
On the existence of asymptotic-l_p structures in Banach spaces

EDUARDO V. TEIXEIRA, *Rutgers University, Piscataway, NJ.*
Regularity theory for differential equations in abstract spaces and applications to nonlinear PDEs

VLADIMIR TROITSKY, *University of Alberta, Edmonton, Canada.*
Norm closed ideals in $\mathcal{L}(\ell_p \oplus \ell_q)$

STANIMIR TROYANSKI, *University of Murcia, Murcia, Spain.*
Banach spaces with modulus of convexity of power type 2

HANS-OLAV TYLLI, *University of Helsinki, Helsinki, Finland.*
Composition operators on vector-valued harmonic functions and Cauchy transforms

M. D. VOISEI, *The University of Texas-Pan American, Edinburg, TX.*
Monotonicity representability and maximality via the Fitzpatrick function

MAREK WÓJTOWICZ, *Casimir the Great University, Bydgoszcz, Poland.*
An isometric form of the theorem of Lindenstrauss and Rosenthal on quotients of $\ell_1(\Gamma)$

GEOFFREY WOOD, *University of Wales, Swansea, UK.*
Transitive norms and unitary Banach algebras

VLADYSLAV YASKIN, *University of Missouri, Columbia, MO.*
The geometry of L_0

MARYNA YASKINA, *University of Missouri, Columbia, MO.*
Centroid bodies and comparison of volumes

JINDRICH ZAPLETAL, *University of Florida, Gainesville, FL.*
Ramsey capacities

ARTEM ZVAVITCH, *Kent State University, Kent, OH.*
On Gaussian measure of sections of dilations and shifts of convex bodies

Bei Fragen zur Produktsicherheit wenden Sie sich bitte an:
If you have any questions regarding product safety,
please contact:

Walter de Gruyter GmbH
Genthiner Straße 13
10785 Berlin
productsafety@degruyterbrill.com